Springer Series in
Electronics and Photonics

33

Edited by A. L. Schawlow

Springer Series in Electronics and Photonics

Editors: I. P. Kaminov W. Engl T. Sugano

Managing Editor: H. K. V. Lotsch

This series was originally published under the title
Springer Series in Electrophysics
and has been renamed starting with Volume 22.

Volumes 1–21 are listed at the end of the book

Nils W. Carlson

Monolithic Diode-Laser Arrays

With 178 Figures

Springer-Verlag
Berlin Heidelberg New York
London Paris Tokyo
Hong Kong Barcelona
Budapest

Dr. Nils W. Carlson

Lawrence Livermore National Laboratories, P. O. Box 808,
Livermore, CA 94551, USA

Guest Editor:

Professor A. L. Schawlow

Department of Physics, Stanford University,
Stanford, CA 94305-4060, USA

Series Editors:

Dr. Ivan P. Kaminow

AT&T Bell Laboratories, P. O. Box 400, Holmdel, NJ 07733, USA

Professor Walter Engl

Institut für Theoretische Elektrotechnik, Rhein.-Westf. Technische Hochschule,
Kopernikusstrasse 16, D-52074 Aachen, Germany

Professor Takuo Sugano

Department of Electrical and Electronic Engineering, Toyo University,
2100 Kujirai Kawagoe Saitama 350, Japan

Managing Editor: Dr.-Ing. Helmut K. V. Lotsch

Springer-Verlag, Tiergartenstrasse 17,
D-69121 Heidelberg, Germany

ISBN-13:978-3-642-78944-1   e-ISBN-13:978-3-642-78942-7
DOI: 10.1007/978-3-642-78942-7

Library of Congress Cataloging-in-Publication Data. Carlson, Nils William. Monolithic diode-laser arrays: Nils W. Carlson. p. cm. – (Springer series in electronics and photonics; vol. 33) Includes bibliographical references and index. ISBN-13:978-3-642-78944-1
1.Semiconductor lasers. I.Title.II. Series: Springer series in electronics and photonics: v. 33. TA1700.C36 1994 621.36'61–dc20 94-11598

© Springer-Verlag Berlin Heidelberg 1994
Softcover reprint of the hardcover 1st edition 1994

Typesetting: Data conversion by Springer-Verlag
SPIN: 10064137   54/3140 - 5 4 3 2 1 0 - Printed on acid-free paper

Preface

The last ten years have witnessed an explosive development in the field of semiconductor-diode lasers. Diode lasers have spread rapidly into the commercial applications areas of telecommunications, optical recording, laser audio/video, laser printing, and pump sources for solid-state lasers. Because of its compact size and high efficiency, the semiconductor-diode laser can offer unparalleled advantages in these and many other optical systems applications. A widespread vision has persisted over the last decade, that more reliable monolithic semiconductor laser sources could eventually be made to replace gas, dye, and solid-state lasers in many applications that would derive great benefit from compact lasers sources with high average power and high-brightness output capabilities. The initial driving force behind this vision has been the notion that both high-power output and high brightness could be increased by coherently coupling an array of single-element diode lasers. With the combined projected advantages of better reliability and high-power output from a very compact device, it is easy to see the attraction of the monolithic semiconductor diode laser array as a particularly user-friendly laser source.

Over the last twenty years, many approaches and concepts for monolithic diode laser arrays have evolved in the quest for a high-power, single-frequency source. But even today, the ultimate monolithic diode laser array has not yet been achieved. Whereas ten years ago or so, there was a diversity of monolithic diode laser designs being actively persued, today the field of concepts and design approaches has narrowed and converged to just a few promising designs. Some approaches such as the unstable-resonator semiconductor laser and the master oscillator power amplifier are well-known from other areas of laser science. But other approaches such as the antiguided-diode laser array and the active-grating surface emitting amplifier are, so far, unique to semiconductor diode lasers.

The goal of this book is to give a unified presention of the physical principles, optical design, operating characteristics, and ultimate performance projections of monolithic diode-laser arrays for single-mode, high-power operation. Emphasis is placed on developing an understanding of the mode discrimination properties of the diode-laser array structures. The book is intended for scientists, engineers, and advanced graduate students working in the field of semiconductor lasers and optoelectronics, as well as those working in other fields of laser science. Phenomenological descriptions have been the preferred mode of presentation. Some detailed mathematical derivations are included to validate certain of the models that are used, and in a few instances, to explain those subjects which cannot be adequately dealt with using the canonical models. The significant developments that have contributed to, and shaped, the present state of the field are reviewed. Numerous references, from the early to the most recent literature, are provided as research in monolithic high-brightness semiconductor laser sources is still an active and expanding field. The book should provide the reader with a perspective on the development of high-power semiconductor lasers and the issues important for mode discrimination. Hopefully, the reader will be left with an appreciation of the potential that semiconductor lasers offer as compact, efficient high-brightness sources of light, as well as the challenging scientific problems that still need to be solved in order for the potential to be realized.

The focus of this book will be limited to the physical principles, design, and performance projections of monolithic diode lasers and diode-laser arrays for high-power single-frequency operation with diffraction-limited quality output beams. In Chaps. 1–4, basic concepts are reviewed that are important for understanding the mode discrimination properties and operating principles of high-power diode-laser devices. Chapter 1 presents a brief history of the diode-laser arrays, as well as some background on the factors that have motivated the development of diode-lasers arrays, and a review of diode-laser array design concepts. In Chap. 2, the physics and design issues related to high-power operation of semiconductor diode lasers and diode-laser arrays are presented. Issues related to mode discrimination (both spectral and spatial) in semiconductor lasers and laser arrays are reviewed in Chap. 3. Theoretical models for diode-laser arrays are the subject of Chap. 4. The remaining four chapters (Chaps. 5–8) discuss the experimental and theoretical analyses of specific types of diode-laser arrays and other novel semiconductor laser designs, where the potential for single-mode, beam-quality operation at high power outputs has been investigated. Basic edge-emitting gain-guided diode-laser arrays are covered in Chap. 5, while Chap. 6 deals with edge-emitting, index-guided diode-laser structures. Surface-emitting diode-laser arrays are presented in Chap. 7, and Chap. 8 covers master-oscillator power amplifier semiconductor lasers (both surface and edge emitting).

I am grateful to Prof. Arthur Schawlow for encouraging me to write this book. Also, I wish to thank Dr Helmut Lotsch of Springer-Verlag for his assistance and encouragement throughout the writing of the book. I am grateful to Prof. Jerome Butler for his encouragement and crtical reviews of parts of the manuscript, as well as for introducing me to TEX. When I began writing this book, I was at the David Sarnoff Research Center; and about halfway through the project, I moved to my present position at the Lawrence Livermore National Laboratory. Hence, I have enjoyed the benefit of technical interactions with colleagues at both laboratories, as well as the use of the fine libraries at each of these institutions.

It is with pleasure that I acknowledge the many scientists whose work has been cited, as their contributions are the fabric of this book. I am especially grateful to Prof. Dan Botez, Prof. Gary Evans, Dr. Robert Amantea, Dr. Ray Beach, Dr. Mark Emanuel, and Dr. Richard Solarz for their critical reviews of parts of the manuscript. I am thankful to Guinevere for her quiet companionship, during many hours spent typing at the computer. Finally, I would like to express my gratitude to my wife, Diane, for her patience, understanding, and encouragement, and I dedicate this book to her.

Danville, California   Nils W. Carlson
April 1994

Table of Contents

1. Introduction and Background

This chapter introduces monolithic diode-laser arrays. Section 1.1 provides some historical background on semiconductor diode-laser arrays, and Sect. 1.2 gives a brief review of work on non-semiconductor laser arrays. Performance limitations of single-element diode lasers, enabling technological advances, and applications that have contributed to continued research and development of monolithic diode-laser arrays are discussed in Sect. 1.3. Concepts and designs of high-power, coherent diode-laser arrays are reviewed in Sect. 1.4. The primary focus is on the geometrical structure of the array and less so on the details of the epilayer structure of the diode.

1.1 History of Diode-Laser Array Development

Not long after the first demonstrations of semiconductor diode-laser oscillation were reported in 1962 [1.1]-[1.4], it was recognized that phase and frequency locking of an array of diffractively-coupled diode lasers could be a useful means for scaling these compact sources to achieve increased power outputs along with an increased brightness in the light output [1.5]-[1.8]. The phenomena of frequency locking of classical nonlinear oscillators in mechanical and electrical systems had been the subject of much research [1.9, 1.10]. Although the laser is a self-sustained quantum-mechanical nonlinear oscillator, frequency locking in systems of laser oscillators has been viewed as being similar to frequency locking in classical oscillator systems. Indeed, the earliest analysis of [1.5]-[1.8], as well as many that followed, have analyzed systems of coupled lasers using classical or semi-classical models.

The diode-laser array design that was the basis for theoretical analysis of *Basov* et al. [1.5], consisted of an open resonator, with the semiconductor diode-lasers element deposited on one of the resonator mirrors. The second mirror formed the resonant cavity and provided diffractive-coupling of light between the laser array elements. This analysis identified the basic stability conditions necessary for fundamental mode operation of a two-element open resonator laser array: 1) the fundamental mode of the coupled resonator must have lower losses than the modes of oscillation for the uncoupled resonators, and 2) the frequency detuning of the two laser oscillators must be within a sufficiently narrow bandwidth, to insure frequency locking of the two lasers.

The first condition is a statment that the mode of the coupled resonators should be the most favored mode of operation based on energy considerations, if the array is to produce a coherent output. Otherwise, each resonator would oscillate independently in it's own mode, and there would be very little, if any, mutual coherence between the output beams. The second condition is a restatement of the well-known condition for frequency locking (also referred to as frequency entrainment) of two nonlinear oscillators [1.9]. *Belenov* and *Letokhov* [1.8] generalized the dynamic stability analysis of a pair of coupled lasers to consider the effects of the time-delay associated with the optical coupling, energy losses resulting from the optical coupling, and the dispersion in the dielectric constant of the gain medium.

The earliest experimental work on external coupling of semiconductor diode lasers was not aimed specifically at studying frequency entrainment, but instead considered quenching of one laser by another. *Fowler* [1.11, 1.12] had observed that quenching of one laser could be induced by injecting light from a second in a perpendicular orientation to the first laser. Shortly, after the observation of [1.11, 1.12], *Kelly* [1.13] studied quenching and cooperative effects in what appear to be the first monolithic pairs of coupled lasers. Figure 1.1 displays the two disitinct geometries that were fabricated [1.13]. The device depicted in Fig. 1.1a was used to study cooperative effects, while that in Fig. 1.1b, where the laser axes are perpendicular to each other, was used for quenching studies. Each homojunction gain element, labeled as 1 and 2 in Fig. 1.1 had an independent contact; and under separate bias, laser oscillation was observed for sufficiently high injection current. When both contacts of the structure in Fig. 1.1a were biased, a significant reduction in the threshold current per contact was observed. It was also noted that the operating wavelength when both lasers operated together was noticeably longer than when either laser section was operated alone. These effects demonstrated that the integrated pair of diode lasers in Fig. 1.1a could be made to act as if it were a single optical cavity. Generally speaking, externally injected light into a laser can compete for the same gain reservoir that is available for the natural oscillation. This can lead to either gain quenching or frequency locking. *Pantell* [1.14] reported an analysis of laser oscillation under the influence of an externally-injected signal. The effects of the frequency pulling were studied, and the intensity level of injected light necessary to quench the natural laser oscillation was also calculated.

A few years later several groups began investigating the properties of broad-area diode lasers and diode-laser arrays in external cavities in an effort to obtain higher power outputs and diffraction-limited beam quality [1.15]-[1.17]. One of the motivations for doing this was that the high-speed modulation characteristics of semiconductor lasers was already recognized as a useful feature for high-data rate, long-range optical space communications [1.15]. However, the power output of single diode lasers was too low for the distances encountered in space communication links. To increase the coherent output power and preserve the high modulation speed, *Vuilleumier* et al.

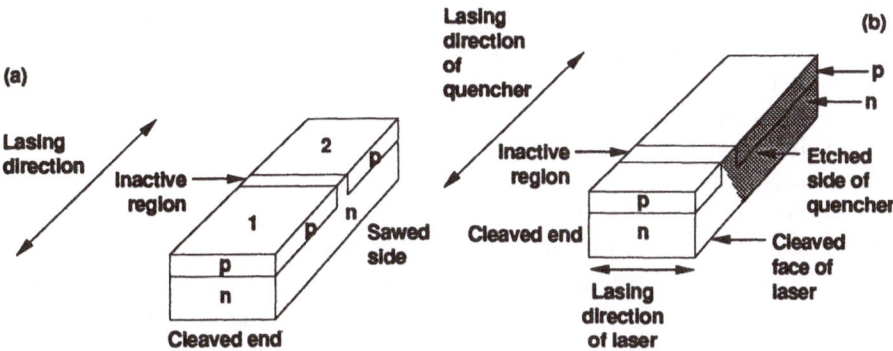

Fig. 1.1. Illustration of the monolithic pairs of coupled diode laser structures with colinear-optic axes (**a**) and orthogonal optic axes (**b**) [1.13].

[1.15] proposed the use of a diode-laser master oscillator to inject an array of adjacent diode-laser amplifiers. Such a laser system was used by [1.15] to demonstrate the feasibility of coherent amplification by an array of diode-laser amplifiers that were injected with light from a common oscillator. Although, this system was operated at cryogenic temperatures, it still provided for a high-degree of optical isolation between the laser oscillator and the amplifiers, as well as a phase correction at the output of each amplifier to compensate for the variations in optical path lengths.

Early experimental demonstrations were also done on diode-laser array oscillators [1.16, 1.17]. Phase locking of GaAs laser arrays in external resonators demonstrated that higher power per element could be obtained from a frequency-locked diode array, if anti-reflection coatings were used to suppress the oscillation modes of each array element [1.16]. A peak output power of 19 W under pulsed operating conditions, along with some narrowing of the angular beam divergence in the plane of the array was observed by [1.16], from a five element array. *Crowe* et al. [1.16] found that better output beam quality was obtained from a coupled diode-laser array comprising five elements of 625 μm width, rather than an array of smaller element size. An improved external resonator design was used by *Philipp-Rutz* et al. [1.17], to demonstrate diffraction-limited operation of a single broad-area GaAs laser with a power output of 2.2 W. The predominantly single-lobed output beam had a far-field angular beam divergence of 0.7 mrad, parallel to the pn junction, and 2.5 mrad perpendicular to the pn junction. This corresponding radiance or photometric brightness was about $150\,\mathrm{MWcm}^{-2}\,\mathrm{sr}^{-1}$ [1.17]. The experiments of *Crowe* et al. and *Philipp-Rutz* et al. pointed out the critical nature of the uniformity of the diode laser material and structure, as well as the phyisical tolerances of the resonator design, that were necessary to obtain a high degree of mutual coherence between array elements.

The power output and output beam quality of compact arrays of uncoupled diode lasers was sufficiently high so that these arrays were attractive sources for illuminator and ranging applications, where single-frequency operation was not a consideration [1.18]-[1.20]. Note that for a given power per element, the power in the far-field output beam of an uncoupled array of diode lasers will scale as the number of elements, though the far-field angular beam divergence will remain a constant. Such uncoupled arrays of InGaAs diode lasers were investigated by *Nuese* et al. as a more efficient alternative alternative to Nd:YAG lasers at 1.06 μm [1.21]. In some external-cavity diode-laser arrays, the uncoupled operation of the array elements produced better output-beam quality, because the array elements could not be placed close enough to reduce the side-lobe structure that occured under frequency-locked operation [1.22, 1.23]. It was recognized that average power outputs of several Watts could be obtained from compact uncoupled diode-laser arrays with sufficiently high-packing densities. With suitable collimation optics, far-field angular beam divergences on the order of a few milliradians were obtained [1.19]. The push for higher packing densities brought on studies of the thermal issues and limitations associated with increased packing density of diode-laser arrays, and progress was made in improving diode-chip mounting and cooling technologies [1.18]-[1.20].

The earliest reported demonstration of what could be considered to be a monolithic diode-laser array, with optical coupling perpendicular to the optic axis, appears to have been by *Kosonocky* et al. [1.24]. This diode-laser array consisted of multiple layers of GaAs in an $p-n-p-n-p-n$ multi-homojunction configuration. When operated at a temperature of 77 K, a device comprising three homojunctions spaced 1.5 μm apart, produced a far-field beam divergence of 0.5°. This narrow far-field output beam indicated a degree of phase locking between the homojunction emitter regions. The possible optical coupling mechanisms proposed by *Kosonocky* et al. in this diode-laser array were higher-order transverse modes and radiation coupling due to the leaky-waveguide characteristics of the multijunctions. It should be noted though, that the waveguiding properties of homojunction lasers have still not been studied in detail [1.25, 1.26].

At about this same time, another concept for an integrated, phase-coherent, semiconductor diode-laser array was proposed by *Pankove* [1.27]. A schematic diagram of this device is shown in Fig. 1.2. The pn junction has been configured in the shape of a cycloid, that also acts as an optical waveguide. A continuous cavity is formed by placing cleaved facets at each apex of the cycloid and high-reflecting facets are placed at the ends. The facets at the apexes act as output couplers in transmission, and in reflection, provide optical coupling between adjacent loops of the cycloid. In this array concept, the elements or cycloids are connected along the longitudinal direction, as opposed to the lateral coupling approach used in the external cavity approaches [1.16, 1.17], and the transverse coupling in the multijunction array [1.24]. The output coupling is distributed along the entire length of the laser-array cavity, unlike most conventional laser designs where the output power is transmitted through one of the end reflectors.

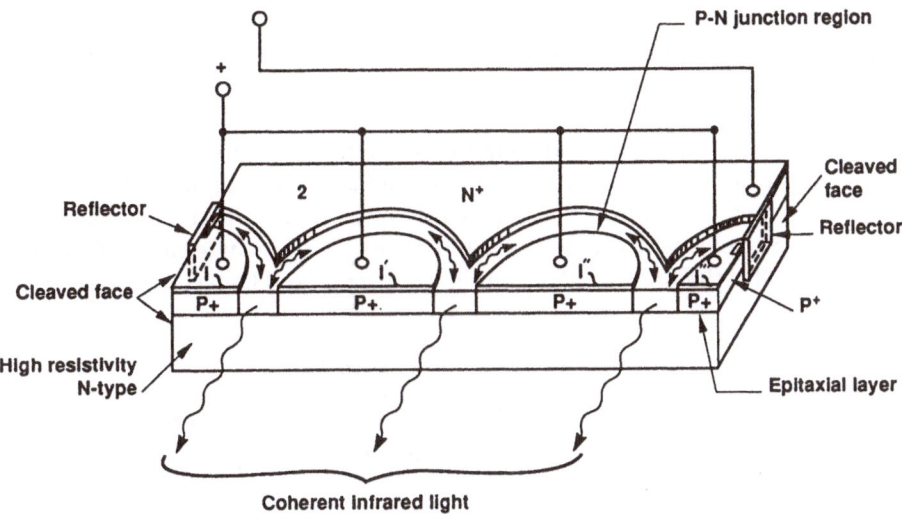

Fig. 1.2. Early concept for an integrated phase coherent laser array [1.27].

Distributed output coupling has a potential benefit in that the power output will scale with the array size (in this case increasing the array size means increasing the number of cycloids) without being limited by the optical losses, as occurs in end-coupled lasers. This distinction is discussed in more detail in Chaps. 2, 7 and 8. However, increasing the cavity length does decrease the separation between adjacent longitudinal modes. This can have an adverse effect on the mode discrimination properties of longer arrays; and if single-frequency operation is desired as well, the cavity length can become a critical parameter. Operation of diode-laser arrays based on the design shown in Fig. 1.2 have not yet been reported. Optically pumped half-ring semiconductor lasers were not demonstrated until 1976 [1.28]. Since then, steady improvements in isotropic etching techniques, photolithography, and processing technology have occured, and recently there have been reports describing the operation and modal properties of circular-shaped ring lasers [1.29] and half-ring geometry lasers [1.30], so perhaps the cycloidal array of *Pankove* was a bit ahead of it's time.

Control of the spatial modes, especially at higher-power outputs, in semiconductor-diode lasers has been, and still is, a very challenging problem. Early experimental studies [1.31, 1.32] encountered difficulties in correlating the near- and far-field output characteristics of diode lasers as cleaved from the wafers. It had been observed by *Dyment* [1.33] that improved control in the spatial mode of homojunction lasers could be obtained by decreasing the stripe width of the diode laser. Obviously, in this approach the improvement in spatial coherence is obtained at the expense of the power output, as the total active area is decreased. A solution for simultaneously obtaining both high-power and high-spatial coherence from a monolithic semiconductor-laser structure was proposed and demonstrated by *Ripper* and *Paoli* [1.34]. Optical coupling at the wafer level was obtained by fabricating two-stripe-geometry lasers in sufficiently close proximity to each other, as shown in Fig. 1.3, so that the optical fields generated in each laser was coupled into the other laser structure. This type of lateral-coupling mechanism was found by *Ripper* and *Paoli* to produce frequency-locked operation (albeit multi-longitudinal mode) for lasers that were fabricated sufficiently close. *Ripper* and *Paoli* suggested that this monolithic array design could be scaled to include many lasers elements, to offer the potential for high-power output and high transverse spatial coherence.

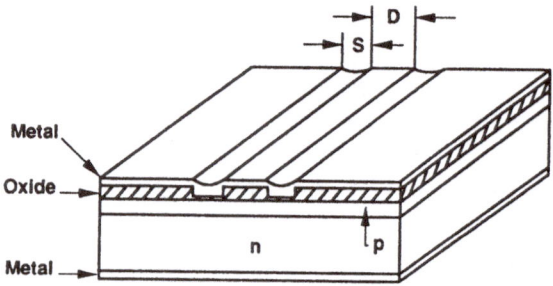

Fig. 1.3. Diagram depicting a two-stripe laser array [1.34].

Other architectures investigated for providing scalable coherent power and narrow output beam divergence were the grating-coupled surface-emitting diode-laser array [1.35] and the two-dimensional distributed feedback laser with grating-output coupling using dye in a solid host material [1.36, 1.37]. Although these early grating-coupled lasers demonstrated beam narrowing from the grating-output coupler, they did not show evidence of highly-coherent operation. *Prozorov* et al. [1.38] fabricated and studied a semiconductor diode-laser structure with quasidistributed output coupling as a means to lift the length restrictions on end coupled diode lasers. When operated near threshold, this monolithic diode-laser structure, which is illustrated in Fig. 1.4, was observed to produce a far-field pattern indicative of that of a coherent array of uniformly space emitters.

Most single-element diode lasers are fabricated as positive-waveguide structures, where the refractive index of the active region is greater than the refractive index of the regions adjacent to the active region. In this case, the electric field outside the active regions is evanescent, and decays exponentially over a distance on the order of a wavelength of light in the material. For this reason, it was believed early on that the laser-array elements must be placed sufficiently close so that the photons generated in each laser array element could "tunnel" into the adjacent laser array element and induce stimulated emission. Similar proposals were reported [1.39] advocating the use of coupled diode-laser waveguide structures as a promising approach for ob-

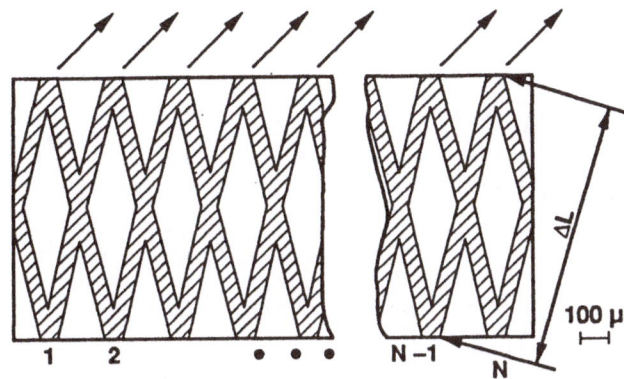

Fig. 1.4. Geometry of a diode laser incorprating quasidistributed energy extraction [1.38].

taining high-power and high-spatial coherence. Other types of single-element diode lasers can be fabricated as a negative-index waveguide, where the gain region has a lower refractive index than adjacent regions. With sufficient gain available, these lasers support modes that radiate light into the regions adjacent to the gain region. In arrays of negative-index waveguide lasers optical coupling is provided by the radiated light or leaky wave from each laser in the array.

The geometry of the coupled diode-laser waveguide structure design shown in Fig. 1.3 served as the basis for many subsequent monolithic diode-laser array designs aimed at obtaining frequency-locked operation. Though, the composition and details of the waveguide design of each laser element, and even the region separating adjacent lasers, can have a significant impact on the nature of the optical coupling, and hence, the ability of the laser elements to frequency lock and oscillate in a mutually coherent fashion. As discussed later in Chaps. 4-6, the original picture of evanescent-field coupling in arrays was incomplete, as it was discovered that it was necessary to consider radiation-field coupling to correctly explain the observed array mode characteristics.

The late 1970's signaled the beginning of a increase in the research and development effort devoted to monolithic diode-laser arrays. The coupled waveguide approach of *Ripper* and *Paoli* was demonstrated in larger gain-guided stripe arrays with five elements by *Scifres* et al., revealing substantial spatial coherence as evidenced by an $\approx 2°$ output-beam divergence [1.40]. Monolithic, index-guided buried-heterostructure arrays were also fabricated and demonstrated by *Tsang* et al. [1.41]. The branched-waveguide laser array illustrated in Fig. 1.5, was proposed and demonstrated by [1.42, 1.43]. In this approach, some of the light generated in each laser-gain element is directly injected by the branched waveguide sections into the adjacent gain element. The branched-waveguide allows direct injection coupling of lasers on

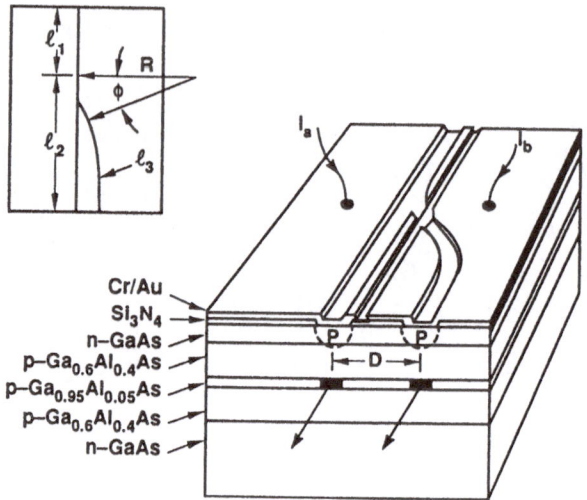

Fig. 1.5. Diagram of a branched-waveguide diode-laser array [1.42].

the same wafer, and provided an alternative approach to evanescently coupling for designing monolithic diode-laser arrays. The feasibility of electronic steering of the output beam of monolthic-diode lasers was demonstrated in both coupled-waveguide diode-laser arrays [1.44] and in branched-coupled, diode-laser arrays [1.42, 1.43]. Electronic beam steering is a characteristic that is desirable in many optical systems applications.

During the 1980's as materials-growth and device-processing techniques became more refined and more widely available, research and development into new monolithic semiconductor diode-laser arrays rapidly spread to laboratories throughout the world. Intensive development and steady improvements in the design, performance characteristics, and reliability of the gain-guided GaAs/AlGaAs diode-laser arrays were reported on in numerous technical publications [1.45]-[1.69]. Much of this work was motivated by the need for high-power diode-laser arrays as efficient pump sources for solid-state lasers.

A considerable amount of work was done in the area of exploring new monolithic array designs for high-power operation with high spatial and temporal coherence. Many applications required power outputs beyond the capabilities of single-element diode lasers. Much of this research was driven by the attraction of monolithic diode-laser arrays as compact, high-power sources for optical systems applications, where there was a need for a compact source capable of producing a least 1 W of continuous-wave power with complete spatial coherence across the output profile. Some applications, such as free-space optical communications, required single-frequency operation, as well, with spectral linewidths in the MHz range. Beam quality, spatial coherence, and temporal coherence became the central issues in the performance of monolithic diode-laser arrays, as the power capabilities had been established.

Since many optical systems required narrow-angular divergence, diffraction-limited, single-lobed output beams, as well as single-frequency operation, these became the performance goals for the coherent monolithic high-power diode-laser array. These system requirements were a primary reason that real index-guided diode-laser arrays were initially investigated in 1983 [1.70, 1.71], as the gain-guided laser arrays did not exhibit stable far-field patterns with drive current [1.43].

The observed tendency of coupled-waveguide diode-laser arrays was to produce a partially coherent output beam, due to oscillation in several spatial modes, with a predominantly double-lobed output beam. The doubled-lobed output beam occured because adjacent array elements would tend to operate with a 180° relative phase shift. In most cases, this 180° or out-of-phase mode of operation has the lowest losses, as most of the field is concentrated in the regions of gain; and the field in the lossy regions in between the gain elements is minimized. The use of integrated-phase shifters on the diode facet, to compensate for the phase differences between elements had been demonstrated as a possible solution for obtaining single-lobed output beams [1.70, 1.72, 1.73]. Though the more immediate challenge was arriving at a method of obtaining single-spatial mode operation from a monolithic diode-laser array at high-power output levels. To this end, many new designs for diode-laser arrays, aimed at producing single-spatial mode operation and single-lobed output beams from the array source, were fabricated and demonstrated.

The array of coupled leaky-mode lasers [1.74] was proposed and demonstrated as a means for providing stronger coupling between array elements. In contrast to most coupled-waveguide array designs, the leaky-mode laser array was designed so that the refractive index of each active region element was less than that of the surrounding material. In this case, each laser element would radiate coherent light at a slight angle to the angle of propagation into the surrounding material. Thus, the degree of optical coupling in an array of leaky-mode (or antiguided arrays as they are often referred to today) diode lasers is much stronger, as the field generated by each element is effectively coupled to all other elements of the array with nearly equal strength.

For a number of years, the ridge and buried-ridge index-guided waveguide laser arrays seemed to be the preferred design for obtaining good beam quality with high spatial coherence, as the spatial mode discrimination was believed to be less sensitive to the drive level than the gain-guided arrays, and lower threshold current densities were possible [1.75]-[1.87]. New geometries of coupled-waveguide laser arrays such as variable-spaced waveguides [1.88, 1.89], nonuniform, lateral-gain distribution [1.77],[1.90]-[1.95] diffraction coupled [1.96]-[1.108], and diode-laser arrays comprised of networks of branched or "Y" configured waveguides [1.109]-[1.116] were all demonstrated with varying degrees of success. The impact of the shape of the laser-waveguide profile on spatial-mode discrimination was also studied, as it became apparent that in some cases this could be controlled by the fabrication and growth procedure [1.71, 1.86][1.117]-[1.125]. Process improvements led to the development of impurity induced disordering in quanutm-well laser structures. This important development made planar processing of index-guided diode-laser arrays possible [1.126]-[1.131] as well as nonabsorbing mirror structures [1.132, 1.133]. The prospect of planar processing was, and still is, quite attractive from a manufacturing viewpoint; and the non-absorbing mirror structures, can be used to raise the facet-damage limit.

The search for scalable monolithic array designs led to the development of monolithic surface-emitting diode-laser arrays. In surface-emitting diode-laser arrays, light is coupled out of the laser structure in a direction normal to the pn junction. Three types of surface emitting laser array were introduced: 1) the vertical cavity surface emitting array Sect. 7.3 [1.134]-[1.158], 2) the etched-45° mirror surface emitting array, [1.159]-[1.170] and 3) the grating-coupled surface-emitting laser array (Sects. 7.1 and 7.2) [1.171]-[1.195]. These surface-emitting laser approaches were also demonstrated as monolithic two-dimensional laser arrays.

Indeed, minaturized integrated-optical versions of larger hybrid laser systems were fabricated and studied, such as monolithic semiconductor-diode-laser versions of the unstable-resonator laser (Sect. 5.3) [1.196]-[1.215] and monolithic injection-locked laser oscillator-array structures Sect. 8.1 [1.216, 1.217]. Monolithic unstable-resonator semiconductor lasers have produced some promising results. With the continued development of novel integrated optical elements such as the waveguide lens and improvements in the methods of fabrication of curved facets, further improvement in the performance of unstable-resonator semiconductor lasers is expected.

It was found that closley-spaced ($\approx 1 - 2\,\mu$m) arrays of antiguided laser based on leaky-wave coupling (Sect. 6.3) could be optimized to operate with low threshold current densities while exhibiting stronger spatial mode discrimination than was possible in evanescently-coupled index-guided diode-laser arrays or leaky-wave coupled gain-guided arrays [1.218]-[1.235]. Antiguided laser arrays have exhibited performance characteristics superior to those of coupled positive-waveguide laser arrays by demonstrating higher-power outputs combined with a high-spatial coherence output. Variations on the basic grating-coupled surface-emitting array approach led to the development of the monolithic surface-emitting *Master-Oscillator-Power-Amplifier* (MOPA) arrays [1.236]-[1.240] and the monolithic active-grating surface emitting amplified laser [1.241]-[1.247] (Sect. 8.2). The active-grating surface emitting amplified laser has demonstrated high spatial coherence, along with single-frequency operation, at power outputs higher than those of coupled-waveguide laser arrays. Significant progress was also made in broad-area edge-emitting MOPAs (Sect. 8.1) and tapered amplifiers (Sect. 8.1.4) as these approaches, too, demonstrated single-frequency operation and spatial coherence at higher-power outputs than coupled-waveguide arrays. Many of the most promising designs (e.g., antiguided arrays and active-grating amplified lasers) for monolithic diode lasers are pushing the limits of growth and processing technologies, so as these technologies evolve further, better performance characteristics can be expected.

Not all efforts were directed towards the development of monolithic diode-laser structures. The need for a compact efficient solid-state laser pump led to the development of hybrid two-dimensional arrays consisting of a stack of wafer bars of edge-emitting diode-laser arrays alternated with heat-sink material. This is referred to as the *rack and stack* diode-laser array) [1.248]-[1.271]. The need for low-cost, efficient, compact, high-power sources of photons to pump solid-state lasers [1.272] has provided much of the driving force behind further improvements in the monolithic arrays of uncoupled gain-guided diode lasers. The development of silicon microchannel cooler technology [1.273] has produced significant improvements in thermal management for rack and stack diode arrays, so that these arrays are useful for applications requiring high-duty cycle, high-power, pump sources [1.274]-[1.276]. Futher improvements in monolithic etched-facet 2-dimensional surface-emitting laser arrays [1.170] and grating-coupled surface emitting laser arrays [1.184, 1.194] have made surface-emitting architectures attractive also as high-power pumps, especially since back-plane cooling can be implemented to allow for greater heat dissapation. The benefits of back-plane cooling have been a major reason for the upsurge in the development of all types of monolithic surface-emitting diode-laser arrays.

Besides GaAs/AlGaAs diode-laser arrays that operated in the $0.78\,\mu$m to $0.88\,\mu$m wavelength region, arrays were fabricated in InGaAsP for the $1.3\,\mu$m to $1.5\,\mu$m wavelength range [1.96, 1.104, 1.159][1.277]-[1.290], and in InGaAlP for the visible wavelength range [1.291, 1.292]. Diode-laser arrays fabricated from Pb salts were demonstrated in the $4-5\,\mu$m wavelength range [1.293]. As strained-layer quantum-well structures became available, diode-laser arrays were fabricated that operated in the $0.9\,\mu$m to $1.03\,\mu$m wavelength range [1.181],[1.294]-[1.296].

1.2 Nonsemiconductor Laser Arrays

The concept of the phased-locked laser array has also been proposed for gas waveguide laser sources that emit in the far-infrared and the submillimeter wavelength ranges such as CO_2 [1.297, 1.298] as well as with solid-state lasers in external cavities [1.299]. Besides higher-intensity outputs offered by arrays, the interest here was also to obtain an electronically steerable output beam from a coupled array of waveguide lasers [1.297, 1.298]. In some applications, the ability to steer the higher intensity output beam of a waveguide-laser array at electronic frequencies would be quite an advantage.

Comparatively little work has been done on laser arrays in gas and solid-state lasers materials. From a performance standpoint, the semiconductor-diode laser has a modal gain of $\approx 100\,\mathrm{cm}^{-1}$, which is much higher than any other laser material. Therefore, the additional losses that are introduced by optically coupling the individual array elements are less likely to degrade the power output of semiconductors. By comparison most solid-state and gas laser materials have much lower gains, typically less than $1\,cm^{-1}$. The coupling losses associated with the array structure will have to be much lower than the available gain, so that the power performance is not degraded. The compact size and methods of fabricating semiconductor lasers are quite conducive to making monolithic laser arrays, as the dimension of the array cavity are determined photolithographically to within a fraction of a micron.

Phased-locked operation of CO_2 waveguide-laser arrays has been demonstrated. The coupling between adjacent CO_2 waveguide lasers was achieved by using transparent partitions between adjacent waveguide lasers [1.300] and with external cavities [1.301, 1.302]. Other approaches employed corrurated electrodes [1.303], as this was easier to implement than placing transparent windows between adjacent waveguide lasers. More recent approaches have focused on tailoring the morphology of the electrodes to obtain phase-locked operation [1.304]-[1.306]. Coherent phasing of an array of neodynium-phosphate glass rods in an external cavity was studied by *Vasil'ev* et al. [1.307]. One of the motivations for using the array was that a group of smaller laser rods could be cooled more effectively than a single larger rod. In situations where high-average power is desirable, thermal stresses degrade the wavefront of the output beam, and can compromise the mechanical integrity of the laser rod as well. Other approaches have involved the use of Nd^{3+}: glass fiber bundles [1.308, 1.309] and the use of multiple laser rods arranged along the axis of a single resonator [1.310, 1.311].

1.3 Contributing Factors to Developing Monolithic Diode-Laser Arrays

The important technological advances that have facilitated the rapid development in monolithic diode-laser arrays over the last ten years are reviewed in Sect. 1.3.1. Of course, these same technological advances have led to parallel improvements in the perfromance characteristics of single-element diode lasers. In Sect. 1.3.2, the physical constraints that have limited the power output of single-element diode lasers are discussed. Section 1.3.3 treats the traditional distinction and terminology which has been used to distinguish between single-mode and multi-mode diode lasers and diode-laser arrays. The need for compact, highly-efficient, coherent light sources with power outputs that well exceed the physical limits of conventional single-element diode lasers has been the primary motivation for developing monolithic diode-laser arrays. This need stems from the many applications that would benefit from using diode-laser array sources. These applications are discussed in Sect. 1.3.4.

1.3.1 Key Technological Advances

Steady advances in epitaxial growth of semiconductor materials [1.312]-[1.319] and processing technologies [1.320]-[1.322] have led to the emergence of the III-V semiconductor-diode laser [1.25, 1.26][1.323]-[1.325] as the most compact and efficient laser source that is presently available. By using different compound semiconductors composed of group-III and group-V elements [1.292, 1.326], operating wavelengths of $\approx 0.78 - 0.88\,\mu m$ from GaAs/AlGaAs, $\approx 0.9 - 1.0\,\mu m$ from InGaAs/GaAs, $\approx 1.3 - 2.0\,\mu m$ from InGaAs/InGaAsP, $\geq 2,\mu m$ from GaInAsSb/AlGaAsSb, and $\approx 0.63 - 0.69\,\mu m$ from InGaAlP are all possible. In the last decade, the development of quantum-well structures and their subsequent application to semiconductor-diode lasers has produced devices with many superior characteristics such as lower-threshold current density, small temperature dependence of threshold current, better efficiency, and good dynamic characteristics [1.327, 1.328]. Such advancements have made it possible to fabricate semiconductor diode lasers with power conversion efficiencies of more than 50 % [1.60][1.329]-[1.331]. More recently strained-layer quantum-well structures of InGaAs/GaAs [1.296, 1.332] and InGaAsP/InGaAs [1.333] have been used in diode lasers to engineer the band structure [1.334]-[1.336] to obtain even lower threshold current densities than are possible with unstrained quantum-well lasers. Strained-layer InGaAs/GaAs diode lasers have demonstrated continuous-wave operation at temperatures in excess of 200° C [1.337]-[1.339]. Although, this may have more to do with better carrier confinement, as lattice-matched GaAs/InGaAlP double-heterostructure diode lasers have recently been observed to operate at 212°C [1.340]. In addition, significant progress has been made in prolonging the lifetime of diode lasers, and reliable operation of over 10,000 hours has been reported by a number of laboratories [1.341]-[1.344].

Under forward-bias operation, quantum-confined diode laser structures effectively contain the injected-carrier densities $\approx 10^{18}\,cm^{-3}$ so that radiative recombination occurs with near-unity quantum efficiency. Small-signal modal gains for quantum-well structures on the order of $100\,cm^{-1}$ or more are possible; while the optical losses are reduced to a few cm^{-1} [1.332],[1.345]-[1.347]. Quantum well structures can provide exceedingly high bulk gains, typically in the range of $3000 - 5000\,cm^{-1}$ [1.348], using relatively-low injected carrier densities of the order of several kA/cm^2. At any given level of injected carrier density, the very thin quantum wells (typically $25 - 100\,Å$ in thickness), support higher carrier densities than conventional double-heterostructure lasers, and this creates a larger population inversion. Another factor that contributes to increasing the population inversion is the lower density of states in the conduction and valence bands associated with the two-dimensional nature of the quantum-well active layer. Even higher gains are achieved by using multiple quantum wells in the active layer [1.327, 1.349]. In unpumped quantum well waveguides, the loss can be saturated to values less than $10\,cm^{-1}$ at optical powers of only a few mW, which means that an unpumped diode laser section can also act as a low-loss waveguide. This saturable loss property of quantum wells has enabled many types of semiconductor laser arrays to be fabricated. The numerous benefits of quantum-well structures have made them the preferred structure for high-power diode lasers.

1.3.2 Performance Limitations of Single-Element Diode Lasers

In spite of the improvements in efficiency, reliable single-mode continuous-wave operation of conventional single-element semiconductor-diode lasers has been achieved only up to power outputs of about one hundred milliWatts, mainly due to the catastrophic facet-damage limitations [1.341, 1.342, 1.350, 1.351]. Here, what is meant by single-mode operation is simultaneous single longitudinal mode (or single frequency) and single spatial-mode operation. One hundred milliWatts does not sound like very much output power, especially when compared to the multi-Watt power-output capabilities of many gas, liquid, and solid-state laser systems, but most conventional single-element diode lasers capable of single-mode operation typically have optical-mode cross-sectional dimensions of only about $1\,\mu$m by $4\,\mu$m. At cw power outputs $\approx 100\,$mW, this small modal area corresponds to a power density of $2.5\,$MW/cm^2 at the facets, which is very close to the material damage limit. Localized heating at the cleaved facets occurs because of absorption of the laser light and nonradiative recombination of the injected carriers [1.341]. Such heating can cause ablation of material from the facet, localized melting of the facet, and generate other material defects that significantly degrade the laser performance characteristics or result in complete failure of the laser diode. The maximum-possible intensity output from a diode laser is considerably more than the that allowed by the catastrophic facet-damage limit, as presented in Sect. 2.8.3.

Through the use of non-absorbing mirror structures [1.341, 1.342],[1.350]-[1.352] as well as specialized facet coating and passivation techniques [1.353, 1.354], significant progress has been made in increasing the power density above the aforementioned $\approx 2.5\,$MW/cm^2 where catastrophic facet damage occurs. Non-absorbing mirror structures have led to an increase in diode-laser projected lifetimes of 100,000 hours or more at 25° C under cw operation at $100\,$mW power outputs [1.341, 1.342],[1.350]-[1.352]. These lasers can produce higher cw power outputs, in the $400 - 500\,$mW [1.342, 1.352, 1.354]. Although, at these higher powers only single spatial-mode operation with a multi-longitudinal-mode spectral output was observed, and the device lifetime was not determined.

There is a clear benefit to be had in developing processes and device designs that increase the catastrophic facet-damage limit and extend device lifetimes at these higher powers. Recently, performance characteristics of lasers with passivated cleaved-facets [1.354] have been reported with maximum cw power outputs of 425 mW. In addition, the power vs current characteristics for this device suggested that the power output was limited by thermal saturation and not catastrophic facet damage. One interesting approach for increasing the single-mode power output involves tapering the single-mode laser-waveguide structure to larger widths in the vicinity of the facet to reduce the power density [1.352]. With the improvements in material quality, cw power outputs as high as 600 mW in both a single spatial and spectral mode have been obtained from single-element diode lasers that were $7\,\mu m$ wide with a cavity length of $800\,\mu m$ [1.355]. This has been attributed to the wider index guide which lowers the intensity at the facet and the longer cavity which reduces the threshold carrier density. Because of the significant potential commercial value of high-power, single-mode diode lasers, many of the techniqes for raising the facet damage limit are regarded as proprietary. Although, the improved performance characteristics of the resulting devices has been reported in the technical literature, the specific details of the processes are usually not disclosed in these reports.

Steady improvements in diode-laser lifetimes and increases in the facet-damage limit intensity are very likely in the future, as work continues in this area. Even if facet damage is eliminated and thermal management is not a limiting factor, the maximum output power density possible from semiconductor-diode lasers with presently available material is $\approx 20\,\mathrm{MW/cm^2}$. This limitation is imposed by the optical losses due to free-carrier absorption in the semiconductor material, which is treated in more detail in Chap. 2. Therefore, the theoretical-maximum cw power output (independent of length) from a single-element laser with cross section dimensions of $1\,\mu m$ by $4\,\mu m$ is $\approx 1\,\mathrm{W}$. Thus, the conventional single-element laser design for single-mode diode lasers is marginal as a source for producing $\geq 1\,\mathrm{W}$ in a single mode under cw operation; and for power outputs of $10\,\mathrm{W}$ or more, other approaches will certainly be required. To obtain single-mode operation at $\geq 1\,\mathrm{W}$ power output from conventional cleaved-facet semiconductor-diode laser sources and operate at an intensity level that will not compromise lifetime, the active layer width will have to be increased well beyond $\approx 5\,\mu m$.

1.3.3 Distinction Between Coherent and Incoherent Diode-Laser Arrays

Many approaches and concepts have been proposed for increasing the lateral extent of the conventionally cleaved facet single-element diode laser to obtain $1\,\mathrm{W}$ or more of cw power output under single-mode operation. In this way, the power density at the facet can be kept below the facet damage limit, as the width is scaled to generate higher-power outputs. If one is not concerned about the spatial- or temporal-coherence properties of the output beam, then simply increasing the active layer width can be the simplest and most effective method for increasing the power output. Although, this approach is limited to widths of about $100\,\mu m$. This limitation occurs because amplified spontaneous emission propagating in the lateral dimension begins to build up and compete for the available gain. This parasitic type of process depletes the available gain for light that is coupled out of the cleaved facets, and can cause reductions in the power output. When the width of the active layer is increased much beyond the several micrometers, laser oscillation in multiple higher-order spatial modes can occur; and the spatial coherence of the output beam is reduced. Hence, the radiometric properties of single-mode and multi-mode diode-laser arrays are often the most important distinguishing characteristic.

(a) Coherent Diode-Laser Arrays

Most approaches for scaling to get higher single-mode power output (in the $1 - 10\,\mathrm{W}$ range) have involved optically coupling many single-element diode lasers at the wafer level, so that they all operate in a mutually coherent fashion (i.e., single-frequency and single-spatial mode, with complete transverse spatial coherence). As is well-known from electromagnetic diffraction theory [1.356], a linear array of M identical coherent sources (for our discussion each source is a diode laser) with each laser emitting light with a wavelength λ, each laser of width w, with a uniform spacing d bewteen adjacent lasers, and each laser emitting light with zero relative phase difference, will have a predominantly single-lobed output beam with a far-field angular divergence of about λ/Md radians at the first nulls in the far field.

An important figure of merit that is used to characterize the energy content of a narrow-divergence beam of light is the *photometric brightness* [1.356] or *radiance* [1.357]. It is defined as the *power per unit area per unit solid angle* that is emitted from a source. Its importance stems from the fact that, the photometric brightness or radiance is conserved as a beam of light propagates through a lossless optical system [1.357]. If each element of the linear array of M identical coherent lasers has a coherent power output per unit area of I, which is independent of the array size, the photometric brightness, will be given by IMd/λ. Thus, the photometric brightness of the coherent array scales with the number of array elements.

Another figure-of-merit that is used to characterize the beam quality of a source, is the *power per unit solid angle*. In the field of radiometry, this is refered to as the *photometric intensity* [1.356] or *radiant intensity* [1.357]. These terms can sometimes be confusing because light intensity is also defined as the power per unit area. In the diode laser community, the term *power-in-the-bucket* is frequently used to refer to the photometric or radiant intensity. The power-in-the-bucket is the power contained within a specific angular beam-divergence of the far-field pattern, that is referred to as the bucket. Typically, the angular beam divergence, or bucket, is selected to correspond to what would be expected were the near-field aperture of the source (a laser array for this discussion) to be illuminated by a plane wave of constant intensity. In the case of a one-dimensional aperture with uniform illumination, the diffraction-limited power contained within the full-width-to-the first zeros of the Airy disk is about 90 %. The beam divergence that corresponds to the full width between the first zeros of the Airy disk is given by $2\lambda/D$, where D is the size of the near field aperture [1.358]. For two-dimensional sources the diffraction-limited powers-in-the- bucket are typically $80 - 86\%$. In actuality, the near field of the laser array is usually not uniformly illuminated, so the power in the dominant-lobe of the far-field beam will be somewhat less than the aforementioned values. Applying the power-in-the-bucket figure of merit to the linear array of M identical coherent lasers, one finds that the photometric intensity or power-in-the-bucket will be proportional to M^2, since the total power output of the array increases as M, and the angular beam divergence decreases as $1/M$. Thus the far field characteristics of the coherent laser array are analogous to those of phased-array radar. In principle, the relative phases between elements of the laser array can be varied in a systematic way to give beam steering.

(b) Incoherent Diode-Laser Arrays

A similar array of M identical lasers operating so that the all elements are mutually incoherent (though each element itself may possess a high degree of temporal coherence) will not give an increase in photometric brightness as the array size is increased. In the case of the incoherent array, the predominantly single-lobed output beam has an angular divergence λ/w, where w is the width of a laser element of the array. The output beam divergence is determined solely by the width of the laser-array element, and is independent of the size of the array. Therefore, the photometric brightness is independent of the size of the array and the photometric intensity or power-in-the-bucket will only be proportional to M.

(c) Terminology

The laser array with mutually coherent elements is a source where the brightness will scale with the array size, whereas for a laser array which has no mutual coherence, the brightness is independent of the array size. This distinction between a coherent and an incoherent array of sources is widely used, and it is serves as the basis for the terminology that is used to distinguish the two types of semiconductor diode laser arrays. Diode-laser arrays that operate as the example of the linear array of identical coherent lasers, will usually produce an output beam with high-temporal coherence and complete spatial coherence across the output beam profile. These arrays are frequently referred to as *coherent diode-laser arrays, phased-locked diode-laser arrays*, or *high-brightness diode-laser arrays* in the technical literature. The ideal coherent diode-laser array operates in a single-spectral and single-spatial mode. In essence, the entire array structure can be thought of as a single laser. However, most types of diode-laser arrays, and even single-element broad-area diode-lasers, operate multi-spectral and multi-spatial mode. Diode-laser arrays with these characteristics are often refered to as *incoherent diode-laser arrays* in the literature, because their far-field output beam characteristics resemble those of an array of lasers that are not mutually coherent. Though each laser array element and spectral mode may have a high degree of temporal coherence, there is little or no temporal or spatial correlation between different laser elements of the array; and there is not complete spatial coherence across the array.

1.3.4 Applications for Diode-Laser Arrays

Because of its compact size, high efficiency, and diversity of wavelengths, the semiconductor-laser array is the preferred source for many applications. Concurrent with the developement of the diode-laser array, there has been a high level of effort devoted to developing innovative optical systems for applications where the diode-laser sources would offer significant benefit over other optical sources. There is a wide variation in the performance requirements and specific design of diode-laser arrays for various applications that have been considered. A single diode-laser array design could not possibly satisfy all application needs, since the requirements for many applications conflict with each other.

Some of the earliest applications considered, such as optical data storage [1.359], data transmission [1.360], and illuminators [1.361], proposed the use of hybrid diode-laser arrays comprised of many discrete diode lasers mounted in a single package. Although, approaches for fabricating monolithic diode-laser arrays were also beginning to be investigated at about this time [1.362]. For these applications, the array elements were not to be optically coupled to each other. In optical recording, the output from a single-mode laser is focused to a diffraction-limited spot size of $\approx 1\,\mu m$ for the purposes of reading data from a storage medium. In this application, the parallelism offered by an individually-addressable N-element array of diode lasers leads to an N-fold increase in the data rate. However, to meet the tolerances on the spatial and angular registration on the array, it was necessary to use monolithic diode-laser array designs, where the photolithographic techniques used in the semiconductor fabrication process could be used to define the array geometry with sufficient precision. It is also necessary that all elements in the array have nearly-identical operating characteristics. Over the last decade, significant progress has occured in developing individually-addressable monolithic diode-laser arrays for optical recording [1.363]-[1.378]. Other applications where individually-addressable diode-laser arrays would be beneficial are displays [1.379]-[1.383], laser printing [1.384]-[1.388], optical interconnects [1.389]-[1.398], and optical processing [1.399]-[1.403].

For some applications, diode-laser arrays are necessary to produce power levels that are well-above the capabilites of single-element diode lasers. The pontential advantages of semiconductor lasers for optically pumping solid-state lasers was realized within a few years of the first demonstration of the GaAs injection laser [1.404, 1.405]. With the improvements in relability, high-power, highly-efficient, diode-laser arrays have become very desireable as pump sources for solid-state lasers. In the last decade, the diode-pumped solid-state laser has received much attention [1.248]-[1.271], and it has provided the impetus for improving the material quality and reliability of high power diode-laser arrays. Other applications that require the high-power capabilities of diode laser are optically-activated switching [1.407]-[1.412] and power transmission [1.413]-[1.415].

The more challenging applications requirements to be met with diode-laser arrays have been those that demand high-power operation, $\geq 1\,$W along with good-beam quality and single-frequency operation. These applications require the high coherence of the laser in addition to the high-power output and high efficiency. The potential advantages of laser-radar systems using phased arrays of lasers had been recognized [1.416, 1.417]. Diode-laser array sources have been investigated for applications in laser-radar systems intended for ranging [1.418, 1.419], as well as, for three-dimensional imaging and velicometry [1.420, 1.421]. Vision systems, based on diode-laser arrays, for short-range applications have also been proposed [1.422, 1.423]. Optical flow sensors for Doppler anemometry using diode lasers and diode-laser arrays have also been demonstrated [1.424, 1.425]. Free-space optical communications is an application that requires a high-power highly-coherent source. Since the required power has been $\geq 1\,$W, diode-laser arrays have been proposed and investigated [1.426]-[1.437] as candidates for the transmitter.

1.4 Design Concepts for Coherent Diode-Laser Arrays

Even with the facet-damage limit on intensity output, one can easily see that the projected power outputs from a semiconductor laser source $\approx 1\,$cm^2 should be in the kW range. Indeed, the prospect of obtaining such high-power levels from a compact source with an efficiency approaching 50 % has provided the primary motivation for research and development of monolithic diode-laser arrays. The precission of photolithography techniques used in semiconductor processing, allow for the fabrication of optical waveguides and other integrated optical structures to sub-micrometer tolerances. The capability to integrate the mode control elements within the semiconductor laser or laser array structure is the basic approach that has been followed for developing coherent monolithic diode-laser arrays. Once properly fabricated, the alignment of all optical elements within the structure of the monolithic semiconductor diode-laser array has been established; and ideally, coherent operation would be obtained without having to realign the mode-selective elements.

Unlike the external cavity designs used with many other types of lasers, the monolithic semiconductor-laser array should be much more robust to pertubations and misalignment due to the external environment. However, monolithic laser-array structures are subjected to the harsh internal environment of the active semiconductor-laser material itself; and in order for a monolithic array structure to operate single mode, it must be robust to this environment. A strong coupling between the carrier density (or gain) and the local refractive index exists in semiconductor-diode lasers (Sect. 2.6). This can cause a significant degradation in the spectral and spatial coherence of the laser array output, especially at the high drive levels required for high cw power output. Also, those carriers that do not contribute to stimulated radiative recombination, recombine non-radiatively and generate heat in the active layer. This heating degrades the electrical-to-optical conversion efficiency. These effects are discussed in more detail in Chap. 2.

In the most general sense, a monolithic semiconductor diode-laser array comprises a group of individual diode-laser gain elements that are fabricated together on the same semiconductor wafer. As with any semiconductor laser, optical gain is established in each diode element by applying an appropriate forward-biased to generate electron-hole recombination radiation within the optical waveguide structure of each structure. This can be accomplished by electrically addressing each gain element individually or by addressing all or various combinations of elements through common contacts. Optical coupling of different gain elements is accomplished by incorporating the diode-laser gain elements within a network of passive optical waveguides that transmit light between some or all of the gain elements on the chip. In essence, monolithic semiconductor diode-laser arrays are a type of integrated optical circuit.

A monolithic diode-laser array may consist of more than just the semiconductor gain elements and the end reflectors. Other integrated optical elements such as gratings [1.194, 1.195], total-internally reflecting etched-facet mirrors [1.170], waveguide lenses [1.215, 1.438], and spatial filters can be incorporated within the waveguide array network to provide the spectral and spatial mode discrimination necessary to obtain coherent operation of a diode-laser array or extended diode-laser source. For the ideal coherent array, this must occur in such a way, that all of the laser-array elements operate at the same frequency, have a high degree of mutual coherence, and emit in-phase with respect to each other. A slightly-less-than ideal design for a coherent array, though still quite acceptable, would relax the in-phase emission requirement, and permit a systematic phase variation over the extent of the array. Such a phase variation can be converted to a uniform phase with an external phase plate [1.70]-[1.73]. Many designs and concepts for such arrays have been proposed. The challenge has been, and still remains, to fabricate the array with sufficient tolerances that will produce a monolithic-array structure where all elements will operate with a high degree of mutual coherence at power outputs, that are significantly higher than those achievable with single-element diode lasers. So far, most monolithic diode-laser array structures, that have been demonstrated, have only operated as coherent arrays at output powers somewhat less than or, comparable to, those obtainable from the best single-element diode lasers. Although, recently monolithic tapered-diode-laser amplifiers [1.246][1.439]-[1.441] and oscillators [1.439, 1.441] have produced power outputs in excess of 1 W cw with good beam quality. In the last few years, the field of approaches seems to be focusing on antiguided-laser arrays (Chap. 6), *Master-Oscillator-Power-Amplifier* (MOPA) arrays (Chap. 8), and unstable resonator semiconductor diode lasers (Chap. 5).

1.4.1 Laterally Coupled Diode-Laser Array Concepts

The structure of the gain-guided laterally-coupled diode-laser array is illustrated in Fig. 1.6. The optical coupling occurs in the direction perpendicular to the optical axis of the cavity, which is defined by the propagation direction of the mode. Here the diode-laser array elements are fabricated in sufficiently close proximity, and are appropriately spaced in the lateral direction, so that the optical field generated by each element is coupled to other elements in the array. Optical coupling can occur either via evanescent-wave coupling or radiation-mode coupling. The nature of the coupling between the waveguides depends strongly on the relative dielectric permitivities of the diode-laser waveguide array elements, and the intervening regions that mediate the optical coupling between array elements. Moreover, the diode-laser array structure imposes a lateral variation in the gain and index of the semiconductor; and the details of this variation are the critical factor in determining the spatial mode discrimination properties of monolithic laser arrays [1.442, 1.443]. This is discussed in more detail in Chap. 3 and in Chap. 4.

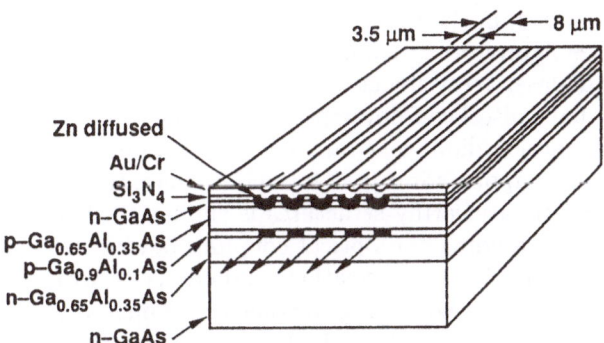

Fig. 1.6. Schematic drawing of the generic gain-guided laterally-coupled diode-laser array. [1.40]

In Fig. 1.7, the Y-guide diode-laser array [1.109]-[1.114],[1.444],[1.445]-[1.448] is depicted. Though similar in structure to the diagram of a branched-waveguide diode laser array that appears in Fig. 1.5, the Y-guide is configured symmetrically to act as a spatial filter to provide mode discrimination. The idea is that only the array mode which is in phase will combine constructivey at the junctions where adjacent laser waveguides come together [1.444, 1.445, 1.449, 1.450]. In this situation, the coupling between adjacent

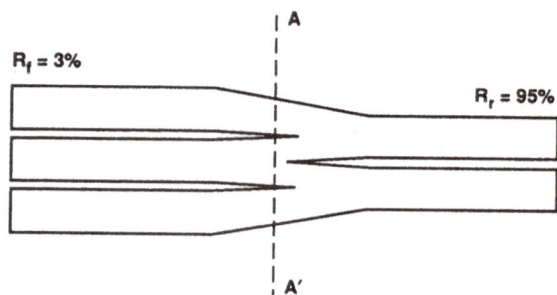

Fig. 1.7. Schematic drawing of a Y-guide diode-laser array. [1.110]

array elements occurs via direct injection at the junctions of the Y guides. Hence, the straight waveguide sections of the array are placed sufficiently far apart so that evanescent coupling does not occur. However, variations on the basic Y-guide approach, incorporating evanesecent coupling between the straight sections of the waveguides, have also been studied [1.451]-[1.455], as well as diffraction coupling [1.456].

Though these device are very interesting in their own right, the benefits of using such multiple-coupling schemes to obtain coherent opertion are not entirely clear. For example, *Mikhal'kov* et al. [1.455] has shown that the presence of weak-evanescent coupling in a Y-guided array can create conditions that favor the excitation of high-order modes at thereshold. Ideally, the waveguide branch is designed so that half of the power traveling in the single guide goes into each branch in the fundamental mode. This is best accomplished if the branch is configured as a symmetric Y-shaped branch.

Several diode-laser array structures incorporating waveguides in branched configurations have also been developed and studied experimentally. As examples, the X-junction diode-laser array [1.115] is indicated in Fig. 1.8, and the tree-array laser appears in Fig. 1.9 [1.116]. The intended mode selection mechanism in these laser-array structures is similar to that of the Y-guide, but with idea of extending the optical coupling beyond adjacent laser elements. Though the initial demonstrations of diode-laser arrays incorporating Y-guide networks were encouraging, further experimental studies indicated these arrays did not operate with a high degree of lateral spatial coherence because of wavelength variations between the array elements [1.457]. Subsequent theoretical analysis of the above-threshold characteristics of Y-guided diode-laser arrays, showed that the projected-mode discrimination properties were severely compromised as a result of non-uniform saturation of the gain [1.458, 1.459]. Given that optical coupling in Y-guide occurs only between adjacent elements, these arrays are very susceptible to any nonuniformity.

Nevertheless, branched-waveguide networks are useful for forming monolithic diode-laser array structures comprised of parallel networks of traveling-wave amplifiers [1.239, 1.460]. Figure 1.10 illustrates an example of a branched-waveguide network structure that was used by *Krebs* et al. [1.460]

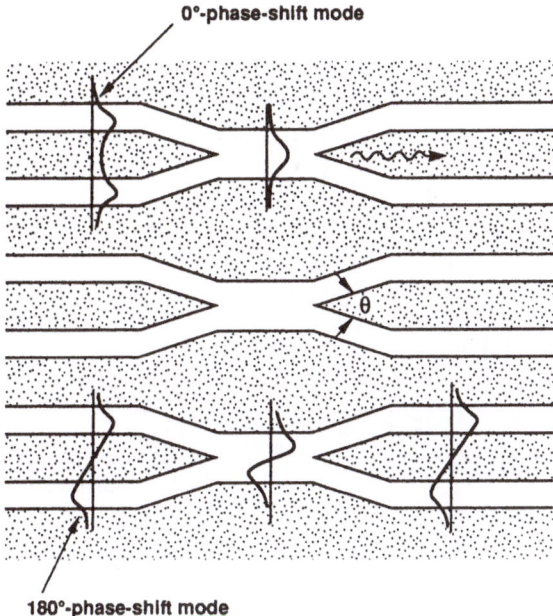

Fig. 1.8. Schematic drawing of an X-guide diode-laser array. [1.115]

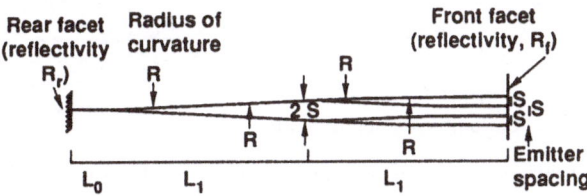

Fig. 1.9. Schematic drawing of a tree-structured diode-laser array. [1.116]

to demonstrate a diode-laser array comprised of 400 traveling-wave amplifiers driven by a single oscillator.

Another method for coupling diode lasers at the wafer level involves injecting light that is not confined in the lateral dimension. Such devices are referred to as *diffraction-coupled* or *interferometerically-coupled* laser arrays. [1.6], [1.96]-[1.105],[1.461, 1.462]. In this approach, the monolithic diode-laser array waveguide elements are all integrated to a common broad-area waveguide section. This broad-area waveguide section still provides waveguiding in the transverse direction; but, it does not provide any waveguiding in the lateral direction.

Schematic illustrations of this type of diode-laser array appear in Figs. 1.11 and 1.12. When the guided-mode of each diode-laser array element enters the broad section of waveguide, it begins to diffract. Alternatively, one could view this as each waveguide mode couples into a superposition of modes of the

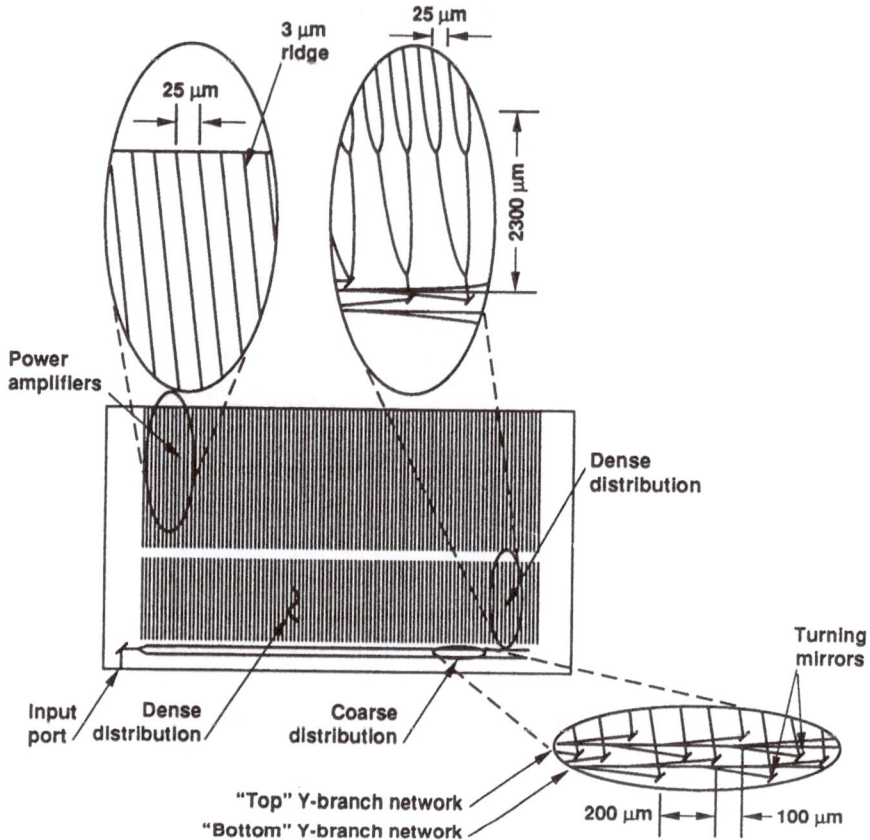

Fig. 1.10. Schematic drawing of the array of 400 traveling-wave amplifiers. [1.460]

broad-area waveguide [1.103]. In any event, as the light continues propagating through the planar-waveguide section towards the facet; and upon reflection and propagation back towards the array, all the light beams corresponding to the array elements will spread and begin to overlap each other. If the diode-laser waveguide structure is resumed at a point where light from adjacent array elements is injected with a 2π phase shift, then the in-phase mode of the array should be reinforced, while other array modes are suppressed [1.97].

Another type of interferometrically-coupled array is illustrated in Fig. 1.13, where the central section of the diode-laser waveguide sections are interupted by a section of broad-active area that is configured to act as a spatial filter [1.106]-[1.107][1.222, 1.230, 1.463, 1.464]. In these types of monolithic diode-laser arrays, the structures have been generally designed to take advantage of the self-imaging properties (or Fresnel imaging) of periodic objects illuminated with coherent light to provide mode-selection in the diffractive coupling [1.465, 1.466]. Moreover, a one-dimensional array of objects of period d_c, illuminated with coherent light will reimage at a distance $Z_T = 2d_c n/\lambda$, where n

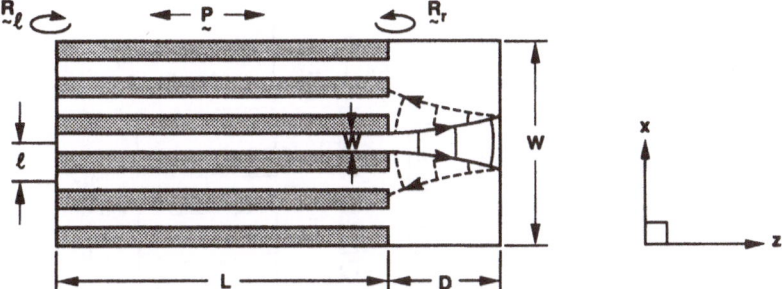

Fig. 1.11. Schematic drawing of a diffraction-coupled diode-laser array with the coupling region at one end of the array. [1.103]

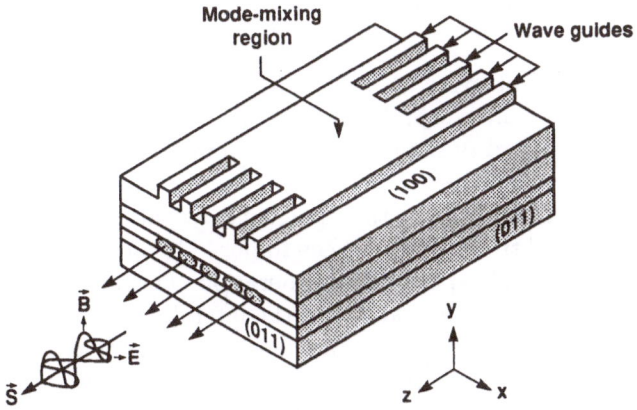

Fig. 1.12. Schematic drawing of a diffraction-coupled diode-laser array with the coupling region in the center of the array [1.99].

and λ are the refractive index and wavelength of light. This property of coherent light can be used to configure an intracavity spatial filter that will block or provide high losses to all array modes save the desired in-phase mode. The use of intracavity spatial filters in combination with the reproductive imaging properties of periodically distributed lasers to produce phase locking has been studied for external cavity lasers [1.467] and demonstrated with semiconductor [1.461, 1.468, 1.469, 1.470] and nonsemiconductor [1.302, 1.307] gain media.

Such spatial filters that operate on the basis of the Fresnel zone self-imaging properties of periodic arrays are often referred to as *Talbot filters*, as it was *Talbot* [1.465] who reported the first observation of self-imaging of gratings illuminated with coherent light. For this reason, the quantity Z_T is often called the *Talbot distance*. The improvements in sub-micrometer fabrication techniques such as focused ion-beam micro-machining [1.471]-[1.473] and e-beam technology have made it possible to fabricate intracavity spatial

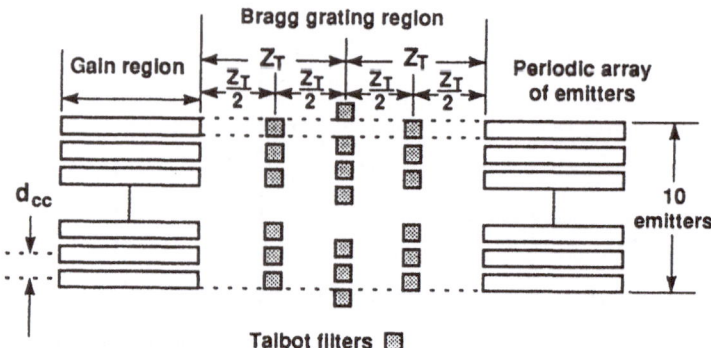

Fig. 1.13. Schematic drawing of diode-laser array that contains an intracavity spatial filter. The non-transmitting elements of the filter are placed at half multiple of the Talbot distance Z_T [1.463].

filters in diode-laser arrays with sufficiently high precision so that insertion losses can be kept at an acceptable level.

In the diode-laser array structures depicted in Figs. 1.6-1.13, the width of the active layer has effectively been increased in the lateral direction to provide more power output; and at the same time, a spatially-periodic variation in the gain and refractive index has been imposed in an effort to obtain mode discrimination and single-spatial mode operation. The question at hand then is how does one engineer a periodicity in the lateral-gain and refractive-index profiles, as well as the boundaries, of the semiconductor laser so that only a single-spatial mode will oscillate over all ranges of operation? Mathematically, it is a complex problem especially in semiconductors where there is a strong coupling between the gain and the index of refraction. The modes of the diode-laser array are solutions to Maxwell's wave equation with the spatially-varying permittivity and boundary conditions that are appropriate to the array structure. Each spatial mode of the array has a different lateral phasefront distribution and operating wavelength. Increasing the width of the laser array by adding to the number of elements will always increase the number of spatial modes, and this is usually accompanied by an increase in the density of spatial modes. For a given array structure to have good mode discrimination properties, all excess spatial modes should have significantly higher threshold gains than the spatial-mode that is desired for operation. Operation in a single-spatial mode does not always guarantee operation in a single longitudinal mode, especially at high drive levels where gain saturation and thermal effects become more important.

1.4.2 Longitudinally Coupled Diode-Laser Array Concepts

Most diode-laser array structures utilize either diffraction or evanescent coupling in the lateral direction, and this approach yields a one-dimensional array. To obtain a two-dimensional array, the diode-laser array elements must be distributed, and optically coupled, in a planar geometry with a suitable optical-coupling arrangement that emits the output beam from the wafer surface. To this end, longitudinally-coupled diode-laser array approaches have also been persued. To get efficient power extraction from a longitudinally-coupled array generally necessitates some form of distributed-output coupling from the surface of the wafer. This can be accomplished by using grating-output couplers [1.194, 1.195, 1.474], as illustrated in Fig. 1.14. Other types of two-dimensional grating-coupled surface-emitting laser arrays are described in more detail in Sects. 7.1 and 7.2, and two-dimensional vertical cavity laser arrays are presented in Sect. 7.3.

In this particular example, the gratings are configured as second-order Bragg reflectors. Then the first diffraction order of the grating is radiated normal to the wafer surface, which is out of the page of Fig. 1.14, to provide the output beam, and the second diffraction order of the grating provides feedback in the longitudinal direction along the waveguide. In this array, each array element or unit cell comprises a *Distributed-Bragg-Reflector* (DBR) laser, as depicted in the inset of Fig. 1.14. The particular 2-dimensional array geometry displayed in Fig. 1.14 is obtained by configuring a linear array of longitudinally coupled DBR lasers in a serpentine structure. Adjacent 1-dimensional arrays are optically coupled by integrated corner-turning mirrors, at each end, that deflect light in the plane of the wafer [1.289, 1.290].

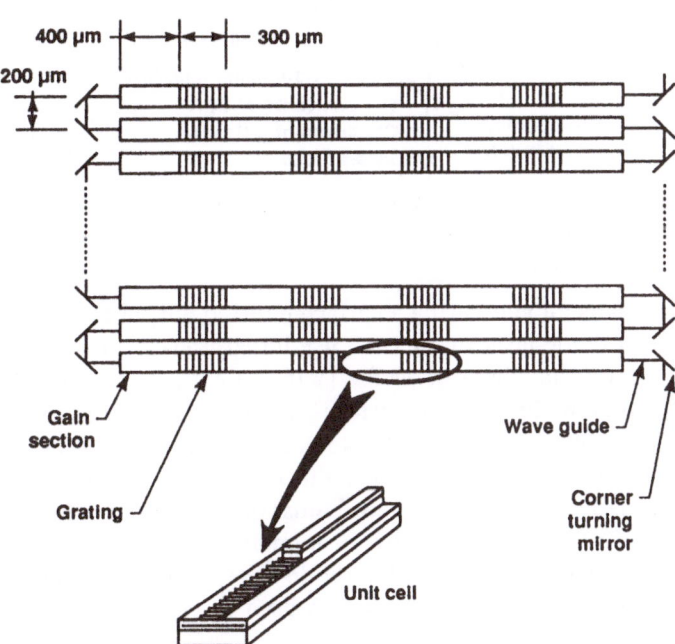

Fig. 1.14. Illustration of longitudinally-coupled grating-surface-emitting diode-laser array. With the serpentine configuration, the entire surface of a wafer can be used as a two-dimensional array [1.194].

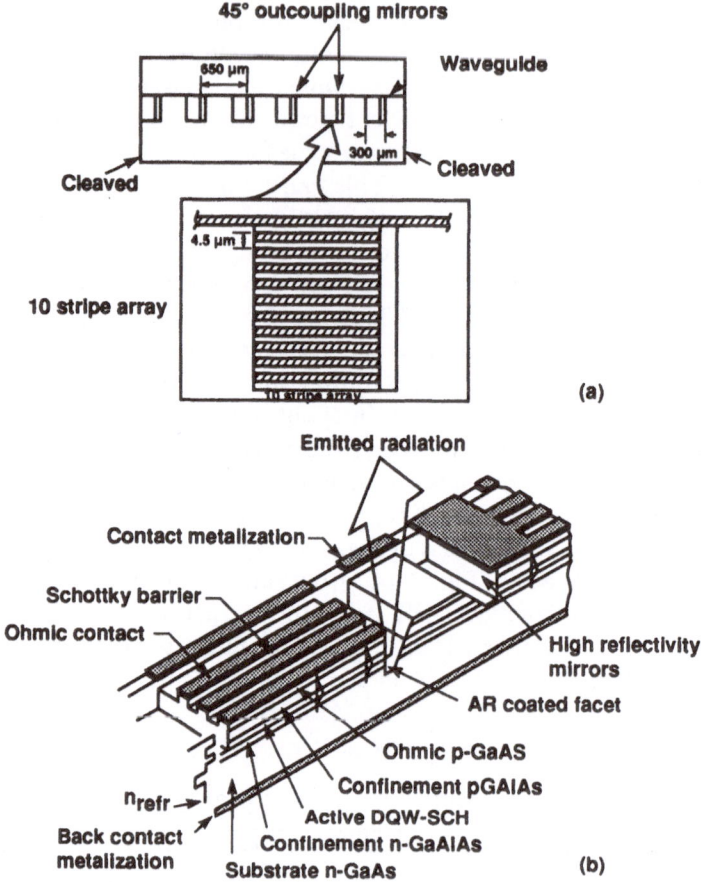

Fig. 1.15. Illustration of a longitudinally-coupled diode laser that employs integrated beam deflectors to provide light emission normal to the wafer surface. [1.475]

Another means for generating a surface-emitted output beam in a longitudinally-coupled diode-laser array is to use integrated-beam deflectors to deflect light out of the plane of wafer [1.161, 1.170, 1.475]. This approach is primarily used in incoherent surface-emitting diode-laser arrays [1.159, 1.160],[1.162]-[1.169],[1.288],[1.476]-[1.480]; however, it can be used in coherent diode-laser array designs. Appearing in Fig. 1.15 is a schematic representation of the design of [1.475], where banks of laterally-coupled diode-laser arrays are longitudinally coupled by a common waveguide on the left side of the structure. Each laterally-coupled diode-laser array emits light from one end, and an integrated-beam deflector is used to redirect the light so it is emitted normal to the surface of the wafer.

1.4.3 Output Coupling Mechanisms

Another consideration in the design of high-power diode laser arrays is the type of output coupling that is used. There are basically two types of output coupling methods available with semiconductor diode lasers, end or boundary-output coupling and distributed-output coupling. The most common type of coupling used though is end coupling, and it is used in most lasers configured as edge-emitting Fabry-Perot lasers. Here the power output is transmitted through a partially reflecting mirror at one end of the laser cavity. This is the type of output coupling that would be used in the laterally-coupled array structures of Figs. 1.6-1.13. The types of output coupling, distributed output coupling and end-coupling, are contrasted in Fig. 1.16. A distributed-output coupler can be segmented along the laser cavity, as illustrated in the example in Fig. 1.2, or it can even occur continuously along the laser cavity, as is seen in Fig. 1.16.

The earliest examples of distributed-output-coupled semiconductor diode lasers were the leaky-waveguide structures [1.24, 1.38, 1.481, 1.482] and intra-cavity gratings [1.35],[1.483]-[1.489]. In recent years, distributed-output coupling in diode-laser arrays and diode-laser amplifier arrays has been realized by fabricating grating-output couplers into the diode-laser array cavity to act as a segmented [1.194, 1.195] or continuous-output couplers in the gain section

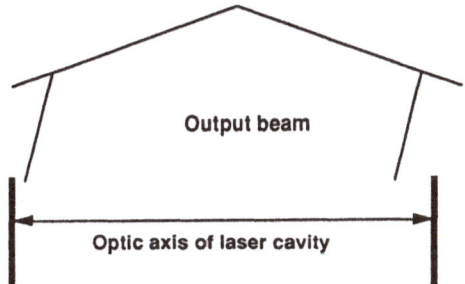

Distributed-Output -Coupled Laser

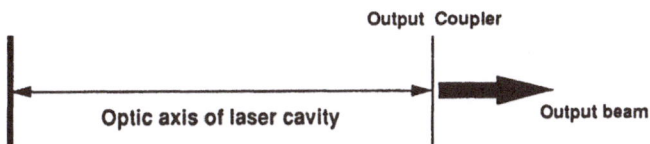

End-Coupled Laser

Fig. 1.16. Schematic representations that depict distributed-output-coupled and end-coupled laser structures.

[1.174],[1.183]-[1.185],[1.242]-[1.245],[1.247, 1.295, 1.490]. Distributed-output coupling has an advantage over end coupling in that as the laser length is scaled to obtain higher-power output, the power-extraction efficiency remains constant. In contrast, the power extraction efficiency of the end-coupled laser decreases as the inverse of the cavity length, as discussed in Chap. 2.

2. Fundamentals of High-Power Operation

The fundamentals and basic concepts that are important for our understanding the physics and technological issues related to high-power, single-mode diode-laser arrays are presented. Principal concepts such as the threshold gain condition (Sect. 2.1), current-gain properties (Sect. 2.2), optimization of the above-threshold operation (Sect. 2.3), and scaling limitations are reviewed (Sect. 2.4). A comparison of the performance characteristics of the semiconductor laser medium with other well-known non-semiconductor laser media (i.e., Nd:YAG, Rhodamine-6G dye, and CO_2) is also included. The effects of injected carriers on the optical properties and spatial-mode discrimination are discussed in Sect. 2.6. Heating and thermal management issues are treated in Sect. 2.8. These are topics that are common to both single-element diode lasers and diode-laser arrays.

2.1 Threshold Characteristics

The conventional, cleaved-facet diode laser, shown in Fig. 2.1, comprises a Fabry-Perot cavity where the mirrors at each end of the laser are formed by cleaving the semiconductor wafer along appropriate crystallographic planes. This type of resonator is widely used both in single-element diode lasers and in multiple-element diode-laser arrays. For reasons of practicality, the semiconductor chip that contains the diode-laser device must have dimensions that are much larger than the active volume. The active volume of a typical diode laser is quite compact.

Figure 2.2 exhibits a schematic diagram of the bandgap energies and layer compositions of four types of quantum-well laser structures. In these structures the gain is contained in the thin quantum-well layers, thickness $\approx 50 - 100$ Å, where the injected carriers and holes undergo radiative recombination. The active layer can be comprised of a *Single Quantum Well* (SQW) or *Multiple Quantum Wells* (MQW). The layers adjacent to the quantum wells act as waveguide cladding layers to confine the optical mode. These cladding layers have higher band-gap energies so that the carrier recombination is confined to the quantum wells. The confinement layers can be configured either as a (SCH) or as a *Graded-Index-Separate-Confinement Heterostructure* (GRINSCH).

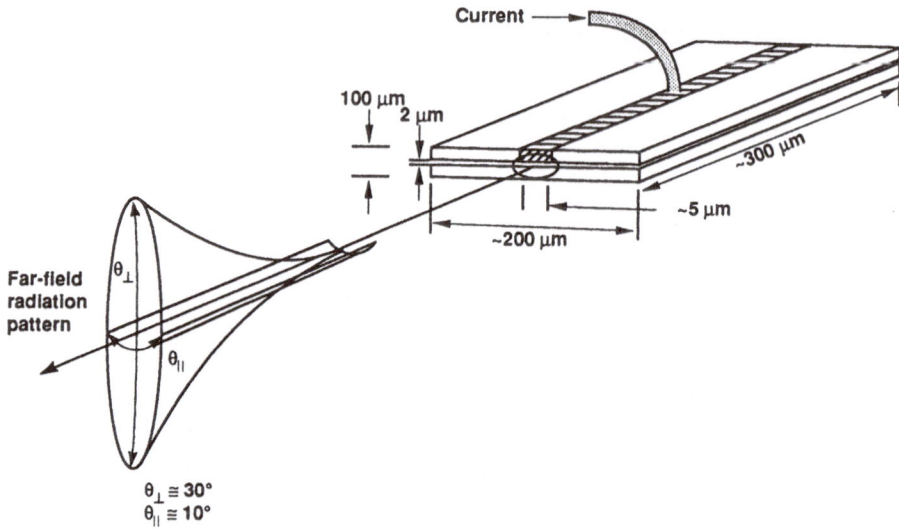

Fig. 2.1. Schematic diagram of a conventional cleaved-facet diode laser

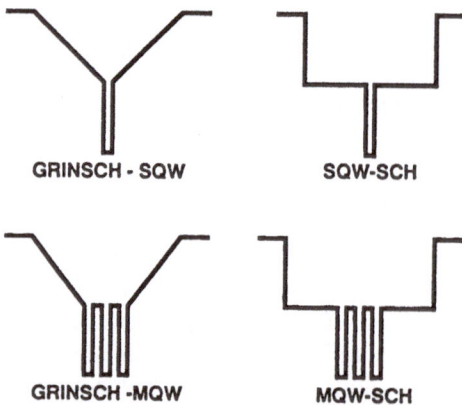

Fig. 2.2. Diagram of GRINSCH-SQW, SQW-SCH, GRINSCH-MQW, and MQW-SCH diode laser structures. The vertical axis represents the band-gap energy of each layer while the horizontal axis corresponds to the layer thickness.

A single-element diode laser comprised of a single-quantum well active layer has typical dimensions on the order of 100 Å or less high, several μm wide, and several hundred μm long, which corresponds to a volume of about 10^{-11} cm^3. A multiple-quantum well structure typically could have an active volume on the order of 10^{-10} cm^3, which is comparable to conventional double-heterostructure lasers, where the active-layer thickness is in the range of $\approx 500 - 800$ Å. Since the thickness of the diode-laser active volume is significantly less than the wavelength of light in the semiconductor material, (≈ 3000 Å), the optical mode will extend well outside the active volume in the direction perpendicular to the pn junction; and in quantum-well structures, only a few percent or even less of the optical mode intensity profile overlaps the active volume. In the plane of the pn junction, a single-element diode laser designed for high-power, single-mode operation will typically have an active volume with a continuous width in the range of $2 - 5\,\mu m$. Therefore, in this direction most of the optical mode interacts with the injected carriers. The aspect ratio of the active volume cross section is about 300 to 1; however, in the waveguide the minimum dimension that the optical mode can be confined to is $\approx \lambda/n$ where n is the refractive index of the active-guide layer. Thus the optical mode in a single-element diode laser has an aspect ratio of about 4 to 1, as indicated schematically in Fig. 2.1. The portion of the mode that is transmitted through the end facet undergoes diffraction as it propagates through free space. The disparate aspect ratio of the mode will give rise to a non-unity aspect ratio in the far field.

The semiconductor structure adjacent to the active volume must act as a low-loss optical waveguide that maximizes the overlap of the optical mode profile with the thin quantum-well active layer. Additional factors that influence the diode-laser chip dimensions are mechanical stability and ease of handling the chip during mounting. These considerations dictate that the semiconductor chip have dimensions of about $100\,\mu m$ high, $100\,\mu m$ wide, and $300\,\mu m$ long. This corresponds to a volume of $\approx 3 \times 10^{-6}\,cm^3$, which is comparable to a grain of sand. Although the diode-laser chip volume is several orders of magnitude larger than the active volume, it is still very small compared to other types of nonsemiconductor lasers. Longer diode-laser cavity lengths, on the order of $1\,mm$ or more, are being used with increasing frequency, as the quality of semiconductor laser material is steadily improving and the corresponding optical losses are decreased. Due to the large refractive index of semiconductors, ≈ 3.4, cleaved facets have reflectivities of about 30 %, which is adequate for sustaining laser oscillation in the high-gain semiconductor material. However, to maximize the power output from such Fabry-Perot type diode lasers, multilayer dielectric coatings are deposited on the cleaved facets so that the rear-facet reflectivity is typically larger than 90 % and the front facet, where the power output is taken from, is around 5 % to 10 %.

Let us turn our attention to the interaction of the optical mode with the active volume. As mentioned above, the optical mode is considerably larger than the active volume where stimulated recombination of injected carriers and holes occurs. Because of the extremely high available gain in the semiconductor active volume, the modal gain can still by quite high. For example, Fig. 2.3 shows the intensity profile perpendicular to the pn junction (also refered to as the transverse-mode profile) superimposed over a diagram of a GRINSCH-SQW of the type that is typically used in high-power diode-laser structures [1.329, 1.331, 1.332][1.345]-[1.347]. This type of structure can be modified to provide more available gain by adding more quantum wells, as indicated in Fig. 2.2. The structure diagram in Fig. 2.3 indicates the refractive

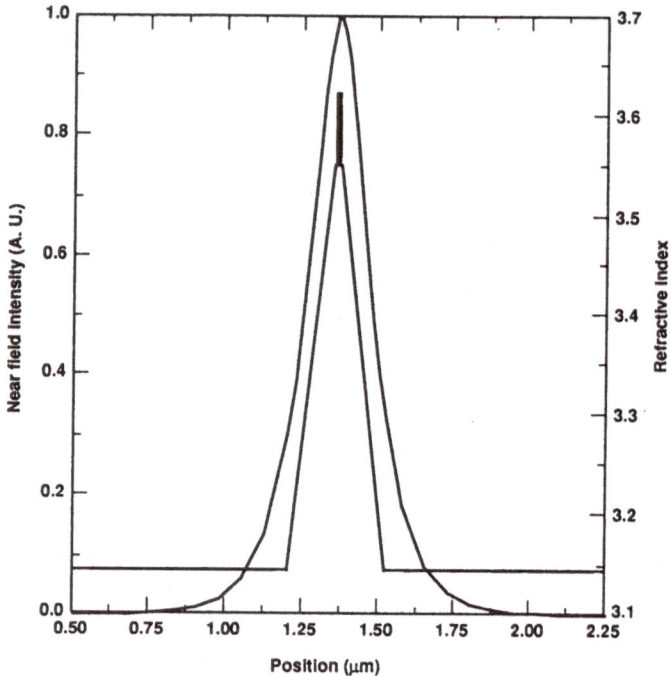

Fig. 2.3. Diagram of the GRINSCH-SQW laser structure, described in [1.332], showing the corresponding intensity profile of the transverse mode.

indices of the epilayers as a function of position. The lower-index material adjacent to the quantum-well active layers acts as the cladding layers of a waveguide that supports the fundamental spatial mode so the field maximum coincides with the active layer. Note that the lower-refractive-index cladding layers must have higher band-gap energies, as well as conduction and valence band-off sets, to confine the electrons and holes so that they can undergo radiative recombination in the active layer [1.25, 1.26, 1.323, 1.324, 2.1].

The interaction of the optical mode with the active layer can be quantified by calculating the cross-sectional overlap of the optical mode with the active volume. This overlap is known as the *confinement factor*. For single-element lasers it is the transverse-mode overlap, perpendicular to the pn junction, that has the largest effect on the modal gain, as the overlap of the optical mode and the active area in the lateral direction is near unity in most cases. The transverse mode confinement factor Γ_y is then defined as

$$\Gamma_y = \frac{\int\limits_{-d/2}^{+d/2} |E_T(y)|^2 dy}{\int\limits_{-\infty}^{+\infty} |E_T(y)|^2 dy} \quad , \tag{2.1}$$

where d is the thickness of the quantum well, and $E_T(y)$ is the transverse optical field. Then the modal gain g and the bulk gain G in the active layer are related by

$$g = \Gamma_y G \ . \tag{2.2}$$

The maximum bulk gain in the quantum well is typically attained for quantum-well thicknesses of $100 \, \text{Å}$ or less. In the example in Fig. 2.3, the thickness is $70 \, \text{Å}$. For maximum power output and efficiency, clearly Γ_y needs to be maximized. Given the limited range of quantum-well thicknesses, it is important to select the other structural parameters so that Γ_y is maximized. Although the waveguide confinement is provided by both the lower-index GRINSCH and n- and p-cladding layers in Fig. 2.3, it is the composition (e.g., Al %), grade, and width of the GRINSCH layers that will have the most effect on Γ_y, as these layers have a strong influence on the optical mode confinement. The n- and p-cladding layers, which are typically $\geq 1 \, \mu\text{m}$, only influence the optical mode properties by changes in composition. The GRINSCH layers must also provide confinement of the injected electrons in the conduction band and holes in the valence band. This is termed *carrier confinement*, and this constraint usually dictates the selection of the composition, so that there is a sufficiently larger band offset between the active layer and GRINSCH layers. For a given composition, the width of the GRINSCH layers should be selected to maximize Γ_y to maximize the gain. As an illustration of this, Fig. 2.4 shows a plot of Γ_y versus w_c, the width of the confinement layers for the structure depicted in Fig. 2.3. It can be seen that the transverse mode confinement reaches a maximum value of about 2.54 % when the width of the confinement layers is $0.15 \, \mu\text{m}$. For confinement layer thicknesses between 0.1 and $0.25 \, \mu\text{m}$ or so, the changes in Γ_y are negligible. Hence the modal gain would be expected to be fairly insensitive to the confinement layer thicknesses in this range. Note that the above discussion applies qualitatively to all the GRINSCH and SCH quantum-well structures displayed in Fig. 2.2, as the SCH layers also provide optical mode and carrier confinement.

The threshold condition for laser operation is obtained in the usual manner by imposing the requirement that the optical field reproduce itself after a round-trip pass through the Fabry-Perot cavity. The amplitude part of the round-trip analysis gives the modal absorption power loss coefficient, α_{th}, at threshold as

$$\alpha_{\text{th}} = \alpha + \frac{1}{2L} \ln \left(\frac{1}{R_1 R_2} \right) \ , \tag{2.3}$$

where α is the modal loss coefficient due to scattering effects, free-carrier absorption, and any other loss process that might be present in the laser waveguide structure, L is the length of the cavity, and R_1 and R_2 are the facet reflectivities. The second term represents the end losses that provide the output coupled light. One can think of (2.3), as an accounting of all the loss mechanisms that must be offset by the gain. The threshold gain for laser

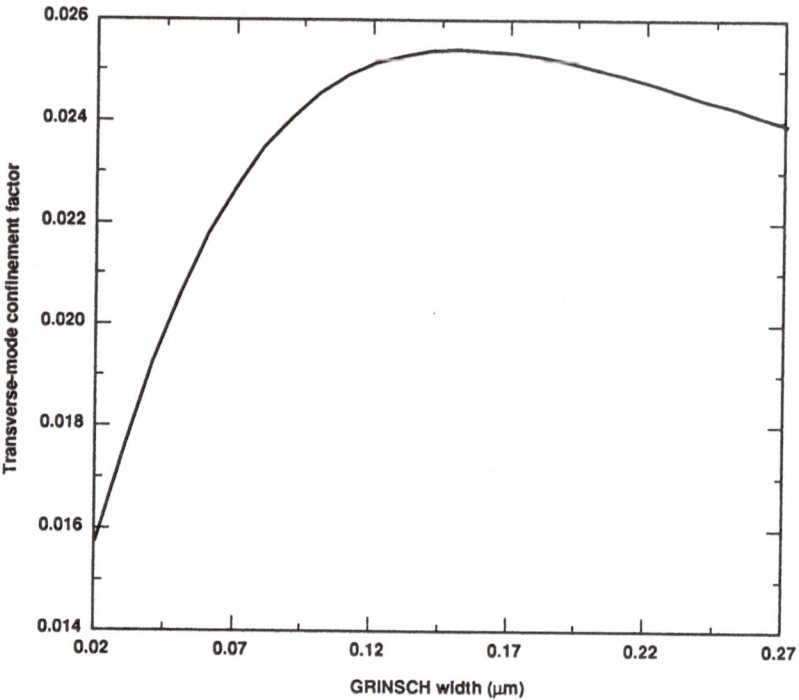

Fig. 2.4. Graph of the calculated Γ_y versus w_c, for the GRINSCH-SQW laser structure described in [1.332].

oscillation occurs when sufficient current is injected so that the modal gain balances the modal losses. This condition can be expressed as

$$g = \Gamma_y G = \alpha_{\text{th}} \; .\tag{2.4}$$

The requirement that the phase of the optical field reproduce itself gives the spectrum of wavelengths where laser oscillation can occur as

$$2n_e k_0 L = 2m\pi \; ,\tag{2.5}$$

where n_e is the effective index of the laser waveguide mode, $k_0 = 2\pi/\lambda$ where λ is the wavelength of light, and m is an integer. From (2.5), the wavelength separation between adjacent longitudinal modes $\Delta\lambda$ can be shown to be [1.26]

$$\Delta\lambda = \frac{\lambda^2}{2Ln_g} \; ,\tag{2.6}$$

where n_g, the group index given by

$$n_{\text{g}} = n_{\text{e}} - \lambda\frac{dn_e}{d\lambda}\tag{2.7}$$

must be used because of the dispersion in the semiconductor material. Typical values for n_e range from 3.2 to 3.5, depending on the laser structure, and

$dn_e/d\lambda \approx -10^{-4}\,\text{Å}^{-1}$. Then n_g falls in the range of 4 to 4.5, depending on the wavelength. The negative dispersion of the semiconductor material has the effect of increasing the apparent cavity length relative to what it would be for the case of zero dispersion in the cavity. The subject of single-longitudinal mode operation will be treated in more detail in Chap. 3.

2.2 Current-Gain Properties

In conventional double-heterostructure diode lasers and diode-laser arrays, the relation between the injected current density J, and bulk optical gain G can be well approximated by a linear relationship of the form $G = A\,(J - J_0)$ [1.25, 1.26, 1.323, 1.324]. Here J_0 is the current density at transparency, which corresponds to $G = 0$, and the coefficient A has units of cm/A. Note that the parameters A and J_0 depend on temperature, as well as the details of the laser structure. In quantum-well diode lasers, the current-gain relationship can be empirically expressed as a logarithmic form. This is a result of the two-dimensional nature of the density of states. The conventional linear current-gain relationship can still be used for quantum-well structures over specific current ranges; however, at higher current ranges reduced values of the A parameter must be used to account for the decreased available gain. In terms of the fundamental structural parameters, A is expressed as

$$A = \frac{\Gamma_y \sigma \tau \eta_i}{ed} \, , \tag{2.8}$$

where σ is the *differential gain* or *gain cross-section*, τ is the carrier lifetime, η_i is the internal efficiency, e is the electric charge, and d is the active-layer thickness. The current dependence of A in quantum wells stems from the nonlinear current-gain relationship; and as seen from (2.8), it can be directly attributed to the carrier dependence of the differential gain σ and the carrier lifetime τ. Generally, values of σ for quantum-well lasers are higher than in double-heterostructure lasers. These trends are predicted by various models for the quantum-well gain [1.327, 1.451] and have been verified experimentally [2.2]-[2.7]. For a given laser structure where the threshold losses are known, the current-gain relationship can be used to determine the threshold current density, differential quantum efficiency, maximum power output, and power conversion efficiency.

To see the utility of the current-gain relationship, let us redirect our attention to the threshold gain condition expressed by (2.4), and the available gain. The net available gain for stimulated recombination is $\Gamma_y G - \alpha_{\text{th}}$. In cases where only the first quantized state (i.e., state with principal quantum number being 1) contributes to the gain, the bulk gain of a single quantum-well active layer can be well approximated by a logarithmic dependence on the injected current density given by [2.8, 2.9]

$$G = G_0 \left[1 + \ln\left(\frac{J}{J_0}\right) \right]$$ (2.9)

or [1.451]

$$G = G_0' \ln\left(\frac{J}{J_0'}\right) ,$$ (2.10)

where J is the injected-current density and G_0, J_0 and G_0', J_0' are constants which depend on temperature, as well as the details of the specific structure. Note that (2.9) and (2.10) are equivalent forms, since $G_0 = G_0'$ and $J_0 = eJ_0'$, where e is the base of the natural logarithm. One should also be aware that the parameters in (2.9 and 2.10) can be defined with respect to the bulk gain [2.8, 2.9] or the modal gain [1.451], where the bulk gain and modal gain are related by (2.2). In any case, it is usually easy to distinguish between the two cases, since bulk gains are several thousand cm^{-1} and modal gains are several tens or hundreds of cm^{-1}. Note that the logarithmic form for the current gain relationship has been shown to be a good approximation by [2.10] for most quantum-well lasers in a variety of semiconductor materials (e.g., AlGaAs/GaAs, InGaAs/GaAs, InGaAs/InP, InGaAsP/InP, and GaInAsSb/GaSb).

From the above form of the current-gain relationship it is seen that the available gain provided by a single quantum well will saturate with injected current. This is refered to as gain flattening, and it is a consequence of the step-like density of states in the quantum well [1.327]. The density of states is largest near the bottom of the well. As more carriers are injected into the well; and the quasi-Fermi energy moves into the conduction band, carriers occupy higher-lying states in the well. This situation is schematically illustrated in Fig. 2.5 [2.11] where the dark shading indicates the occupied states, and the quasi-Fermi energy is denoted by E_F. At sufficiently high current densities, the carriers begin to fill states in the second quantized state of the well, i.e. the state with principal quantum number being 2, as shown in Fig. 2.5, and the gain doubles. This is a primary reason that quantum well lasers with short cavity lengths, short meaning $< 200\,\mu$m, exhibit a significant increase in threshold current [2.12]. Although the two-dimensional density of states is the main factor that influences the gain of quantum-well lasers, there are other physical processes which must be considered to accurately model experimental observations. As discussed in [2.13], carrier overflow from the active layer into the barrier and optical waveguide layers, as well as nonradiative recombination in the active layer, are processes which also affect the current-gain relationship in quantum-well lasers.

When the gain contribution comes from the second quantized state, the operating wavelength of the laser will be at a shorter wavelength or higher energy, as reported by *Mittelstein* et al. [2.11]. Figure 2.6 displays the calculated modal gain vs wavelength from [2.11]. The solid curves denote gain spectra at different quasi-Fermi energies. Hence, each curve represents a different level

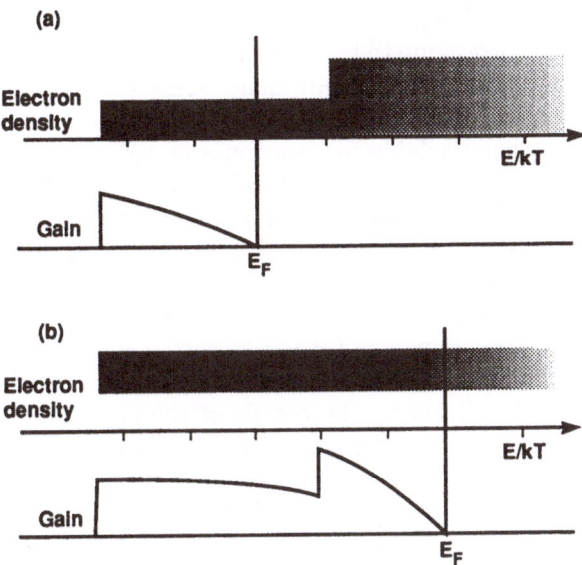

Fig. 2.5. Calculated gain spectra and energy level diagram from [2.11] depicting the occupation of states at, a current density corresponding to gain flattening (a); and (b), at a higher current density, where the maximum gain is shifted higher in energy to the second quantized state.

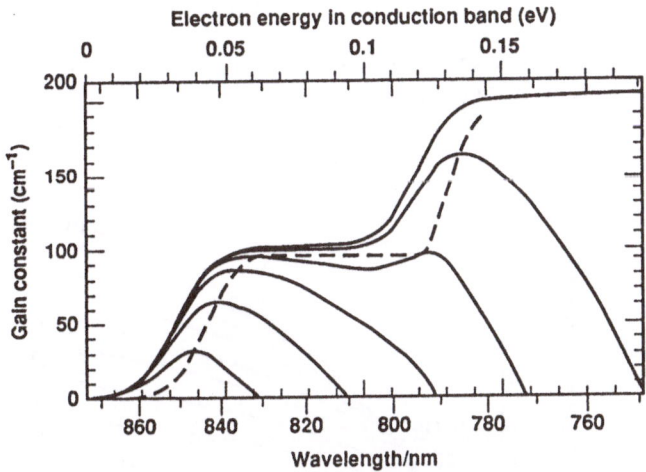

Fig. 2.6. The solid lines are calculated gain spectra from [2.11], and each curve corresponds to a different quasi-Fermi energy level. The broken curve represents the dependence of the maximum available gain on the quasi-Fermi level energy.

of carrier injection. The upper curve corresponds to the highest level; and the lower-lying curves correspond to lower carrier-injection levels. The dotted line relates to the peak-gain variation with quasi-Fermi energy. The dotted curve represents the maximum available modal gain versus injected current. From this curve, we see that available gain saturates at $100\,cm^{-1}$. After this point, increasing the current level will not provide any more gain, until the injection level of carriers is increased to the point where there is enough gain for laser oscillation to occur at $\approx 780\,nm$, from the second quantized state in the well. This curve also illustrates the strong dependence that σ has on gain in quantum-well structures.

Experimental measurements of σ versus carrier density in quantum-well lasers have verified that σ decreases with increasing carrier density or gain [2.5, 2.7]. As mentioned above, the details of the current-gain relationship depend on the structure. For example, GRINSCH structures usually exhibit lower threshold-current densities than SCH structures due to a lower density of states in the confining layer of the GRINSCH [2.14].

For laser structures with sufficiently high modal losses, gain flattening can severely limit the power output. In these situations, it is necessary to use more quantum wells in the active layer in order to get high power, or in some cases any power. For a multiple quantum-well structure the available bulk gain can be expressed empirically as [2.9]

$$G\left(n_{\mathrm{w}}\right) = n_{\mathrm{w}}G_0 \left[1 + \ln\left(\frac{J}{n_{\mathrm{w}}J_0}\right)\right] , \qquad (2.11)$$

where n_{w} is the number of wells. The form of (2.11) assumes that the coupling between wells is negligible and that the injected current density is distributed uniformly between all the wells [1.327]. An example of a study of the dependence of the modal gain on well number by *Whiteaway* et al.

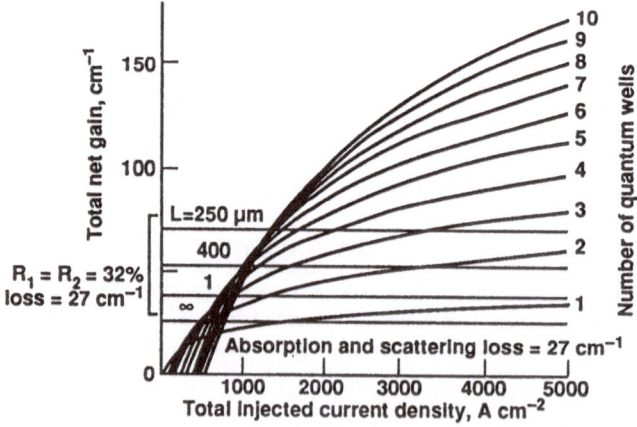

Fig. 2.7. Calculated dependence of the modal gain on injected current density for active layers with different numbers of quantum wells. [2.15]

[2.15] is exhibited in Fig. 2.7. Here the calculated available modal $\Gamma_y G$ is plotted versus the injected current density for $n_w = 1 - 10$. This analysis was done for an InGaAs/InGaAlAs/InP SCH multiple quantum-well laser structure with an operating wavelength of $1.5\,\mu m$. Similar studies have been reported for GaAs/AlGaAs by *Kurobe* et al. [2.9]. The parameters for the InGaAs/InGaAlAs/InP structure used to calculate the threshold gain were, $\alpha = 27\,cm^{-1}$, $\Gamma_y = 0.016$, and $R_1 = R_2 = 0.32$. As the threshold gain (2.4) also depends on length, the horizontal lines in Fig. 2.7 indicate the threshold modal gain required for cavity lengths of $250\,\mu m$, $400\,\mu m$, $1\,mm$, and $L \rightarrow \infty$. For the shortest cavity length, the minimum number of quantum wells that will support laser oscillation is three. But this requires a very high current density of $\approx 3\,kA/cm^2$, where there is significant gain flattening. To obtain low thresholds of $\leq 1\,kA/cm^2$ even with the longer cavity lengths, at a point well below the onset of gain flattening, it is necessary to use at least three wells. As more wells are added, the available gain increases and the threshold-current density increases.

Increasing the well number provides more available gain so higher power outputs are possible, because the gain flattening is pushed out to higher current densities. But the current density corresponding to transparency, and hence the threshold current density, will increase with the number of quantum wells. Higher threshold current densities will also increase the free-carrier absorption losses (free-carrier effects are discussed later in Sect. 2.6). In addition, more thermal energy is dissipated in the active layer as the current density is increased. As the active-layer temperature of a diode laser is increased, the internal quantum efficiency for radiative recombination decreases, and this will reduce the overall efficiency. This is discussed in more detail in Sect. 2.8. The point here is that heat dissipation, thermal management issues and efficiency considerations will ultimately limit the maximum available gain for high power operation; therefore, one may need to account for these effects when using the current-gain relationship to model performance characteristics of high power diode-laser arrays [2.16].

2.3 Optimizing Operation Above-Threshold

To be of maximum utility for many applications, a diode-laser array that is intended for high-power operation should be designed to operate at or very near the peak efficiency for converting the electrical input power into useable optical output power. Such an efficiency specification will impose constraints on the cavity length, facet reflectivites, and operating current.

2.3.1 Differential Quantum Efficiency

Generally, a diode laser that is operating above threshold will exhibit a linear relationship between the light output power and the injected drive current.

However, at higher drive levels the power-current characteristic will become sublinear as the effects of heating decrease the internal quantum efficiency. This is true for quantum-well lasers and double-heterostructure lasers alike. In practice, one usually avoids operating at current levels where thermal roll-off in the power-current characteristic becomes noticeable. In the linear region, then the light output power P_{out} can be written as [1.323]

$$P_{out} = \eta_e \left(\frac{E}{e} \right) (I - I_{th}) \ , \tag{2.12}$$

where E is the photon energy, e is the electron charge, I is the injected current, I_{th} is the threshold current, and η_e is the differential quantum efficiency. The *differential quantum efficiency* is defined as the ratio of the change in optical-power output to the corresponding change in electrical-power input. From (2.12), η_e is given by

$$\eta_e = \left(\frac{e}{E} \right) \frac{\partial P_{out}}{\partial I} = \frac{\lambda}{12400} \frac{\Delta P_{out}}{\Delta I} \ , \tag{2.13}$$

where λ is the wavelength in Å, ΔP_{out} is the power change in Watts, and ΔI is the corresponding current change in Amperes. It is also referred to as the *slope efficiency*, since it can be directly determined from the slope of the power output versus current characteristic of the laser. At threshold, η_e is given by

$$\eta_e = \eta_i \left(\frac{\alpha_{out}}{\alpha_{th}} \right) \ , \tag{2.14}$$

where η_i is the internal efficiency in the stimulated emission regime (above threshold), α_{out} is the loss-coefficient that corresponds to output coupling, and α_{th} is the threshold gain. The internal efficiency η_i corresponds to the fraction of injected carriers that recombine to give photons. It is introduced to account for any incomplete gain clamping and other processes, such as carrier leakage over the heterojunction confinement layers and carrier diffusion [2.17], that remove injected carriers that would otherwise be available for stimulated recombination. Clearly there is much physics contained in η_i, and it would be extremely difficult to calculate from first principles; however, it is straight forward to determine η_i empirically from experiment, as discussed below. Note that the internal quantum efficiency part of η_i can be calculated from the analytic expression (4.61) (as discussed on page 159), which is derived from (4.49) under the assumption that carrier diffusion can be neglected.

For a cleaved-facet, Fabry-Perot diode-laser array, it is the end losses, which correspond to the second term of (2.3), that generate the power output. Therefore, α_{out} is given by

$$\alpha_{out} = \frac{1}{2L} \ln \left(\frac{1}{R_1 R_2} \right) \ . \tag{2.15}$$

Then from (2.3) and (2.14), it is seen that η_e can be expressed as [2.18]

$$\eta_e = \frac{\eta_i}{1 + \frac{\alpha}{\alpha_{\text{out}}}} = \frac{\eta_i}{1 - \frac{2\alpha L}{\ln(R_1 R_2)}} \; . \qquad (2.16)$$

This expression for η_e is valid provided that η_i and α are independent of the cavity length and carrier concentration, which is usually the case for GaAs/AlGaAs quantum-well lasers with cavity lengths greater than $300\,\mu m$ [1.347, 2.12, 2.19]. From (2.16), it is seen that η_e will be largest when α is as small as possible and η_i is as close to unity as possible. This is also beneficial for reducing the threshold current. Decreasing L, the length of the cavity, and decreasing the facet reflectivities, R_1 and R_2, will increase the output coupling, which also results in an increase in η_e. However, this occurs at the expense of an increased threshold current, as the total losses are increased. The form of (2.16) is the basis for the experimental technique for measuring η_i and α. For a given structure with fixed facet reflectivities, the power-current characteristic is measured for several different cavity length lasers. According to (2.13), the differential quantum efficiency for each cavity length can be determined from the slope of the power-current characteristic. A plot of the reciprocal of the observed differential quantum efficiency $1/\eta_e$ versus cavity length, L, can be fit to a straight line. Then applying (2.16), the intercept with the vertical axis corresponds to the reciprocal of the internal quantum efficiency $1/\eta_i$, and the slope of the line is equal to

$$\frac{-2\alpha L}{\eta_i \ln(R_1 R_2)} \; . \qquad (2.17)$$

In some InGaAs/InGaAsP multi-quantum well structures, weak length dependences of η_i and α have been observed in lasers with longer cavity lengths, i.e. $L > 300\,\mu m$, [2.20]. This has been attributed to the length-dependent contributions of free-carrier and intervalence band absorption to the internal optical losses. To account for this, *Koren* et al. [2.20] have expressed α that appears in (2.16) as

$$\alpha = \alpha_{\text{sc}} + (1 - \Gamma_y)\alpha_c + \Gamma_y(\alpha_0 + bN) \; , \qquad (2.18)$$

where α_{sc} is the loss coefficient due to scattering in the waveguide, α_c is the loss coefficient in all of the confinement and cladding layers that surround the active layer, and $\alpha_0 + bN$ represents the carrier dependent loss coefficient in the active layer. This form for the carrier dependent losses has a linear dependence on the carrier density N plus a constant background term α_0. The linear approximation is valid for free-carrier absorption [2.21] and intervalence band absorption [2.22]. As the cavity length is varied, the threshold carrier density and losses, given by (2.18), will also vary. When the modal gain is approximated as $g = \sigma\Gamma_y(N - N_{\text{th}})$, the length-dependent losses at threshold can be expressed as [2.20]

$$\alpha = \frac{1}{1 - b/\sigma}\left[\tilde{\alpha} + \frac{b}{2\sigma L}\ln\left(\frac{1}{R_1 R_2}\right)\right] \; , \qquad (2.19)$$

where

$$\tilde{\alpha} = \alpha_{sc} + (1 - \Gamma_y)\alpha_c + \Gamma_y(\alpha_0 + bN_{th}) \tag{2.20}$$

and N_{th} is the carrier density at threshold.

As the cavity length is increased, it is seen from (2.19) that α decreases because the end losses decrease, which leads to a reduction in the carrier density at threshold. Since the active-layer losses, $\alpha_0 + bN$, are generally larger than the losses outside the active layer, α_c, reducing Γ_y will reduce α, as is seen from (2.19 and 2.20). This will result in a reduction in the threshold current and an increase in differential quantum efficiency. Another benefit that is derived from reducing α is that the maximum conversion efficiency will occur for longer cavity lengths, where heat removal is more efficient. When α is given by (2.19), the form for the differential quantum efficiency is given by

$$\eta_e = \frac{\tilde{\eta}_i}{1 + \dfrac{2\tilde{\alpha}L}{\ln\left(\dfrac{1}{R_1 R_2}\right)}}, \tag{2.21}$$

where

$$\tilde{\eta}_i = \eta_i \left(1 - \frac{b}{\sigma}\right). \tag{2.22}$$

From (2.21) it is seen that a plot of $1/\eta_e$ vs cavity length will give values for an effective internal quantum efficiency (2.22) and an effective internal loss (2.20).

In general, the use of either (2.16 or 2.21) to extract values of the parameters η_i and α from power-current characteristics is most accurate for longer length devices, where $L > 300\,\mu$m, where the threshold carrier density is minimized. Furthermore, this measurement should be done on an ensemble of lasers that are identical in all respects except for the cavity length. However, internal structural variations in lasers from the same wafer can produce large uncertainties in the measured values of η_i and α. These uncertainties can be more pronounced in quantum well structures where cavity lengths of several mm or more are encountered, internal losses are low, and the internal quantum efficiency approaches unity. To remedy this situation, improved techniques for independently measuring the internal loss of diode lasers are beginning to be developed [2.23].

2.3.2 Conversion Efficiency

In designing high-power diode-laser arrays, it is very important that the total efficiency for the conversion of electric power input to optical power output be maximized. Higher-power outputs can be obtained by increasing the cavity length of the diode-laser array. However, for cavity lengths that are longer than the an *absorption length*, which is defined as $1/\alpha$, the power output saturates rapidly as the cavity length is increased. For diode-laser arrays with cavity lengths less than one absorption length, there is a range of lengths where the conversion efficiency will be nearly maximized. The *power conversion efficiency* or *conversion efficiency* η_T is defined as the ratio of optical output power to the electric input power and can be expressed as [1.323]

$$\eta_T(I) = \frac{P_{out}}{IV(I)} = \frac{\eta_e(E/e)(I - I_{th})}{I(V_0 + IR_s)} \ , \tag{2.23}$$

where I is the *drive current*, P_{out} is given by (2.12), $V(I)$ is the *current-voltage characteristic* of the diode, V_0 is the *turn-on voltage* of the diode, and R_s is the *series resistance*, which is given by,

$$R_s = \frac{\rho_s}{wL} \ . \tag{2.24}$$

Here ρ_s is the *sheet resistivity* of the diode-laser structure, which is due to both the material resistivity of the epilayers and the resistivity of the metal contacts, and w is the width of the active region. All the laser parameters in (2.23) can be determined by a direct measurement of the diode-laser characteristics. The numerator of (2.23) is the optical power output expressed in terms of the drive current and the differential quantum efficiency. The denominator of (2.23), which corresponds to the electric input power, is comprised of the power corresponding to the voltage drop across the band gap, IV_0, and the power dissipated due to Ohmic heating $I^2 R_S$. The form of the denominator in (2.23) assumes that the diode is biased well-above the turn-on voltage so that $V(I)$ of can be approximated by [1.323]

$$V(I) = V_0 + IR_s \ . \tag{2.25}$$

In using (2.25) to approximate $V(I)$, V_0 should be determined by an extrapolation of $V(I)$ measured at high-current levels back to $I = 0$ [2.24]. An example of this is shown in Fig. 2.8 [2.24], where the $V(I)$ characteristic of two diode lasers of differing cavity lengths are shown. In these data that were measured by *Bour* and *Rosen* [2.24], the diode lasers were fabricated from the same wafer and have an operating wavelength of $0.93\,\mu$m. The voltage that corresponds to this photon energy is given by $h\nu/e$, and it has a value of about 1.5 V. As seen in Fig. 2.8 the diode is just beginning to pass current at a bias of 1.5 V. However, the measured $V(I)$ is super-linear up to about $V = 1.8$ V; and for $V > 1.8$ V, the linear form of (2.25) is approached. The dotted lines show the linear extrapolation of high-current operation back to $I = 0$, where it is found that $V_0 = 1.7$ V. In studying the use of (2.25) to approximate diode-laser $V(I)$ characteristics, the typical range for values of V_0 is bounded below by the photon energy and bounded from above by the band-gap energy of the cladding layers [2.24].

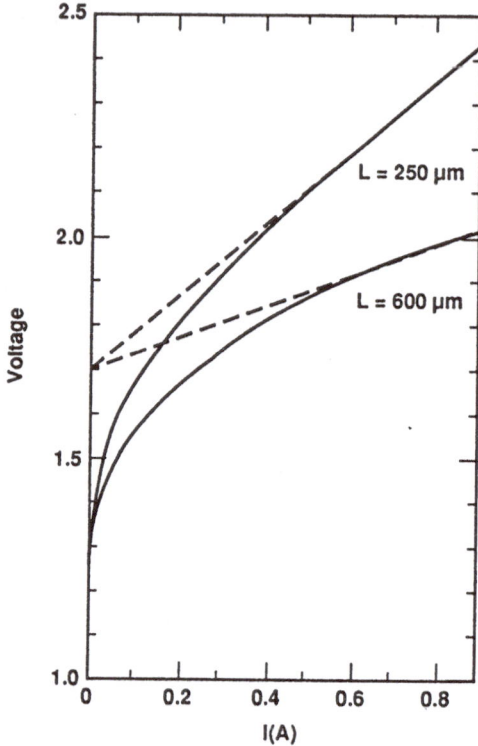

Fig. 2.8. Example of the current-voltage characeteristic of diodes of similar width $90 \, \mu\text{m}$ but different lengths. The linear fit was obtained using $V_0 = 1.7 \, \text{V}$ and $\rho_s = 1.8 \times 10^{-4} \, \Omega \, \text{cm}^2$. [2.24]

2.3.3 Maximizing Conversion Efficiency

Since the optical power output increases with drive current and the electrical power dissipated in the laser is quadratic in the current, the conversion efficiency (2.23) will have a peak only at a specific current value. For a linear power-current characteristic given by (2.12), this current value I_0 is found by maximizing η_T as given by (2.23) with respect to drive current I [1.330, 2.24]. Note that the peak conversion efficiency $\eta_T \, (I_0)$ and corresponding power output $P_{out} \, (I_0)$ are functions of the parameters η_e, R_s, and I_{th}, all of which depend on the laser cavity length L. For η_e and R_s, the length dependencies are given by (2.16 and 2.24) respectively and the conversion efficiency versus current density, where the *current density* is defined as $J = I/\,(wL)$, can be written as

$$\eta_T \, (J) = \frac{\eta_e \, (E/e) \, (J - J_{th})}{J \, (V_0 + J\rho_s)} \, . \tag{2.26}$$

From (2.26), it is seen that the conversion efficiency can be maximized with respect to the operating current density. The optimum current density J_{opt}

is found by setting the first derivative of (2.26) with respect to J to zero, to yield

$$J_{\text{opt}} = J_{\text{th}} \left(1 + \sqrt{1+v}\right) , \qquad (2.27)$$

where $v = V_0/(I_{\text{th}} R_{\text{s}}) = V_0/(J_{\text{th}} \rho_{\text{s}})$ is a dimensionless parameter and J_{th} is the threshold current density. For AlGaAs diode-laser arrays, typically $V_0 = 1.7\,\text{V}$, $J_{\text{th}} \approx 100\,\text{A/cm}^2$, and $\rho_s \approx 2 \times 10^{-4}\,\Omega\,\text{cm}^2$, so according to (2.27) the maximum conversion efficiency is obtained when the laser is operated at a current density $J \approx 10 \times J_{\text{th}}$. In order to express J_{th} in terms of the threshold losses, the current-gain relationship must be known. For the sake of example, if the logarithmic form of (2.10) is used, then J_{th} is given by

$$J_{\text{th}} = J_{\text{o}}' \exp\left(\frac{\alpha + \alpha_{\text{out}}}{G_0'}\right) = J_{\text{o}}' \exp\left\{\frac{1}{G_0'}\left[\alpha + \frac{1}{2L}\ln\left(\frac{1}{R_1 R_2}\right)\right]\right\} . \qquad (2.28)$$

The corresponding peak conversion efficiency is given by

$$\eta_{\text{T}}^{\text{peak}} = \eta_{\text{T}}\left(I_0\right) = \eta_{\text{e}} \frac{E}{eV_0} \frac{v}{\left(1 + \sqrt{1+v^2}\right)^2} . \qquad (2.29)$$

From (2.29), it seen that the length dependence of the peak conversion efficiency is determined by η_e and J_{th}, which is due to the dependence of these parameters on the end losses, (2.15). Note that for a given diode-laser structure, (2.27 and 2.29) are functions of the end losses. Therefore, different cavity length structures can have the same peak conversion efficiency by appropriately adjusting the facet reflectivities, so that the value of output-coupling coefficient (2.15) is the same for each structure. Using (2.12), the power output at peak conversion efficiency is given by

$$P_{\text{out}} = \eta_e \left(\frac{E}{c}\right) w L J_{\text{th}} \sqrt{1+v^2} . \qquad (2.30)$$

Note that (2.30) represents an extrapolation of the threshold characteristics to obtain an estimate of the power output at an operating level well above threshold. It does not account for modifications of the current-gain characteristic due to heating effects or nonuniform saturation of the gain along the cavity. These effects are discussed in Sect. 2.8.

In many of the applications that use single-element and multiple-element diode-laser sources, it is desirable to maximize the total conversion efficiency, while generating as high a power output as possible. *Bour* and *Rosen* [2.24] used the logarithmic current-gain relationship reported by *Chinn* et al. [1.451], which is given by (2.10) with $G_0' = 40\,\text{cm}^{-1}$ and $J_0' = 100\,\text{A/cm}^2$. Figure 2.9 shows a plot of the calculated peak conversion efficiency versus cavity length for the parameters that are indicative of InGaAs/AlGaAs diode lasers containing a single strained-layer quantum-well active layer. The two curves that are shown each correspond to rear-facet reflectivities $R_2 = 1$ and front-facet reflectivities $R_{\text{F}} = R_1 = 0.1$ and 0.3. For short cavity lengths,

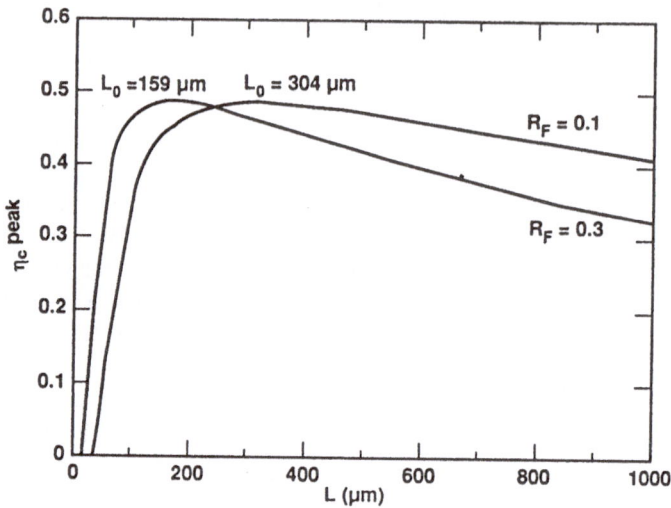

Fig. 2.9. Calculated peak-power conversion efficiency as a function of cavity length for a broad-area laser. [2.24]

where $L < 100\,\mu m$, the peak conversion efficiency is low because this is where the series resistance R_s is largest. As the cavity length is increased, the peak conversion efficiency rapidly increases to a maximum value, and then decreases gradually as L is increased further. Longer cavity lengths result in a decrease in both R_s and J_{th}, however, the differential quantum efficiency will decrease, for fixed facet reflectivities, because the output-coupling losses decrease as $1/L$. The maximum peak-conversion efficiency of ≈ 0.49 is the same for the 0.1 and 0.3 front-facet reflectivities. However, for the larger 0.3 front-facet reflectivity, the peak efficiency is maximized at a shorter cavity length. This occurs because the shorter length is necessary to increase the differential quantum efficiency (2.16) by increasing output-coupling coefficient α_{out}. As mentioned above, the peak conversion efficiency is a function of α_{out}, so that for any length L, one can readjust the front-facet reflectivity R_1 so that the peak conversion efficiency is again maximized. Except for very-short cavity lengths, the power output at the peak conversion efficiency (2.30) increases monotonically with increasing cavity length, but it eventually levels off and saturates as L is increased beyond $1/\alpha$. This occurs because at large L, we have $\eta_e \sim 1/L$, $I_{th} \sim L$, and $v = \text{constant}$.

Although, there is more current available in a longer cavity to generate higher optical power, the energy that is dissipated by the distributed absorption losses α also increases, so that P_{out} approaches a constant for very large L. However, (2.30) overestimates the asymptote of P_{out}, as well as the maximum extent for L. The aforementioned analysis used to determine (2.29) does not account for the nonuniform intensity distribution in the laser or nonlinear gain saturation. The model employed an above-threshold extrapolation of the round-trip threshold gain analysis given by (2.3), and the optical intensity was assumed to be uniform along the cavity. This is a good approximation when the facet reflectivities are equal or nearly equal. For the unequal facet reflectivities considered above for maximizing the conversion efficiency, the intracavity intensity is highest at the output facet where the reflectivity is lowest [2.25]-[2.28]. When these effects are included, the calculated power output saturates more rapidly as the cavity length approaches $1/\alpha$, resulting in lower conversion efficiencies.

2.3.4 Nonlinear Gain Saturation and Conversion Efficiency

To accurately model the effects of nonuniform gain saturation in the laser cavity, it is necessary to use a self-consistent nonlinear model of the laser to account for the spatial dependence of the optical intensity and carrier density, as well as carrier diffusion effects [2.29]. Of course, these more complicated models must be solved numerically. An example of a more manageable non-linear, self-consistent laser model is the model developed by *Rigrod* [2.30]. In this section, the *Rigrod* model is described and used to optimize the Fabry-Perot diode-laser structure for maximum power output and conversion efficiency.

An improved understanding of the cw high-power output characteristics of semiconductor laser arrays can be obtained by using a well-known self-consistent model calculation that was originally developed by *Rigrod* [2.30]. This type of steady-state model calculation is often refered to as the *Rigrod* analysis, and it has been discussed extensively in numerous books devoted to non-semiconductor lasers [1.272, 2.31, 2.32]. Since this analysis incorporates nonlinear gain saturation, it can be used to calculate the power output for a Fabry-Perot type laser. It is particularly well suited for analyzing the continuous-wave power-output characteristics of homogeneously broadened laser systems such as semiconductor-laser sources. Consider the schematic diagram of a Fabry-Perot laser cavity that is shown in Fig. 2.10. The oppositely propagating intensities as a function of position, $I_+ (z)$ and $I_- (z)$, are

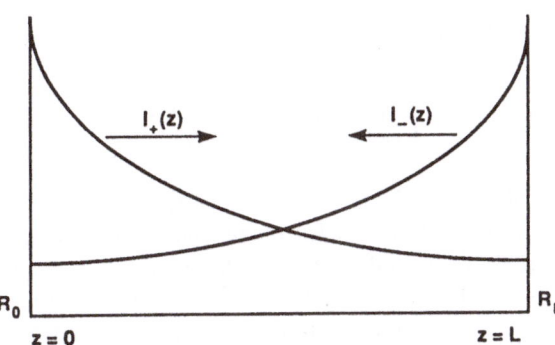

Fig. 2.10. Schematic Fabry-Perot laser cavity illustrating the two traveling-wave intensities that are use in the *Rigrod* model.

assumed to satisfy the following set of coupled first-order nonlinear differential equations,

$$\frac{\partial I_+(z)}{\partial z} = [g(z) - \alpha] I_+(z) \tag{2.31}$$

and

$$\frac{\partial I_-(z)}{\partial z} = [g(z) - \alpha] I_-(z) \ , \tag{2.32}$$

where α is the modal loss coefficient, z is the position coordinate along the axis of the laser cavity, and $g(z)$ is the saturable gain coefficient which is given by

$$g(z) = \frac{g_0}{1 + \dfrac{I_+(z) + I_-(z)}{I_S}} \ , \tag{2.33}$$

where g_0 is the unsaturated or small-signal gain coefficient and I_S is the saturation intensity, which is the intensity where $g(z)$ is reduced to half the value of the unsaturated gain g_0.

For semiconductor laser media, the unsaturated gain coefficient and the saturation intensity are given by [2.29]

$$g_0 = \Gamma_y \sigma \left(\frac{\tau \eta_i J}{ed} - N_{tr} \right) \tag{2.34}$$

and

$$I_S = \frac{E}{\tau \sigma} \ , \tag{2.35}$$

where Γ_y is the fractional overlap of the optical mode with the active layer of the semiconductor laser structure given by (2.1), σ is as before the differential gain or stimulated emission cross-section, τ is the electron-hole recombination time, d is the thickness of the diode laser active layer, and N_{tr} is the carrier density where the active layer is transparent. The saturable modal gain coefficient, $g(z)$, has the following dependence on $N(z)$, the position-dependent carrier density,

$$g(z) = \Gamma_y \sigma [N(z) - N_{tr}] \ . \tag{2.36}$$

From a steady-state analysis of the carrier density rate equation [2.29], it can be shown that

$$N(z) = \tau \left(\frac{\eta_i J}{ed} - \frac{g(z)[I_+(z) + I_-(z)]}{\Gamma_y E} \right) \ . \tag{2.37}$$

By combining (2.34 and 2.37), the form for $g(z)$ given by (2.33) is obtained. It is important to note that the forms of the positional gain dependence given by (2.37 and 2.33) are only valid for the linear current-gain relationship given by (2.34), as the detailed behavior of the gain saturation depends on the differential gain σ. The aforementioned logarithmic current-gain relationship, (2.9, 2.10, and 2.11) cannot be used. These cases effectively exhibit a gain-dependent or current-dependent σ, which leads to a gain-dependent I_S. This is an example where one needs to use an appropriately selected linear current-gain relationship to approximate the logarithmic gain of the quantum well.

The *Rigrod* analysis neglects interference of the oppositely propagating waves. Such interference effects can give rise to a strong spatial dependence in the gain saturation, which causes spatial-hole burning on the scale of the wavelength of light. In semiconductors, this corresponds to a distance of $\approx 0.3\,\mu m$. At the carrier densities where laser oscillation occurs in quantum-well semiconductor lasers, which is $\approx 10^{18}\,\mathrm{cm}^{-3}$, the carrier-diffusion lengths are typically $\approx 1\,\mu m$, so that the spatial-hole burning in the gain is not established [2.33]. Even in cases of homogeneous laser media where spatial-hole burning does occur, the *Rigrod* analysis has been shown to be a good approximation in the case of high-gain lasers [2.34].

Consider the laser cavity depicted in Fig. 2.10, which has reflectivities of R_0 and R_L respectively at the $z = 0$ and $z = L$ ends of the gain section. The intensity output I_{out} at the $z = L$ end is given by

$$I_{\mathrm{out}} = (1 - a - R_\mathrm{L})\, I_+ (L) \ , \tag{2.38}$$

where a are the losses that occur due to transmission through the end mirror at $z = L$. Integration of (2.31 and 2.32) yields a transcendental equation for I_{out} of the form [2.30, 2.35]

$$\alpha L - \ln \sqrt{R_0 R_\mathrm{L}} = \frac{g_0}{\sqrt{(g_0 - \alpha)^2 - (2\alpha p)^2}} \ln \left(\frac{F\left(p, p\sqrt{R_0}\right)}{F\left(p, p/\sqrt{R_L}\right)} \right) \ , \tag{2.39}$$

where the function $F(x, y)$ given by,

$$F(x,y) = \frac{\sqrt{(g_0 - \alpha)^2 - (2\alpha x)^2} + (g_0 - \alpha - 2\alpha y)}{\sqrt{(g_0 - \alpha)^2 - (2\alpha x)^2} - (g_0 - \alpha - 2\alpha y)} \ . \tag{2.40}$$

Equation (2.39) can be evaluated numerically to obtain p, which is related to the laser-power output at $z = L$ by

$$I_{\mathrm{out}} = I_\mathrm{S} \frac{(1 - R_\mathrm{L})}{\sqrt{R_\mathrm{L}}} p \ , \tag{2.41}$$

where it has been assumed that $a = 0$. The two terms on the left-hand side of (2.39) corresponds to the total absorption losses and end losses in the cavity of length L. The term on the right-hand side of (2.39) represents the total gain in the cavity with the accumulated saturation due to the spatial dependence of $I_+ (z)$ and $I_- (z)$. In order to determine the intensity output I_{out}, as given by (2.41), (2.39) must be solved numerically to determine p.

There are several limiting cases where approximate analytic expressions for (2.39) can be found [2.31]. These cases are near threshold, where the gain saturation along the cavity can be neglected, and for saturated operation far-above threshold where the effects of absorption losses are negligible, when $L \ll 1/\alpha$ and $g_0 \gg \alpha$. As a primary concern of this volume is the scaling limitations of high-power diode-laser arrays, the I_{out} vs L characteristics, which are ultimately limited by the absorption losses, will be determined numerically. The scaling characteristics of the *power extraction efficiency*, η_{Ex}, as the cavity length is increased is also of interest. The power extraction efficiency η_{Ex} is defined as

$$\eta_{Ex} = \frac{1}{g_0 L} \frac{(1 - R_L)}{\sqrt{R_L}} p = \frac{I_{out}}{g_0 L I_S} \; . \qquad (2.42)$$

In practice, (2.42) is not at all convenient to use because one must first determine p by numerically solving (2.39). Note that power extraction efficiency (2.42) and conversion efficiency η_T, given by (2.23), are related by a scale factor that depends on the injected current. This can be seen by substituting (2.42) into (2.23), and we find that

$$\eta_T (I) = \frac{P_{out}}{IV (I)} = \frac{g_0 L I_S w d}{IV (I) \Gamma_y} \eta_{Ex} \; , \qquad (2.43)$$

where

$$P_{out} = \frac{w d}{\Gamma_y} I_{out} \; . \qquad (2.44)$$

Comparing (2.43) with (2.23), it is seen that η_{Ex} is analogous to the differential quantum efficiency η_e. The power extraction efficiency, η_{Ex}, is more general, because it accounts for the intensity variation at the output-facet due to nonuniform gain saturation along the laser cavity. It is important to note that η_{Ex} can only be equated with η_e, when the gain is well saturated along the entire cavity length. This occurs for operation well-above threshold for any facet reflectivities or when the end-facet reflectivities are equal. In these regimes, η_{Ex} is nearly equal to the slope efficiency definition of the differential quantum efficiency (2.13).

It should be pointed out that thermal effects were not considered in the nonlinear *Rigrod* model presented in this section. To accurately model thermal effects, one needs to resort to the self-consistent calculation which is presented in Chap. 4. However, it is possible to approximate thermal effects in the *Rigrod* analysis by accounting for the temperature dependence of σ and N_{tr}. As discussed by *Katz* [2.16], if the thermal resistance of the laser is known, then the temperture rise to the active layer can be calculated. The resulting temperature rise can be used to calculate the changes in σ and N_{tr} due to heating of the active-layer. The active layer temperature increases as the thermal resistance increases, causing σ to decrease and N_{tr} to increase. This results in an increase in the threshold current density, and in turn more heat is dissipated in the active layer. This chain of events can lead to thermal runaway where the device heats up to the point where it will no longer operate.

2.3.5 Maximizing Power-Extraction Efficiency

A useful parameterization method, developed by *Shindler* [2.35] can be applied to (2.39 and 2.41) so that the power output extraction efficiency η_{Ex} can be numerically evaluated as a function of the output coupling mirror reflectivity R_L. The only constraint is that $R_0 = 1$. This is not at all restrictive, as in most practical applications requiring high-power output, one would like to couple all of the power out of one end, call it the end at $z = L$, of the laser. Therefore, the reflectivity of the back mirror at $z = 0$ should be as high as a possible. The parameter λ (not to be confused with the wavelength of light) that is introduced is defined as

$$\sin(2\lambda) = \frac{\sqrt{R_L}I_+(L)}{I_S} \frac{2\alpha_0}{g_0 - \alpha_0} \ . \tag{2.45}$$

Using (2.45), the transcendental equation (2.39) can be expressed in the form

$$\alpha L - \ln\left(\sqrt{R_L}\right) = \frac{g_0}{(g_0 - \alpha)}\frac{1}{\cos(2\lambda)}\ln\left(\frac{1 - \sqrt{R_L}\tan(\lambda)}{\sqrt{R_L} - \tan(\lambda)}\right) \ . \tag{2.46}$$

and the power extraction efficiency can be expressed as

$$\eta_{Ex} = \frac{g_0 - \alpha_0}{2\alpha_0 g_0 L}\frac{1 - R_L}{\sqrt{R_L}}\sin(2\lambda) \ , \tag{2.47}$$

From (2.45, 2.47), it is seen that λ is confined to the range $0 < \lambda < \pi/2$, since both $I_+(z)$ and η_{Ex} must always be positive and less than one to be physically meaningful. The range of λ can be restricted further to $0 < \lambda < \pi/4$, since the argument of the natural logarithm on the right-hand side of (2.46) must be real for the solution to be physically meaningful. By subdividing the range of allowable values of λ into an appropriate number of values, λ_n, (2.46) can be evaluated numerically to give R_L as a function of λ_n. Then η_{Ex} as a function of λ_n is easily calculated from (2.47). Either a sorting routine or graphical technique can then be used to find the maximum value for $\eta_{Ex} = \max[\eta_{Ex}(\lambda_n)]$, and the corresponding value for the optimized output-coupler reflectivity R_L is designated as R_{opt}. The corresponding value of the intensity output is then simply given by

$$I_{out} = (g_0 L)\max(\eta_e) \ . \tag{2.48}$$

The aforementioned solution method offered by *Schindler* is very useful because it can be used as a design tool for determining the optimum performance characteristics of any laser. With the rapid improvements that have occured in personal computers, the transcendental equation (2.46 or 2.39) can be solved in a straightforward manner using any one of a number of available mathematical software packages.

As an example Fig. 2.11 shows the result of a numerical calculation of R_{opt} versus L. For this calculation, $\alpha = 5\,cm^{-1}$ and $I_S = 0.5\,MW/cm^2$ were

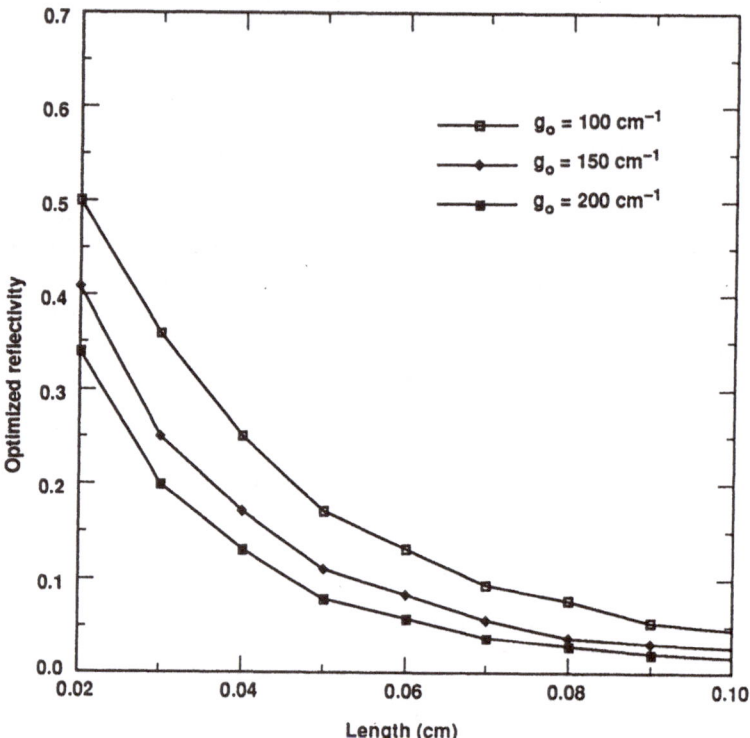

Fig. 2.11. Calculated R_{opt} versus L using typical parameters (see text for actual values) for high-power GaAs diode lasers.

used; and the three curves correspond to available gains of $g_0 = 100$, 150, and $200\,cm^{-1}$. These parameters are representative of the operating range of high-power diode lasers and arrays. The values for $\max(\eta_{Ex})$ corresponding to $g_0 = 100$, 150, and $200\,cm^{-1}$ were, respectively, found to be 0.6, 0.66, and 0.7. The length dependence of $\max(\eta_{Ex})$ was negligible over the length range considered in Fig. 2.11, because $L \ll 1/\alpha = 2\,mm$.

In Sect. 2.4, which deals with the scaling limitations of high-power lasers, the case when $L > 1/\alpha$ is considered; in this regime, $\max(\eta_{Ex})$ decreases with increasing length. The range of values of $R_{opt} \approx 0.1 - 0.35$ for $L \approx 0.03 - 0.05$ that are shown in Fig. 2.11 are consistent with those that are typically used in fabricating high-power, high-efficiency diode-laser arrays [1.329, 1.332, 1.346, 1.347, 2.24].

In Fig. 2.12, the normalized intensity output I_{out}/I_S versus length corresponding to the parameter set of Fig. 2.11 is plotted. The horizontal line in Fig. 2.12 represents the intensity level where catastrophic facet damage can start to occur. For the cases of $g_0 = 150$, and $200\,cm^{-1}$, it is seen that catastrophic facet damage places a limitation on the maximum intensity output that can be extracted. This illustrates the potential for increasing the power

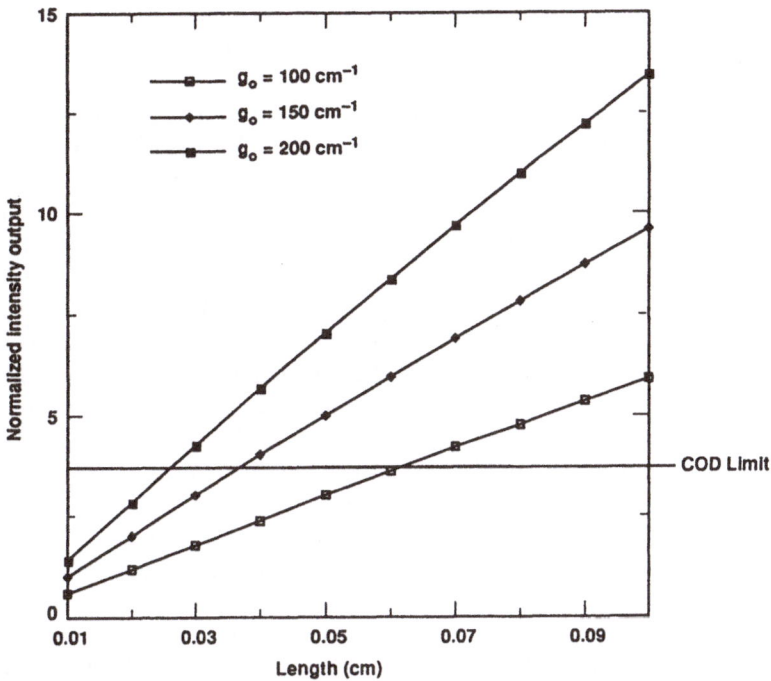

Fig. 2.12. Normalized intensity (I_{out}/I_s) vs L plot that corresponds to the R_{opt} versus L plot appearing in Fig. 2.11.

output of diode lasers further by finding methods for raising the catastrophic facet damage limit. The conversion efficiency that corresponds to the curves in Figs. 2.12 and 2.11 can be directly calculated from (2.42), if the parameters for the current-gain relationship (2.34) and the current-voltage characteristic (2.25) of the diode-laser structure are known. This is examined in the next subsection.

2.3.6 Efficiency and Power-Current Characteristics: Nonlinear Calculation

The optimum values of R_{opt} that were calculated in the preceeding subsection, maximize the extraction efficiency only for the specific operating current that corresponds to g_0. This would seem to suggest that the output-facet reflectivity should be selected for a specific operating level. Actually, this is not the case. In many situations, a primary concern is that the conversion efficiency be maximized. Fortunately, the extent of the conversion efficiency maximum is quite broad with respect to output-facet reflectivity and operating level. To illustrate this, the current-gain relationship (2.34) is used to calculate the values of g_0. This set of values of g_0 are then used as input to (2.39, 2.41-2.43). In this case the threshold current will have the form

$$I_{th} = wL \left(\frac{ed\alpha_{th}}{\Gamma\sigma\tau\eta_i} + J_{tr} \right) , \tag{2.49}$$

where α_{th} is given by (2.3). Other parameters that are necessary for calculating the conversion efficiency from (2.23) are V_0, ρ_s, E, w, and L.

Figure 2.13 illustrates the calculated conversion efficiency versus power output for the indicated output-facet reflectivities. The parameters that were used to generate these curves were as follows; $\Gamma_y = 0.02$, $\sigma = 6 \times 10^{-16}$ cm^2, $\tau = 0.75$ ns, $\eta_i = 0.95$, $d = 50$ Å, $N_{tr} = 5 \times 10^{17}$ cm^{-3}, $\alpha = 5$ cm^{-1}, $w = 120\,\mu$m, $L = 500\,\mu$m, $V_0 = 1.5$ V, and $\lambda = 0.84\,\mu$m. This particular set of parameters is indicative of some of the high-power, high-efficiency, SQW-GRINSCH diode-laser structures that have been reported [1.329, 1.332, 1.346, 1.347, 2.24]. As seen in Fig. 2.13, the maximum of the conversion efficiency is very broad, extending over more than two thirds of the indicated range of power outputs. The maximum conversion efficiencies range between 0.52 at 1.16 W for $R_0 = 0.3$ and 0.58 at 1.64 W for $R_0 = 0.05$. However, for this structure the catastrophic facet damage limit would be expected to be at a power output ≈ 1 W, where the conversion efficiency is not quite maximized.

Given that the power output is limited to ≈ 1 W by catastrophic facet damage, the maximum conversion efficiency can by shifted to better coincide with the available operating range by reducing the cavity length. This is illustrated in Fig. 2.14 where the conversion efficiency vs power output is shown for the same set of output-facet reflectivities. Here, a maximum conversion efficiency of 58 % occurs at 1 W power output for output facet reflectivities of 0.1, 0.2, and 0.3. The data in Figs. 2.14 and 2.13 serve to illustrate the broad range over which near-optimized performance can be obtained from diode lasers and diode-laser arrays. In addition, Figs. 2.14 and 2.13 provide power projections that indicate the potential for still higher-power diode lasers, being limited at present by the catastrophic facet damage limit.

The nonlinear method of calculating conversion efficiency by using (2.43) is not nearly as convenient as using (2.29), which is based on an extrapolation of a linear analysis of the threshold characteristics. In many of the cases that are of interest, the use of (2.29) is a good approximation for the conversion efficiency. This is illustrated in Figs. 2.15 and 2.16 that depict comparisons of the peak conversion efficiency versus cavity length as calculated using (2.43 and 2.29). In Fig. 2.15 where $R_L = 0.05$, it is seen that the threshold extrapolation increasingly overestimates the conversion efficiency as the cavity length is increased. Whereas in Fig. 2.16 where $R_L = 0.3$, the two models show a negligible difference in the conversion efficiency over the indicated length range. Eventually as L is increased beyond $1/\alpha$, (2.29) increasingly overestimates the conversion efficiency. When $L < 1/\alpha$, the agreement is not as good when the difference in the front and back facet reflectivities is larger, because the optical intensity has a larger variation over the length of the cavity. This inturn causes non-uniform saturation of the gain along the cavity. The optical intensity will be largest at the end of the cavity where the facet

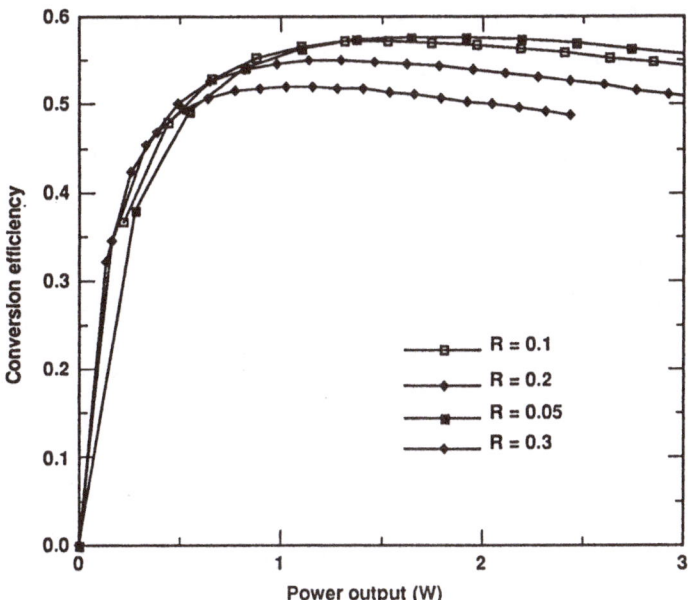

Fig. 2.13. Calculated conversion efficiency versus power output for a cavity of length $500\,\mu$m and output-facet reflectivities of 0.05, 0.1, 0.2, and 0.3.

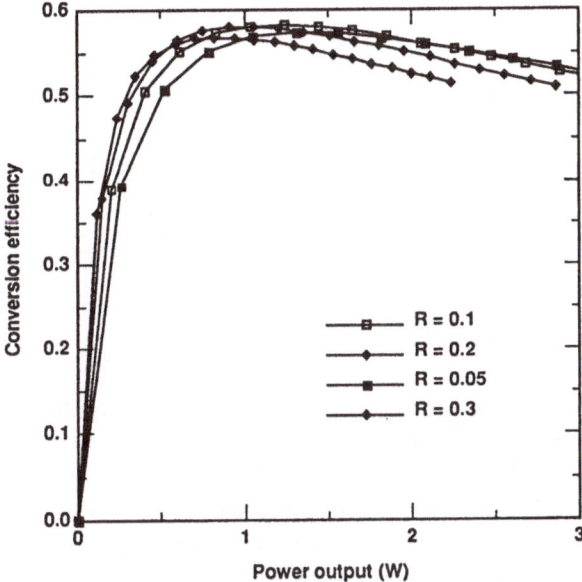

Fig. 2.14. Calculated conversion efficiency versus power output for a cavity of length $300\,\mu$m and output-facet reflectivities of 0.05, 0.1, 0.2, and 0.3.

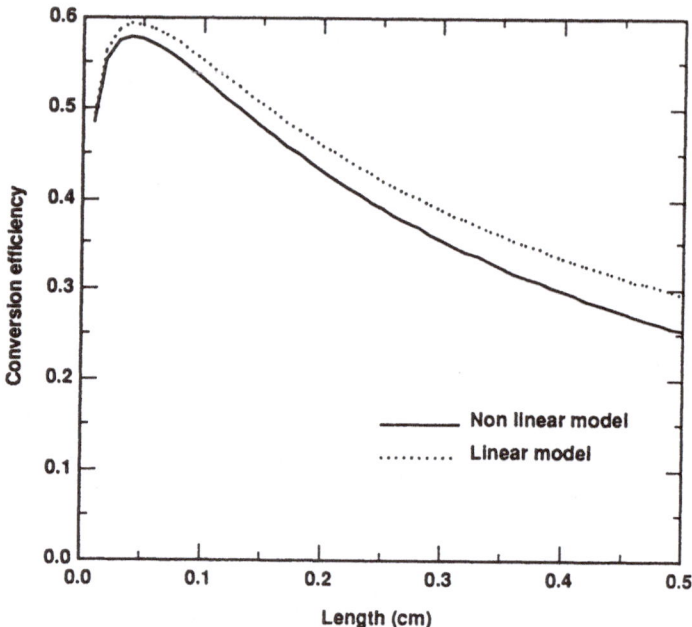

Fig. 2.15. Calculated conversion efficiencies versus cavity length for an output-facet reflectivity of $R_L = 0.05$. The dotted line was calculated using the threshold extrapolation of (2.29) and the solid line was calculated using the nonlinear above-threshold model of (2.43).

reflectivity is lowest [2.25]-[2.28]. Therefore, the gain is saturated at a lower level in the vicinity of the output facet. The linear threshold analysis used to obtain (2.29) does not account for these effects, so it overestimates the conversion efficiency in cases where the optical intensity has a larger spatial variation along the cavity.

2.4 Scaling Properties of Semiconductor and Non-Semiconductor Lasers

The form of the solution given by *Schindler* will be applied to illustrate some general scaling properties of the laser-power output as a function of the cavity length and mirror reflectivites for semiconductor lasers, as well as other nonsemiconductor, homogeneously broadened laser media. The calculated continuous-wave intensity output from (2.48) at maximum η_{Ex} vs cavity length L are shown on a log-log plot in Fig. 2.17 for GaAs, R6G dye, Nd:YAG, and CO_2 lasers. For this calculation, as well as subsequent ones in this section, the set of parameters given Table 2.1 were used. These selected values are indicative of cw operating conditions [1.272, 2.32][2.36]-[2.38].

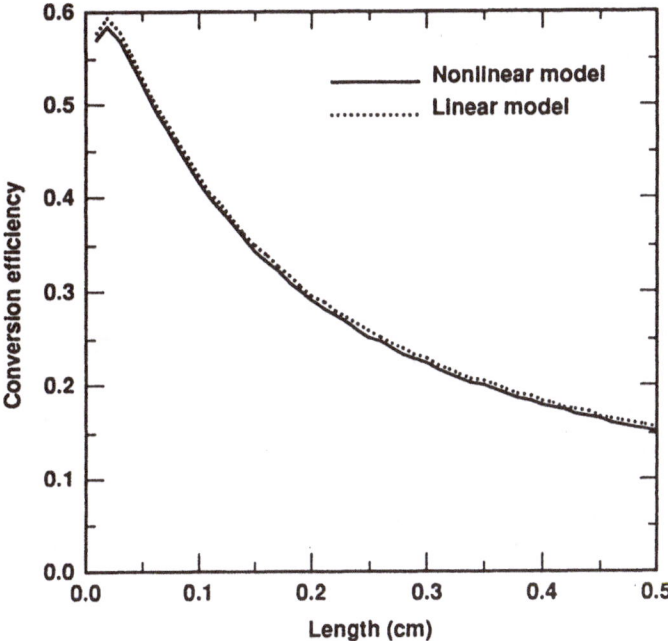

Fig. 2.16. Calculated conversion efficiencies versus cavity length for an output-facet reflectivity of $R_L = 0.3$. The dotted line was calculated using the threshold extrapolation of (2.29) and the solid line was calculated using the nonlinear above-threshold model of (2.43).

Table 2.1. Parameters in the scaling calculation

Laser Type	$g_0 \left[\text{cm}^{-1}\right]$	$\alpha \left[\text{cm}^{-1}\right]$	$I_s \left[\text{W/cm}^2\right]$
GaAs	200	5	0.5×10^6
R6G dye	5	0.2	1×10^6
Nd:YAG	0.35	0.01	2×10^3
CO_2	1	0.01	5

The calculated results are displayed as follows: the intensity output is in Fig. 2.17, the optimum reflectivity of the output-coupling mirror R_{opt} vs cavity length L is represented by the log−log plot in Fig. 2.18, and the corresponding maximum power extraction efficiency $\eta_{Ex} = \max\left[\eta_{Ex}\left(\lambda_n\right)\right]$ vs L, is displayed as a semi-log plot in Fig. 2.19. In comparing, the intensity output vs length characteristics of GaAs, R6G, Nd:YAG, and CO_2 shown in Fig. 2.17 one sees each is useful over a different range of lengths. The GaAs semiconductor laser can be scaled up to lengths of several millimeters, R6G dye to nearly 10 cm, and both Nd:YAG and CO_2 to many tens of centimeters. For all laser types the log−log plot in Fig. 2.17 is linear over a wide range

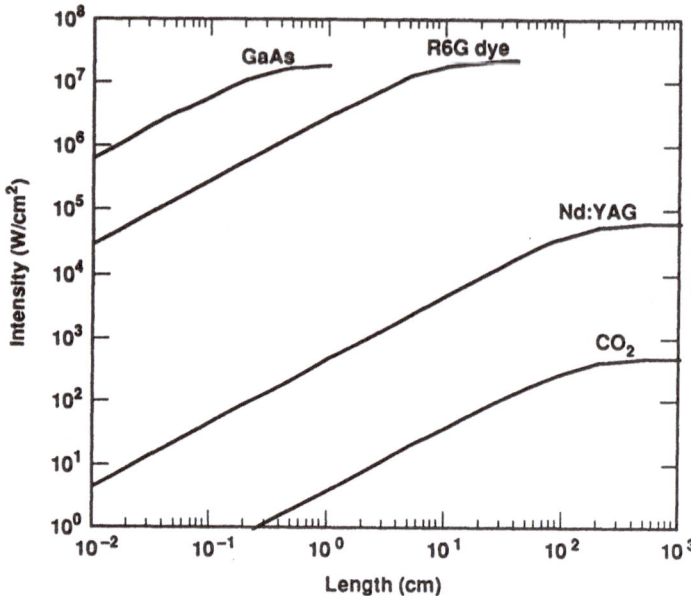

Fig. 2.17. Calculated intensity at maximum η_{Ex} versus cavity length for GaAs, R6G dye, Nd:YAG, and CO_2 laser gain media.

of shorter cavity lengths, indicating an exponential increase of the intensity output (at maximum power extraction efficiency) versus cavity length. As the cavity length is increased outside this range, for each laser type there is a rapid saturation of the intensity output with length. In all cases, the intensity output versus cavity length levels-off to a constant over about a one decade span in cavity length; it is further seen that the center of each of these length spans corresponds to the absorption length $1/\alpha$ for each of the laser types. Thus Fig. 2.17, illustrates how the sizes of different laser media are limited by the cavity losses.

In comparing the different types of lasers, it is seen that GaAs and R6G dye produce much higher intensity outputs than Nd:YAG and CO_2. This occurs because GaAs and R6G dye have higher values of g_0, and I_S. However, in actual laser systems the diameter of the mode in Nd:YAG and CO_2 can be as large as several mm or even cm, while still maintaining good beam quality at very-high cw power outputs [1.272, 2.32, 2.37]. The lower values of I_S in Nd:YAG and CO_2 are due to the longer upper-state lifetimes, μs to ms range, in these laser systems. This long lifetime makes it possible to operate both of these non-semiconductor laser materials in either a regenerative mode or a Q-switched mode, to obtain very high peak powers under pulsed operating conditions [1.272, 2.32, 2.37]. In semiconductor lasers the carrier lifetimes ≈ 1 ns or less are not long enough to be useful for producing high-peak power and high-pulse energy under Q-switched operation. The semiconductor laser is most useful as a compact source of high-average power.

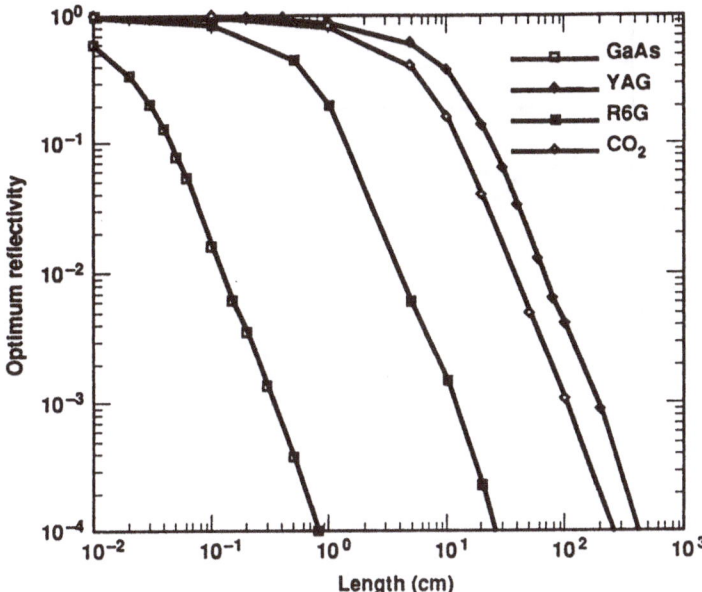

Fig. 2.18. Optimum output facet reflectivities corresponding to the intensity at maximum η_{Ex} versus cavity length curves in Fig. 2.17 for GaAs, R6G dye, Nd:YAG, and CO_2 laser gain media.

The optimum reflectivity of the output coupling mirror $R_L = R_{opt}$ versus L in Fig. 2.18, shows that as the length is increased, the value of R_{opt} decreases. As the laser length is increased, the end loss that produces optimum output coupling decreases steadily. Therefore, smaller values of R_{opt} are necessary for longer cavity lengths, in order to keep the output intensity at the level necessary to maximize the power extraction efficiency η_{Ex}. For very long cavity lengths, R_{opt} becomes so small that the laser is essentially a single-pass device, and the gain is saturated along almost the entire length of the laser. This is referred to as the loss-limited saturated operating regime [2.39]. As was seen in Fig. 2.17, this occurs for cavity lengths larger than $1/\alpha$, where the intensity output of the laser rapidly approaches a constant given by

$$I_{Lim} = I_S \left(\frac{g_0}{\alpha} - 1 \right) . \qquad (2.50)$$

This limit on the intensity output is a characteristic of all lasers and optical amplifiers where the output coupling is taken from the end reflectors. It is a consequence of the distributed optical losses that are present in the laser cavity.

From Fig. 2.19, the semi-log plot of $\eta_{Ex} = \max [\eta_{Ex}(\lambda_n)]$ vs L, it is seen for all laser types that the maximum-power-extraction efficiency is constant over the range of cavity lengths where the intensity output increases exponentially with cavity length. This is followed by a very sharp decrease as the cavity

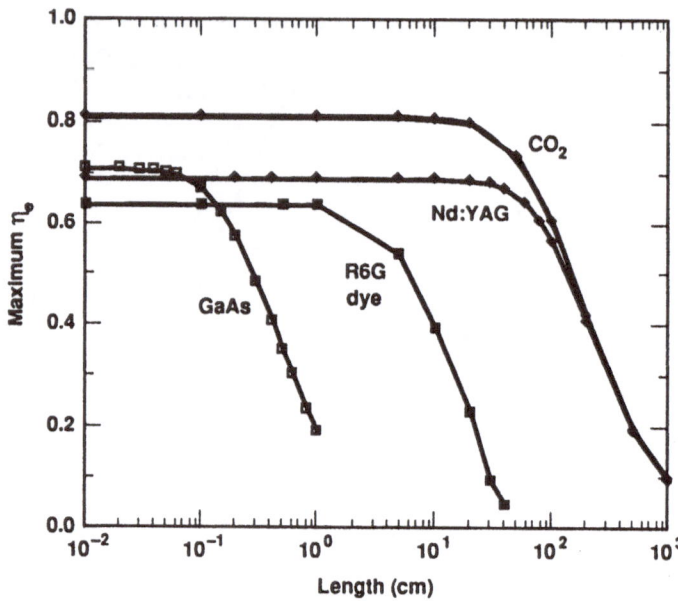

Fig. 2.19. Maximum η_{Ex} versus cavity length associated with Fig. 2.17 for GaAs, R6G dye, Nd:YAG, and CO_2 laser gain media.

length approaches $1/\alpha$. For cavity lengths greater than $1/\alpha$, $\max[\eta_{Ex}(\lambda_n)]$ will decrease exponentially with increasing L. Clearly then for maximum power-extraction efficiency, the cavity length should satisfy $L \ll 1/\alpha$. If one is interested in maximum power output, then the cavity length should be comparable to $1/\alpha$; however, as discussed earlier, one pays a price in terms of a reduction in conversion efficiency. Hence, it is not necessarily beneficial to have L comparable to the absorption length, since the conversion efficiency decreases as the cavity length is increased.

If we relax the constraint that the power-extraction efficiency be maximized, and select a larger reflectivity, R_L, for the output-coupling mirror, the maximum power intensity output will be lower, even though the end losses have been reduced. This is illustrated for the GaAs laser in Fig. 2.20, where the intensity output normalized to I_S is plotted as a function of cavity length, which has been normalized to the losses, for three different values of R_L. The curves in Fig. 2.20 were calculated numerically using (2.39 and 2.41). Note that $R_L = 0.3$ is the facet reflectivity of a cleaved GaAs chip without any dielectric coating. The values of $R_L = 0.05$ and 0.1 correspond to those typically used in high-power diode laser and diode-laser array designs; they are obtained by depositing an antireflection coating on the cleaved facet. For larger values of R_L, a lower maximum intensity output is obtained. This occurs because the power extraction efficiency, given by (2.42), decreases monotonically as the output coupler reflectivity R_L is increased. Since the

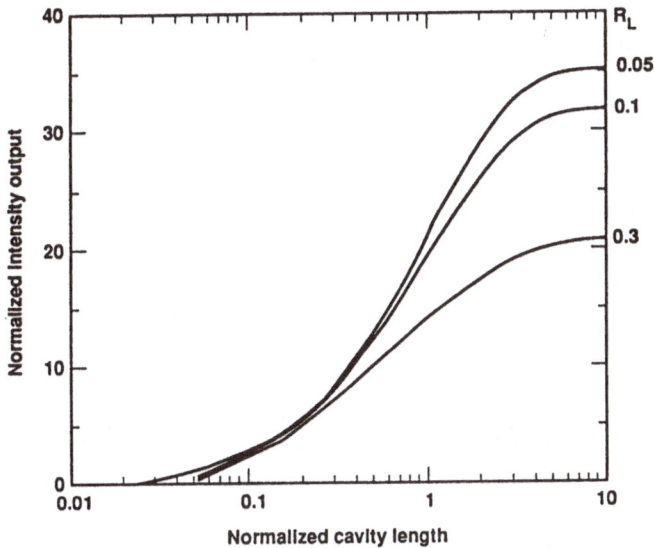

Fig. 2.20. Calculated intensity (normalized to I_S) is plotted as a function of the cavity length (normalized to $1/\alpha$) for a GaAs laser with output-facet reflectivities of 0.05, 0.1, and 0.3.

end losses are lower, the intensity output saturates at smaller values of the cavity length (Fig. 2.20).

2.5 Scaling Limitations on Power Performance of Diode Lasers

The major obstacle in realizing the maximum-projected performance levels presented in Sect. 2.4 for diode lasers is the catastrophic optical facet damage problem that was alluded to in Sect. 1.3.2. When the optical intensity approaches the $\approx 1\,\mathrm{MW/cm^2}$ level at the output facet, the risk of facet damage is increased. Moreover, operation at levels near the damage limit causes a considerable reduction in the operating lifetime and reliability. Figure 2.21 presents a graph of the intensity output versus product of the absorption coefficient and cavity length αL for selected experimental results (denoted by the numbered squares), as well as a theoretical curve for $g_0/\alpha = 500$ and $I_S = 0.5\,\mathrm{MW/cm^2}$ (representative of the best reported semiconductor material at the highest operating levels) that was calculated by the method presented in Sect. 2.4.

The experimental points are representative of diode-laser devices that have operated at high-output power levels while maintaining a predominately single-lobed output beam of reasonable quality. Point 1 is the 700 mW cw result of [1.355] for a 7 μm wide single-element index guided laser. Note that this laser displayed single-spatial and single-spectral mode behavior to 600 mW, but failed suddenly upon reaching 700 mW as indicated in the power-current characteristic in [1.355]. Presumably, the sudden failure was due to optical damage at the facet. Point 2 is a single-element index-guided laser with a *Non-Absorbing Mirror* (NAM) structure at the facets to raise the optical damage limit. This NAM-structure laser exhibited improved reliability over lasers without the NAM structure [1.351]. Point 3 is yet another single-element index guided laser, but with a width of 4 μm, that has demonstrated reliable operation at cw power outputs of 100 mW, and operated to as high 450 mW where sudden failure was observed to occur [1.342]. The details of the index-guided structure in [1.342] have not yet been revealed. However, there has been some speculation (perhaps motivated by the relibility data) that this laser design may contain a NAM-type of structure. Point 4 is a tapered semiconductor-laser amplifier (Sect. 8.1.4) with a 325 μm wide output facet that produced 4 W cw power output in a single-lobed beam [1.441], while point 5 is also a tapered semiconductor amplifier, with a 450μm wide output facet, that gave a cw power output of 4.5 W [2.40]. Note that the maximum power output of tapered-semiconductor lasers (Sect. 5.3.4) and amplifiers (Sect. 8.1.4) is not constrained by the loss-limited saturation as it is in non-tapered laser amplifiers when $\alpha L > 1$. Therefore, the theoretical maxiumum intensity output of the tapered amplifiers is not indicated by the theoretical curve that appears in Fig. 2.21, and in fact lies well above it, as illustrated in Fig. 8.11. Point 6 is an antiguided-laser array of width 120 μm that produced 1 W of cw power output [2.41], and point 7 is a 600 μm wide broad-area semiconductor amplifier that demonstrated 3.3 W of cw power output. Finally, point 8 is a 3 μm wide single-element index-guided laser that produced 425 mW, and was limited to this level by thermal roll over [1.354]. For this particular laser, facet passivation techniques were employed to raise the optical damage limit [1.353]. An important conclusion that can be drawn from Fig. 2.21 is that the maximum potential power output of high-power semiconductor laser arrays is greater, by more than an order of magnitude, than what is presently being found in experiments. To the extent that the optical damage limit can be raised, or even eliminated, significant improvements in the performance of diode lasers and diode-laser arrays can still be expected in the future.

In lasers where the output coupling is generated by a distributed loss, rather than an end loss, the power output will scale with the laser length, even when the laser is operating in the loss-limited saturated regime. Gratings can be fabricated into the structure of a semiconductor laser to act as distributed output couplers, and this will be discussed in detail in Chaps. 7 and 8. The intensity output of a distributed output coupled laser is given by

$$I_{\text{out}} = \alpha_{\text{o}} \int_0^L |E_-(z) + E_+(z)|^2 \, dz \; , \qquad (2.51)$$

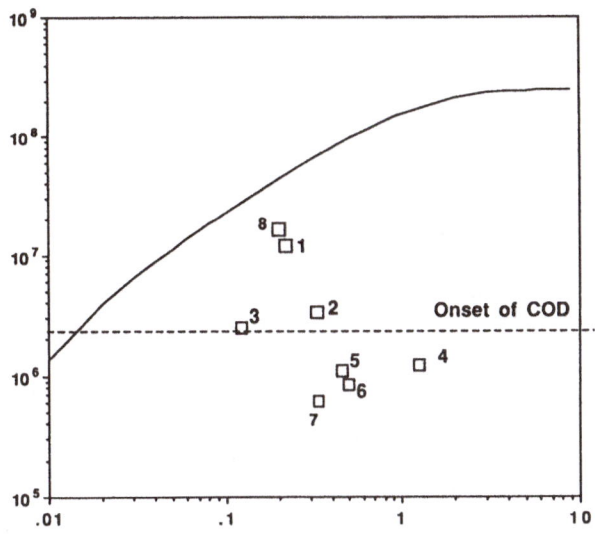

Cavity Length*Absorption Coefficient

Fig. 2.21. Calculated intensity is plotted as a function of the product αL for experimental observations of 1) [1.355], 2) [1.351], 3) [1.342], 4) [1.441], 5) [2.40], 6) [2.41], 7) [2.42], and 8) [1.354].

where α_o is the distributed-output coupling coefficient and $E_-(z)$ and $E_+(z)$ are the electric field of the traveling waves. The intensity expression (2.51) is used to account for interference effects due to the coherence between the oppositely-propagating traveling waves. The ramifications of such coherent losses are presented in Chap. 7. In the limit of large L, the maximum intensity is given by (2.50). For the case of a distributed output coupler, the loss due to the distributed-output coupling must be added to the absorption and scattering losses, α. In the large L limit of (2.51), the gain is saturated down to the level of the losses, so we have $|E_-(z) + E_+(z)|^2 \to I_{\mathrm{Lim}}$ where I_{Lim} is given by (2.50); therefore (2.51) is approximated well by

$$I_{\mathrm{out}} = \alpha_o L I_S \left(\frac{g_0}{\alpha_o + \alpha} - 1 \right) . \qquad (2.52)$$

The maximum intracavity intensity is reduced because of the larger losses due to the introduction of the distributed-output coupler. However, I_{out} has a linear dependence on L, that extends into the loss-limited saturated regime. The intensity output scales linearly with the cavity length, even though the gain medium is completely saturated, until a critical length (discussed in Chap. 8), where amplified spontaneous emission noise quenches the gain. When the effects of spontaneous emission are negligible, the power extraction efficiency is given by

$$\eta_{\mathrm{Ex}} = \frac{\alpha_o}{\alpha_o + \alpha} - \frac{\alpha_o}{g_0} , \qquad (2.53)$$

The power extraction efficiency for the distributed-output coupled laser is independent of the cavity length L. However, from the form of (2.53) it can be seen that there is an optimum value for the distributed-output coupling coefficient α_o, where the power extraction efficiency due to distributed-output coupling is maximized. By setting the first derivative of η_{Ex} with respect to α_o equal to zero in (2.53), the optimum value is found to be

$$\alpha_o = \sqrt{\alpha g_0} - \alpha \ . \tag{2.54}$$

In Sect. 8.2.2, an example is considered where a high level of amplified spontaneous emission is present in the laser cavity along with the coherent light. In this situation, it becomes necessary to use a more general form for (2.52), because (2.54) will not give the optimum value of α_o for maximizing the power output. Substituting (2.54) into (2.53), the maximum power extraction efficiency (when spontaneous emission is negligible) of the distributed-output coupled laser is found to be

$$\eta_{Ex} = \left(1 - \sqrt{\frac{\alpha}{g_0}}\right)^2 \ . \tag{2.55}$$

2.6 Effect of Carriers on Optical Properties

In semiconductor-laser material, there is a strong dependence of the local refractive index on the carrier density. The coupling between the refractive index and the gain can severely affect the mode discrimination and output-beam quality of the diode laser. The refractive index of the active layer decreases as the carrier density is increased owing to free-carrier absorption, band-filling, intraband absorption and scattering, and band-gap shrinkage effects. The carrier-induced index changes are all basically a consequence of the Kramers-Kronig relation [1.26, 2.43, 2.44]. In regions of higher optical intensity, where the gain is reduced due to saturation, the refractive index will be higher than in regions of lower optical intensity where the gain is higher. This results in a self-focusing effect on the optical beam in the laser. In wide laser stuctures, this effect can cause uncontrolled filamentation of the beam, resulting in a low or no mutual coherence between filaments, which severely degrade the spatial and temporal coherence of the output beam [1.26, 2.43]. As early as the first demonstrations of the semiconductor-diode laser, it was realized that increasing the width of the junction degraded the temporal and spatial coherence of the light output of the laser. Today, about thirty years later, the problem of how to scale semiconductor lasers to larger size to obtain high power while maintaining a high degree of spatial and temporal coherence still challenges the scientific community. At present this is one of the most important problems in the area of high-power semiconductor lasers. Below we will review the mechanisms that introduce coupling between the gain and refractive index in semiconductor lasers.

Stimulated recombination in all lasers is generated by creating a population inversion on a transition between energy states. In such dissipative materials, the frequency dependent complex dielectric permittivity will exhibit anamolous dispersion for frequencies that are near resonance with the laser transition [2.31]. The damping rate (or growth rate when gain is present) of electromagnetic energy in dissipative material is related to the imaginary part of the complex dielectric permittivity and the refractive index is determined by the real part. The causality principle requires that the real and imaginary parts of the dielectric permittivity be related to each other by the Kramers-Kronig relations [2.45, 2.46]. This coupling between the refractive index and gain (or loss) in dispersive materials is expressed as

$$n\left(E\right) = 1 + \frac{2ch}{\pi e^2} \mathrm{P} \int_0^\infty \frac{G\left(E'\right) dE'}{E'^2 - E^2} \tag{2.56}$$

where E is the photon energy, $n\left(E\right)$ is the index of refraction, $G\left(E\right) > 1$ is the gain or $G\left(E\right) < 1$ is the loss coefficent, c is the speed of light, h is Planck's constant, e is the electric charge, and P indicates that the principal value of the integral is to be taken. Most gas and solid-state laser materials comprise dilute systems of atoms or molecules that do not interact with each other, and the laser transition usually involves discrete levels. In this case, the energy integral in (2.56) extends only over the bandwidth of a single atomic or molecular transition, and increasing the number of atoms or molecules in the population inversion has no effect on $G\left(E\right)$ or $n\left(E\right)$ because the density of states, which is contained in $G\left(E\right)$, does not change. This is true as long as the optical intensity is low enough so that the transition is not saturated and nonlinear effects can be neglected.

In contrast, quantum-well semiconductor-laser systems comprise high concentrations of carriers, $\approx 10^{18}$ cm^{-3}, that are interacting Fermions; and obey the Pauli-exclusion principle. The stimulated recombination occurs on a transition between the conduction and valence energy bands of a semiconductor. Therefore, when electrons and holes are injected into the semiconductor-active layer, the density of states will change significantly because no two carriers (or holes) can occupy the same quantum state in the conduction (or valence band). In addition, the injected carriers and holes interact via Coulomb forces, and this can result in a carrier-concentration dependence in the form of the density of states. At high optical intensities where gain saturation occurs, nonlinear contributions will also contribute to the index changes. The combined effects of the dependence of the density of states on pump level and the large available gain are the reasons that gain-induced index changes are more significant in semiconductor lasers that other types of lasers. Consider (2.56) for the case of a semiconductor laser. As mentioned above, both $n\left(E\right)$ and $G\left(E\right)$ are dependent on carrier concentration, N, so we may write,

$$n\left(E, N\right) = n\left(E, 0\right) + \Delta n\left(E, N\right) \tag{2.57}$$

and

$$G(E, N) = G(E, 0) + \Delta G(E, N) \quad . \tag{2.58}$$

Then the changes in the gain or loss spectrum, $\Delta G(E, N)$, that are due to an injected carrrier concentration N will induce a change in the index of refraction, $\Delta n(E, N)$, that is given by

$$\Delta n(E, N) = \frac{2ch}{\pi e^2} P \int_0^\infty \frac{\Delta G(E', N)\, dE'}{E'^2 - E^2} \quad . \tag{2.59}$$

From (2.59) it is seen that gain-induced changes to the dispersion of the refractive index occur only in spectral regions where the injected carriers cause the gain spectrum to deviate from $G(E, 0)$. For semiconductor-diode lasers, this spectral region spans an energy range from about the band edge to the quasi-Fermi energy (which depends on the carrier concentration). At energies outside this range, gain-induced index changes will still occur, but they will be attenuated by the denominator of (2.59). The effects of band-gap shrinkage and free-carrier absorption all have significant contributions to $\Delta G(E, N)$ in the vicinity of the bandedge of semiconductors, where diode lasers tend to operate [2.44].

Band filling occurs when carriers fill the lowest lying states in the conduction band causing a shift in the apparent position of the band edge to higher energies. This causes a reduction of the absorption that begins at the band-gap energy and decreases monotonically with increasing energy. The contribution of band filling to the gain-induced changes causes a decrease in the refractive index, for energies at or below E_g the band-gap energy. For energies that are sufficiently larger than E_g, the gain-induced index change can be positive. At carrier concentrations in the range of $10^{16} - 10^{18}$ cm^{-3} the contribution to $\Delta n(E, N)$ due to band filling has a linear dependence on N the carrier concentration. For higher carrier concentrations, the contribution to $\Delta n(E, N)$ due to bandfilling exhibits a sublinear dependence on N [2.44].

At carrier concentrations $\approx 5 \times 10^{16}$ cm^{-3} or more, band-gap shrinkage effects become important. At these higher concentrations, the carriers interact via Coulombic repulsion. The resulting many-body interactions cause a decrease in the carrier energy in the conduction band. In the valence band, it results in an increase in the energy of the holes. The net effect is a decrease in the band-gap energy. Band-gap shrinkage causes an increase in the absorption that is largest near the band edge, and decreases rapidly at higher photon energies. In contrast to band filling, the gain-induced changes due to band-gap shrinkage result in an increase in the refractive index. In bulk active layers, the effects of band filling are usually dominant. This is because the band-gap shrinkage scales as the cube root of the carrier density, whereas, the band filling scales linearly with the carrier density. However, in quantum-well structures, where higher-carrier concentrations are encountered, the effects of band-gap shrinkage and intraband scattering can become important [1.451],[2.47]-[2.51]. Intraband scattering results in a spectral broadening of the gain profile, and therefore it also causes a shift of the band edge to lower energies.

Free-carrier absorption causes changes in the refractive index because a free carrier in the conduction band, or hole in the valence band, can be excited to a higher unfilled energy level by absorption of a photon. This process is modeled as a plasma effect using Drude theory [2.52], where both the carriers and holes are treated using the effective-mass approximation. The absorption coefficient for carriers in semiconductors can be expressed as [2.21]

$$\alpha_{fc} = \frac{N\lambda^2 e^3}{4\pi^2 \varepsilon_0 m_c^2 c^3 n\mu_c} \ , \tag{2.60}$$

where N is the carrier concentration in the conduction band, λ is the wavelength of light, e is the electronic charge, μ_c is the mobility, n is the refractive index, m_c is the electron effective mass, and ε_0 is the permittivity of free space. The absorption coefficient corresponding to free holes is obtained from (2.60) by substituting suitable values for m_h and μ_h, the hole effective mass and mobility [2.21]. It is seen from (2.60) that the free-carrier absorption increases with longer-wavelength radiation. Note that free-carrier absorption is an important loss mechanism in semiconductor-diode lasers. At carrier densities of 10^{18} to 10^{19} cm^{-3} that occur in quantum-well diode lasers and diode-laser arrays, the bulk free carrier absorption is in the range of 10 cm^{-1} to 100 cm^{-1} for AlGaAs/GaAs quantum-well active layers [1.347]. The actual contribution to the internal optical loss is reduced by the confinement factor to be ≈ 0.1 cm^{-1} to 1 cm^{-1}. For GaAs, the free-carrier contribution to the internal absorption can be thought of as the lower limit on the internal loss. For the semiconductors materials such as InGaAsP/InP that operate at longer wavelengths, there are other mechanisms such as intra-valence band absorption [2.22],[2.53]-[2.55] that also contribute to the internal loss; and in these materials, the lower limits on internal losses will be higher than in AlGaAs/GaAs lasers.

The contribution to the index change that corresponds to free-carrier absorption, when modeled as a plasma effect, can be written as [2.56]

$$\Delta n = -\frac{r_0}{2\pi n\left(E,0\right)E^2}\left(\frac{N}{m_e} + \frac{P}{m_h}\right) \ , \tag{2.61}$$

where $r_0 = 2.82 \times 10^{-13}$ cm is the classical radius of the electron, m_e is the effective mass of a conduction band electron, m_h is the effective mass of a hole in the valence band, and P is the concentration of holes in the valence band. The free-carrier contribution will decrease the refractive index, and sum with the bandfilling contribution for photon energies less than the bandgap. As the energy is decreased relative to the bandgap, the free-carrier effect increases due to the E^2 in the denominator of (2.61).

The two parameters that are used to characterize the net effect of injected carriers on the optical properties of a diode laser structure are the *differential gain* $dg\,(E,N)\,/dN$ and the *carrier-induced index change* or *differential index* $dn\,(E,N)\,/dN$. Note that the differential gain has units of length squared, and it is identical to the gain cross-section σ. Both the differential gain and carrier-induced index change can be determined in a straight forward manner from experiment [2.57]-[2.59]. From (2.61), the carrier-induced index change of the active layer can be expressed as

$$n\,(N) = n\,(0) + \frac{dn}{dN} N \ . \tag{2.62}$$

Note that the carrier-induced index change in the active layer dn/dN and the carrier-induced index change of the effective index of the optical mode dn_e/dN are related by

$$\frac{dn}{dN} = \frac{1}{\Gamma_y} \frac{dn_e}{dN} \tag{2.63}$$

where Γ_y is transverse-mode confinement factor given by (2.1). The carrier-induced index change in active layers of double-heterostructure diode lasers has been measured to be about $-1.2 \times 10^{-20}\,\mathrm{cm^3}$ for AlGaAs and about $-2.8 \times 10^{-20}\,\mathrm{cm^3}$ for InGaAsP [2.60]. Measurements of the carrier-induced index change in quantum-well active layers have yielded values that range from slightly lesser magnitude [2.5, 2.61, 2.62] to a about the same magnitude [2.5, 2.62, 2.63] and even slightly greater magnitudes [2.4, 2.5, 2.62, 2.64] than those of conventional double-heterostructure lasers. Some of this spread is likely due to the differences in compositions and thicknesses in the quanutm-well structures that were studied in each case. However, some of these apparent discrepancies in the reported measurements of the carrier-induced index change in quantum-well lasers might also be attributed to the method used to determine the carrier density in the experiment. As pointed out by *Rideout* et al. [2.5], the more accurate measurements are those that employ techniques that are insensitive to non-radiative recombination and leakage currents. Another factor that must be considered in interpreting such data is the wavelength and carrier dependence of the carrier-induced index change, as discussed in [2.4, 2.5],[2.62]-[2.64].

2.7 Linewidth Broadening Factor

The coupling between the gain and index that occurs through the carrier density can be described by a single parameter, α, called the *linewidth-broadening factor*, the *linewidth-enhancement factor*, the *anti-guiding factor* or the α-*factor*. An excellent review on the linewidth-broadening factor and its importance to the behavior of semiconductor lasers has been written by *Osinski* and *Buus* [2.65]. It is defined as

$$\alpha = -\frac{d\mathrm{Re}\,\{\chi\,(E,N)\}\,/dN}{d\mathrm{Im}\,\{\chi\,(E,N)\}\,/dN} \tag{2.64}$$

where $\chi(E, N)$ is the dielectric susceptibility, which depends on both E, the photon energy, and N, the carrier density. Note that the linewidth enhancement factor is dimensionless. Although it is designated by α, which is also used to denote absorption loss, one can usually infer the correct interpretation from the context or by doing a simple dimensional analysis. The linewidth-enhancement factor can also be expressed in terms of the differential gain and differential index as

$$\alpha = -\frac{4\pi}{\lambda} \frac{dn(E, N)/dN}{dg(E, N)/dN} , \qquad (2.65)$$

where λ is the wavelength corresponding to the photon energy E. The definitions of the linewidth-enhancement factor given by (2.64, 2.65) are with respect to the bulk properties of the active layer. As the optical mode of diode lasers is usually characterized in terms of the *effective index* or *modal index* n_e and modal gain g_e, it is more convenient to define the α-factor as

$$\alpha = -\frac{4\pi}{\lambda} \frac{dn_e(E, N)/dN}{dg_e(E, N)/dN} . \qquad (2.66)$$

For most diode-laser structures, (2.66 and 2.64, 2.65) will give nearly identical values for the α factor. Exceptions to this occur when the carrier density varies significantly over the extent of the optical mode [2.65].

The linewidth-enhancement factor for double-heterostructure lasers typically ranges between about 4 and 8 [2.65]. For quantum-well lasers lower values of α are found. This occurs because the differential gain in quantum-well structures is significantly larger than that of double heterostructures, while the differential index is comparable. Even lower values of α have been measured, 0.5 to 3 in strained-layer quantum-well structures [2.5, 2.66]. Recently, though there have been some reports [2.67] of values of α as high as 30 in InGaAsP/InGaAs MQW lasers, which has been attributed to carrier overflow into the adjacent separate confinement layers.

Many of the unique properties of semiconductor-diode lasers can be attributed to the strong coupling between the gain and index. The broadening of the spectral linewidth, dynamic-response characteristics, and waveguide properties of diode lasers and diode-laser arrays can all be characterized in terms of α. Here, we will mainly be concerned with the impact of the gain-index coupling on the waveguide properties of a laser array. In this context, the term *antiguiding factor* is most commonly used. For optimum mode discrimination and stability under high-power operation, it is desirable to minimize the antiguiding factor. The role of the antiguiding factor in laser array modeling is discussed in Chap. 4.

2.8 Thermal Effects in High-Power Arrays

Besides the strong coupling of gain and the refractive index, one also must deal with the dissipation of thermal energy in the diode-laser structure, especially at high operating powers where large injection current densities are required. Given the compactness of semiconductor diode-laser structures, the thermal characteristics of the monolithic diode-laser array structure are the most important limiting factor in obtaining high-average power output both for single and multi-mode operation. The electrical input power that is dissipated as heat within the diode-laser structure is given by the product of the optical power output and $(1 - \eta_T)/\eta_T$, where η_T is the electrical-to-optical conversion efficiency. The best electrical-to-optical diode laser efficiencies are in the $30-60\%$ range [1.329, 1.332, 1.346, 1.347, 2.24]. This means a narrow-stripe single-element laser producing $0.5\,\mathrm{W}$ optical power output with $50\,\%$ conversion efficiency will dissipate $0.5\,\mathrm{W}$ of the electrical input power as heat in the laser structure. For typical dimensions of $5\,\mu\mathrm{m}$ wide by $400\,\mu\mathrm{m}$ long, this corresponds to a thermal power per unit area of $\approx 25\,\mathrm{kW/cm^2}$. In single-element lasers, the active area is small enough so that this magnitude of heat dissipation can be managed. However, for broad-area lasers where the width is larger than $10\,\mu\mathrm{m}$, the thermal powers that must be dissipated are considerably higher, as the active-area where the heat is generated is larger. The rate at which heat can be removed from the diode-laser chip becomes the limiting factor in wider devices, because of the sublinear dependence of the thermal resistance on the width of the laser stripe, which is illustrated in Fig. 2.23. Since heating of the semiconductor active layer has a deleterious effect on the performance characteristics of diode lasers, thermal considerations are very important in the design of any high power monolithic diode-laser array.

2.8.1 Effect of Heating on Diode-Laser Characteristics

In the narrow-stripe, single-element, semiconductor-diode laser, the dominant source of heat generation is the non-radiative recombination of the carriers in the active layer. In addition, Joule heating occurs across the entire device because of the electrical resistance of the epilayers and the contacts. Another source of heat generation that can be important is the absorption of spontaneous emission in the regions of the diode laser outside of the active layer. In high power diode-laser arrays and broad-area lasers, that are operated at high current level, the Joule heating caused by Ohmic losses in the structure can be dominant. There are several temperature-dependent phemonena which will modify the optical properties of the semiconductor-laser structure. As the temperature is increased, the band-gap energy will decrease. In addition, the transition rates associated with nonradiative recombination processes will increase, and the form for the density of states can also change. These combined effects cause the threshold current density for laser oscillation to increase, the peak gain to decrease, and the internal quantum efficiency to decrease [2.22, 2.53, 2.68]. In addition, the temperature dependence of the index of refraction can cause thermal waveguiding to occur in narrow-stripe single-mode diode lasers [2.69] and thermal lensing to occur in broad-area diode lasers and diode-laser arrays (see Sects. 4.3 and 5.1).

In the AlGaAs quantum-well active layers that are typically used in high-power laser designs, the threshold current density increases with temperature because of an increase in the intraband carrier-carrier scattering and current leakage over the barriers. Intraband absorption and scattering causes a broadening of the gain spectrum and a reduction in the value of the peak gain [1.451, 2.48, 2.54, 2.55, 2.70]. The increased carrier leakage over the confinement layer energy barriers results in a decrease of the internal quantum efficiency at elevated temperatures [2.71]. At higher temperatures, a larger threshold current density will be required to offset the same losses.

Because of the numerous factors which affect the threshold current density and current-gain relationship, there is no single equation that is applicable to all devices and temperature ranges. To accurately model the effects of temperature on gain, one must use detailed numerical models such as those described in [1.451, 2.48, 2.54, 2.55, 2.70]. The temperature dependence of the threshold current density is characterized by a temperature parameter T_0 which is defined as [2.53]

$$\frac{1}{T_0} \equiv \frac{1}{J}\frac{dJ}{dT} \ .$$

(2.67)

The parameter T_0 is called the *characteristic temperature*. From experimental observation it has been empirically determined that the temperature dependence of threshold current density is usually well approximated by

$$J(T) = J(T_r)\exp\left[\frac{(T - T_r)}{T_0}\right]$$

(2.68)

over certain temperature ranges, where T_r is an arbitrary reference temperature. Although the above expression (2.68) for $J(T)$ does not appear to have any theoretical basis, other than being consistent with the definition (2.67), comprehensive models of quantum-well diode lasers have been used to calculate values of T_0, and its variation with temperature range [1.451, 2.10], and a T-dependent character similar to (2.67) was found. For larger values of T_0, the threshold current density will be less sensitive to temperature. In general, quantum-well diode lasers exhibit larger T_0 values $> 200\,\mathrm{K}$, whereas the best double-heterostructure diode lasers have $T_0 \approx 170\,\mathrm{K}$, and this is yet another reason why quantum-well structures are preferred for high-power operation. Under certain circumstances, T_0 in quantum-well diode lasers can become small, $< 100\,\mathrm{K}$, resulting in increased threshold-current density and lower available gain.

As the operating temperature is increased in quantum-well lasers, it has been found experimentally that T_0 decreases [1.329, 2.72]. Furthermore, it has been observed that increasing the cavity length in quantum-well structures can result in an increase in T_0 [2.72, 2.73]. Abrupt changes of T_0 in quantum-well lasers have been observed in short-cavity structures where higher end losses cause laser oscillation to occur on the n = 2 transition [2.12]. This results in a significant lowering of T_0, to about 60 K, accompanied by a shift to shorter laser operating wavelength. At cavity lengths > 500 μm, larger values of T_0 prevail, and the quantum-well diode lasers sensitivity to thermal effects will be minimized.

To summarize, heating of the diode-laser active layer reduces the available gain at a given current density. Therefore, higher threshold current densities are required to provide sufficient gain to offset the losses, and higher operating current densities are required to maintain a given power output level. As the carrier density is increased to compensate the deleterious effects of active layer heating, additional heating occurs; without proper heat sinking, thermal runaway can occur and the diode laser will cease to oscillate [1.26].

2.8.2 Heat Dissipation in Diode-Laser Arrays

Given that the net effect of heating on the diode-laser active layer is to decrease the output power and conversion efficiency, it is important to engineer high-power diode-laser arrays and associated packaging in such a way as to allow for maximum heat removal, so that the active-layer temperature is maintained at an acceptable level. The basic approach used to model the temperature distribution within a semiconductor diode laser has been to consider the two-dimensional heat flow from a uniform stripe source that represents

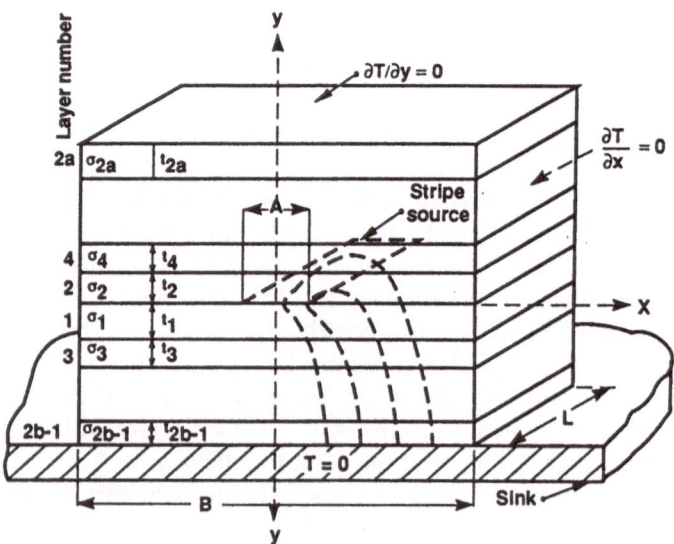

Fig. 2.22. Schematic diagram of the two-dimensional heat flow in the epilayer structure of a diode laser. [2.74]

the active layer [2.74]-[2.76]. This is depicted schematically in Fig. 2.22. Heat flow through the side, end, and top boundaries of the laser is neglected, as it is assumed that the bottom of the laser is in contact with a highly-conductive heat sink which is held at a fixed temperature T_{hs}. However, for some laser structures, such as proton implanted lasers, it is important to consider the radiative transfer of energy from the active layer to the other layers [2.75],[2.77]-[2.79], as well as to the ambient atmosphere [2.80]. The basic approach described above has been extended for analyzing oxide-stripe lasers [2.81], the effect of lateral current spreading [2.82]-[2.84], lateral current spreading in Zn-diffused oxide-stripe lasers [2.85, 2.86], and buried-heterostructure lasers [2.87]. In diode-laser arrays, it is usually the size of the array that is the most important factor in determining the heat-dissipation properties. Therefore, in the following discussion on the relation between active layer temperature and scaling of the array size, we can use the basic approach of [2.74] without loss of generality.

Following the treatment of *Joyce* and *Dixon* [2.74], the temperature within the ith layer of the laser, $T_i(x,y)$ is expressed as

$$T_i(x,y) = \beta_{i,0}(1 - \gamma_{i,0}y) + \sum_{n=1}^{\infty} \beta_{i,n}[\cosh(k_ny) - \gamma_{i,n}\sinh(k_ny)]\cos(k_nx)$$

$$(2.69)$$

where $\beta_{i,n}$ and $\gamma_{i,n}$ are constants that are determined by the boundary condition at the top of the laser and the continuity of temperature and normal heat flow between the layers that do not contain heat sources. The constant k_n is equal to $2\pi n/B$ because of the assumption that no heat escapes from the sides located at $x = \pm B/2$, where B is the substrate width.

The thermal properties of a diode-laser structure can be represented by the *thermal resistance* $R_{th}(x)$. The thermal resistance is defined as

$$R_{th}(x) = \frac{T_1(x,0)}{J_hLA} = \frac{1}{J_hLA}\sum_{n=0}^{\infty} \beta_{1,n}\cos(k_nx) , \qquad (2.70)$$

where J_{th} is heat generation rate per unit active layer area, L is the length of the laser active layer, A is the width of the active layer, and the lateral temperature profile of the active layer $T_1(x,0)$ is found by setting $y = 0$ and $i = 1$ in (2.69). From (2.70), we see that $R_{th}(x)$ is the quantity which when multiplied by the dissipated power, J_hLA, gives the temperature at a point x in the active layer. The dependence on the lateral position x can be averaged by using the mean-thermal resistance of the active layer $\langle R_{th}\rangle$ which is calculated as

$$\langle R_{th}\rangle = \frac{1}{A}\int_{-A/2}^{-A/2} R_{th}(x)\,dx = \frac{\beta_{1,0}}{J_hLA} + \frac{2}{J_hLA^2}\sum_{n=1}^{\infty} \frac{\beta_{1,n}}{k_n}\sin\left(\frac{k_nA}{2}\right) . \quad (2.71)$$

Thus, the mean temperature of the active layer is found by multiplying the mean-thermal resistance by the dissipated power. By applying the afore-mentioned boundary conditions, analytic forms for the constants $\gamma_{m,n}$ and $\beta_{m,n}$ can be calculated. In the top layer, indicated as $i = 2a$ in Fig. 2.22, where it is assumed that no heat can escape, we find

$$\gamma_{2a,n} = \tanh\left(k_n t_{2a}\right) \tag{2.72}$$

where t_{2a} is the thickness of the top layer. In the bottom layer, indicated by $i = 2b - 1$ in Fig. 2.22, that interfaces with the heat sink

$$\gamma_{2b-1,n} = \coth\left(k_n t_{2b-1}\right) \ , \tag{2.73}$$

where t_{2b-1} is the thickness of the bottom layer. In the intervening layers, where it has been assumed that the interfaces are source-free, we have

$$\gamma_{i,n} = \frac{\tanh\left(k_n t_i\right) + \left(\sigma_{i+2}/\sigma_i\right)\gamma_{i+2,n}}{1 + \left(\sigma_{i+2}/\sigma_i\right)\gamma_{i+2,n}\tanh\left(k_n t_i\right)} \ , \tag{2.74}$$

where t_i is the thickness of the layer i, and σ_i is the thermal conductivity of layer i. By considering the boundary conditions on the heat flow from the sides of the active layer, we obtain

$$\beta_{1,n} = \frac{4\sin\left(k_n A/2\right)}{B k_n^2 \left(\sigma_1 \gamma_{1,n} + \sigma_2 \gamma_{2,n}\right)} \tag{2.75}$$

and

$$\beta_{1,0} = \frac{A}{B}\sum_{j=1}^{b}\frac{t_{2j-1}}{\sigma_{2j-1}} \ . \tag{2.76}$$

For a given diode-laser structure, (2.71-2.76) or (2.70, 2.72-2.76) can be used then to calculate the thermal resistance. In most applications of this thermal analysis it is necessary to calculate about 100 terms in (2.71 or 2.70) to obtain a result that is accurate to three significant figures.

For a given epilayer structure, it has been shown [2.74] that the laser length L and the active layer width A are the parameters that will have the greatest impact on the thermal properties of laser. From the forms of the thermal resistance (2.70,2.71) and the expressions for constants (2.75,2.76) it is clear that $\langle R_{th}\rangle$ and $R_{th}(x)$ will scale as $1/L$. Also, it is seen that the thermal resistance will experience only a limited decrease as the active layer width A is increased. The dissipated power scales linearly in both L and A. The spreading resistance causes R_{th} to decrease more slowly than $1/A$. This means that for constant-power dissipation, the active-layer temperature will increase as the width of the active layer is increased. This situation is quite common in monolithic diode-laser arrays, and it is a significant factor that limits the maximum-operating power and efficiency. Moreover, the tempera-ture gradient that can occur between elements in diode-laser array structures can be large enough so that coherent operation is no longer possible [2.88].

To consider this further, we will use a simpler form for the thermal resistance that was derived by *Liau* et al. [2.89] to model thermal resistances of devices mounted junction-side-up on the heat sink. In this analysis, it was assumed that only a single layer of substrate material was between the active layer and the heat sink. The considerably simpler analytic expression

$$\langle R_{\mathrm{th}} \rangle = \frac{1}{\pi \sigma_{sub} L} \sinh^{-1} \left(\frac{\sinh (\pi d / B)}{\sin (\pi A / 2B)} \right) \tag{2.77}$$

for the thermal resistance was obtained. In (2.77), d is the thickness of substrate material between the active layer and the heat sink and σ_{sub} is the thermal conductivity of the substrate. Figure 2.23 displays the result of using (2.77) to calculate the thermal resistance and active layer temperature as a function of the active layer width with an InP substrate. Note that these values are much higher than would be found for the case of junction-down mounting, so this represents a worst case example. The smallest width used of $A = 3\,\mu$m corresponds to $\langle R_{\mathrm{th}} \rangle \approx 50\,°$C/W in Fig. 2.23. The corresponding active layer temperature of $\approx 25\,°$C was calculated assuming that a power of 0.5 W being dissipated in the $L = 500\,\mu$m long by $A = 3\,\mu$m active layer.

As the width A is increased, the power per unit area in the active layer is fixed at the value assumed for $B = 3\,\mu$m. This corresponds to the situation that occurs when the width of the active layer is scaled to larger values to obtain higher power outputs. The thermal resistance decreases steadily towards an apparent asymptotic value of $\approx 15\,°$C/W. However, the active-

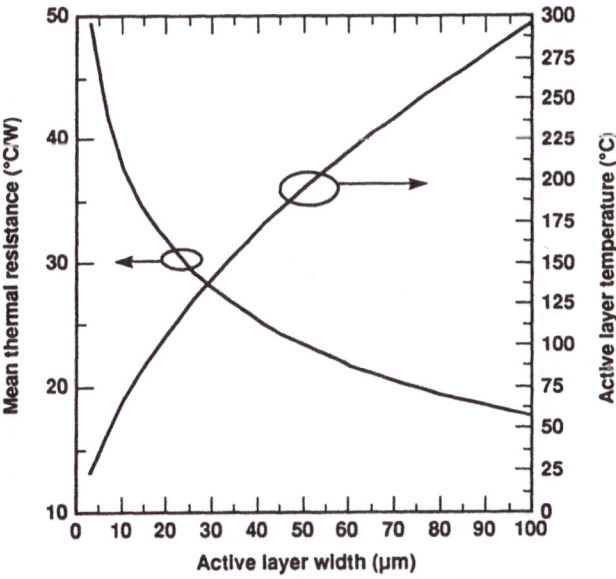

Fig. 2.23. Calculated results for the thermal resistance and active layer temperature as a function of the active layer width using (2.77).

layer temperature exhibits a rapid increase with increasing width. A doubling of the width from $3\,\mu$m to $6\,\mu$m results in nearly a doubling of the active-layer temperature. At a width of $40\,\mu$m where the thermal resistance has decreased by a factor of two, relative to the $3\,\mu$m width, the active-layer temperature is 173 °C. Although, this may seem like an unrealistic operating temperature, diode lasers have been operated at temperatures as high as 200 °C [1.337], as there are some applications where performance in high-temperature environments is required. However, active-layer temperatures of this magnitude will lead to a significant degradation in the conversion efficiency and the power output.

Another parameter that is frequently used to characterize the cooling capacity is the thermal impedance, which is obtained by multiplying the thermal resistance by the area of the diode-laser active area. The thermal impedance is particularly useful in comparing the cooling capacities of diode lasers of different sizes, since it accounts for any size differential. This is depicted in Fig. 2.24 where the thermal impedance corresponding to the data in Fig. 2.23 is plotted as a function of the active layer width.

Increasing the active layer width results in a monotonic increase in the thermal resistance. It can be seen that the thermal impedance is more reflective of the general trend that the active-layer temperature increase as the width of the active layer is increased. The illustrative example of Figs. 2.23 and 2.24 emphasize the importance of thermal management as diode-laser array sizes are scaled to achieve higher-power ouputs.

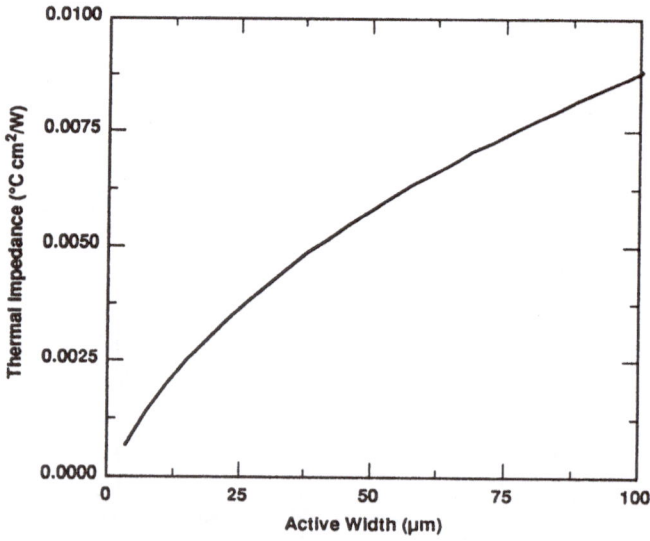

Fig. 2.24. Thermal impedance versus active-layer width plot corresponding to the thermal resistance plotted in Fig. 2.23.

2.8.3 Thermal Management and Performance Limitations

Although not discussed above, the manner in which the diode laser is bonded
to the heat sink, as well as the details of heat-sink design and overall packaging architecture, play a critical role in determining the thermal performance
of diode-laser arrays. Indeed, significant progress has occurred in improving heat-sink and packaging technology for high power diode-laser arrays
[1.275, 1.276, 2.90]. Much of this work has been motivated by the significant
advantages that hybrid diode-laser arrays offer as pump sources for a variety
of solid-state lasers [1.272]. Successful heat sink architectures for high power
diode laser operation have been developed using copper impingement coolers [2.90, 2.91], silicon microchannel coolers [1.275, 1.276], diamond coated
submounts [1.275, 1.69], and copper microchannel coolers [2.92]. Thermal
impedances of $0.014°C\,cm^2/W$ [1.275, 1.276] have been reported with silicon microchannel coolers. Copper impingement coolers have demonstrated
thermal impedances of $\approx 0.075°C\,cm^2/W$ [2.91] (as estimated from thermal
resistance data in [2.91]), while copper microchannel coolers have exhibited
$0.026°C\,cm^2/W$ [2.92] (as estimated from thermal resistance data in [2.92]).
All of this advanced heat-sinking technology can be implemented in monolithic surface-emitting diode-laser arrays using backplane cooling architectures. A monolithic monolithic surface-emitting laser array architecture has
the potential of greater cooling capacity than conventional rack-and-stack
edge-emitting array architecture.

It is section will consider the silicon microchannel cooler is assessing
the thermal limitations that are placed on high power operation. However, as already mentioned there are other approaches for thermal management. In the case of the copper impingement cooler, a detailed review has
been given by [2.90]. The silicon microchannel heat sink [1.274, 1.276][2.93]-
[2.99] has demonstrated the capability for dissipating heat levels as high as
$2000\,W/cm^2$, which occur in both hybrid and monolithic high-average power
diode-laser arrays. In addition, the modular construction, light weight, and
compact size of the silicon microchannel heat sink can be easily scaled to fabricate large two-dimensional diode-laser arrays, while minimizing the system
complexity [2.100]. The combination of heat-removal capacity with small size
and low weight is desirable for scaling to larger array sizes to increase power
output. Silicon microchannel coolers have been demonstrated with monolithic
etched-facet surface-emitting laser arrays [1.184, 2.99] and surface-emitting
DFB lasers [1.184], but have not yet achieved the performance levels that have
been demonstrated in hybrid edge-emitting diode-laser arrays [1.275, 1.276].

It is useful to consider the limits placed by heat removal on the scaling properties of high-power diode-laser arrays. The principle issue in diode-
arrays that are used as pump sources for solid-state lasers is minimizing the
cost per Watt of output power [1.276, 2.100], so beam quality and spectral
purity are not considerations. Therefore, scaling the length of these arrays,
as discussed in Sect. 2.4, should lead to increase power output, and is a

promising approach for reducing the cost per Watt of power output [2.100]. Figure 2.25 presents the calculated theoretical-maximum power outputs (left-hand vertical axis) for a 1 cm wide diode laser-array bar of the type described in [1.275, 1.276, 2.100]. Also, displayed on the same plot is the thermal power removed by a silicon-microchannel heat sink assuming a power conversion efficiency of 50 %, a temperature rise of 28°C between the active layer and the heat sink, and a thermal impedance of $0.014°C\,cm^2/W$. Note that the temperature rise number is rather conservative, but is representative of some of the most demanding requirements that are encountered. On the other hand, the power conversion efficiency is representative of the best performing diode-laser arrays.

We see from Fig. 2.25 that in scaling to beyond mm cavity lengths to increase power-output levels, the limiting factor is no longer the heat-removal capacity, but the facet damage limit. Increasing the length of a laser array also increases the size of the area over which the heat is dissipated. Since the thermal resistance (2.70) scales as the reciprocal of the length of the device, the thermal power removed will scale linearly with cavity length for a fixed

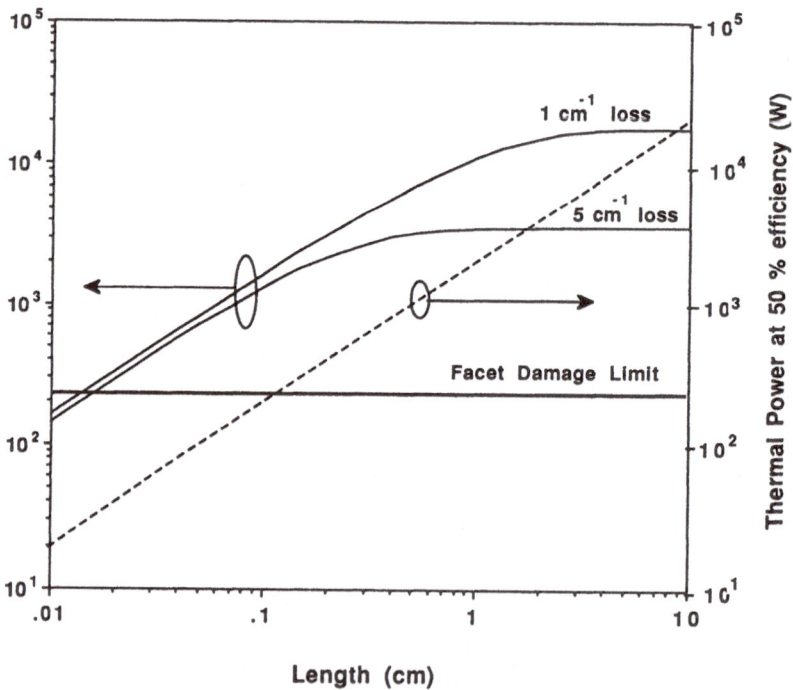

Fig. 2.25. The calculated power output (left-side vertical axis) of a 1 cm wide diode-laser array bar is plotted as a function of the cavity length of the bar. The curves for $\alpha = 5\,cm^{-1}$ and $1\,cm^{-1}$ were both calculated using $g_0 = 500\,cm^{-1}$. The horizontal line corresponds to the optical facet damage limit. The dotted line (right-hand side vertical axis) is the heat removal capacity in W.

temperature rise. To maintain a constant temperture rise as the length is increased, it will be necessary to increase the flow rate in the micro-channel coolers to compensate for caloric self-heating of the coolant, which accounts for $\approx 1/3$ of the thermal resistivity of the package [1.275, 1.276]. As already mentioned in Sect. 2.4, the optical damage limit of the diode-laser facet is the principle obstacle in the path towards further improvement in the high-power performance of diode-laser arrays. To this end, facet passivation techniques that raise or eliminate optical damage would be of great benefit. Also, the grating-coupled surface-emitting laser arrays presented in Chaps. 7 and 8 represent an alternative structure that should be immune to facet damage, but more development of grating-output coupled laser arrays is required as the routine performance levels of these devices is not yet up to the standard of conventional edge emitters.

3. Spatial and Spectral Mode Discrimination

The previous chapter dealt with topics that were primarily related to optimizing the power output and conversion efficiency of diode-laser arrays and single-element diode lasers. This chapter is devoted to principal concepts related to spatial and spectral mode discrimination in semiconductor diode-laser arrays. We begin in Sect. 3.1 by reviewing the symmetric three-layer planar waveguide model. This model is central to our understanding of the spatial-mode properties of both diode lasers and diode-laser arrays, as it is often used to elucidate the waveguide and spatial-mode characteristics. Analysis of specific multilayer diode-laser array structures is dealt with in Chap. 4.

Section 3.2 discusses spatial-mode discrimination in diode-laser arrays comprised of a periodic variation of the dielectric permittivity along the lateral direction. A mode confinement factor approach is used for this discussion, since it illustrates mode discrimination and scaling in diode-laser arrays from a geometrical perspective. For this section, the coherent diode-laser array is is analyzed as if it were a large-cavity single-element laser with a periodic variation in the dielectric permittivity.

Fundamental principles of single-frequency diode lasers and scaling characteristics are reviewed in Sect. 3.3. Some fundamental aspects of electromagnetic modes in a cavity are reviewed in Sect. 3.3.1. Section 3.4 discusses injection-locking and entrainment of laser oscillators. In contrast to Sects. 3.2 and 3.3, for this discussion the diode-laser array is viewed as an ensemble of laser oscillators that can operate independently. The conditions for frequency locking or entrainment to produce coherent single-frequency operation of the complete diode-laser ensemble are considered.

3.1 Spatial Modes of Planar Semiconductor Waveguides

In this section, the symmetric three-layer planar waveguide model for semiconductor diode lasers is reviewed. The three-layer waveguide is very useful for providing an intuitive understanding of the waveguiding mechanisms in diode laser and diode-laser array structures. Although, both diode lasers and diode-laser arrays are actually two-dimensional waveguide structures, the waveguide characteristics in the two directions perpendicular to the optic

axis can usually be well-approximated by effective planar-waveguide structures, and the two-dimensional waveguide can be reduced to an effective
one-dimensional waveguide. These topics are discussed in depth in Chap. 4.
Rather than present a detailed analysis, the aim of this section is to elucidate
the various types of modes (e.g., index-guided modes, gain-guided modes, and
radiation or leaky-wave modes) that are encountered in semiconductor diode-
laser waveguides and the conditions necessary to support them. Numerous
volumes have been devoted to the subject of waveguiding in semiconductor
lasers and layered optical media. More detailed treatments of complicated
waveguide structures can be found in [1.25, 1.26, 1.323, 1.325][3.1]-[3.8].

There are several important characteristics that distinguish semiconductor laser waveguides from passive waveguides. Some of these were touched
on in Chap. 2, where the effects of carrier injection and temperature on the
semiconductor dielectric coefficient were described. In semiconductor diode-
laser array structures, the waveguide properties and coupling characteristics
can vary significantly with drive level, which often results in a degradation of
the mode discrimination properties. In most cases, the gain or loss will have
a significant effect on the semiconductor laser-waveguide properties. Therefore, in this discussion of the three-layer planar waveguide, we will follow
the approach of *Schlosser* [3.9] where each layer is described in terms of its
complex-dielectric permittivity as opposed to the usual treatment of lossless
waveguides, where only the real part of the dielectric permittivity is used.
In order to describe the gain-guided modes and leaky-wave modes that are
often found in diode-laser arrays, it is necessary to use the complex dielectric permittivity of each layer. Figure 3.1 shows a diagram of a symmetric
three-layer slab waveguide that will be used. The active or guide layer has a
dielectric permittivity ε_1 and the adjacent cladding layers have a dielectric
permittivity ε_2. The dielectric permittivity ε_j, where $j = 1, 2$ of each layer
will be of the form,

$$\varepsilon_j = (n_j + i\kappa_j)^2 \ , \tag{3.1}$$

where n_j is the index of refraction of layer j and κ_j is the extinction coefficient
of layer j,

$$\kappa_j = \frac{\lambda \alpha_j}{4\pi} \ , \tag{3.2}$$

where λ is the wavelength of light in vacuum and α_j is the loss coefficient of
layer j when $\alpha < 0$ or the gain coefficient of layer j when $\alpha > 0$. For this
discussion the center layer, or guide layer in Fig. 3.1, will represent the active
layer of the laser; and the direction of propagation along the laser cavity lies
along the z axis, which is defined as the *longitudinal direction*.

The electromagnetic modes of interest are those which will experience the
most gain and undergo laser operation. These will be modes that propagate
along the z axis, with the field energy primarily localized in the active-guide
layer of the laser. Therefore, in solving the wave equation we need only consider solutions where the field decreases to zero with increasing distance,

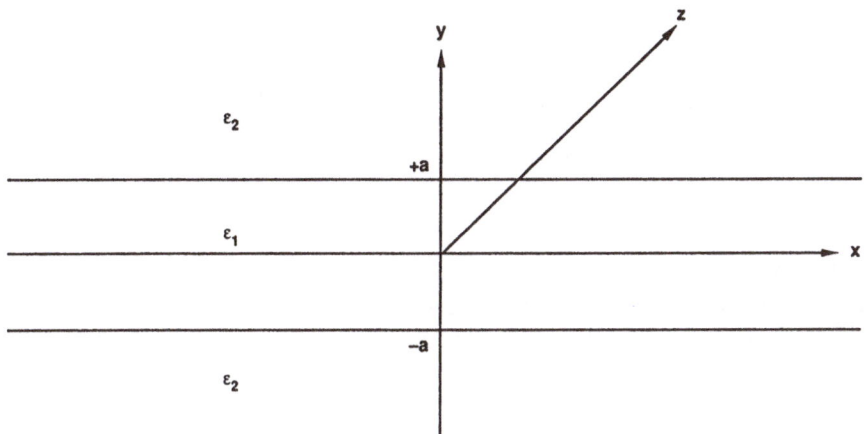

Fig. 3.1. A drawing of the symmetric three-layer dielectric-waveguide structure.

along the y or *transverse direction*, from the active layer. Also, field solutions corresponding to waves propagating towards the guide structure from infinity can be eliminated. Without loss of generality, we can restrict our solutions to the case of the *transverse polarized* or *TE modes*, where the electric field is polarized only along the x axis or *lateral direction*. These solutions for the electric field E_x of the TE modes will be of the form

$$E_x = A_x(y)\, exp\left[i\left(\omega t - \beta z\right)\right] \; , \tag{3.3}$$

where ω is the angular frequency of light, β is the propagation constant of the waveguide mode, and $A_x(y)$ satisfies the following wave equation

$$\frac{\partial^2 A_x}{\partial y^2} + \left(k^2\varepsilon - \beta^2\right) A_x = 0 \; , \tag{3.4}$$

where $k = \omega c$ is the free-space wavenumber. The remaining field components, H_y and H_z can be determined from E_x and Maxwell's curl equations, as shown in the next section. When the solutions of (3.4) that satisfy the boundary conditions at large values of y are combined with (3.3) the electric field of the TE modes can be expressed as,

$$E_y = A_0 cos\left(\frac{uy}{a}\right) \exp\left(-i\beta z\right) \text{ for } |y| \leq a \tag{3.5}$$

and

$$E_x = A_0 \cos(u)\, exp\left[-i\beta z + w\left(|y/a| - 1\right)\right] \text{ for } |y| \geq a \; , \tag{3.6}$$

where

$$u^2 = \varepsilon_1(ka)^2 - (\beta a)^2 \tag{3.7}$$

and

$$w^2 = (\beta a)^2 - \varepsilon_2(ka)^2 \; . \tag{3.8}$$

Note that (3.6) satisfies the aforemention requirement that the field goes to zero at large distances y from the laser waveguide layer, and that there are no waves propagating into the waveguide from infinitely large y distances. The boundary conditions require continuity of $A_x(y)$ and $\partial A_x(y)/\partial y$ at $y = \pm a$, the interface of the guide and cladding layers, which gives the secular equation,

$$w = u \tan(u) \ . \tag{3.9}$$

The propagation constant β for the allowed modes of the waveguide are those values of β which satisfy the secular equation (3.9). Usually, this gives a discrete spectrum of allowed modes β_m which must be determined numerically, since (3.9) is a transendental equation. Because both ε_1, the dielectric coefficient of the active layer, and ε_2, the dielectric coefficient of the cladding layers, are complex, the resulting spectrum of modal propagation constants β_m are also complex, and can be expressed in the form,

$$\beta_m = k n_{\text{eff}}^{(m)} + i \alpha_{\text{eff}}^{(m)} \ , \tag{3.10}$$

where $k = 2\pi/\lambda$. The imaginary part of β_m corresponds to the *modal loss* or *gain coefficient*, $\alpha_{\text{eff}}^{(m)}$, of the mth mode. The real part of β_m corresponds to the propagation constant of the mth mode, and $n_{\text{eff}}^{(m)}$ is defined as the *effective index of refraction* of the mth mode. The concept of the effective index of refraction is useful because it enables one to view the allowed modes in terms of geometric optics and ray-tracing. The vector relationship between the waveguide-propagation constants and the modal propagation constant is found by substituting (3.10) into (3.7) and taking the real part to give

$$\text{Re}\{\beta_m\} = k n_{\text{eff}}^{(m)} = k n_1 \cos(\theta_m) \ . \tag{3.11}$$

The vector relationship that illustrates (3.11) is depicted in Fig. 3.2; and it is seen from the viewpoint of geometric optics, that the effective index is related to the cosine of the bounce angle of the rays comprising the guided mode.

For a guided-mode, we have the following condition on $\text{Re}\{\beta_m\}$:

$$k n_2 \leq \text{Re}\{\beta_m\} \leq k n_1 \ . \tag{3.12}$$

For a mode to be able to propagate in the waveguide without decay, other than that due to dissipative optical losses such as absorption and scattering, the total phase change of the plane wave as it travels between the two interfaces must be an integer multiple of 2π. If φ represents the phase shift at the interface for the symmetric 3-layer waveguide, then it follows that

$$4akn_1 \sin(\theta_m) - 4\varphi = 2m\pi \tag{3.13}$$

where $2a$ is the thickness of the waveguide layer. The term 4φ represents what is know as the Goos-Hänchen shift [3.3, 3.10, 3.11]. The effect of the Goos-Hänchen shift is to cause the apparent position of the deflection point of the

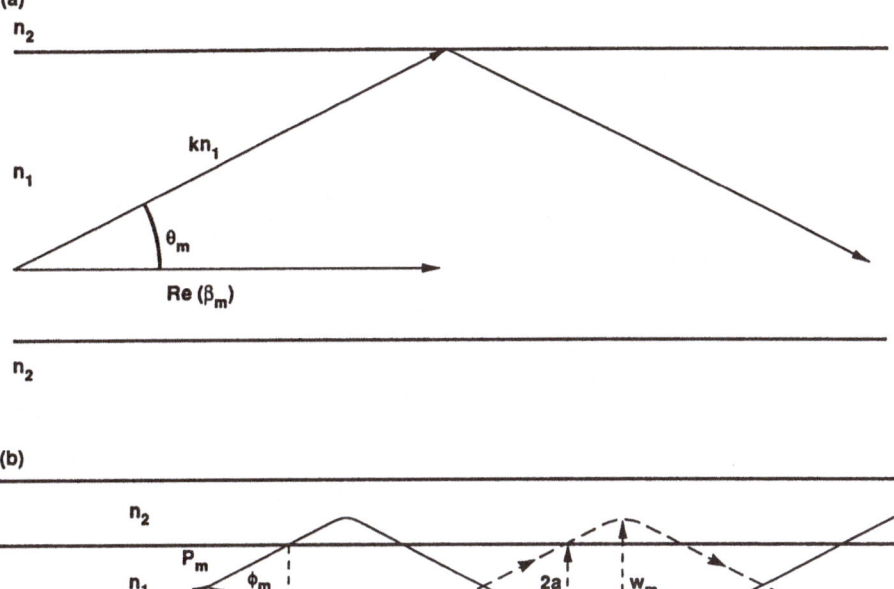

Fig. 3.2. Diagrammatic representation of the vectorial relationship between the waveguide-layer propagation constants and the modal-propagation constant (**a**). Effective optical path length due to the Goos-Hänchen shift (**b**).

ray to be displaced from the physical position of the interface into the lower-index cladding region, as illustrated in Fig. 3.2b. Hence, the electric field will penetrate some distance into the cladding material, causing the physical width of the optical mode w_m to be larger than that of the waveguide $2a$. The effective width of the optical mode can be approximated as

$$w_m \approx 2a + 2\frac{a}{w} \; , \tag{3.14}$$

where $a/2w$ is the exponential-decay rate of the electric field in the cladding layer (3.6). For the case of an asymmetric waveguide the effective mode width w_m is expressed as

$$w_m \approx 2a + \frac{a}{w_u} + \frac{a}{w_l} \; , \tag{3.15}$$

where a/w_u and a/w_l are respectivly, the exponential-decay rates of the electric field in the upper and lower cladding layers.

The Goos-Hänchen shift is a very important effect in semiconductor lasers, because the active-waveguide layer is always much less than the wavelength of light so that most of the electric field resides in the cladding layers adjacent to the active layer, as illustrated in the example displayed in Fig. 2.3. Therefore, to minimize the optical losses of the laser-waveguide structure, the losses associated with the cladding layers and other layers, where there is significant field penetration, should be as low as possible. Although a bounce angle can be associated with a given mode according to (3.11), to accurately model diode-laser waveguide structures it is really necessary to use the therory of physical optics rather than the geometric optics approximation. The geometric optics picture provides an intuitive explanation of the optical confinement of the modes, but not the details of the modal fields.

In the presence of loss or gain, we have a more general condition for the confined optical modes of the waveguide. The conditions on the waveguide layers necessary to support confined modes can be found by using (3.7, 3.8, and 3.9) and the requirement that the field be exponentially damped as y goes to infinity. It can be shown that [3.9]

$$\text{Im}\left\{\varepsilon_1 - \varepsilon_2\right\} > 0 \tag{3.16}$$

is a necessary, though not sufficient, condition for a solution to the secular equation (3.9) to exist. Using (3.1), the condition (3.16) can be expressed as

$$n_1\kappa_1 - n_2\kappa_2 > 0 \ . \tag{3.17}$$

There are three special cases of (3.17) that are of particular relevance to semiconductor diode-laser arrays. These cases are identified and discussed below.

3.1.1 Index Guiding

When $\kappa_1 = \kappa_2$, the inequality (3.17) reduces to $\Delta n = n_1 - n_2 > 0$. This is identical to the well-known condition for refractive-index waveguiding in a lossless structure. It is usually refered to as *index guiding*. The index of refraction of the guide layer must be larger than that of the cladding layers. Here, we see that this condition is more general, and it is applicable to the case when a uniform loss or gain is present throughout the entire waveguide structure. When $n_1 > n_2$, light propagating in the guide layer, within a narrow range of wave vectors, will undergo total internal reflection at each interface of the guide and cladding layers, and be reflected back into the waveguide. This is shown schematically in Fig. 3.2. In the terminology of geometric optics, the total internal reflection balances the diffractive losses of the waveguide layer for certain wave vectors of light. This is the mechanism that confines the optical field to the guide layer. For index-guided modes there is a discrete spectrum of allowed modes within the range $kn_1 > \text{Re}\left\{\beta_m\right\} > kn_2$. In many cases, there is also a *cutoff condition* for each mode. A mode is said to be at cutoff when the optical field is no longer confined to the guide layer of the waveguide. This occurs when $\text{Re}\left\{\beta_m\right\} = kn_2$. At this point, the attenuation coefficient in the y direction w, which is given by (3.8), goes to zero; and as a result E_x, which is given by (3.6), is no longer exponentially damped at infinitely large values of y. Such modes, which are not confined to the guide layer, are referred to as *radiation modes* [3.1, 3.2]. At cutoff, the secular equation (3.9) for the three-layer symmetric wavguide, with $\kappa_1 = \kappa_2$, can be simplified to give [3.3]

$$n_1^2 - n_2^2 > \frac{m^2 \lambda^2}{8a^2}; \quad \text{where } m = 0, 1, 2, \dots \ , \tag{3.18}$$

where m is the mode order, which corresponds to the number of nulls in the transverse-field amplitude $A_x(y)$. The lowest-order even mode, also called the *fundamental mode*, is given by $m = 0$; and it will usually provide the maximum confinement of light in the active-guide layer. From (3.18), it is seen that there is no cut-off condition for the $m = 0$ mode. Therefore, for a given refractive index difference between the guide and cladding layers, all modes with $m \geq 1$ can be cut-off so that the only allowed mode is the $m = 0$ or fundamental mode. From (3.18), it is seen that this occurs for waveguide structures where the guide layer thickness a and indices of refraction for the guide and cladding layers, n_1 and n_2, satisfy the condition

$$\frac{2a^2 k^2}{\pi^2} \left(n_1^2 - n_2^2 \right) \leq 1 \ . \tag{3.19}$$

To design an effective single-mode index-guided diode-laser structure it is best if the layer compositions and active-layer thickness can be selected so that all modes but the $m = 0$ mode are beyond cutoff for all anticipated operating conditions of the laser. This applies both to the transverse and lateral modes. It is seen from (3.19) that for a given set of guide and cladding-layer compositions, this can be accomplished by making the guide layer a sufficiently thin. The cutoff condition for the symmetric three-layer waveguide can be applied to both the transverse and lateral modes of diode lasers and diode-laser arrays.

3.1.2 Gain Guiding

Now consider the case when $n_1 = n_2$. The inequality (3.17) reduces to $\Delta\kappa = \kappa_1 - \kappa_2 > 0$. This is what is referred to as pure *gain guiding* or *loss guiding*, wherein the guide layer has lower optical losses or higher gain, in the case of active structures, than the adjacent cladding layers. The optical field is confined to the more transparent guide layer, since the optical field is more strongly absorbed in the cladding layers. The fundamental mode, or lowest order even mode, of a gain-guided waveguide structure with $n_1 = n_2$ is not cutoff. The cutoff condition for the next mode, which is the lowest-order odd mode, has been determined numerically by *Schlosser* [3.9], and can be expressed by,

$$2ak \left(\kappa_1 - \kappa_2 \right) \leq 1.877 \ . \tag{3.20}$$

Qualitatively this is very similar to the condition (3.19) for a pure index-guided structure to support only the fundamental mode. Though, in diode lasers pure gain-guiding does not actually occur because the index of refraction is dependent on the local carrier density. Therefore, spatial variations in the gain will produce a corresponding variations in the refractive index according to (2.59).

3.1.3 Mixed Guiding

In semiconductor-diode lasers, the waveguide will always be a combination of index and gain guiding; and depending on the details of the structure, one type of guiding can be made to dominate over the other. More general cases of optical waveguiding can by considered by substituting $\Delta n = n_1 - n_2$ and $\Delta\kappa = \kappa_1 - \kappa_2$, into (3.17) to obtain,

$$\frac{\Delta\kappa}{\kappa_1} + \frac{\Delta n}{n_2} > 0 \ . \tag{3.21}$$

It is seen that in situations where either Δn or $\Delta\kappa$ is negative, guided modes will exist if whichever of Δn or $\Delta\kappa$ that is positive, is sufficiently large that (3.21) is satisfied. For example, waveguide structures that do not satisfy the condition for purely gain- or loss-guided modes, will support guided modes if the refractive index difference between the guide and cladding layers is sufficiently large to offset the negative difference in the imaginary part of the dielectric coefficient. For diode-laser arrays, it is the opposite case, when $\Delta n < 0$ and $\Delta\kappa > 0$ which is of more importance. When there is a negative index difference, $\Delta n < 0$, between the guide and cladding layers, confined modes can be supported if the gain of the active layer is large enough so that (3.21) is satisfied.

Waveguides with a negative-index difference are often referred to as *antiguided structures, leaky-wave structures* or *negative-index guides* . For these structures, the propagation constant of the waveguide satisfies $\beta < kn_2$, and there is no total internal reflection at the interface between the guide and cladding layers to confine the mode. Therefore, the confined modes supported by a $\Delta n < 0$ structure will radiate light from the guide layer into the adjacent cladding layers. For this reason, this type of confined mode is often called a *leaky mode*. This additional radiative loss mechanism is compensated by the gain of the active layer in a diode-laser waveguide structure. Generally, low-loss leaky modes can only be supported in semiconductor waveguide structures where the guide-layer thickness is significantly larger than the wavelength of light [3.12] or where there is sufficient gain present in the active layer [3.13].

Although optical confinement can be achieved with gain guiding and mixed guiding, strongly index-guided laser structures are usually preferable in single-mode diode-laser arrays. This is because index-guided modes have plane wave fronts, which do not depend on the operating conditions of the laser. In contrast, purely gain-guided waveguide or mixed-guided modes possess curved wave fronts, and the degree of wave front curvature depends on $\Delta\kappa$ or the gain step which in turn depends on the local-carrier density. Therefore, as the injection current is varied, the strength of the waveguide can vary, resulting in a change in the output-beam divergence. In actual devices, the contribution of both the gain and the index changes with drive current must be considered. Carrier-induced changes in the dielectric properties of the laser-waveguide depend on the details of the laser structure. This is illustrated next in Sect. 3.1.4 using the parabolic waveguide model for a stripe laser.

It should also be mentioned that actual diode lasers usually have waveguide structures with more than three layers; and very often, the layers are not configured in a symmetric fashion. In asymmetric waveguides, the fundamental mode can also have a cut-off condition [3.3], so it is possible to design a waveguide that will not support any guided modes. Obviously, one needs to be cognizant of this fact, as it is a situation that is undesireable in designing a semiconductor-diode laser or laser array. In practice, more accurate numerical methods are used to design and optimize the semiconductor waveguide structure for fundamental mode operation. These are discussed in Chap. 4.

3.1.4 Parabolic Dielectic-Profile Waveguide Model

An approach which is very useful for illustrating the waveguide characteristics of gain-guided stripe diode lasers, as well as the effects of mixed guiding and gain-induced index effects, is to model the gain distribution within the active-layer volume of the laser-waveguide structure as a smoothly-varying parabolic gain and index profile in the plane of the pn junction [3.14, 3.15]. In the direction perpendicular to the pn junction, the dielectric permittivity is modeled as in the index-guided symmetric three layer slab waveguide. In this case, the waveguide modes are solutions to the wave equation with a two-dimensional complex permittivity of the form,

$$\varepsilon\left(x,y\right) = \begin{cases} \varepsilon_0\left(0\right) - a^2 x^2 & \text{for } x \text{ within the active layer} \\ \varepsilon_1 & \text{for } x \text{ outside of the active layer} \end{cases} , \qquad (3.22)$$

where $a = a_r + i a_i$ is a complex-valued number. Following the analysis of *Paoli* [3.15], the electric fields of the modes are assumed to satisfy the reduced scalar wave equation

$$\nabla^2 E + \varepsilon\left(x,y\right) k^2 E = 0 \ , \qquad (3.23)$$

where $k = 2\pi/\lambda$. For this analysis, the solutions to (3.23) are approximated by

$$E = E\left(x\right) E\left(y\right) exp\left(-i\beta z\right) \ . \qquad (3.24)$$

Note that $\varepsilon\left(x,y\right)$ is expected to vary much more slowly in the plane of the pn junction (x direction) than in the direction perpendicular to the pn junction (y direction). Hence, the assumption can be made that $E\left(y\right)$ is not effected by the optical confinement in the plane of the pn junction, and therefore $E\left(y\right)$ satisfies

$$\frac{d^2 E(y)}{dy^2} + \beta_y^2 E(y) = 0 \ , \tag{3.25}$$

where β_y is the propagation constant or eigenvalue for the symmetric three-layer slab waveguide in the y direction. It can be shown then that when $\varepsilon(x, y)$ is given by (3.22), then the modal electric fields in the plane of the pn junction satisfy

$$\frac{d^2 E(x)}{dx^2} + \left[\left(\varepsilon_0 - a^2 x^2 \right) k^2 \Gamma_y^2 + \varepsilon_1 k^2 (1 - \Gamma_y) - \beta_y^2 - \beta^2 \right] E(x) = 0 \ , \tag{3.26}$$

where Γ_y is the transverse mode confinement factor for E_y which is defined by (2.1). The solutions to (3.26) which correspond to the modal-field distributions are Hermite-Gaussian functions of the form

$$E_p(x) = H_p \left(x \sqrt{ak} \right) \exp \left(-\frac{\sqrt{\Gamma_y}}{2} kax^2 \right) \ , \tag{3.27}$$

where H_p is the Hermite polynomial of order p which is given as [3.16]

$$H_p(s) = (-1)^p \exp \left(s^2 \right) \frac{\partial^p \left[\exp \left(s^2 \right) \right]}{\partial s^p} \ . \tag{3.28}$$

Note that the fundamental mode corresponds $H_p = 1$, and higher-order mode solutions are obtained from values of $p > 1$.

The parabolic model for the dielectric permittivity can, in some cases, serve as an acceptable approximation to the carrier-density distribution in the plane of the pn junction of stripe lasers [3.14]. Following the approach of *Cook* and *Nash* [3.14], the corresponding parabolic profile of the refractive index is

$$n(x) = n(0) - \frac{4\Delta n}{S^2} x^2 \ , \tag{3.29}$$

and the parabolic gain profile is

$$g(x) = g(0) - \frac{4\Delta g}{S^2} x^2 \ , \tag{3.30}$$

where $\Delta g > 0$ and $\Delta n < 0$. The real and imaginary parts of a are related to Δg and Δn by

$$a_r = \sqrt{ \frac{4n(0)\Delta n}{S^2} \left\{ 1 \pm \left[1 + \left(\frac{\lambda \Delta g}{4\pi \Delta n} \right)^2 \right]^{1/2} \right\} } \ , \tag{3.31}$$

and

$$a_i = \frac{\lambda n(0) \Delta g}{\pi S^2 a_r} \ . \tag{3.32}$$

Then from (3.28) it is seen that the fundamental mode E_y in the plane of the pn junction is a Gaussian beam with a full-width at the $1/e^2$ intensity point w_0 given by

$$w_0^2 = \frac{1}{ka_r\sqrt{\Gamma_y}} \tag{3.33}$$

with a wave front radius of curvature R given by

$$R = \frac{1}{a_i\sqrt{\Gamma_y}} \ . \tag{3.34}$$

Note that the mode confinement in the transverse or y direction is very strongly index guided because of the large-refractive index differences in the epilayers of the pn junction of the diode laser-waveguide structure. Hence, the effect of the gain guiding in the plane of the pn junction is to produce a cylindrical wave front, as illustrated in Fig. 3.3. From (3.31 and 3.32), it is seen that w_0 and R are not independent, but are both determined by the width of the stripe S. In laser structures where the stripe width varys along the length of the laser, it is possible to independently control w_0 and R [3.17].

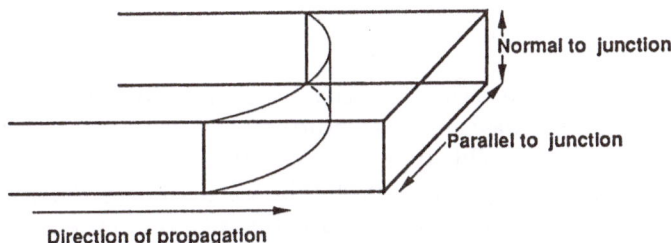

Fig. 3.3. Wavefront coresponding to a mode which is index guided normal to the pn junction and gain guided parallel to the pn junction [3.14].

This sort of deviation of the wave front from that of a spherical wave is refered to as *astigmatism* [1.356]. In the case of a diode laser it means in effect that when imaging the output beam, an observer will find that the beam will have an apparent beam waist, w_0, at a distance D behind the output facet. The parameter D is known as the *astigmatic focal distance* [1.356]; and for the parabolic permittivity waveguide, it is given by [3.14, 3.18]

$$D = R\left(1 + \frac{\lambda R}{\pi w_0^2}\right)^{-1} \ , \tag{3.35}$$

where R is expressed by (3.34) and w_0^2 by (3.33). By substituting (3.33 and 3.34) into (3.27), we find that the fundamental mode, $p = 0$ corresponds to the Gaussian beam

$$E_0(x) = A_0 \exp\left[-\frac{x^2}{2w_0^2}(1 + i\chi)\right] \ , \tag{3.36}$$

where the astigmatism factor χ is given by

$$\chi = \frac{w_0^2 k}{R} \ . \tag{3.37}$$

In diode-laser array structures where gain or mixed guiding is the dominant confinement mechanism, the Gaussian beam model for the fundamental mode can be very useful for understanding the influence of the waveguide confinement on the output-beam properties.

An interesting example, contrasting waveguiding in conventional GaAs/AlGaAs double-heterostructure and multi-quantum-well lasers, has been reported by *Hausser* et al. [2.63]. Both types of lasers exhibited nearly the same values for the differential index, but the differential gain in the multi-quantum-well laser is much stronger than that of the conventional double-heterostructure laser. In this case, the multi-quantum-well laser produced a narrow output beam with a 7° angular divergence in the plane of the pn junction, whereas a double-heterostructure of similar dimensions produced an output beam with a 35° angular divergence.

3.2 Lateral-Mode Discrimination and Modal Gain

We will now consider the basic modal characteristics of diode-laser arrays that are based on periodic-waveguide structures. The description will be from a geometrical perspective. This type of approach will be helpful in developing an intuitive understanding of the effect of lateral scaling on diode-laser array mode discrimination. Later in Sect. 3.3.1, the relation between the number of modes and the overall size of the diode-laser array cavity is discussed.

Now we will derive a fundamental relationship between the bulk gain and modal gain in a laterally-coupled diode-laser array structure. A diode-laser array that comprises a periodic structure can be thought of as a waveguide with a periodic variation in the dielectric permittivity. The modal gain can be represented using a simple model that considers only the overlap of the intensity profile of the optical mode with the cross-sectional area of the array active volume. This mode confinement description is very intuitive because it allows one to view the mode discrimination problem from a geometric perspective. It does, however, have its limitations. The effects of gain saturation are neglected, so it is valid only near threshold. Nevertheless, this sort of threshold analysis is very useful in assessing the mode discrimination characteristics of large-laser arrays, as good-mode discrimination at threshold is very often a prerequiste for maintaining mode discrimination at high-power output levels.

Because of the incomplete overlap of the optical field and the active volume in diode lasers, the gain that a given mode experiences will be less than the bulk gain that is present in the active volume of the laser cavity. This is illustrated in Fig. 3.4, where a schematic drawing of the end view of a generic

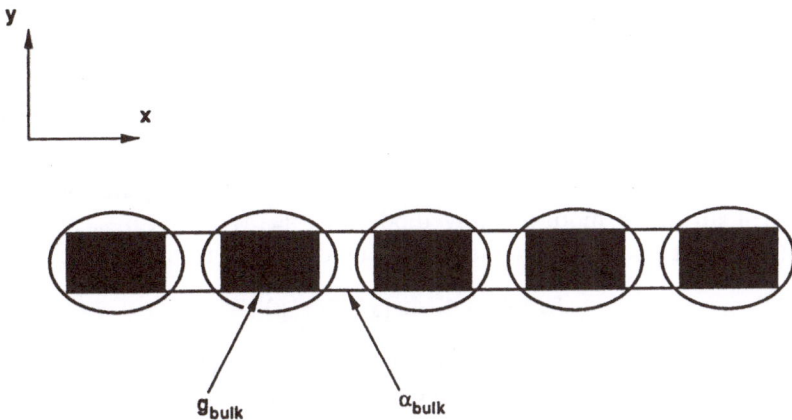

Fig. 3.4. Illustration of the lateral mode profile overlap with the periodic dielectric structure of a laser array.

laterally-coupled diode laser array is shown. The dark regions represent the regions where gain, are denoted by g_{bulk}, is present; and those areas where there is no gain (or less gain), will have loss (or less gain), denoted by α_{bulk}. Superimposed over the end view of the array, is the cross-sectional intensity profile of nth mode, $E_n(x,y)$, where y corresponds to the direction perpendicular to the plane of the pn junction (also referred to as the transverse direction) and x corresponds to the lateral direction which is in the plane of the pn junction, and the z direction corresponds to the axis of the laser cavity. For laser oscillation, it is the round-trip gain that is pinned at a level that must balence the round-trip losses; therefore as discussed in Chap. 2, the local gain can vary as a function of z subject to the round-trip constraint. Only the portion of $E_n(x,y)$ that interacts with the injected carriers will experience gain. The fractional overlap of the optical mode with the active region in the $x-y$ plane is given by

$$\Gamma_{xy:n} = \frac{\iint_{\text{Active}} |E_n(x,y)|^2 dxdy}{\iint_{-\infty}^{+\infty} |E_n(x,y)|^2 dxdy} \ , \tag{3.38}$$

where $\Gamma_{xy:n}$ is the 2-dimensional mode confinement factor for the nth array mode. The net gain of the nth mode or modal gain g_n, is given by [1.324]

$$g_n = \Gamma_{xy:n} g_{\text{bulk}} - (1 - \Gamma_{xy:n}) \alpha_{\text{bulk}} \ . \tag{3.39}$$

The concept of the mode-confinement factor, or confinement factor as it is usually termed, is very useful because it provides an intuitive relationship between the available gain in the active-area and the net gain that an array mode will actually experience. Though it should be noted that the form of (3.39) does not account for positional-dependent gain saturation, hence it is valid only near threshold. The array mode that will oscillate first will, according to (3.39), have the largest mode confinement factor, see more of the bulk gain and less of the bulk loss. This also follows from simple energy considerations. Array modes with lower values of the mode confinement factor will be suppressed. Then the problem of desigining a single spatial-mode laser array is seen to be one of engineering a lateral structure that will maximize the confinement factor for one array mode, while minimizing the confinement factors of all other modes. Many diode-laser array structures, satisfy the conditions that allow the cross-sectional mode profile to be separated as

$$E_n(x, y) = A_n(x) F(x, y) \ , \tag{3.40}$$

where $A_n(x)$ is the lateral-mode profile and $F(x, y)$ is the transverse-mode profile (exceptions are considered in Chap. 4). Note that the transverse-mode profile can have a parametric dependence on lateral position in the array structure. We will see that such a lateral dependence in the transverse-mode profile is necessary to establish good lateral mode discrimination. The transverse and lateral profiles in (3.40) are caluclated using a method called the *effective index-approximation*. This is discussed in more detail in Chap. 4. Using (3.40) we can define separate confinement factors for the transverse and lateral modes. From (3.38) it is seen that

$$\Gamma_{xy:n} = \frac{\int_{\text{Active}} |A_n(x)|^2 \Gamma_y(x) \, dx}{\int_{-\infty}^{+\infty} |A_n(x)|^2 dx} \ , \tag{3.41}$$

and the transverse-confinement factor, $\Gamma_y(x)$, is given by

$$\Gamma_y(x) = \frac{\int_{\text{Active}} |F(x, y)|^2 dy}{\int_{-\infty}^{+\infty} |F(x, y)|^2 dy} \ . \tag{3.42}$$

From (3.42) it is seen that when the transverse confinement factor $\Gamma_y(x)$ is independent of lateral position, x the two-dimensional confinement factor $\Gamma_{xy:n}$ can be separated as follows

$$\Gamma_{xy:n} = \Gamma_{x:n} \Gamma_y \ , \tag{3.43}$$

where

$$\Gamma_{x:n} = \frac{1}{C_n} \int_{\text{Active}} |A_n(x)|^2 dx \tag{3.44}$$

and the normalization constant, C_n, is given by

$$C_n = \int_{-\infty}^{+\infty} |A_n(x)|^2 dx \ . \tag{3.45}$$

This case where $\Gamma_y(x)$ is independent of x would only apply to laser arrays that have a uniform lateral structure such as a broad-area stripe laser. Generally, broad-area stripe lasers do not offer good spatial-mode discrimination. This occurs because (3.44) approaches unity for all values of n, as the laser array is scaled to larger lateral dimensions. Thus (3.43) will approach Γ_y and all lateral modes will see the same gain. One way to get around this problem is to vary the gain or loss in the lateral direction. This can be done by applying a nonuniform lateral-current distribution [1.91, 1.92, 1.462][3.19]-[3.22] or by engineering a spatial variation of the modal losses into the epilayer structure [1.71, 1.93, 1.94, 1.95, 1.117, 1.120, 1.123, 1.125, 3.23].

Generally speaking, large laser-array structures that are designed with the intent of operating single-mode must have transverse confinement factors that are dependent on lateral position. This will be shown below. A more detailed analysis in Chap. 4 uses the Bloch analysis to show that for a laser array to operate in a single-spatial mode, the gain, or imaginary part of the dielectric permittivity, must vary periodically in the lateral direction. The case represented by (3.43) is not usually encountered when one is dealing with single-spatial mode array structrues.

For a single-mode array, the two-dimensional confinement factor will have a more complicated form than shown in (3.43). To illustrate this situation, consider the generic laterally-coupled array structure depicted in Fig. 3.4. This periodic structure in the x-direction is a reasonable representation for many laterally-coupled laser array structures. Let each array element of width d_1, indicated by the regions in Fig. 3.4 with gain g_{bulk}, have a transverse confinement factor Γ_{1y}. Let each array element of width d_2, indicated by the regions in Fig. 3.4 with loss or lower gain α_{bulk}, have a transverse confinement factor of Γ_{2y}, then from (3.41) the two-dimensional confinement factor will be

$$\Gamma_{xy:n} \doteq \Gamma_{2y}\Gamma_{2x:n}^{(-\infty)} + \Gamma_{1y}\sum_{j=1}^{N}\Gamma_{1x:n}^{(j)} + \Gamma_{2y}\sum_{j=1}^{N-1}\Gamma_{2x:n}^{(j)} + \Gamma_{2y}\Gamma_{2x:n}^{(+\infty)} , \qquad (3.46)$$

where

$$\Gamma_{2x:n}^{(-\infty)} = \frac{1}{C_n}\int_{-\infty}^{-(N-1)\Lambda/2}|A_n(x)|^2 dx , \qquad (3.47)$$

$$\Gamma_{2x:n}^{(+\infty)} = \frac{1}{C_n}\int_{(N-1)\Lambda/2}^{+\infty}|A_n(x)|^2 dx , \qquad (3.48)$$

$$\Gamma_{1x:n}^{(j)} = \frac{1}{C_n}\int_{\Lambda(2j-N-1)/2}^{d_1+\Lambda(2j-N-1)/2}|A_n(x)|^2 dx \qquad (3.49)$$

and

$$\Gamma_{2x:n}^{(j)} = \frac{1}{C_n}\int_{-d_1+\Lambda(2j-N-1)/2}^{\Lambda(2j-N-1)/2}|A_n(x)|^2 dx . \qquad (3.50)$$

Here N is the total number of gain elements of width d_1, hence there are a total of $N-1$ loss elements of width d_2, $\Lambda = d_1 + d_2$ is the period of the of the array, and the array extends from $y = -\Lambda(N-1)/2$ to $y = +\Lambda(N-1)/2$ and by symmetry the end contributions satisfy

$$\Gamma_{2x:n}^{(-\infty)} = \Gamma_{2x:n}^{(+\infty)} . \tag{3.51}$$

In the limit as N becomes very large, the intensity profile of all the array modes will be become translationally invariant with respect to a lateral displacement through a distance Λ, as the end contributions become negliglible, and $\Gamma_y(x)$ approaches a constant. The integrals in the numerator and denomenator of (3.41) can be expressed as integrals over a period or unit cell length, Λ, of the array multiplied by the total number of array elements. Realizing this, we can express (3.41) to a good approximation as

$$\Gamma_{xy:n} \cong \frac{N \int_{-\Lambda/2}^{+\Lambda/2} |A_n(x)|^2 \Gamma_y(x)\, dx}{N \int_{-\Lambda/2}^{+\Lambda/2} |A_n(x)|^2 dx} = \Gamma_{\Lambda y:n} . \tag{3.52}$$

The two-dimensional confinement factor of the infinite extent array is denoted by, $\Gamma_{\Lambda y:n}$, and it corresponds to the fractional intensity overlap of the nth array mode with the active-area in a unit cell or one period of the array structure. The normalization constant, C_n, that is used in defining (3.47-3.50) will have the following large N dependence

$$C_n = \int_{-\infty}^{+\infty} |A_n(x)|^2 dx \rightarrow N \int_{-\Lambda/2}^{+\Lambda/2} |A_n(x)|^2 dx . \tag{3.53}$$

Then substituting (3.47-3.50, 3.52, and 3.53) into (3.46) one finds that in the limit of an infinitely large array, the two-dimensional confinement factor can be expressed as

$$\lim_{N \to +\infty} \Gamma_{xy:n} = \Gamma_{\Lambda y:n} = \Gamma_{1y}\Gamma_{1x:n} + \Gamma_{2y}\Gamma_{2x:n} , \tag{3.54}$$

where

$$\Gamma_{\ell x:n} = \frac{\int_{-d_\ell/2}^{+d_\ell/2} |A_n(y)|^2 dy}{\int_{-\Lambda/2}^{+\Lambda/2} |A_n(y)|^2 dy} . \tag{3.55}$$

In (3.55) when $\ell = 1$, $\Gamma_{\ell x:n}$ is the fraction of the nth array mode intensity that is contained within a gain element of width d_1; and when $\ell = 2$, $\Gamma_{2x:n}$ is the fraction of the nth array mode intensity that is contained within an element of width d_2. Note that the end contributions, (3.47 and 3.48), go to zero in the large N limit, so they do not appear in (3.54). From (3.52 and 3.55), it is seen that the lateral confinement factors $\Gamma_{\ell x:n}$ and $\Gamma_{2x:n}$ are simply related by

$$1 = \Gamma_{1x:n} + \Gamma_{2x:n} . \tag{3.56}$$

Substituting (3.56) into (3.54), we find that the two-dimensional confinement factor for the infinite array example is given by

$$\Gamma_{Ay:n} = \Gamma_{1x:n} \left(\Gamma_{1y} - \Gamma_{2y} \right) + \Gamma_{2y} . \tag{3.57}$$

The form of (3.57) has been used (with the Bloch analysis of Sect. 4.2.3) to derive a closed-form expression for the two-dimensional confinement factor of very-large arrays of anti-guided diode lasers [1.234]. Equation (3.57) is an interesting result, because it shows that for very-large arrays, it is necessary that the lateral variation of the transverse confinement factor satisfy $\Gamma_{1y} > \Gamma_{2y}$, in order to have lateral mode discrimination. Recall that there are N-gain elements with Γ_{1y} and $N - 1$ loss or lower gain elements with Γ_{2y}, and this is why $\Gamma_{1y} > \Gamma_{2y}$ for a spatial mode to be energetically favored by (3.39). If there is no lateral variation in the transverse confinement factor, then $\Gamma_{1y} = \Gamma_{2y}$ and $\Gamma_{Ax:n} = \Gamma_{2y}$ is independent of n, and according to (3.39) all lateral modes have the same gain. For cases where $\Gamma_{1y} < \Gamma_{2y}$, there will also be lateral mode discrimination, but these modes are energetically less favored according to (3.39) because the mode-intensity profile is maximized in the loss or lower gain regions. If we interchange the gain elements with the loss elements, then the energetically favored modes will be those with $\Gamma_{1y} < \Gamma_{2y}$, and the least favored modes will be those with $\Gamma_{1y} > \Gamma_{2y}$.

The confinement-factor analysis demonstrates an important scaling constraint in laterally-coupled laser arrays. That is, in very large laterally-coupled laser arrays, a periodic variation in the local gain in the lateral direction is necessary to obtain spatial mode discrimination. However, by itself it is not a sufficient condition for obtaining spatial-mode discrimination. The simple confinement factor analysis does not provide any information on the waveguide properties (real part of the dielectric permittivity) of the laterally-coupled array. The only assumption was that all array modes exhibit the translational symmetry of the uniform laterally-coupled array of infinite extent. This is a valid assumption because Bloch's theorem for periodic lattices also applies to the periodic laser waveguide structures. In Chap. 4., the Bloch analysis of laterally-coupled arrays is presented and the necessary and sufficient conditions for single spatial-mode operation are established.

3.3 Single-Frequency Operation

In the two preceeding sections, we considered spatial-mode characteristics and scaling issues related to the lateral waveguide structure of diode-laser arrays. In this section, we will consider some fundamental topics that pertain to single-frequency operation and longitudinal structure in diode lasers and diode-laser arrays. Recall that in Chap. 2, we discussed the power-scaling characteristics of diode-laser arrays for the case when the spectral output properties were not a concern. Here we will consider the issues related to single spectral mode operation. Beginning with Sect. 3.3.1, the picture of the diode-laser resonator as a dielectric slab is introduced and used to estimate the number of confined modes that are contained within the gain bandwidth. In Sect. 3.3.2, the effects of spontaneous emission on spectral purity, mode discrimination, and waveguide properties are discussed. Issues related to logitudinal-mode discrimination are reviewed in Sect. 3.3.3. Section 3.3.4 deals with the spectral linewidth.

3.3.1 Number of Modes in Diode-Laser Arrays

To gain better insight into the topic of mode discrimination in diode-laser arrays, we will begin by considering the number of modes that are contained within the gain-bandwidth of the semiconductor medium. For this initial discussion on the number of modes, we will assume that the diode-laser array cavity can be modeled by a rectangular slab of refractive index n with dimensions L in the longitudinal direction, dimension w in the lateral direction, and dimension t in the transverse direction. Here the waveguide-volume of the diode-laser array can be thought of as the relevant cavity volume. The standing-wave modes confined within this volume will have maximum overlap with the smaller volume of the active layer; therefore, these modes will compete most effectively for the available gain.

Of course, the rectangular-slab model is an approximation, because only the boundaries comprised by the end reflectors of the diode-laser array represent rigid side walls. In the lateral and transverse directions, the boundaries, though comprised of abrupt interfaces between materials of differing dielectric coefficients, will not be quite as well localized. When a forward bias is applied across the pn junction of the diode-laser array, there will be a diffusion of the injected carriers from the active volume into the adjoining passive regions. Hence, the boundary interface is smoothed over a distance of $\approx 1\,\mu\text{m}$. Therefore, the dimension w and height t should be thought of more as effective-cavity dimensions. Since active-layer thicknesses are typically much less than the wavelength of light, the transverse dimension t corresponds to d/Γ_y, where d is the active-layer thickness and Γ_y is the transverse mode confinement factor defined by (2.1).

In most diode-laser arrays, the width of the active layer is much larger than the wavelength, so the lateral dimension w corresponds to the width of the active layer. Note that standing-wave modes are not the only modes that will be excited. When gain is present in the cavity, spontaneous emission is radiated into a continuum of plane-wave radiation modes that propagate away from the active volume in all directions. Usually, these radiation modes do not compete effectively for the gain.

The standing-wave modes that can be supported by a rectangular slab have a discrete spectrum of frequencies, $\nu\,(\ell_L, \ell_w, \ell_t)$, that is established by the boundary conditions. These allowed modal frequencies are related to the magnitude of the wavevector k by

$$\nu^2\,(\ell_L, \ell_w, \ell_t) = \left(\frac{c}{2n}\right)^2 \left[\left(\frac{\ell_L}{L}\right)^2 + \left(\frac{\ell_w}{w}\right)^2 + \left(\frac{\ell_t}{t}\right)^2\right] = \left(\frac{ck}{2\pi n}\right)^2 , \quad (3.58)$$

where ℓ_L, ℓ_w, and ℓ_t are the integer-number of half-wavelengths along the respective slab dimensions. When the slab dimensions are all much larger than the wavelength of light, there are many combinations of the integers ℓ_L, ℓ_w, and ℓ_t that correspond to modes within the same narrow wavelength interval. Then the number of modes $N\,(\nu)$ between 0 and a given frequency ν is given by [1.323]

$$N\,(\nu) = \frac{8\pi^3 V_{\text{slab}} n^3 \nu^3}{c^3} , \quad (3.59)$$

where $V_{\text{slab}} = Lwt$ is the volume of the laser resonator. Note that (3.59) is a general result that applies to all wave-like phenomena. An additional factor of 2 appears for light waves to account for the number of polarization states associated with each mode. Corrections to (3.59) that occur when the slab dimensions are not necessarily much larger than the wavelength have been analyzed for the case of sound waves in a room by *Morse* and *Bolt* [3.24]. Another quantity of interest is the density of modes per unit frequency per unit volume, $\rho\,(\nu)$. The form of $\rho\,(\nu)$ that corresponds to (3.59) is given by

$$\rho\,(\nu) = \frac{1}{V_{\text{slab}}} \frac{dN\,(\nu)}{d\nu} = \frac{8\pi\nu^2 n^3}{c^3} . \quad (3.60)$$

Equations (3.59 and 3.60) are a general result; and as long as the slab dimensions are much greater than the wavelength of light, $N\,(\nu)$ and $\rho\,(\nu)$ are independent of the shape of the cavity [2.31]. From (3.59, 3.60) it is seen that for sufficiently large cavity size, the number of modes per-unit-volume and the mode density are independent of cavity size.

When one is dealing with waveguide structures such as single-element diode lasers or diode-laser arrays (3.59 and 3.60) are no longer accurate expressions for the confined modes, because at least one of the cavity dimensions is comparable to the wavelength of light. The active layer, which corresponds to t for this discussion, can almost always be made thin enough so that, depending on the structure, either (3.19 or 3.20) is satisfied. Therefore, the number of modes in a diode-laser array corresponds to the number of modes of the two-dimensional sheet of length L and width w. However, (3.59, 3.60) are accurate expressions when applied to spontaneous emission, as spontaneous emission is emitted into the entire volume of the larger semiconductor chip that contains the active layer.

To derive expressions for the number of modes in diode-laser arrays, we will begin by considering a single-element single-spatial mode diode laser. Such lasers comprise waveguide structures that usually support only one guided mode. Using the nomenclature of (3.59), this corresponds to the constraint that $\ell_w = 1$, and $\ell_t = 1$ for all allowed modes. The confinement provided by the single-mode waveguide effectively produces a one-dimensional structure for guided modes. Then the number of modes in a single-spatial mode laser, N_L, is just the number of longitudinal or Fabry-Perot modes contained within the gain bandwidth, which is given by

$$N_L = \text{In}\left\{\frac{\Delta\lambda_{\text{Gain}}}{\Delta\lambda_{FP}}\right\} = \text{In}\left\{\frac{2\Delta\lambda_{\text{Gain}}Ln_g}{c}\right\} , \qquad (3.61)$$

where In is the integer part of the expression in parenthesis, $\Delta\lambda_{\text{Gain}}$ is the gain bandwidth, $\Delta\lambda_{FP}$ is the wavelength separation of adjacent longitudinal modes given by (2.6). It is seen from (3.61), that the number of modes is increased for longer lasers. Though, for a typical single-spatial mode diode laser, $\Delta\lambda_{\text{Gain}} \approx 200\,\text{Å}$ and $\Delta\lambda_{FP} \approx 2\,\text{Å}$, which gives ≈ 100 available modes.

Generally speaking, in diode-lasers where the lateral dimension is extended much beyond $\approx 2\,\mu m$, one needs to be concerned with the possibility of multi-spatial mode operation. The number of guided-lateral modes versus lateral width, N_w can be estimated using (3.18), which is the secular equation for a symmetric three-layer index-guide. The rectangular slab model being used in this discussion corresponds to a symmetric three-layer index-guide in both the lateral and transverse dimension. Therefore, when only a single transverse mode is supported, the number of lateral modes is given by

$$N_w = \text{In}\left\{1 + \frac{w}{\lambda}\sqrt{8\left(n^2 - n_c^2\right)}\right\} \cong \text{In}\left\{1 + \frac{4w}{\lambda}\sqrt{n\left(n - n_c\right)}\right\} , \qquad (3.62)$$

where n_c is the refractive index of the material outside of the rectangular slab. Since $n \approx n_c$ in most cases, the approximate expression given in (3.62) is usually valid.

Each of the lateral modes that is counted in (3.62) has associated with it an index m given by (3.18) and an effective refractive index, $n_{\text{eff}}^{(m)}$, that is determined by (3.9 and 3.10). Therefore, each lateral mode can be thought of as a plane wave propagating in a one-dimensional Fabry-Perot cavity of length L and refractive index $n_{\text{eff}}^{(m)}$. In this case, the wavelength of each plane wave representation must satisfy the round-trip phase condition that is given by (2.5), which gives the spectrum of allowed wavelengths $\lambda_{\ell,m}$ as

$$\lambda_{\ell,m} = \frac{2Ln_{\text{eff}}^{(m)}\left(\lambda_{\ell,m}\right)}{\ell} , \qquad (3.63)$$

where ℓ is the longitudinal mode index. Note that (3.63) must be solved numerically to find the mode spectrum, because of the dispersion in $n_{\text{eff}}^{(m)}\left(\lambda_{\ell,m}\right)$.

The values of the effective indices typically satisfy $n_c < n_{\text{eff}}^{(m)} < n$ for the guided lateral modes. Then each value of ℓ, consistent with the gain bandwidth, will consist of N_w distinct wavelengths. In this case, the number of modes is simply given by the product $N_w N_L$.

Thus, it is seen that mode discrimination in diode-laser arrays can be viewed as being separable into two distinct problems; one being lateral or spatial mode discrimination and the other being longitudinal mode or spectral discrimination. Note that true single-frequency requires both single lateral and longitudinal-mode operation. To appreciate this, consider the case of single longitudinal-mode operation, where ℓ will only have a single value. Then from (3.63), operation in a number of lateral modes will give rise to a similar number of frequencies whose relative separations are determined by the dependence of $n_{\text{eff}}^{(m)}(\lambda_{\ell,m})$ on m and dispersion effects. The one exception to this would occur when there are two spatial modes, say m_1 and m_2 that have indentical forms of $n_{\text{eff}}^{(m)}(\lambda_{\ell,m})$ over a region of the gain bandwidth. In this situation, both spatial modes would oscillate at the same frequency.

3.3.2 Spontaneous-Emission Factor

In diode lasers and diode-laser amplifiers, spontaneous emission has a significant influence on the spectral properties, dynamic behavior, mode discrimination, and power-conversion efficiency. The level of spontaneous emission in semiconductor-diode lasers is considerably larger than that of non-semiconductor laser media because the gain in semiconductors is much higher. In Fig. 3.5, we see pictured a generic diode laser or diode-laser array that shows the active-layer volume embedded within the greater volume of the diode-laser resonator.

Spontaneous emission noise is generated throughout the gain or active-layer volume and radiated nearly-isotropically into the surrounding medium. Some of the spontaneously emitted light will be coupled into all the modes of the diode-laser resonator. Note that spontaneous light is emitted as plane

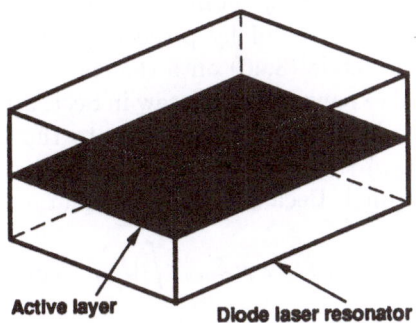

Active layer   Diode laser resonator

Fig. 3.5. Schematic drawing that shows the configuation of the active-layer and diode-laser resonator volume.

waves, but the modes of the laser cavity are generally not plane waves. As a result, the level of spontaneous emission coupled into the laser cavity modes varys for different modes. Therefore, all modes within the gain-bandwidth experience different levels of gain, depending on the spectral distribution of the gain and the level of spontaneous emission coupled into each mode. As discussed by [3.25, 3.26], this situation causes multimode operation in semi-conductor lasers and other homogeneously-broadened lasers operating near threshold. Also, amplification of broadband incoherent spontaneous emission occurs at the expense of a reduction in the power of the light in the coherent mode. Spontaneous emission will be coupled into the oscillating mode of the laser and will contribute to the spectral linewidth of the laser.

The fraction of spontaneous emission power that is coupled into a mode is refered to as the *spontaneous emission factor*, γ, and is defined as

$$\gamma = \frac{\text{rate of spontaneous emission into oscillating mode}}{\text{total rate of spontaneous emission}} . \qquad (3.64)$$

It is evident that the value of (3.64) will depend on the specific mode of oscillation. At a given point or location in the active layer, the total spontaneous emission noise power that is generated is proportional to the local-carrier density. So generally, γ, is influenced by the active characteristics of the laser-waveguide structure.

Many analyses calculate γ using a classical model where the electric field of the spontaneous emission noise is generated by an ensemble of randomly oriented, uncorrelated, point-like electric dipoles that are incorporated in the active-layer volume of the laser [3.27, 3.28]. Moreover, the active layer represents a thin sheet of current density J_a within the rectangular slab that corresponds to the diode-laser resonator. To calculate γ using this model, one solves the wave equation for the complex electric field where the current density sheet acts as a source term [3.27]-[3.31]. The polarization source term due to the current density is expressed as a superposition of the modes of the diode laser structure, $E_n(x,y)\exp(i\beta_n z)$. Recall that $E_n(x,y)$ is the two-dimensional mode profile given by (3.40) and β_n is the propagation constant (3.10). In this form, the rate of spontaneous emission power that is coupled into the oscillating mode is found by projecting the two-dimensional field profile of the oscillating mode (3.40) onto the spontaneous emssion field, as expressed by the modal expansion. As we saw in Sect. 3.1 the dielectric profile of a semiconductor-diode laser is complex due to the presence of gain and loss. As a consequence of this fact, the modes of a diode-laser array are not generally power orthogonal. Because the wave equation is not Hermitian

$$\int E_m(x,y)\,E_n^*(x,y)\,dxdy \neq \delta_{m,n}\int |E_m(x,y)|^2 dxdy , \qquad (3.65)$$

where $\delta_{m,n}$ is the Kronecker delta function. However, the field orthogonality relation given by

$$\int E_m\left(x,y\right) E_n\left(x,y\right) dxdy = \delta_{m,n} \int \left[E_m\left(x,y\right)\right]^2 dxdy \; , \qquad (3.66)$$

is valid. It should be pointed out that in diode lasers, where index guiding is predominate, the modal fields have plane phase fronts, and can therefore be described by a set of power-orthogonal modes. Gain- or mixed-guided modes that posses curved phase fronts do not satisfy power orthogonality, consistent with (3.65), but still satisfy field orthogonality (3.66).

The nature of the phase-front curvature of the diode-laser modes can have a strong influence on the value of γ. Spontaneous emission, being generated by an ensemble of point-like sources, will couple more strongly into modes that have curved-phase fronts than those that have plane-phase fronts. The result of the afore-mentioned proceedure for calculating γ yields

$$\gamma = \frac{c^3 g\left(\lambda_0\right) K}{8\pi n_a n n_g V_{\mathrm{eff}}} \; , \qquad (3.67)$$

where $g\left(\lambda_0\right)$ is the normalized lineshape of the spontaneous emission, n_a is the refractive index of the active layer, n is the refractive index of the material outside the active layer volume, n_g is the group index, and V_{eff} is the effective volume of the active layer that is defined below. The factor K is often referred to as the *excess spontaneous emission factor* [3.28], K-factor, or *excess noise factor* [3.30], and it is given by

$$K = \frac{\left[\int |A_n\left(y\right)|^2 dy\right]^2}{\left|\int A_n^2\left(x\right) dx\right|^2} \; , \qquad (3.68)$$

where $A_n\left(x\right)$ is the lateral-mode profile, as used in (3.40). The effective volume is defined $V_{\mathrm{eff}} = Lw_{\mathrm{eff}}d/\Gamma_y$ [3.28, 3.29], where w_{eff} is given by

$$w_{\mathrm{eff}} = \frac{\int |A_n\left(x\right)|^2 dx \int I_{sp}\left(x\right) dx}{\int |A_n\left(x\right)|^2 I_{sp}\left(x\right) dx} \qquad (3.69)$$

where $I_{sp}\left(x\right)$ is the shape of the spontaneous emission distribution along the lateral dimension of the active layer. From (3.69), it is seen that w_{eff} corresponds to the overlap width of the oscillating mode and the spontaneous emission intensity. Thus V_{eff} that appears in (3.66) can be interpreted as a type of modal volume.

In index-guided structures $A_n\left(x\right) = A_n^*\left(x\right)$ so $K = 1$. In purely gain-guided structures $K = \sqrt{2}$. For laser-waveguide structures with mixed guiding, larger phase curvatures are present in the modes, leading to $K > \sqrt{2}$. Hence, γ in either gain-guided or mixed-guided laser-waveguide structures will be larger than in index-guided laser structures. For the example of the Gaussian-beam fundamental mode given by (3.36), we find $K = \sqrt{1+\chi^2}$.

(a) Quantum Mechanical Description of K-Factor

Note that a values of $K > 1$ does not imply that more total spontaneous emission is generated in gain- and mixed-guided structures. Though it does imply that the spontaneous emission rate into a mode is larger when $K > 1$. As discussed by *Yariv* and *Margalit* [3.32], this is not inconsistent with the quantum-mechanical description of spontaneous emission. A Gaussian-beam mode of the type given by (3.36) with $\chi \neq 0$ is not a mode of the quantum-mechanical Hamiltonian for the electromagnetic field. Modes such as those given by (3.36) with curved-wave fronts only exist when carriers are present. As such, these have been refered to as *improper modes*. The *proper modes* of the quantum-mechanical Hamiltonian for the electromagnetic field are plane waves of the sort described in Sect. 3.3.1, where the electromagnetic field is confined to a large rectangular enclosure. These proper modes exist without carriers being present. To determine the spontaneous emission rate into an improper mode, one needs to express the improper mode in terms of an expansion of the proper modes of the electromagnetic field. For the Gaussian-beam mode given by (3.36), this gives [3.32]

$$\exp\left[-\frac{x^2}{2w_0^2}(1+i\chi)\right] = \sum_{-\infty}^{\infty} A_l \exp\left(il\frac{2\pi x}{L_x}\right) , \qquad (3.70)$$

where L_x is the enclosure dimension, and A_l is given by

$$A_l = \frac{w_0}{L_x}\sqrt{\frac{2\pi}{1+i\chi}} \exp\left(-\frac{2l^2\pi^2 w_0^2(1-i\chi)}{L_x^2(1+\chi^2)}\right) . \qquad (3.71)$$

The effective number of proper modes that comprise the Gaussian beam mode $N = 2l$ is determined by the value of l where $|A_l| = e^{-1}|A_0|$, and is given by $N = L_x\sqrt{2(1+\chi^2)}/(\pi w_0)$. The spontaneous-emission rate into the improper mode is then seen to be given by the spontaneous emission rate into one proper mode of the large rectangular enclosure multiplied by the effective number of the proper modes N that are used in (3.70) for the improper mode. Thus, the ratio of the spontaneous-emission rates into an improper Gaussian-beam mode to that of an index-guided Gaussian-beam mode each with the same beam size w_0 is given by $\sqrt{(1+\chi^2)}$.

In non-power orthogonal-mode systems, more spontaneous emission is coupled into the oscillating mode within the diode-laser cavity. However, this does not imply that one will necessarily measure more spontaneous emission noise in the output beam. As discussed by [3.30, 3.33], the laser output is usually measured with an optical system, *external to the laser cavity*, that is represented by a power orthogonal set of modes (e.g., free space, optical fiber, or other passive structure). In projecting the non-power orthogonal mode of the laser onto the power orthogonal set of modes of the measurement system, mode filtering occurs and the apparent excess noise can be effectively removed. In cases where there are gain inhomogenities present within the laser cavity, such as found in long amplifiers or lasers with significantly different facet reflectivities, then higher levels of spontaneous-emission noise occur because the noise can effectively compete for the available gain [3.30, 3.34].

(b) K-Factor for Gaussian Modes

For diode-laser structures that are well approximated by the parabolic-profile waveguide model of Sect. 3.1.4, there is an analytic relation that relates the K-factor (3.68) to the modal parameters w_0 and R, of the fundamental Gaussian-beam mode (3.36) [3.18]

$$K = \sqrt{1 + \left(\frac{\pi w_0^2}{\lambda R}\right)^2} . \tag{3.72}$$

Equation (3.72) elucidates the influence of the modal parameters on the K-factor. It is seen that modes with smaller beam waists and larger radii of curvature will have a K-factor closer to unity. The K-factor for a Gaussian-beam mode can also be expressed in terms of the near-field and far-field sizes as [2.63]

$$K = \frac{\pi w_0 \phi_y}{2\lambda \ln 2} , \tag{3.73}$$

where ϕ_y is the full-beam divergence at half-maximum in the plane of the pn junction. Note that in lasers where the parabolic-gain model applies, (3.73) can be used to determine the K-factor directly from measurements of w_0 and ϕ_y.

The relation between the K-factor and the parameters Δg and Δn that appear in the quadratic terms of (3.29 and 3.30) in the parabolic-gain model is given by

$$\Delta g = (2\ln 2)^2 \frac{\lambda S^2}{\pi w_0^4 n(0) \Gamma_y} \sqrt{K^2 - 1} , \tag{3.74}$$

and

$$\Delta n = (2\ln 2)^2 \frac{\lambda^2 S^2}{8\pi w_0^4 n(0) \Gamma_y} (2 - K^2) . \tag{3.75}$$

Dividing (3.75) by (3.74) and using (2.66), the expression [3.28]

$$\alpha = \frac{4\pi}{\lambda} \frac{\Delta n}{\Delta g} = \frac{2 - K^2}{2\sqrt{K^2 - 1}} \tag{3.76}$$

is found that relates the K-factor to the antiguiding factor α (Sect. 2.7). Note that (3.76) is only valid for the parabolic-dielectric-profile waveguide model. We find that in the case of pure gain-guiding where $K = \sqrt{2}$, we have $\alpha = 0$. For pure index guiding where $K = 1$, α is not defined in the parabolic dielectric profile waveguide model. In actual diode lasers, typical values for α fall in the range between 4 and 8. In this case, (3.76) can be approximated by $K \simeq 2\alpha$. In general, K and its relation to α will differ for other types of laser-waveguide structures.

An interesting experimental study of the K-factor and its relation to waveguiding in conventional GaAs/AlGaAs double-heterostructure and multiquantum-well stripe lasers of equal widths, has been reported by *Hausser* et al. [2.63]. By measuring the near- and far-field beam sizes and using (3.73), *Hausser* et al. found that $K = 3$ for the multiquantum-well laser and $K = 12$ for the double-heterostructure laser. The smaller value of K observed for the multi-quantum-well laser was attributed to the stronger gain-guiding being present than in the double heterostructure. Both types of lasers exhibited nearly the same values for the differential index, but the differential gain in the multi-quantum-well laser was much stronger than that of the conventional double-heterostructure laser.

3.3.3 Longitudinal Mode Structure

In Sect. 3.3.1 we saw that even in a laser that operates in a single-lateral mode, there are ≈ 100 longitudinal modes that lie within the typical semiconductor laser bandwidth. Although semiconductor-diode lasers and laser arrays are homogeneously broadened, they quite often operate in multiple-longitudinal modes. Below, we review several topics that are important for our understanding of the issues that relate to single-frequency operation.

(a) K-Factor and Longitudinal-Mode Spectra

The spontaneous emission rate into different longitudinal modes can vary, and this will directly influence the number of laser modes that will oscillate above threshold [3.25, 3.26]. This is especially true in gain and mixed guided structures. The modes of these structures have curved wave fronts, which means that the plane-wave-like spontaneous emission light will not be coupled uniformly into the modes of the laser cavity. Generally, those modes with a larger K-factor will couple spontaneous emission more strongly, hence these modes will reach threshold first.

Streifer et al. have analyzed the case of a semiconductor laser operating in a single TE_{00} spatial mode [3.26]. They found that the full-width half-maximum spectral envelope of the longitudinal-mode spectrum, λ_s, could be expressed as

$$\lambda_s = \lambda_h \left(\sqrt{\frac{P_t^2}{4P_s^2} + 1} - \frac{P_t}{2P_s} \right) , \qquad (3.77)$$

where λ_h is the homogeneous linewidth of the spontaneous emission, P_t is the total power output from the front facet, and P_s is a reference power which is given by

$$P_s = \left(\frac{(1 - R_1)^2}{2R_1 \ln(1/R_1)} \right) \frac{hc\lambda K}{4\pi n_e n_a \mathcal{A}} , \qquad (3.78)$$

where R_1 is the reflectivity of the facet through which the output is transmitted and it has been assumed that the reflectivity of the other facet R_2 is equal to unity. The parameter $\mathcal{A} = \sigma_{th}\tau$, where σ_{th} is the differential gain at threshold and τ is the carrier-recombination time. Note that the value P_s is on the order of the power output in the spatial mode just below threshold, so that well-above threshold where $P_t > 2P_s$, (3.77) is approximated as

$$\lambda_s \approx \lambda_h \frac{P_s}{P_t} \ . \tag{3.79}$$

As the power output P_t is increased, the number of longitudinal modes above threshold will decrease. From (3.77 and 3.78) it seen that larger K-factors result in a larger P_s, which, in turn, increases λ_s and the number of longitudinal modes above threshold at a given power output. This analysis illustrates the influence that the waveguide characteristics can have on the longitudinal mode spectrum. We see from (3.78 and 3.79) that λ_s is minimized when $K = 1$; therefore, the optimum longitudinal-mode discrimination should occur for the case of pure-index guiding. Although for pure gain guiding where $K = \sqrt{2}$, the longitudinal mode envelope is only about 40 % larger than for pure index guiding. In the case of mixed guiding where $K \gg 1$, the longitudinal-mode envelope will be considerably broader; and consequently, the longitudinal mode discrimination would be expected to be poorest in these structures.

(b) Carrier Diffusion and Spatial-Hole Burning

There are other factors besides the spontaneous emission rate that influence the longitudinal mode spectrum. Effects such as *spatial-hole burning* [3.35, 3.36] as well as *gain non-linearities* through beating of the oscillating and nonoscillating modal fields [3.37, 3.38] have been shown to have an important influence on the longitudinal-mode structure and stability in semiconductor lasers. In particular, *Kazarinov* et al. [3.37] explained longitudinal mode self-stabilization in index-guided diode lasers as a modulation of the gain due to beating of the oscillating mode with the modes below threshold. This results in a gain nonlinearity that tends to suppress modes below threshold and stabilize the oscillating mode.

The standing-wave field within the laser cavity gives rise to a periodic depletion of the gain along the optic axis. At the nulls of the standing wave, where the optical intensity is a minimum, the population inversion and hence the gain will be a maximum. The converse of this is true at the intensity peaks of the standing wave. This phenomena is referred to as *spatial-hole burning*. Other longitudinal modes within the gain bandwidth can effectively use the maxima of the spatially-periodic population inversion to obtain enough gain to undergo laser oscillation. This can cause multi-longitudinal mode operation even in a homogeneously-broadened gain medium.

In semiconductor lasers, the phenomena of carrier diffusion can result in a smoothing of the periodic inversion. If a carrier moves through an axial distance of one wavelength, or more, in the material during the carrier lifetime, then the effects of the standing wave will effectively be averaged out, and gain would not be available for other longitudinal modes. The diffusion length of an injected carrier is $\sqrt{D_e/\tau}$, where D_e is the diffusion constant and τ is the carrier lifetime. An optical wavelength in the semiconductor material is $\lambda/n \approx 0.3\,\mu m$, which is typically smaller than the $\sqrt{D_e/\tau} \approx 3\,\mu m$ distance a carrier travels within its lifetime. Using the approach of [3.35], *Streifer* et al. have done an analysis of the longitudinal mode structure that accounts for the combined effects of spatial hole burning and carrier diffusion [3.36]. They found that that the spectral width of longitudinal modes that can oscillate, $\Delta\lambda_l$, can be modeled as

$$\Delta\lambda_l = \Delta\lambda_g \sqrt{\frac{\lambda^2\,\sigma\,P}{32\,\pi^2\,n^2\,w\,t\,T\,E\,D_e}} \,, \qquad (3.80)$$

where E is the photon energy, w is the width of the active layer, and t is the thickness of the active layer. When $\Delta\lambda_l$ is less than or equal to the wavelength separation between adjacent Fabry-Perot modes, single-frequency operation is expected. The analysis used to derive (3.80) does not account for carrier diffusion in the lateral dimension.

In diode-laser arrays, spatial-hole burning has an additional dimension because the spatial variation of the lateral-mode intensity also causes a nonuniform depletion of the carriers that results in a spatially-periodic gain distribution. This causes a spatially-periodic variation in the refrative index. In regions of higher gain the index is smaller; and in regions of lower gain, the index is higher. This gain-induced index variation is superimposed on the built-in index profile of the array structure. Also, lateral variations in the temperture can influence the gain distribtution. However, in most diode-laser arrays the resulting periods in the spatial variation of the gain are usually several μm, and carrier diffusion is not as effective at smoothing out the spatial-hole burning, as it is for the longitudinal modes. Therefore, spatial-hole burning can be an important factor in determining the lateral mode structure of a diode-laser array. Specific examples of this in antiguided-laser arrays are presented in Sect. 6.3.5. A proper analysis of the lateral-mode structure, that accounts for spatial-hole burning, requires a self-consistent analysis of the carrier-transport equation and the wave equation, as described in Chap. 4.

(c) Defects and Nonuniformities

In actual diode lasers, material and structural nonuniformities, as well as defects and damage sites, that effect the intracavity optical characteristics have been shown to have a significant effect on the spectral-output characteristics [3.39]-[3.41]. Nonuniformities and defects correspond to well-localized random-spatial variations in the dielectric permittivity of the diode laser that can act as scattering sites. Such a scattering center will both reflect and transmit light, so in effect there can be multiple cavities within the Fabry-Perot cavity comprised by the diode-laser chip. This picture has been shown to be consistent with the observation that lasers showing strong internal scattering tend to exhibit modulation or pertubations the in spectral envelope [3.39]. Experimental studies [3.40, 3.41] have also revealed a correlation between the characertistics of multi-mode spectra and the location of the damage site within the diode-laser structure. Similar effects should occur in diode-laser arrays with the additional complication that the lateral-mode characteristics will also be perturbed. It is seen then that developing a clear understanding of the modal characterisics of diode-laser arrays can be quite a challenge when the possibility of random structural variations exist. One would need material of sufficient uniformity and low enough defect density, so that random structural variations have a negligible influence on the modal characteristics. With improved quality in laser material and processing controls, this may becoming a reality.

3.3.4 Spectral Linewidth

In Sect. 3.3.2 we have seen how the spontaneous emission rate into the diode-laser spectral and spatial modes can have a significant effect on the spectral output of a diode laser. Now we will consider the spectral linewidth of a single oscillating mode in a semiconductor diode laser. In semiconductor-diode lasers the spectral linewidth behavior is typically modeled using a modified *Schawlow-Townes* formula [3.42]-[3.45]. Because of the strong coupling between the carrier density and refractive index in semiconductor lasers, random fluctuations in the carrier density produce a random phase modulation of the coherent light [3.45]-[3.48] which acts to broaden the spectral linewidth of the laser. The above-threshold lineshape is a Lorentzian with a spectral width $\Delta\nu$ given by [3.48]

$$\Delta\nu = \frac{R\left(1+\alpha^2\right)}{4\pi I} \; ,$$

(3.81)

where R is the spontaneous emission rate into the oscillating mode, I is the intensity of light, and α is the linewidth-enhancement factor which is discussed in Sect. 2.7. The contribution of the linewidth broadening due to carrier-induced index fluctuations is given by the α^2 term. For most semiconductor-diode-laser structures $\alpha > 2$, so it is usually the dominant source of spectral linewidth broadening. In contrast, non-semiconductor lasers usually have $\alpha \ll 1$ so that the spontaneous emission at the coherent signal frequency dominates the linewidth. The spontaneous emission rate, represented by R in (3.81), also plays a very important role in determining the linewidth. As we saw in Sect. 3.3.2, the details of the lateral waveguide structure in diode lasers has a strong influence on the spontaneous emission rate; and this is accounted for in the calculation of R. From the analysis of *Henry* [3.46], it has been shown that R can be expressed as

$$R = KGn_{sp}F_R \; ,$$

(3.82)

where G is the saturated gain, F_R is a structural-dependent factor discussed below, and n_{sp} is the average mode occupation number which is expressed as

$$n_{sp} = \frac{1}{1 - \exp\left(\frac{h\nu - eV}{k_B T}\right)} \;, \tag{3.83}$$

where V is the bias voltage and ν is the optical frequency. Note that eV is the energy separation of the quasi-Fermi levels in the conduction and valence bands. The band of optical emission frequencies is such that the photon energy $h\nu$ continuously spans the range $E_{\text{gap}} < h\nu < eV$. Here E_{gap} is the band-gap energy. When $h\nu < eV$, we see that $n_{sp} \approx 1$. As $h\nu \to eV$, we see from (3.83) that $n_{sp} \to \infty$; however, $G \to 0$ so that R is continuous and with a finite-positive value. Similarly when $h\nu > eV$, we have $n_{sp} < 0$, but also $G < 0$ so that R is still positive. Laser oscillation can only occur for photon energies where $h\nu < eV$. The parameter n_{sp} characterizes the degree of the population inversion associated with the energy levels that are involved in the laser transition. For most diode lasers, experiments indicate that $n_{sp} \approx 2$ [3.47]. A complete population inversion corresponds to $n_{sp} = 1$. This can only occur when $h\nu \approx E_{\text{gap}} \ll eV$; and in this frequency range there is less available gain, so the power output would be compromised.

Both the K and the F_R factors that appear in (3.83) characterize the structural contribution to the spontaneous-emission rate. The essence of these contributions is the spatial dependence in the local spontaneous-emission rate due to the spatial variation of the optical-mode overlap with the active-layer volume. As we saw in Sect 3.3.2, the K-factor gives a quantitative measure of the influence of the lateral-waveguide structure on the spontaneous-emission rate. The factor F_R accounts for the contribution of the details of the longitudinal structure to the spontaneous-emission rate. For a Fabry-Perot laser cavity, F_R can be shown to be given as [3.48]

$$F_R = \left(\frac{(r_1 + r_2)(1 - r_1 r_2)}{2 r_1 r_2 \ln(r_1 r_2)}\right)^2 \;, \tag{3.84}$$

where r_1 and r_2 are electric field facet reflectivities. For a high-power diode-laser array, we would expect to have $r_2 = \sqrt{R_2} = 1$ and $r_1 = \sqrt{R}$, and (3.84) can be expressed as

$$F_R = \left(\frac{(1 - R)}{\sqrt{R}\ln(R)}\right)^2 \;. \tag{3.85}$$

Note that (3.85) is indentical to the case of equal facet reflectivites $r_1 = r_2 = \sqrt{R}$ that has been discussed in [3.46]. Recall from Chap. 2 that to maximize power extraction efficiency with respect to the length of the Fabry-Perot laser cavity requires that the output-facet reflectivity be decreased as the cavity length is increased. This is illustrated in Fig. 3.6, where (3.85) is plotted versus R for the reflectivity range that corresponds to the output-facet reflectivity that maximized the power-extraction efficiency, as discussed in Sect. 2.3.5.

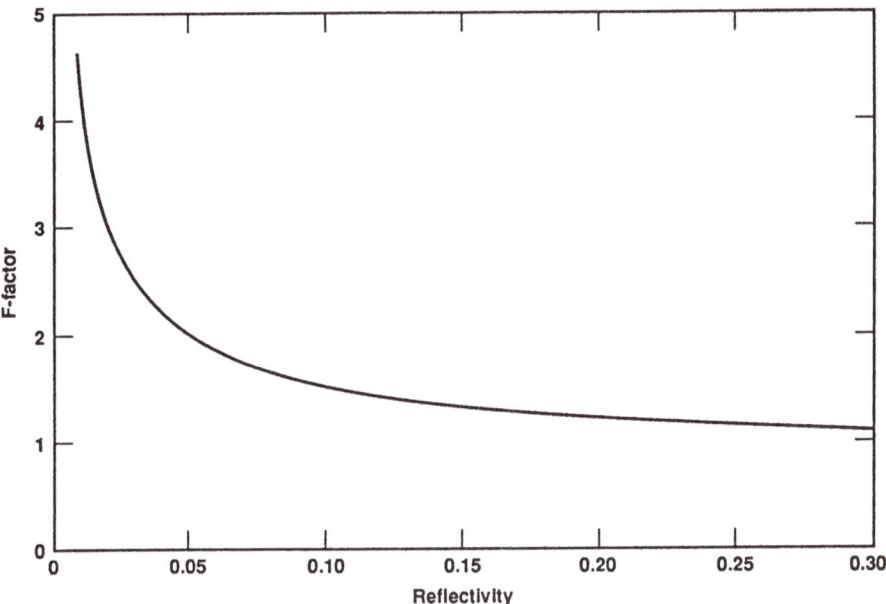

Fig. 3.6. Graph of F_R as given by (3.85) as a function of output reflectivity R.

As R is decreased, the end losses are increased and the carrier density necessary for supporting laser oscillation must also be increased to provide enough gain to offset the higher losses. Therefore, at the same intensity level I that appears in (3.81), a higher carrier density is needed in a laser with lower R so the spontaneous emission rate will be larger.

(a) General Linewidth Calculation

Since the spontaneous emission rate is largely determined by the local-carrier density, diode-laser arrays that operate with inhomogeneous-gain distributions would be expected to exhibit spectral linewidth characteristics not accounted for by the standard linewidth formulae (3.81 and 3.82). A more general analysis of the spectral linewidth for an arbitrary 3-dimensional laser cavity has been developed by *Arnaud* [3.49]. For isotropic media where the dielectric permittivity is a scalar, the linewidth formula found by *Arnaud* can be expressed as

$$P\Delta\nu = \frac{8\pi \left[\int \nu \mathrm{Im}\left\{ \varepsilon \right\} \left| E_m \right|^2 dV \right]^2}{\left| \int \left[\left(\frac{\partial \nu \varepsilon}{\partial \nu} \right) E_m^2 - \mu_0 H_m^2 \right] dV \right|^2} . \qquad (3.86)$$

where E_m and H_m are the modal fields associated with the active cavity, ε is the complex dielectric permittivity, and P is the number of photons per second. As we saw in Sect. 3.2, in laterally-coupled diode-laser arrays it is usually necessary to have a periodic variation in the gain in order to get mode discrimination, especially when the size of the array is large. Therefore, $\text{Im}\{\varepsilon\}$ will vary spatially for some diode-laser array structures and it may be necessary to use the more general linewidth formula (3.86). In [3.49], it is shown that when $\text{Im}\{\varepsilon\}$ does not vary much spatially, (3.86) reduces to the usual defintion of the linewidth (3.81). In this case, the influence of the lateral dimension is accounted for by the spontaneous emission factor (3.64), which from (3.67) is seen to depend on the reciprocal of the lateral dimension. Note that the relation between the spectral linewidth of an array and the details of the array structure have not been studied extensively either theoretically or experimentally. However, recently a linewidth of 203 kHz was measured for a single-freqeuncy 5-element ridge-guide distributed-feedback array [3.50]. This is about a factor of five narrower than the spectral linewidths that are typically measured for single-element distributed-feedback lasers. The observed narrowing of the array linewidth was attributed to the reduction in the spontaneous emission factor due to the larger cavity dimension associated with the lateral structure of the array [3.50]. Control of the spectral linewidth under single-frequency operation through the design of the array structure could be an interesing area for future studies.

3.4 Frequency-Locking of Diode-Laser Oscillators

Another way to view a diode-laser array is that of an ensemble of coupled-laser oscillators that can operate independently of each other. A portion of the electrical field output of each array element is coupled in some manner into some or all of the other elements in the array. This picture allows for array modes where one or more of the laser elements are not undergoing synchronous single-frequency oscillation; and therefore, not operating coherently with respect to each other. The analyses of Sects. 3.2 and 3.3 were restricted to determining the modes of a perfectly-uniform array structure at threshold. Structural nonuniformities, gain saturation, and even thermal effects can induce a diode-laser array to break up into sub-arrays that oscillate independently. For this to occur, the available gain in the array must be sufficiently high so that the modes of the sub-arrays, which could even be single elements, are at or above threshold. This situation is frequently encountered when operation at the highest powers is desired. Coherent single-frequency operation of the ensemble of lasers requires synchronous operation of all elements at the same optical frequency. Frequency-locking in coupled oscillators is a subject that has been well studied in mechanical, acoustical, and electrical systems [1.9, 1.10]. The phenomena of frequency locking of coupled oscillators was perhaps first recognized by C. Huygens in 1665. An interesting account of Huygen's original observations, as well as the latter observations of Lord Rayleigh in 1907, has been given by *Sargent* et. al. [3.51]. The question of synchronous frequency-locking of diode lasers for the purpose of creating large-scale coherent operation has several unique aspects to it which place constraints of the structural tolerances, coupling strength, and array size.

3.4.1 Frequency Locking of Independent Laser Oscillators

The subject of synchronous frequency-locking of a pair of mutually-coupled semiconductor lasers for the purpose of forming a coherent laser array was perhaps considered first by *Basov* [1.5]. Although at about the same time, other analyses [1.14, 3.52] considered the frequency-locking characteristics of a laser under the influence of an externally-injected signal, where the external source was not coupled back to the laser. Numerous other studies have since been reported on the properties of discrete coupled lasers [3.53]-[3.65], as well as diode-laser arrays [3.66]-[3.70].

The aforementioned treatment that allows mutual coupling of oscillators is an appropriate treatment for diode-laser arrays that can be modeled as mutually-coupled oscillators. However, the case where an externally-injected signal is coupled into a laser is relevent to *Master-Oscillator Power-Amplifier* (MOPA) arrays and array structures that use non-reciprocal coupling. Diode-laser array amplifiers are the subject of Chap. 8.

Consider a pair of single-frequency laser elements, operating at frequencies ν_1 and ν_2 respectively, and suppose the electric field of each laser is coupled into the cavity of the other. When the lasers are not operating sychronously there will be an instantaneous beat note generated at the difference frequency $|\nu_1 - \nu_2|$. As the frequency of one laser is tuned closer to the other, the beat frequency decreases, and eventually a point will be reached where the beat frequency completely disappears. The range of frequency differences where synchronous operation at the same frequency will occur is given by [1.5]

$$|\nu_1 - \nu_2| \leq \frac{c\Gamma_{1,2}}{nL} \ , \qquad (3.87)$$

where $\Gamma_{1,2}$ is the fraction of optical field that is coupled between the two lasers and L is the cavity length of the lasers. It is seen that the optical coupling between the two lasers must be sufficiently large in order to establish synchronous single-frequency operation. This is also refered to as phase-locked operation. Note that (3.87) is valid for weak reciprocal-coupling between the two lasers. What is meant by weak coupling in this case is that the coupling does not perturb the mode structure of the isolated laser. This is not always the case, and it may then become necessary to use a more general analysis (see, for example [3.60]). Also, the actual characteristics of the locking bandwidth that are observed in semiconductor lasers are more complicated than can be accounted for by (3.87). The presence of carrier-induced changes in the refractive index causes the locking bandwidth to become asymmetric and introduces regions of instability. These topics are discussed in more detail in [3.54, 3.57, 3.58, 3.62, 3.64].

(a) Time-Delay Effects

Another important parameter that can effect the frequency-locking condition is the time delay associated with the coupling of the fields between the laser oscillators. *Belenov* and *Letokov* [1.8] pointed out that the analysis of [1.5] could be extended to obtain a frequency-locking condition for a pair of oscillators that accounted for propagation delays in the mutual coupling of the fields. Also, the effect of time-delayed feedback into a single laser has been considered [3.54, 3.60]. The effect of temporal delays on spatially-separated oscillators is an important topic for our understanding of the potential scaling limitations of single-frequency operation in diode-laser arrays. More recently, *Dente* et al. [3.71] have done a comprehensive theoretical and experimental analysis of time-delay effects on frequency-locking of spatially-separated, mutually-coupled oscillators. They found, that for a given separation distance, that there was a maximum value of the coupling strength beyond which it was not possible to achieve good frequency locking. This interesting phenomena is particularly relevant to diode-laser arrays, as it may be an important factor in limiting the maximum size of a coherent array. The analysis developed in [3.71] is reviewed below.

In their analysis, *Dente* et al. modeled the pair of coupled oscillators as a large laser of length L_{ext} with regenerative gain regions of length L_D at each end. The distance between the two lasers corresponds to a time delay of $\tau_{ext} = L_{ext}/c$. The coupled-laser system is illustrated in Fig. 3.7. The coupler is modeled as a beam splitter with a power transmissivity of ϵ^2. For this coupled-laser system, frequency locking requires that only one mode of the coupled-laser system undergo laser oscillation.

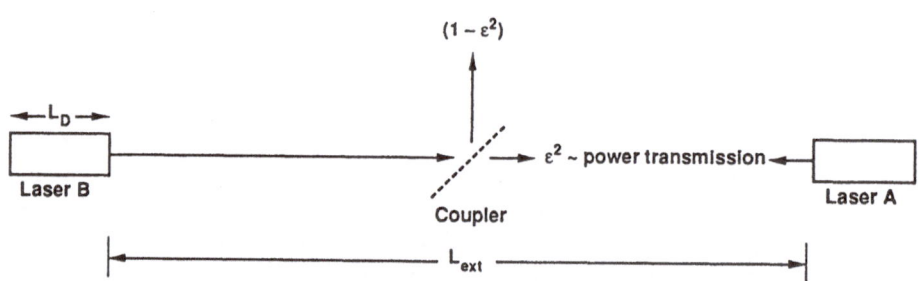

Fig. 3.7. Schematic diagram of a pair of mutually-coupled oscillators. [3.71]

In Fig. 3.8, we see a more detailed illustration of a laser oscillator section. As depicted, light that is incident on one of the mirrors of the Fabry-Perot cavity makes multiple passes through the gain region experiencing amplification on each pass. The regenerated signal that is transmitted out through the same mirror will be increased in power; and also, it will have aquired a phase shift, relative to the input light, due to the multiple passes in the amplifier.

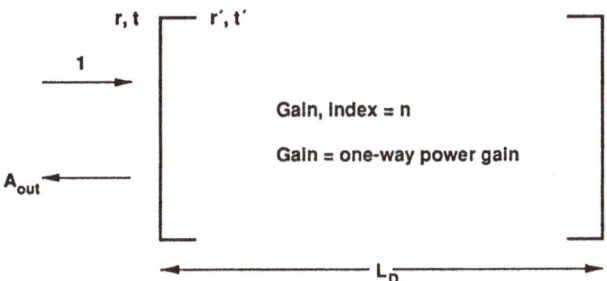

Fig. 3.8. Detailed view of single-laser oscillator section. [3.71]

For this configuration, the near-resonance regenerative gain of each laser section, $G_{\text{eff}}(\Delta \nu)$ can be expressed as

$$G_{\text{eff}}(\Delta \nu) = \frac{G_{\text{eff}}(0)}{1 - G_{\text{eff}}(0) \left(\frac{\Delta \nu}{\Delta \nu_{cc}}\right)^2} \quad , \tag{3.88}$$

where $\Delta \nu$ is the frequency detuning of the input light from the resosnant frequency of the cavity. The saturated regenerative gain at resonance $G_{\text{eff}}(0)$ is

$$G_{\text{eff}}(0) = \frac{R(1-G)^2}{(1-RG)^2} \quad , \tag{3.89}$$

where R is the power reflectivity of the laser mirror or facet and G is the saturated single-pass power gain. The parameter $\Delta \nu_{cc}$ correponds to the half width of the output coupling resonance, and it is given by

$$\Delta \nu_{cc} = \frac{c}{2nL_D} \frac{1-R}{2\pi\sqrt{R}} \quad . \tag{3.90}$$

For frequency locking of the two laser oscillators to occur, the oscillation condition for the complete coupled-laser system must be satisfied. For typical diode lasers, $L_D \approx 300\,\mu m$, $n \approx 3.3$, and $R \approx 0.3$, which gives $\Delta \nu_{cc} \approx 30\,\text{GHz}$. For the two oscillators designated as A and B in Fig. 3.7, the round-trip gain condition corresponds to

$$G_{\text{eff}}^A(\Delta \nu_A) G_{\text{eff}}^B(\Delta \nu_B) \left(\varepsilon^2\right)^2 \geq 1 \quad . \tag{3.91}$$

where the frequency detunings are referenced to the mode of the coupled cavity with the highest gain. From (3.91 and 3.88) and assuming non-zero frequency detunings, the minimum coupling strength requirement is given by [3.71]

$$\epsilon_{\text{min}}^2 \geq \left(\frac{1}{2\Delta \nu_{cc}}\right)^2 \left|(\nu_A - \nu_B)^2 - \gamma^2\right| \quad , \tag{3.92}$$

where γ is the frequency detuning of the coupled-cavity mode with respect to the frequency $\nu = (\nu_A + \nu_B)/2$, which is at the midpoint between ν_A and

ν_B. When the coupled-cavity resonance is tuned to the midpoint frequency ν, then $\gamma = 0$ and (3.92) reduces to the oscillator-locking condition that had been derived earlier by *Adler* [1.9]. For the case when lasers A and B are nearly resonant, $|\nu_A - \nu_B| \leq 100\,\text{MHz}$, and the power coupling strength for frequency locking is then $\epsilon_{\min}^2 \approx 10^{-6}$, which is quite small.

The frequency separation between the longitudinal modes of the coupled laser system of Fig. 3.7, $\Delta\nu_L$, has been calculated to be [3.71]

$$\Delta\nu_L = \frac{2\pi}{\dfrac{4\pi L_{\text{ext}}}{c} + \dfrac{\sqrt{G_{\text{eff}}^A (0)}}{\Delta\nu_{cc}} + \dfrac{\sqrt{G_{\text{eff}}^B (0)}}{\Delta\nu_{cc}}} . \tag{3.93}$$

The second and third terms in the denomenator of (3.93) represent the phase-shift contributions of from the two regenerative oscillators A and B. As expected, the frequency separation of modes of the coupled-cavity decreases as the separation distance L_{ext} is increased. In the case of near-resonant coupling, $\nu_A \approx \nu_B$, and (3.93) reduces to

$$\Delta\nu_L = \frac{\dfrac{c}{2L_{\text{ext}}}}{1 + \dfrac{c}{2\pi\epsilon L_{\text{ext}}\Delta\nu_{cc}}} . \tag{3.94}$$

Frequency locking will degrade when the coupled-laser system can oscillate in more than one longitudinal mode. This will occur for sufficiently large coupling strengths. To see this, consider the locking condition (3.92) for the case when the two oscillators are very nearly resonant with a mode of the coupled system. This corresponds to $\Delta\nu_L \gg |\nu_A - \nu_B|$. To determine the maximum coupling strength, *Dente* et al. [3.71] evaluated the condition (3.92) for locking to an adjacent mode of the coupled system, so that $\gamma \approx \Delta\nu_L \gg |\nu_A - \nu_B|$. Then the maximum coupling strength $\epsilon_{\max}$ beyond which frequency locking will not occur is given by

$$\epsilon_{\max} = \frac{\left(1 - \frac{1}{\pi}\right)\left(\frac{c}{2L_{\text{ext}}}\right)}{\Delta\nu_{cc}} . \tag{3.95}$$

For frequency locking discrete diode lasers in a laboratory experiment as in [3.71], one typically has $L_{\text{ext}} \approx 10\,\text{cm}$, which corresponds to a near-resonant maximum power coupling strength of $\epsilon_{\max}^2 \approx 10^{-4}$.

For a fixed value of the coupling strength, the condition (3.95) defines a critical distance L_{crit} between oscillators

$$L_{\text{crit}} = \frac{\left(1 - \frac{1}{\pi}\right)c}{2\epsilon\Delta\nu_{cc}} . \tag{3.96}$$

For the values of $\Delta\nu_{cc}$ and $\epsilon_{\max}^2$ given above, we find that the maximum allowable distance between the oscillators is $L_{\text{ext}} \approx 30\,\text{cm}$. This is quite a large distance in comparison to the $\approx 1\,\text{cm}$, or less, size of semiconductor wafers chips that comprise monolithic diode laser arrays. For the monolithic case ϵ^2_{max} would be about two orders of magnitude larger. The difficulty enters in fabricating lasers on the same wafer with sufficiently high uniformity, so that they can be made to operate in a near-resonant condition. As discussed below, a departure from near-resonance operation results in degradation of the frequency lock.

A criterion for assessing the quality of frequency-locking between two laser oscillators was introduced in the analysis of [3.71]. In addition, to phase locking it is also desireable for both lasers to have nearly equal power outputs. This is particulary relevant if the oscillators are part of an array, because large disparities in the power outputs of the emitting elements can lead to a broadening of the far-field beam divergence. In the case of the coupled-laser system of Fig. 3.7, *Dente* et al. [3.71] used the fringe visibility of the interference pattern resulting from the combination of the two laser output beams as a measure of the quality of the frequency lock [3.72]. The fringe visiblity $V(\gamma)$ due to an imbalence in the output powers P_A and P_B can be expressed in terms of the frequency detunings as

$$V(\gamma) < \frac{2\sqrt{P_A P_B}}{P_A + P_B} = \frac{2\sqrt{|\Delta\nu_A \Delta\nu_B|}}{|\Delta\nu_A| + |\Delta\nu_B|} . \tag{3.97}$$

In Fig. 3.9, we see a graph of the visibility $V(\gamma)$ versus detuning γ for the cases $\Delta\nu_A \approx \Delta\nu_B$ and $|\Delta\nu_A - \Delta\nu_B| = 200\,\mathrm{MHz}$. It is seen that the maximum visibility is obtained when the two oscillator frequencies are nearly equal to each other. The case of $|\Delta\nu_A - \Delta\nu_B| = 200\,\mathrm{MHz}$ is not a large detuning by diode laser standards, yet the visibility decreases rapidly as γ deviates on either side of zero.

(b) Intermodal-Injection Locking

We have seen that for mutually-injection coupled-laser oscillators to operate with a maximum degree of mutual coherence requires that the free-running frequencies are all nearly resonant. In practice, it is usually very difficult, even in monolithic arrays, to fabricate multiple diode lasers that operate at nearly-resonant frequencies. Material and structural nonuniformities, as well as variations in the frequency versus temperature characteristics, make it very difficult to get a group of diode lasers to oscillate within the narrow locking bandwidths, exhibited in Fig. 3.9. For this reason, the possibility of using unidirectional or nonreciprocal coupling of laser oscillators has been considered as a means for increasing the locking bandwidth [3.56].

Goldberg et al. have used nonreciprocal injection of light from a master os-cillator operating at a frequency that corresponds to one of the non-oscillating Fabry-Perot modes of the slave laser, so that the slave laser will act as a res-onant amplifier [3.56]. As the power of the injected signal is increased, the power at the free-running frequency decreases because of gain competition. A sufficiently high power of the injected signal will saturate the gain and quench

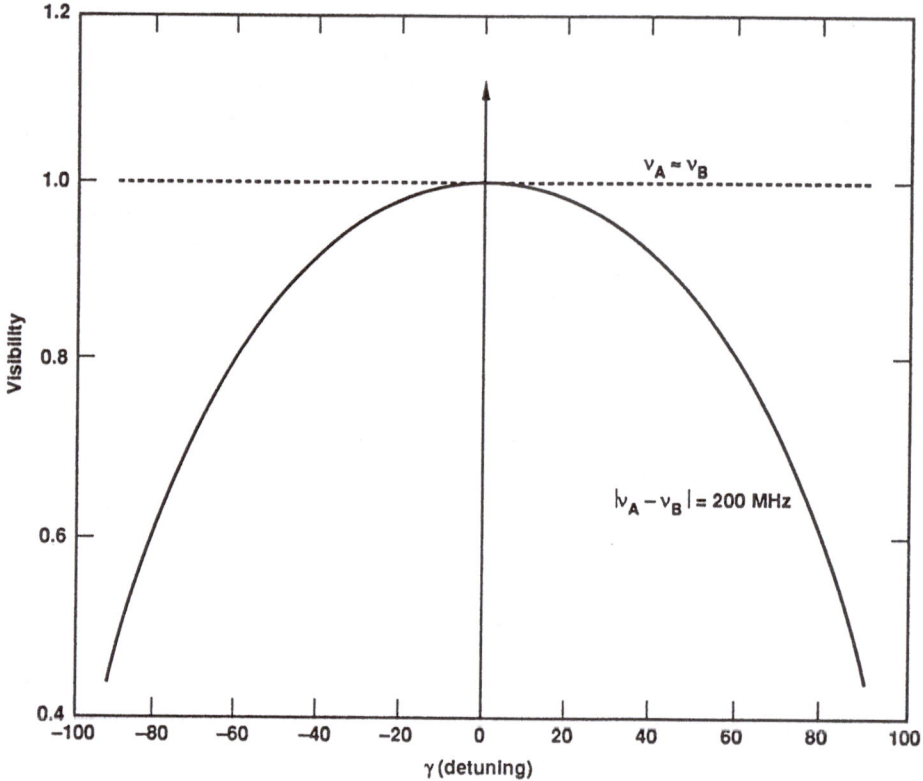

Fig. 3.9. Graph of $V(\gamma)$ versus detuning γ for the case of $\Delta\nu_A \approx \Delta\nu_B$ and $|\Delta\nu_A - \Delta\nu_B| = 200\,\mathrm{MHz}$ [3.71].

the laser oscillation at the free-running frequency. This method is known as *intermodal-injection locking*, and it has been used to demonstrate injection locking of a slave laser to the frequency of a master laser that was detuned $60\,\text{Å}$ from the free-running frequency [3.56]. This illustrates a potential advantage offered by uni-directional coupling, in that a common frequency may be found for a master oscillator that can be used to injection lock a group of free-running slave lasers.

(c) Quantum-Noise Effects in Coupled Lasers

The coupled laser analysis of the previous subsection established the range of coupling strengths ϵ where stable frequency locking will occur. However, experimental observations of *Bossert* et al. [3.73, 3.74] indicate that the measured fringe visibility as a function of ϵ show a gradual increase in the visibility from 0 to 1 as the coupling strength was tuned througth $\epsilon_{\min}$. This soft entry into the frequency-locked condition was observed even when the two lasers were very near resonance. To explain this behavior, it is necessary to anaylze the frequency locking of two lasers using quantum mechanics. A basic understanding of the influence of quantum fluctuations on frequency locking can be obtained by using the uncertainty principle. When both lasers are operating at the same power level P, the average number of photons injected into one laser during a time interval t_{lock} will be given by

$$N_{lock} = \frac{\epsilon^2 P t_{lock}}{h\nu} \quad . \tag{3.98}$$

Here t_{lock} is then called the *locking-control interval*, which is defined below. If the lasers are both above threshold then the photon statistics can be represented by a *Poisson* process, then the fluctuations in the number of photons that are injected in the time interval t_{lock} are given by $\Delta N = \sqrt{N_{lock}}$. From the uncertainty relation, photon-number fluctuations are related to phase fluctuations as $\Delta N \, \Delta\varphi \geq 1/2$. This corresponds to a phase uncertainty of $(\Delta\varphi)^2$ given by

$$(\Delta\varphi)^2 \geq \frac{1}{4N_{lock}} \quad . \tag{3.99}$$

As ϵ is decreased, N_{lock} also decreases and this results in an increase in phase fluctuations during the locking interval t_{lock}. The locking control interval t_{lock} is the recovery time for a fluctuation in the relative phase. It can be thought of as the accumualted time it takes for the an injected signal to reach a steady-state with the regenerative oscillator. Using this approach, *Dente* et al. [3.73] calculated t_{lock} in the near resonance case to be

$$t_{lock} = \frac{1}{2\pi\epsilon\Delta\nu_{cc}} \quad . \tag{3.100}$$

The afore-mentioned argument illustrates, in an intuitive way, the impact of quantum-noise fluctuations on the realtive phase of coupled oscillators. To obtain an analytic form for the fringe visibility requires a more exact treatment which can be found in [3.74]. The resulting form for the visibility that accounts for quantum fluctuations is

$$V(\epsilon) = \exp\left(\frac{-\pi h\nu\Delta\nu_{cc}\sqrt{1+\alpha^2}}{4\pi P\sqrt{\epsilon^2 - \epsilon_{min}^2}}\right) \quad . \tag{3.101}$$

The above form for the visibility has been found by *Bossert* [3.74] to give agreement with experiment.

3.4.2 Influence of Nonuniformities on Frequency Locking

In the previous subsection, it has been established that near-resonance operation is critical for a pair of coupled lasers to exhibit good frequency-locking characteristics. This subsection will consider the topic of frequency locking large (large being much greater than 2 oscillators) ensembles of lasers. Viewing the diode-laser array as a collection of mutually-coupled lasers will allow

us to assess the impact of random structual and material nonuniformities on
the spectral characteristics of diode-laser arrays.

The phase-locking condition (3.87) has been generalized to a uniform
array of N coupled diode-laser elements [3.75],

$$\left| \sum_{i=1}^{N} E_i^2 \left(\nu_0 - \nu_i \right) \right| \leq \frac{c\Gamma}{nL} E_{N+1} E_N \; ; \text{ where } N = 2, 3, \cdots, (N-1) \; , \quad (3.102)$$

where E_i is the field amplitude of element, Γ is the fractional coupling between adjacent lasers, and ν_0 is the frequency corresponding to phase-locked
operation of the N element array which is given by

$$\left| \sum_{i=1}^{N} E_i^2 \left(\nu_0 - \nu_i \right) \right| = 0 \; . \quad (3.103)$$

The condition expressed by (3.102 and 3.103) contains N equations. Each
equation comprises a sum of weighted frequency differences that must be
less than a specific locking bandwidth. For a typical diode-laser array the
free-running laser frequencies $\nu_i \approx 3 \times 10^{14}$ Hz. The locking bandwidth on
the right hand side of (3.102) is $c\Gamma / (nL) \approx 3 \times 10^{10}$ Hz, which corresponds
to only about a one Ångstrom difference in wavelength. Therefore, the slight
changes in the operating wavelengths, which can occur due to local variations
in environment, will drive the lasers outside of the locking condition (3.102).

It was pointed out by *Nishi* and *Lang* [3.76], that small nonuniformities in multi-waveguide diode-laser structures would cause localization of the
spatial mode into a single waveguide of the array. Such localization of the
mode causes spatial hole-burning, which induces multi-lateral mode operation, and phase-locked operation is lost. *Garmire* [3.77] has analyzed the
tolerances of the locking condition (3.102) for the case of two weakly-coupled
positive-waveguide diode lasers coupled in the lateral dimension by the overlap of their evanescent electric fields. The phase-locking condition dictates
the uniformity tolerance requirements for the composition, dimensions, temperature, and electrical properties of the ridge-guide type diode-laser array
(Sect. 6.2). For the case analyzed by [3.77], which is representative of many
evanescently-coupled index-guided diode laser arrays, the tolerance requirements were found to be very demanding. For example, the locking tolerance
dictates that the refractive index, averaged along the length of two lasers,
differ by less than 10^{-4}. *Garmire* [3.77] has calculated that the difference in
average thickness of the epilayers must be less than 0.2 %. Although some
of the highest uniformity MOCVD-grown laser structures [3.78] and MBE-
grown laser structures come close to these requirements, maintaining such
stringent tolerances throughout the subsequent fabrication and processing of
the complete laser structure can be difficult [1.474]. With the continuing improvements in growth and technology, refractive index uniformities of index-
guided diode lasers that extrapolate to less than 10^{-4} over a several hundred

micron distance have been inferred from the high-power operating characteristics of some single-element diode lasers [1.355]. The temperature tolerance
on phase-locking requires that the temperature difference between two lasers
be less than 0.04 °C. Thus, a suitably uniform heat sink must be used to
insure that the locking requirement is met. As discussed by [2.88, 3.77], there
are several feasible heat-sinking approaches that can satisfy the aforementioned temperature uniformity requirement. A more recent analysis of the
locking requirement [3.65] corroborates the stringent tolerance requirements
on the variation in laser parameters in [3.77].

The consequence of having a diode-laser array structure that does not
satisfy the locking requirement is the break down of phase locking that results in incoherent operation of the array. The uniformity tolerances discussed
in [3.65, 3.77] have been a contributing factor towards the lack of coherent
operation found in many types of diode-laser structures. Experimental studies of the time-resolved output characteristics of some types of diode-laser
arrays done by *Elliott* and *DeFreeze* [3.79]-[3.82] have shown a rich dynamical behavior and temporal instabilities in many types of diode-laser arrays.
Theoretical analyses that model laser arrays as a system of coupled nonlinear oscillators attribute the observed dynamical behavior in laser arrays to
operation outside the locking condition [3.65, 3.68, 3.83].

The question of phase-locked operation in very-large two-dimensional
laser arrays with random frequency variations among the array elements has
been analyzed by *Golubentsev* et al. [3.69, 3.70]. In this treatment, the laser
array was modeled as an ensemble of nearest-neighbor coupled, oscillators
with a homogeneously saturable gain medium. Under the following set of
assumptions: 1) the random variations in operating frequencies of the array
elements were small enough so that the locking requirements (3.87 and 3.102)
were satisfied, 2) the available gain was sufficiently large that field amplitudes
did not differ significantly between array elements, and the nearest-neighbor
distance was uniform over the array; *Golubentsev* et al. claimed that the spatial dependence of the phase distribution could be modeled with a diffusion
equation of the form

$$\frac{\partial \varphi}{\partial t} = \frac{\Gamma a^2 c}{2L} \nabla_\perp^2 \varphi + \delta \nu (x, y) \ , \qquad (3.104)$$

where φ is the phase distribution of the electric field across the array, a is the
nearest-neighbor distance, and $\delta \nu (x, y)$ gives the random frequency variation
between laser elements across the array. It is seen from (3.104) that the random frequency variations act as sources and sinks of phase. *Golubentsev* et
al. did a numerical study of arrays with between 100 and 1000 laser elements
and found, that for small levels of disorder, a steady-state phase distribution
with a domain type of structure would exist over the array. The phase difference between adjacent domains was found to differ by about π. The size of a
domain expressed in terms of the number of laser elements was found to be

$$N_{1-D} = \left(\frac{\Gamma^2 c^2}{\langle \delta\nu^2 \rangle n^2 L^2} \right)^{1/3} , \tag{3.105}$$

for a one-dimensional array, and for a two-dimensional array the domain size was

$$N_{2-D} \ln (N_{2-D}) = \frac{\Gamma^2 c^2}{\langle \delta\nu^2 \rangle n^2 L^2} , \tag{3.106}$$

where $\langle \ \rangle$ corresponds to an average over the random frequency variations in the array. By considering higher levels of disorder in their numerical study, *Golubentsev* et al. were able to infer the following locking requirement

$$\langle \delta\nu^2 \rangle \leq \frac{\Gamma^2 c^2}{n^2 L^2 \ln (N_A)} , \tag{3.107}$$

where N_A is the total number of lasers in the array. The locking-condition expressed by (3.107) can be thought of as a condition on the domain size N_{2-D}. From the definition of N_{2-D} given by (3.106), it is seen that (3.107) is equivalent to the requirement that $N_{2-D} \geq 1$ in order for a steady-state phase distribution to exist across the array, which agrees with ones intuition. It is also seen that when $N_{2-D} = 2$, (3.107) essentially reproduces the locking condition for two lasers given by (3.87). Note that in arrays where all laser elements are more-or-less coupled together (i.e., parallel or global coupling) such as devices that operate on the principle of leaky-wave coupling (Sect. 6.3), it may be possible to tolerate a larger variation in $\langle \delta\nu^2 \rangle$, than that given by (3.107) for nearest-neighbor coupling. However, it appears as if a quantitative analysis of the locking condition for a parallel-coupled laser array has not yet been done.

The effects of an externally-injected signal on very-large coupled-laser arrays were also analyzed by *Golubentsev* et al., and they found that a small external signal was sufficient to lock the phase across all domains in the array. This situation is analogous to what happens when an external magnetic field is applied to a ferromagnet and produces long-range order. Other experimental and theoretical studies of injection locking of diode-laser arrays to an external laser signal have, in fact, shown that with appropriate control of the frequency and angle of incidence of the injected signal, it is possible to select a specific array mode [3.59],[3.63],[3.84]-[3.88]. This method of active-mode control is especially useful because it can be used to electronically control the characteristics of the output beam. Experimental examples of mode control with externally-injected signals are presented in Sect. 8.1.

4. Theoretical Models for Monolithic Diode-Laser Arrays

This chapter will review some of the theoretical models that have been used to analyze the modal properties and simulate the operating characteristics of various types of diode-laser array designs. The starting point for all of these models is Maxwell's equations. The details of a particular diode-laser array structure will specify the boundary conditions and the spatial dependence of the dielectric function which corresponds to the gain and refractive-index profiles that comprise the monolithic laser-array structure. We will start out in Sect. 4.1 by reviewing the basic properties of the multi-layer waveguide problem, as this model is basic to our understanding of more complex analyses. Section 4.2 will cover methods specifically used in analyzing laterally-coupled arrays such as the effective-index approximation, coupled-mode theory, the Bloch-function method, and the two-dimensional Helmholtz wave equation. In Sect. 4.3, self-consistent models of diode-laser arrays are presented. Finally, a section (Sect. 4.4) has been included to present the fundamentals of optical coherence theory that are important for understanding and interpreting the output-beam characteristics of diode-laser arrays.

4.1 Multi-Layer Waveguide Structures

In analyzing the modal properties of diode-lasers arrays, it is usually necessary to resort to a numerical calculation in order to find the allowed modes. An approach that is commonly used to find the lateral modes of diode-laser arrays is to model the lateral structure as an effective, multi-layer dielectric stack. Maxwell's curl equations (along with a suitable set of boundary conditions) can then be used to find the lateral modes associated with the array structure. A periodic multi-layered dielectric structure of the sort that is used to model arrays is illustrated in Fig. 4.1. Here we are interested in modes that propagate along the z-direction of the array cavity, so the general solutions will be of the form

$$\boldsymbol{E}\left(x,y\right)\exp\left[\mathrm{i}\left(\omega t - \beta z\right)\right] \text{ and } \boldsymbol{H}\left(x,y\right)\exp\left[\mathrm{i}\left(\omega t - \beta z\right)\right] \; . \qquad (4.1)$$

The dielectric layers stacked along the x-direction represent the lateral structure of the array; and in the plane of each layer, the composition is assumed

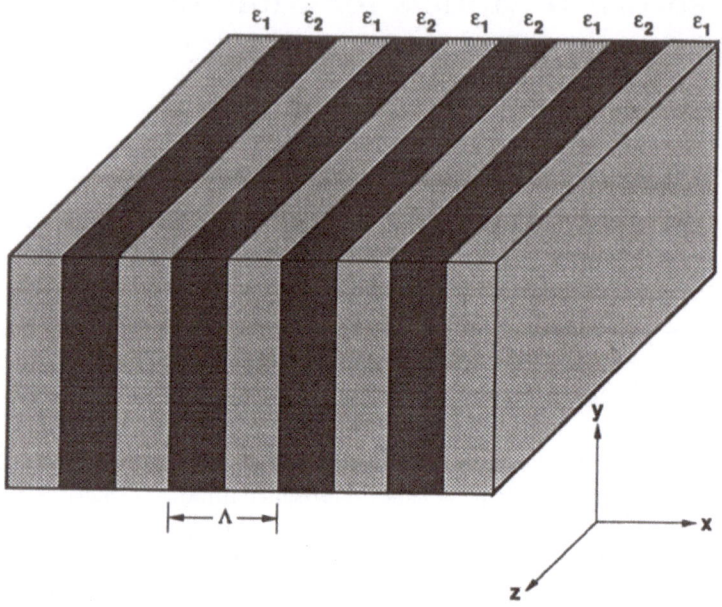

Fig. 4.1. Geometry of the planar multi-layer dielectric structure used to represent diode-laser arrays.

to be uniform and of infinite extent. Note that in semiconductor diode-laser array structures the x (lateral) and z (longitudinal) dimensions of the cavity are much greater than the y (transverse) dimension. However, the waveguide structure in the y-direction typically supports only one mode. As will be shown in Sect. 4.2.1 that follows, the y dependence can be removed and replaced by an effective one-dimensional dielectric profile along the x direction. Then the electromagnetic fields in the dielectric stack model will be independent of y, and applying Maxwell's curl equations to (4.1) gives the following set of six equations for the x, y, and z components of the electromagnetic field

$$\beta E_y = -\omega\mu H_x , \qquad\qquad \beta H_y = \omega\varepsilon E_x ,$$

$$\frac{\partial E_z}{\partial x} + i\beta E_x = i\omega\mu H_y , \quad \frac{\partial H_z}{\partial x} + i\beta H_x = -i\omega\varepsilon E_y , \qquad (4.2)$$

$$\frac{\partial E_y}{\partial x} = -i\omega\mu H_z , \qquad \frac{\partial H_y}{\partial x} = i\omega\varepsilon E_z ,$$

where ε is the dielectric permittivity, and μ magnetic permeability within a layer. At all boundaries between layers, the tangential field components E_y, E_z, H_y, and H_z in (4.2) are all continuous. As is well-known, the six equations of (4.2) have independent sets of modes that can be solved for separately [1.356]. These modes correspond to the following polarization states of the electromagnetic field: 1) *Transverse Electric* (TE) where $E_z = 0$ and $H_z \neq 0$, and 2) *Transverse Magnetic* (TM) where $H_z = 0$ and $E_z \neq 0$. Thus, the TE modes are completely specified by E_y, H_x, and H_z, while the TM modes are completely specified by H_z, E_z, and H_y. For both the TE and TM modes, the system of equations given by (4.2) can be reduced to a second-order wave equation in a single field component [1.356]. Alternatively, (4.2) can be reduced to the following system of coupled first-order differential equations for the TE set of modes,

$$\frac{\partial E_y}{\partial x} = -i\omega\mu H_z \text{ and } \frac{\partial H_z}{\partial x} = \frac{iE_y}{\omega\mu}\left(\beta^2 - \varepsilon\mu\omega^2\right) \qquad (4.3)$$

and for the TM modes,

$$\frac{\partial H_y}{\partial x} = -i\omega\varepsilon E_z \text{ and } \frac{\partial E_z}{\partial x} = \frac{iH_y}{\omega\varepsilon}\left(\varepsilon\mu\omega^2 - \beta^2\right) . \qquad (4.4)$$

The sets of coupled first-order differential equations in (4.3 and 4.4) are a more convient representation to use when the solutions are to be obtained by numerical methods. Such systems of coupled first-order differential equations can be transformed directly into a matrix reperesentation for the solutions. For multi-layered structures, the chain-type of matrix representation, that is well-known in thin-film optics, is what is commonly used [1.356, 3.4, 4.1, 4.2]. Imposing the field boundary conditions at the layer interfaces, along with the condition that the modal fields go to zero as x → ±∞, gives rise to a eigenvalue problem in β, and the complex propagation constants of lateral laser array modes can be found using standard algorithms [4.3]-[4.5]. The effective one-dimensional multi-layer waveguide model is widely used for analyzing both the transverse (perpendicular to the plane of the pn-junction), as well as the lateral (in the plane of the pn-junction and perpendicular to the direction of propagation) waveguide characteristics of diode lasers and diode-laser arrays. In the next section, it will be shown how the effective-index approximation is used to reduce the two-dimensional waveguide problem associated with the x and y dimensions of the diode-laser array structure to a simpler effective one-dimensional waveguide that can be analyzed with the aforementioned matrix techniques.

4.2 Analysis of Diode-Laser Waveguide Structures

The ideal, monolithic diode-laser array structure would operate in a single spectral and spatial mode over its entire range of operation. Such an array would be expected to comprise a waveguide structure that is very robust to the carrier and thermally induced changes in the dielectric profile across the array. A desirable way to accomplish this is to fabricate a spatial variation in the dielectric coefficient along the lateral dimension of the waveguide structure, such that the fabricated dielectric variation is significantly larger than either the carrier or thermally induced changes in the dielectric profile. In such structures, the built-in spatial variation of the dielectric coefficient across the waveguide is intended to play the dominant role in determining the modal properties. For this reason, diode-laser array structures are frequently modeled as a fixed-periodic dielectric permittivity.

This section is organized into subsections that are devoted to the theoretical models that treat the diode-laser array structure as a fixed dielectric permitivitty that varies periodically in the lateral direction. The effects of injected carriers and heating of the structure are not included in these models. Section 4.2.1 will review the effective-index method for finding the propagating solutions to Maxwells' wave equation for laser-array structures with a fixed spatial dependence in the dielectric coefficient along the lateral direction. This will provide a basis for later discussions on more complicated models. Section 4.2.2 will review coupled-mode analysis of diode-laser array structures, and Sect. 4.2.3 will review the more exact Bloch-wave analysis. The two-dimensional Helmholtz model laser arrays is discussed in Sect. 4.2.4.

4.2.1 Effective-Index Method

Diode-laser arrays are three-dimensional structures, but in many cases it is possible to make simplifying approximations that reduce the problem to an effective one-dimensional analysis, which is considerably easier to solve. One example of this type of approximation, the parabolic dielectric-waveguide model, was presented in Sect. 3.1.4. The approach that was used there is quite general and has been employed to model most diode-laser array structures. Here will we review in more detail this method which is known as the *effective-index approximation.*

The electric field inside the cavity of a diode-laser array, $\mathbf{E}\,(\mathbf{r}, t)$, satisfies the three-dimensional wave equation

$$\nabla^2 \mathbf{E}\,(\mathbf{r}, t) - \frac{\varepsilon\,(\mathbf{r})}{c^2} \frac{\partial^2 \mathbf{E}\,(\mathbf{r}, t)}{\partial t^2} = 0 \ , \tag{4.5}$$

where c is the speed of light in vacuum, $\mathbf{r} = (x, y, z)$ are the coordinates, and $\varepsilon\,(\mathbf{r})$ is the dielectric coefficient which depends on the refractive-index and gain profile that is fabricated into the array structure, as well as the carrier density and thermal profile. The most common approach used for solving (4.5) is to apply the effective-index approximation [3.15][4.6]-[4.11]. For simplicity, we will also restrict this discussion to the case of TE-polarized modes. Most quantum-well active layers that are used in diode-laser array structures only have significant gain for modes polarized in the plane of the pn junction, so this is not a restrictive assumption. Then the electric field solutions to (4.5) can be written as follows

$$\mathbf{E}\,(\mathbf{r}, t) = \hat{x} E_x\,(x, z)\,F\,(x, y) \exp\left[\mathrm{i}\,(\beta z - \omega t)\right] \ , \tag{4.6}$$

where $E_x(x, z)$ describes the lateral variation in the mode, which is of primary interest for the diode-laser array problem, and $F(x, y)$ gives the transverse dependence of the mode, which is allowed to a have a parametric dependence on x. This assumption is not valid in diode laser and diode-laser array structures that contain regions that support more than one transverse mode [4.12, 4.13](e.g. such as some antiguided laser-array structures [1.235]), and in these cases the effective-index method is not accurate. Note that the z-dependence of the lateral-field distribution has been retained so that longitudinal or axially-nonuniform arrays can be modeled. Recall from Chap. 2 that axial intensity and gain nonuniformities are expected in laser structures with significantly different facet reflectivities, which is usually the case for high-power, high-efficiency lasers. For many semiconductor-laser arrays, the fundamental transverse mode is strongly index-guided and all higher-order modes are beyond cut-off. There are however important exceptions, notably the antiguided diode-laser array presented in Sect. 6.3, which can require a two-dimensional waveguide treatment (Sect 4.2.4) to accurately model the modal properties. In general, $F(x, y)$ can have a dependence on x, as the shape of the transverse mode profile may exhibit changes corresponding to the lateral variations that have been fabricated into the dielectric coefficient. If the lateral variation in $\varepsilon(x, y)$ is slow enough, then the transverse mode confinement is not significantly perturbed. Following the approach of [1.26], (4.6) is substituted into (4.5) to give

$$\frac{1}{F(x, y)} \frac{\partial^2 F(x, y)}{\partial y^2} + S(x, z) - \beta^2 + k^2 \varepsilon(x, y, z) = 0 , \qquad (4.7)$$

where $S(x, z)$, which is independent of y, represents the terms involving the partial derivatives of $E_x(x, z)$, and is given by

$$S(x, z) = \frac{2i\beta}{F(x, y)} \frac{\partial E_x(x, z)}{\partial z} + \frac{1}{E_y(y, z)} \frac{\partial^2 E_x(x, z)}{\partial y^2} . \qquad (4.8)$$

The terms containing $\partial F(x, y)/\partial x$ and $\partial^2 F(x, y)/\partial x^2$ have been neglected in (4.7) because the x variation of $F(x, y)$ has been assumed to be small. In most cases this is a valid approximation; however, in some cases it is not, and it is then necessary to numerically solve the two-dimensional wave equation, as discussed in Sect 4.2.4. Since the longitudinal (z-direction) variations in the field are expected to be small compared to the wavlength of light, the paraxial approximation [4.14, 4.15] has been invoked and the term containing $\partial^2 E_x(x, z)/\partial z^2$ has been neglected. Equation (4.7) corresponds to a one-dimensional equation for a multilayer waveguide in the transverse direction, where the propagation constant has been modified to be

$$\beta_{\text{eff}}^2(x, z) = k^2 \varepsilon_{\text{eff}}(x, z) = \beta^2 - S(x, z) . \qquad (4.9)$$

The effective propagation constant, $\beta_{\text{eff}}(x, z)$ in (4.9) has a parametric dependence on the lateral and longitudinal positions in the array. This can be associated with an effective dielectric permittivity, $\varepsilon_{\text{eff}}(x, z)$, hence the name *effective index*. Note that in some cases the longitudinal dependence is weak enough to be negligible; however, we have retained it here because it will be required in the self-consistent model that is discussed in Sect. 4.3.

To calculate an effective one-dimensional dielectric permittivity in the lateral direction, the array structure is subdivided, as shown in Fig. 4.2. Also, the z dependence can be neglected, without loss of generality. Within each slice, the dielectric permittivity is assumed to be independent of x. In effect, the x dependence of dielectric permittivity is approximated as a piecewise function. The size of the slices should be small enough so that the first and second derivatives of $F(x, y)$ with respect to x can be neglected in each subdivision. Solving the transverse wave equation (4.7) for each slice gives $\beta_{\text{eff}}(x)$. Then using (4.9), the effective lateral permittivity, $\varepsilon_{\text{eff}}(x_i)$, that models the lateral refractive index and loss variation in the array structure is obtained. Combining (4.9 and 4.8), we obtain the following wave equation for the lateral modes $E_x(x, z)$

$$\frac{\partial^2 E_x(x, z)}{\partial x^2} + 2i\beta \frac{\partial E_x(x, z)}{\partial z} + \left[k^2 \varepsilon_{\text{eff}}(x, z) - \beta^2\right] E_x(x, z) = 0 . \quad (4.10)$$

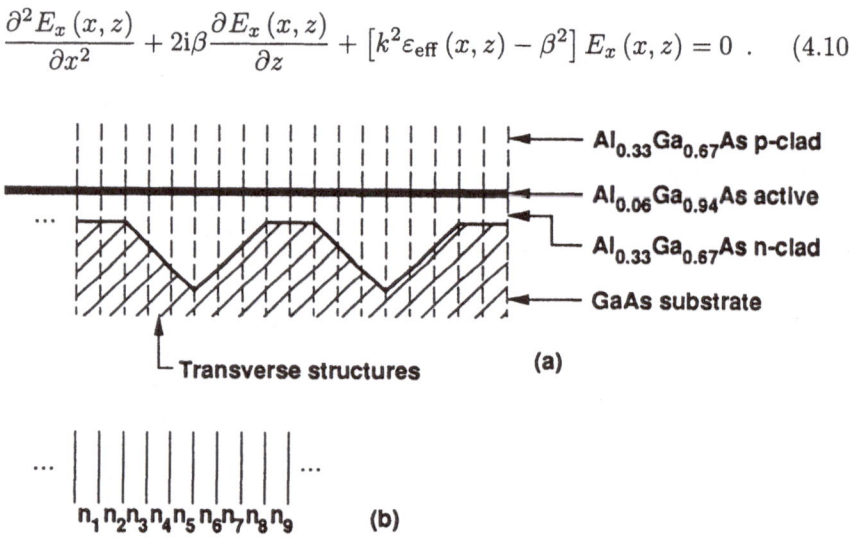

Fig. 4.2. Diagram of x-y plane of a generic diode-laser array structure that has been subdivided into regions where the effective-index approximation can be applied.

Thus, it is seen that in the effective-index approximation, the problem of solving for the lateral modes of a laser-array structure is reduced to the one-dimensional paraxial wave equation which contains an effective lateral dielectric permittivity that characterizes the array structure. Note that initially we restricted this analyses to the TE modes where the electric field is polarized in the plane of the pn junction, as this is the usual operating mode for most diode-laser arrays. This means that the corresponding lateral mode solutions will be TM solutions of the lateral wave equation (4.10). Though for diode-laser array structures the characteristics of the TE and TM modes are nearly identical because the dielectric permittivity step $\Delta\varepsilon$ is very small compared to the magnitude of the background permittivity ε. The largest value for $\Delta\varepsilon/\varepsilon = 2\Delta n/n$ that is usually encountered in diode-laser arrays is ≈ 0.1. In many cases, the TE mode analysis is simpler and gives the same modal characteristics as the TM modes, so it is very often used in solving the lateral wave equation (4.10) even though it is not rigorously correct.

The effective-index method is a powerful result because all of the techniques and algorithms that have been developed for solving for the modes of planar-multilayer waveguides can be used for two-dimensional laser-waveguide structures. Also, the pair of one-dimensional wave equations (4.7 and 4.10), resulting from the effective-index approximation, are much easier to solve than the three-dimensional equation (4.5). Note that (4.10) accounts for propagation effects due to axial variations in the dielectric function. Numerical solutions to the lateral-wave equation (4.10) for diode-laser array waveguide structures (consistent with the assumptions of the effective-index method) can be found by using the *beam propagation methods* that have been used to solve the paraxial wave equation in modeling large, high-power laser systems [4.14, 4.16], as well semiconductor waveguides [4.15][4.17]-[4.19]. In a few special cases where there is no z dependence in the dielectric function, analytic solutions have been possible. These cases include the parabolic lateral variations [1.26, 3.15], described in Sect. 3.1.4, and linear lateral variations of the lateral-gain profile [4.20], which is presented in Sect. 5.2. Diode-laser structures of this type fall more into the category of broad-area devices. As already remarked, accurate modeling of diode-laser array waveguide structures that violate the assumptions made in the effective-index approximation requires the two-dimensional waveguide analysis presented in Sect. 4.2.4.

4.2.2 Coupled-Mode Approximation

Traditionally, diode-laser arrays have been thought of as a group of identical laser elements that are fabricated in close proximity to each other so that optical coupling occurs between elements in such a way that all laser elements in the array are forced to oscillate at the same frequency and with a high-degree of mutual coherence. This point of view that diode-laser arrays were comprised of distinct laser-waveguide elements may have had something to do with why many early analyses of diode-laser arrays [4.21]-[4.23] were done using a coupled-mode type of analysis similar to what had been used earlier to model systems of multiple-coupled passive waveguides [3.1, 3.2][4.24]-[4.26]. Extensive work on the coupled-mode theory, as it applies to diode-laser arrays, has been reported [4.27, 4.28], and some of these analyses have been quite detailed.

In essence, the coupled-mode theory assumes that the modal fields of the entire array can be expressed as a linear superposition of the modal fields of the individual waveguide structures that comprise the array. For an array of weakly-coupled lasers, the modal field in each laser element can be approximated by the field associated with the single-isolated laser element [4.21]-[4.23]. This type of weak coupling has also been termed *nearest-neighbor coupling*. For N-element laser-array structures described by the nearest-neighbor coupling approximation, the coupled-mode theory of *Butler* et al. predicted that the frequency separation between adjacent array modes decreased as $1/N$ (for fixed coupling strength) [4.21]-[4.23]. This indicated that mode discrimination for large numbers of nearest-neighbor coupled lasers would be compromised. Another coupling scheme that was theoretically explored with coupled-mode theory was one where each laser element was coupled to every other element in the array [4.29]. In this case, *Fader* and *Palma* [4.29] found that the frequency separation of adjacent modes decreased as $N/(N-1)$, which gives improved mode discrimination (relative to the nearest-neighbor coupled array) for large N. The nature of the coupling in diode-laser arrays has a strong influence on the mode discrimination properties. Therefore, to accurately model diode-laser array structures one should use a theory that does not make any approximations on the optical coupling. The Bloch function analysis presented in the next section, as well as the the paraxial-wave equation presented in Sect. 4.2.1 and the two-dimensional waveguide model of laser arrays (Sect. 4.2.4) are models that incorporate the full range of coupling that can occur in one- and two-dimensional array structures.

In a sense, coupled-mode theory of weakly-coupled laser arrays is analogous to the tight-binding method used in solid-state physics, where the Bloch states are expressed as linear combinations of atomic orbitals [2.52]. Therefore, coupled-mode analysis of laser arrays can suffer from many of the limitations that occur when the single-element eigenstates, that are used to express the solutions of the coupled system, do not satisfy the boundary conditions of the coupled system. For example, as discussed in Sect. 3.1, when finding the guided modes of a waveguide structure used to model a single-element laser, the boundary condition that the field goes to zero at large distances eliminates the radiation modes corresponding to energy propagating into or out of the laser waveguide. In strongly-coupled array structures where the elements are sufficiently close (within a few wavelengths), the guided-modes of the individual elements may no longer exist. Instead, the field between the laser elements is best described by plane-wave radiation modes, rather than the exponentially-decaying evanescent fields of the guided modes. However, this can be remedied by including the appropriate radiation modes of each element in the superpostion solution for the array [4.27].

In principle, one can always carry out the coupled-mode expansion to include enough modes to give an accurate solution. However, this can become quite complicated; and if care is not taken in selecting the proper expansion basis, inaccuracies can occur. For example, using the matrix-method solution of Maxwell's equations to analyze index-guided laser arrays, it was shown by [4.3] that the coupled-mode array analysis of [4.30] had not found all of the possible modes that could effectively compete for the available gain.

In the case of gain-guided laser arrays, direct analysis methods [3.86, 4.4, 4.31] gave additional modes that had been observed experimentally [4.32, 4.33], but had not been predicted from earlier coupled-mode analyses [4.21]-[4.23]. Basically, the direct analyses of arrays [3.86, 4.3, 4.4, 4.31] accounts for strong coupling or *parallel coupling*, as it has also been termed, between array elements via radiation modes, termed *leaky-wave coupling*, which is a more complete description of the optical-coupling mechanism present in diode-laser arrays. In contrast, coupled-mode analyses [4.21]-[4.23][4.30] that only account for the weaker evanescent-mode coupling neglect leaky-wave coupling, and therfore, do not always give a correct description of the array modes. As discussed in Chaps. 5 and 6, many diode-laser arrays structures operate on the principle of leaky-wave coupling, and so the direct analysis or Bloch-function analysis (Sect. 4.2.3) is required to accurately calculate the modal characteristics.

By applying a coupled-mode theory where the basis set comprised plane waves counter-propagating in the lateral dimension of the array, *Mehuys* and *Yariv* [4.34] were able to model the higher-order modes that had been observed in gain-guided laser arrays. These additional array modes were attributed to distributed-feedback resonances associated with the periodic characteristic of the lateral dielectric permitivitty. Such array modes could not be accounted for using a superposition of the modes of the individual array elements. Note that the plane-wave basis set selected in the coupled-mode model of [4.34] corresponds to a direct Bloch-function analysis (Sect. 4.2.3). Generally speaking, it is often most efficient to use the exact multilayer approach described in Sect 4.1 to find the allowed modes of a laser-array cavity under steady-state conditions. Though, coupled-mode theory is useful for analyzing dynamic behavior of arrays where the elements are not necessarily frequency-locked. For example, see [4.35] and references cited in Sect. 3.4.

Coupled-mode theory can be generalized to be useful for some strongly-coupled arrays. This can be done when the coupled-mode expansion is comprised of modes that are exact solutions to a broad-area laser structure, and the periodic variation of the dielectric permittivity can be regarded as a small pertubation. *Verdiell* et al. [4.36, 4.37] have applied this type of analysis successfully to model experimental observations on diode-laser arrays with a weak built-in lateral periodic-index variation. We now turn our attention to the broad-area coupled-mode approach for modeling diode-laser arrays. This analysis uses the effective-index approximation (4.10) for finding the array modes, but neglects the axial dependence of the field, hence $\partial E_x(x, z)/\partial z = 0$ in the paraxial wave equation (4.10). The effective dielectric permittivity, which is $\varepsilon_{\mathrm{eff}}(x, z)$ in (4.10), is separated as follows,

$$\varepsilon_{\mathrm{eff}}(x) = n_0(x) - \mathrm{i}\frac{g_0(x)}{k} + \varepsilon_{\mathrm{pert}}(x) \ , \qquad (4.11)$$

where $n_0(x)$ and $g_0(x)$ correspond, respectively, to the background effective index and background gain of the broad-area laser. The pertubation term is $\varepsilon_{\text{pert}}(x)$, and it contains any periodic lateral variation fabricated into the laser-array structure, as well as any other lateral variations such as thermal effects.

The method of solving for the modes of the perturbed broad-area structure parallels the pertubation theory used in quantum mechanics. First, one solves (4.10) neglecting the pertubation term $\varepsilon_{\text{pert}}(x)$ to find the unperturbed modes of the broad-area structure $E_m^{(0)}(x)$, where m is the mode index. The first-order solution for the modes in the presence of the pertubation are then given by

$$E_m^{(1)}(x) = E_m^{(0)}(x) + \sum_{n \neq m} c_n^m E_n^{(0)}(x) \ . \tag{4.12}$$

The coupling coefficients, c_n^m, in (4.12) are expressed as

$$c_n^m = \frac{\int E_m^{(0)}(x) W(x) E_n^{(0)}(x) \, dx}{\beta_0^2(m) - \beta_0^2(n)} \ , \tag{4.13}$$

where $W(x)$ is the pertubation operator which is given by

$$W(x) = 2k^2 n_0(x) \varepsilon_{\text{pert}}(x) \ , \tag{4.14}$$

and $\beta_0(m)$ denotes the mode propagation constants for the modes of the unperturbed broad-area structure. Note that for the form for $W(x)$ given by (4.14) it has been assumed that $n_0 \gg g_0/(2k)$, which is a very-good approximation for semiconductor lasers. In order for the pertubation approach to be accurate, the coefficients, c_n^m, that appear in (4.13) must all be much less than unity.

The modal gains and wavelength separations of the perturbed broad-area structure, are found by calculating the eigenvalues of (4.10), which correspond to $\beta_0^2(m)$, to second order with the following expansion

$$\beta^2(m) = \beta_0^2(m) + \beta_1^2(m) + \beta_2^2(m) \ , \tag{4.15}$$

where $\beta_1^2(m)$ is the first-order correction which is given by

$$\beta_1^2(m) = \int E_m^{(0)}(x) W(x) E_m^{(0)}(x) \, dx \ , \tag{4.16}$$

and the second-order correction $\beta_2^2(m)$ is expressed as

$$\beta_2^2(m) = \sum_{n \neq m} \frac{\left[\int E_n^{(0)}(x) W(x) E_m^{(0)}(x) \, dx \right]^2}{\beta_0^2(m) - \beta_0^2(n)} \ . \tag{4.17}$$

The effective indices of the perturbed modes, $n_{\text{eff}}(m)$, to second order are given by

$$n_{\text{eff}}(m) = \text{Re}\left\{\frac{\beta}{k}\right\} = \text{Re}\left\{\frac{\beta_0(m)}{k}\right\}\left(1 + \frac{\text{Re}\left\{\beta_1^2(m)\right\} + \text{Re}\left\{\beta_2^2(m)\right\}}{2\text{Re}\left\{\beta_0^2(m)\right\}}\right)$$

$$(4.18)$$

and the modal gains, to second order, are expressed as,

$$g(m) \approx \frac{\text{Im}\left\{\beta^2(m)\right\}}{n_0 k} = \frac{\left(\text{Im}\left\{\beta_0^2(m)\right\} + \text{Im}\left\{\beta_1^2(m)\right\} + \text{Im}\left\{\beta_2^2(m)\right\}\right)}{n_0 k}.$$

$$(4.19)$$

Equation (4.18) can be used to find the wavelength separations of the array modes, (4.19) can be employed to calculate the gain discrimination; and from (4.12) and (4.13), the near field pattern can be determined. The broad-area coupled-mode approach can be useful for analyzing periodic-array structures, because in effect it models the array structure as a Bragg grating [4.34, 4.38]. The modes of the periodic array structure, (4.12), are then linear combinations of the unperturbed broad-area structure modes, which have non-zero coupling coefficients, i.e. $c_n^m \neq 0$. This picture can be useful for analyzing laser arrays when there are several structural pertubations or inhomogeneities with differing periodicities present.

An interesting example of this appears in the analysis of [4.36, 4.37], where a gain-guided laser was modeled as a broad-area laser perturbed by a periodic modulation of the gain and index, and a temperature-dependent refractive index. Although, this model provides a simple picture of the nature of coupled modes in gain-guided diode-laser arrays, it does so at the expense of making some oversimplified assumptions [4.39]. Following the approach of [4.36, 4.37], the broad-area waveguide was modeled as having a constant permittivity in the gain stripe of half width x_0, with $n_0(x) = n_0$ and $g_0(x) = g_0$ for $|x| \leq x_0$. Outside of the gain stripe where $|x| > x_0$, the absorption was assumed to be infinite, meaning $g_0(x) = -\infty$. This assumption is equivalent to the infinite square-well potential model that is often used in quantum mechanics. The approximation of infinite absorption outside the gain region as infinite is not accurate for the lateral-waveguide structure of most diode-laser arrays, though it can provide an intuitive explanation of some array mode behavior.

In most laser arrays, the absorption outside the gain region is low enough so that the optical field can extend many wavelengths into the unpumped regions at the ends of the array. Since this field penetration is neglected in the infinite absorption approximation, significant errors can occur in calculating the gains of the higher order modes [4.39]. The effect of the infinite absorption approximation outside the gain region is to introduce a significant, but artificial loss guide, so all modes are necessarily confined completely within the lateral extent of the gain region. This is clear from the unperturbed modes which in this case are given by

$$E_m^{(0)}(x) = \frac{1}{\sqrt{x_0}}\sin\left(\frac{m\pi x}{2x_0} + \frac{m\pi}{2}\right) \quad \text{for } |x| \leq x_0,$$

$$E_m^{(0)}(x) = 0 \quad \text{for } |x| > x_0.$$

$$(4.20)$$

The mode propagation constants corresponding to these eigenfuctions are given by,

$$\text{Re}\left\{\beta_0\left(m\right)\right\} = k\sqrt{n_0^2 - \frac{m^2\pi^2}{4k^2x_0^2}} \quad , \tag{4.21}$$

and

$$\text{Im}\left\{\beta_0\left(m\right)\right\} = \frac{kn_0g_0}{\text{Re}\left[\beta_0\left(m\right)\right]} \quad , \tag{4.22}$$

The pertubation corresponding to the periodic modulation of the gain and index of the gain-guided laser array is modeled as

$$\varepsilon_{\text{pert,gain}}\left(x\right) = \left(-1\right)^N \frac{\Delta g}{2k}\left(\alpha + \text{i}\right)\cos\left(\frac{\pi Nx}{x_0}\right) = \varepsilon_{\text{gain}}\left(0\right)\cos\left(\frac{\pi Nx}{x_0}\right) \quad , \tag{4.23}$$

where α is the anti-guiding factor, Δg is the amplitude of the gain pertubation, and N is the number of gain elements. The pertubation due to heating of the active region is of the form

$$\varepsilon_{\text{pert,temp}}\left(x\right) = \frac{\Delta T}{2}\frac{dn}{dT}\cos\left(\frac{\pi x}{x_0}\right) = \varepsilon_{\text{temp}}\left(0\right)\cos\left(\frac{\pi x}{x_0}\right) \quad , \tag{4.24}$$

where ΔT is the temperature difference between the center and the ends of the array, and dn/dT is the index variation with temperature. With the gain pertubation of (4.23), and the temperature pertubation of (4.24), the first-order correction to the eigenvalues, $\beta^2\left(m\right)$, are given as

$$\begin{aligned}
\beta_1^2\left(m\right) &= 0 \text{ for } m \neq N \text{ and } m \neq 1 \;, \\
\beta_1^2\left(1\right) &= -n_0k^2\varepsilon_{\text{temp}}\left(0\right) \;, \\
\beta_1^2\left(N\right) &= \left(-1\right)^{N+1}n_0k^2\varepsilon_{\text{gain}}\left(0\right) \;.
\end{aligned} \tag{4.25}$$

Since $\varepsilon_{\text{temp}}\left(0\right)$ is real and $\varepsilon_{\text{gain}}\left(0\right)$ is complex, we see from (4.25) and the expression for the modal gain (4.19), that only the mode with $m = N$ has an enhanced gain in the first-order correction. This arises simply from the fact that the maxima in the near-field intensity of the $m = N$ mode coincide with the gain stripes. It was found that the second-order correction to $\beta^2\left(m\right)$ due to (4.23 and 4.24) could be well-approximated by

$$\begin{aligned}
\beta_2^2\left(m\right) &= \frac{x_0^2n_0^2k^4}{\pi^2N\left(N-m\right)}\varepsilon_{\text{gain}}^2\left(0\right) \text{ for } n \neq N \;, \\
\beta_2^2\left(N\right) &= 0 \;.
\end{aligned} \tag{4.26}$$

We see from (4.26) that in second order, the gain pertubation increases the modal gain for $m > N$ and decreases the modal gain for $m < N$. Note also that the temperature pertubation again has no effect on the gain discrimination. This occurs in the present model calculation because of the assumption of infinite absorption outside the gain stripe, which acts as a strong loss guide and confines all the lateral modes completely within the gain region.

The effect of the temperature rise at the center of the array increases the refractive index which also strengthens the waveguide under the current stripe by increasing the lateral-mode confinement. However in this model, the difference between the loss in the gain region and the loss outside the gain region is infinite; whereas, the thermal-induced index difference is a small finite pertubation. If we think of the lateral structure of the array in terms of the three-layer waveguide discussed in Sect. 3.1.3, then we see that the lateral mode confinement is determined almost entirely be the infinite absorption outside the active region, and the small contribution due to heating of the active region acts as a weak pertubation. In effect, the inifinite absorption assumption introduces waveguide confinement characteristics that are similar to the those associated with uniform heating in the active region with finite absorption outside the gain region.

Note that the effects of the temperature pertubation influence the secondary-lobe structure in the far-field pattern. Due to the different spatial dependencies of the gain pertubation, (4.23), and the temperature pertubation, (4.24), these pertubations do not have the same non-zero coupling coefficients. Hence each pertubation can be associated with a unique set of the unperturbed broad-area modes, (4.20), that appear in the near-field solution (4.12). In the analysis of [4.36, 4.37], (4.23) couples only $E_{2N-m}^{(0)}(x)$ into (4.12), and the temperature pertubation only couples $E_{m+2}^{(0)}(x)$ and $E_{m-2}^{(0)}(x)$ into (4.12). Since each of the modes, (4.20), comprises a sinusoid of finite spatial extent, it will produce a far-field pattern with dominant lobes at angles $\pm\theta_m = \sin^{-1}[(m\lambda/(4x_0)]$ with respect to the optic axis. Therefore, the gain pertubation, (4.23), produces secondary lobes at $\pm\theta_{2N-m}$; and the temperture pertubation, produces additional lobes at $\pm\theta_{m+2}$ and $\pm\theta_{m+2}$.

As shown by [4.36, 4.37], the lobe structure due to the temperature pertubation is sufficiently close to the dominant lobes so that interference occurs. The $E_{m+2}^{(0)}$ component interferes constructively with the main mode $E_m^{(0)}$, causing an apparent shift of the dominant lobes to an angular position between θ_m and θ_{m+2}. The $E_{m-2}^{(0)}$ component interferes destructively with $E_m^{(0)}$. This creates a secondary lobe at an angle slightly less than θ_m and adjacent to the dominant lobe. This additional side-lobe structure is consistent with experimental observations (Chap. 5). As mentioned above, the infinite absorption approximation is not completely accurate. As shown by [4.39], even though the calculated near- and far-field patterns may be insensitive to the value of the loss outside the gain region, the modal gains of the higher-order modes area greatly influenced. Furthermore, when realistic values for the loss outside the gain region area considered, thermal effects have been found to have a significant influence on the modal gain characteristics of diode-laser arrays. This is also discussed more in Chap. 5.

4.2.3 Bloch-Function Analysis of Laser Arrays

In many diode-laser array structures, the periodic lateral variation in the dielectric permittivity is actually obtained by fabricating a laterally-stratified structure. Thus, the laser array is mathematically equivalent to a one-dimensional lattice. Therefore, the Bloch-wave analysis that is used in solid-state physics has been used to find the modes of layered-optical media [4.2] and laser-array structures [4.40]-[4.42]. The periodic variation in the dielectric permittivity is analogous to the periodic potential enocuntered in solid-state physics. Note that the form of the scalar wave equation (4.10) for the modes of a laser array, is identical to the Schrödinger equation for a single electron moving through a one-dimensional potential. Although in the case of the laser array, the resulting eigenvalues will generally be complex because of gain and loss which are represented by the imaginary part of the dielectric permittivity. Another distinguishing feature is that most practical laser-array structures will have a finite extent, and so end effects can be important in determining the mode structure; whereas in solid-state physics, the crystal lattice can often be treated as having infinite extent. For sufficiently large arrays, the modal characteristics obtained from the Bloch analysis of infinite extent structures are applicable to finite structures [4.38]. Although, recent progress in the Bloch-function analysis of antiguided diode-laser arrays (Sect. 6.3) has resulted in an analytic theory of finite-extent arrays [4.43].

The Bloch-wave analysis is a very powerful tool for analyzing the modal properties of laser arrays because it is a complete description that gives closed form expressions for the modal propagation constants and near fields of the array modes. This is very useful given that the problem one is faced with when designing a laser array is to engineer a semiconductor structure that will give the desired modal characteristics. With closed form expressions, the parametric dependence of the mode discrimination can be studied in a systematic way so that array designs can be more readily optimized [1.234, 4.41, 4.42].

Consider the diode-laser structure shown in Fig. 4.1. It is comprised of periodic variation along the x-direction that alternates between permittivities of ε_1 and ε_2, where Λ is the period, and the mode propagates along the z-direction. Such a stratified structure corresponds to the effective-permittivity profile resulting from a lateral variation due to either ridges or channels with straight side walls. Bloch's theorem tells us that the modal field solutions will be plane waves that are modulated by a function that has the same periodicity as the stratified-dielectric permittivity. The field solutions, $E_K(x, z)$, for the wave equation (4.10) will have the form

$$E_K(x, z) = A(x) \exp(i\beta z) = U_K(x) \exp[i(Kx + \beta z)] , \qquad (4.27)$$

where K is the Bloch wave number, and $U_K(x)$ is the periodic function that satisfies,

$$U_K(x + \Lambda) = U_K(x) . \qquad (4.28)$$

The real part of β corresponds to the modal propagation constant, and the imaginary part of β corresponds to the array mode loss or gain, and $|U_K(x)|^2$ corresponds to the near-field intensity of the array. The phase variation of the electric field across the array structure is determined by K, the Bloch wave number. For example when $K = 0$, there is no phase variation in the near field of the laser array. This corresponds to the in-phase or $0°$ phase mode, and it is the most desired mode of operation, since it will give rise to a predominantly single-lobed far-field with the largest percentage of the power in the dominant lobe. When $K = \pi/\Lambda$, the relative phase between adjacent emitting elements of the array will be $180°$. This is the out-of-phase mode that produces a double-lobed far-field pattern. Because we are considering an infinite extent array, K will be continuous and so modes with other types of phase variations can exist. For a finite extent array, the boundary conditions at the ends of the array would impose a discrete set of allowed values for K. However, for sufficiently large arrays, end effects will be negligible, and the modal characteristics of the infinite structure is a good approximation.

From (4.27) it is seen that the array modes are designated by a spectrum of propagation constants that will be specified by both the frequency of light ω and the Bloch wave number K, so $\beta = \beta(\omega, K)$. In effect each propagation constant represents a branch of modes in $\omega - K$ space; since K must be real, $\beta(\omega, K)$ will have a band-like structure similiar to the energy bands in a solid. To completely characterize the modes of the laser array, we need to determine both $\beta(\omega, K)$ and $U_K(x)$. A general method for solving this problem has been described by *Yeh* et al. [4.2], and more recently been applied to the problem of finding all the modes of a laser-array structure by *Eliseev* et al. [4.41, 4.42]. Using the solution of *Eliseev* et al., the Bloch-function representation of the electric field in the laser array can be expressed as a superposition of an incident and reflected plane wave in each layer that satisfy (4.27 and 4.28) to give

$$
\begin{aligned}
U_m(x) \quad = \quad & \exp\left(iKn\Lambda\right)\{C_m\exp\left[ik_m(x - n\Lambda)\right] + \\
& D_m\exp\left[-ik_m(x - n\Lambda)\right]\}
\end{aligned}
\tag{4.29}
$$

where

$$
k_m^2 = \left(\frac{\omega}{c}\right)^2 \varepsilon_m - \beta^2 ,
\tag{4.30}
$$

and $m = 1$ corresponds to the layers with permittivity ε_1 when $n\Lambda - d_1 < x < d_2$ and C_1 and D_1 are given by

$$
C_1 = -k_1 \sin\left(k_2 d_2\right) + ik_2 \exp\left[i\left(K\Lambda + k_1 d_1\right)\right] - ik_2 \cos\left(k_2 d_2\right)
\tag{4.31}
$$

and

$$
D_1 = -k_1 \sin\left(k_2 d_2\right) - ik_2 \exp\left[i\left(K\Lambda - k_1 d_1\right)\right] + ik_2 \cos\left(k_2 d_2\right) .
\tag{4.32}
$$

$m = 2$ corresponds to the layers with permittivity ε_2 where $n\Lambda < x < n\Lambda + d_2$ and C_2 and D_2 are given by

$$C_2 = \exp\left(iK\Lambda\right)\left[ik_1\cos\left(k_1d_1\right) - k_2\sin\left(k_1d_1\right)\right] - ik_1\exp\left(-ik_2d_2\right) \quad (4.33)$$

and

$$D_2 = -\exp\left(iK\Lambda\right)\left[ik_1\cos\left(k_1d_1\right) + k_2\sin\left(k_1d_1\right)\right] + ik_1\exp\left(ik_2d_2\right) . \quad (4.34)$$

The dispersion relation that relates K, β, and ω is found by matching the electromagnetic field boundary conditions at the layer interfaces, and it is expressed for TE polarized modes by

$$K\left(\omega,\beta\right) = \frac{1}{\Lambda}\cos^{-1}\left[\cos\left(k_1d_1\right)\cos\left(k_2d_2\right) - \right.$$
$$\left.\left(\frac{1}{2}\right)\left(\frac{k_1}{k_2} + \frac{k_2}{k_1}\right)\sin\left(k_1d_1\right)\sin\left(k_2d_2\right)\right] \quad (4.35)$$

and for TM polarized modes by

$$K\left(\omega,\beta\right) = \frac{1}{\Lambda}\cos^{-1}\left[\cos\left(k_1d_1\right)\cos\left(k_2d_2\right) - \right.$$
$$\left.\left(\frac{1}{2}\right)\left(\frac{\varepsilon_2 k_1}{\varepsilon_1 k_2} + \frac{\varepsilon_1 k_2}{\varepsilon_2 k_1}\right)\sin\left(k_1d_1\right)\sin\left(k_2d_2\right)\right] . \quad (4.36)$$

Note the close similarity between the dispersion relations for the TE and TM modes. The only difference being the $\varepsilon_2/\varepsilon_1$ and $\varepsilon_1/\varepsilon_2$ factors that appear in the second term of the TM dispersion relation (4.36). For most diode-laser array structures, the maximum variation in the permitivitty is $|\varepsilon_2 - \varepsilon_1| \approx 0.01$, so both TE and TM modes will to a good approximation satsify the same dispersion relations. In order for K to be real, the argument of the inverse cosine function in (4.35 or 4.36) must be less than one. Otherwise, K would be complex and the Bloch wave would be damped. Therefore, Bloch wave solutions will not exist for all combinations of ω, β, d_1 and d_2 that one might encounter for a given structure. Note that k_1 and k_2 are directly related to ω and β through (4.30). Just as in the case of a crystal lattice in solid-state physics, there will be a band structure in the allowed modes of a laser-array structure with bandedges occuring when $K\Lambda = N\pi$, with N being an integer (see for example [4.38]).

Eliseev et al. have applied the Bloch-wave analysis to a periodic structure that is representative of many laterally-coupled diode-laser array designs. It is worthwhile to review this particular example of Bloch-wave analysis because it describes all types of modes, including evanescently-coupled and leaky-wave coupled modes, that can occur in a laser array. This will be very helpful in providing additional insight and an improved understanding of the mode discrimination properties of diode-laser arrays. Figure 4.3 shows a schematic diagram of the periodic variation of the dielectric profile. The laser array is modeled as an infinite structure of alternating $3\,\mu m$ wide layers of real permittivities $\varepsilon_1 = 12.45$, corresponding to a refractive index $n_1 = 3.5285$, and $\varepsilon_1 = 12.25$, corresponding to a refractive index $n_1 = 3.5$.

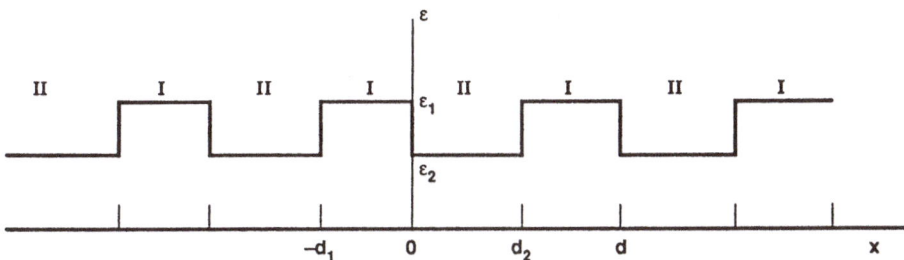

Fig. 4.3. Diagramatic representation of periodic-array structure that was analyzed by [4.41, 4.42].

In Figure 4.4 six mode branches for the array are represented by plotting the real part of the normalized modal propagation constant $Re\{2\pi\beta(K)/\lambda\}$ versus $K\Lambda/\pi$, the normalized Bloch wave number. Recall that the real part of the normalized modal propagation constant is equal to the mode effective index. Also, the Bloch wave number K is a continuous variable for the infinite extent array that is the subject of this analysis. In the case of a finite array, K could only take on take on discrete values. There are four branches where $n_2 < Re\{2\pi\beta(K)/\lambda\} < n_1$, labeled as $m = 1$, 2, 3, and 4; and two branches where $Re\{2\pi\beta(K)/\lambda\} < n_2$, labeled by $m = -1$ and -2. The significance of the labeling can be understood by looking at the corresponding near-field intensity pattern $|U_K(x)|^2$.

In Fig. 4.5 the near-field intensities at $K = 0$ and $K = \pi/\Lambda$ have been plotted for several selected modes, indicated by the mode index m. Those modes that are labeled with positive indices $m = 1$, 2, 3, and 4 have $m - 1$ nulls in the higher-index layers; and those modes labeled by negative indices $m = -1$, have nulls in the lower-index layers. The modes with $m = 1 - 3$ and $K = 0$, which correspond to the highest values of the mode effective index, all have most of the intensity confined to the higher-index wavguide layers and show no dispersion in $Re\{2\pi\beta(K)/\lambda\}$ versus $K\Lambda/\pi$, as can be seen from Fig. 4.4. These mode branches correspond to superpositions of the eigenmodes of the individual waveguide regions. This is the type of mode solution that is typically found when using coupled-mode theory to analyze the modes of an array structure. The $m = 1$, 2, 3, and 4 mode branches correspond to the modes that are found by using the coupled-mode theory that was discussed in Sect 4.2.2. In the lower-index region between the waveguides, the field falls off exponentially and is very nearly zero at the center of each lower-index layer. For the $m = 1$ mode, the field-intensity profile in each higher-index layer corresponds to the fundamental mode of the single high-index waveguide layer; whereas the intensity profiles of the $m = 2, 3,$ and 4 modes all correspond to higher-order confined modes of each high-index waveguide layer. These are the evanescently coupled modes that motivated much of the early work in diode-laser arrays.

The lack of K dependence in $Re\{2\pi\beta(K)/\lambda\}$ for the $m = 1$, 2, and 3 mode branches indicates that there is very little interaction or coupling

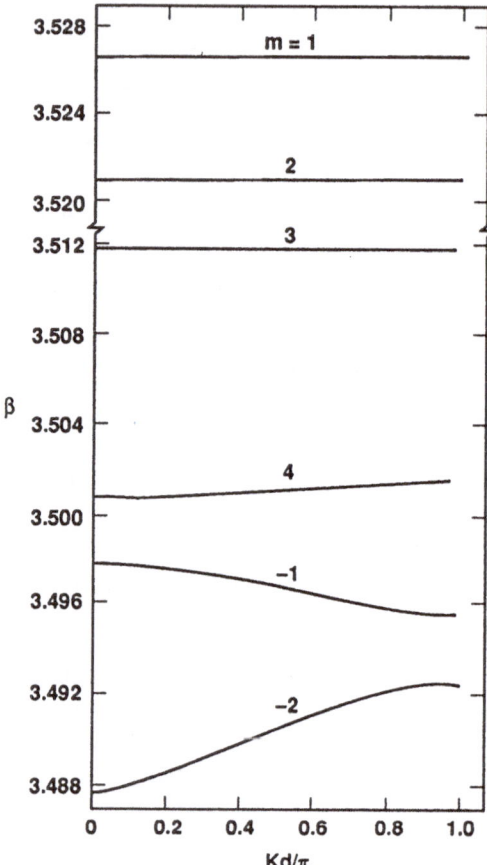

Fig. 4.4. The real part of $\beta(K)$ is plotted as a function of $K\Lambda/\pi$, [4.41, 4.42].

of the optical fields between high-index waveguide regions. Hence, the near-field intensity is also independent of K, and all modes on these branches have identical intensity profiles independent of the phase profile across the array. However, this is not the case for the $m = 4$ mode branch. Of all the positive integer-mode branches, the $m = 4$ branch has the lowest optical-field confinement in the higher-index waveguide layers; and the intensity profile in the lower-index layers does not fall off exponentially at $K = 0$, as can be seen from Fig. 4.5. In addition, we see from Fig. 4.4 that the dispersion in Re $\{2\pi\beta(K)/\lambda\}$ increases slightly with increasing K. This increase in the mode effective index with K, is accompanied by a change in the near-field intensity profile in the lower-index region; and when $K = \pi/\Lambda$, the near field profile in the lower-index regions is attenuated. This behavior is consistent with the positive dispersion of Re $\{2\pi\beta(K)/\lambda\}$, since the near-field profile indicates that the optical-field confinement in the high-index waveguide layer is larger at $K = \pi/\Lambda$ than at $K = 0$. Note that the $m = 4$ mode branch

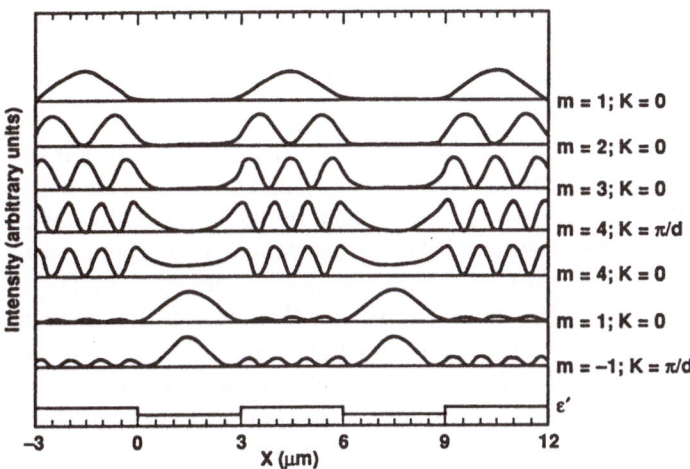

Fig. 4.5. Near-field intensities for Bloch modes for the periodic array structure shown in Fig. 4.3 [4.44].

has the weakest field confinement in the high-index waveguide layer, the field extends further into the adjacent lower-index layers so that there is a greater interaction of the optical fields in these layers. As K is varied, and the relative phase of the over-lapping fields changes and interference effects occur that change the near-field profile and give rise to the dispersion in $\mathrm{Re}\left\{2\pi\beta\left(K\right)/\lambda\right\}$.

The modes that are labeled with negative integers show stronger dispersion in $\mathrm{Re}\left\{2\pi\beta\left(K\right)/\lambda\right\}$ than the positive integer modes. As mentioned above, for these modes $\mathrm{Re}\left\{2\pi\beta\left(K\right)/\lambda\right\} < n_2$, so that there are no confined mode solutions for the high-index waveguide layers. Therefore, these mode branches will have a sinusoidal type of dependence for the optical field in the low-index waveguide layers, as opposed to the exponentially-attenuated field dependence for the positive integer m branches. It is seen from near-field profiles (Fig. 4.5) at $K = 0$ and $K = \pi/\Lambda$ that the near-field intensity in both the high and low index layers is dependent on K.

Perhaps the most important property of a laser-array structure is the mode discrimination. For the Bloch analysis we have modeled the array as a passive structure, so it will be necessary to determine the relative losses of the array modes, which are given by $\mathrm{Im}\left\{\beta\left(K\right)\right\}$. In the Bloch wave analysis, mode discrimination can be determined by doing a stability analysis of the Bloch solution with respect to small variations in the imaginary parts of the dielectric permittivities of the layers. Consider the dispersion relation (4.35) for $K\left(\omega,\beta\right)$. The introduction of gain, in the form of injected carriers, can be treated phenomenologically by making incremental changes to ε_1 and ε_2 in (4.35). Such small changes in the permittivities will cause small changes in K and β that, to first order, are related by

$$\delta K = \left(\frac{\partial K}{\partial \varepsilon_1}\right) \delta \varepsilon_1 + \left(\frac{\partial K}{\partial \varepsilon_2}\right) \delta \varepsilon_2 + \left(\frac{\partial K}{\partial \beta}\right) \delta \beta \ , \tag{4.37}$$

where the functional form of K to be used in calculating the derivatives in (4.37) is given by (4.35). In order for the Bloch-wave solution to remain as such when small changes are made in ε_1 and ε_2, δK must be real. Therefore, the imaginary part of (4.37) must be equal to zero, and this can be used to derive the relation between the modal gain change $\delta \beta''$, and the gain changes $\delta \varepsilon_1''$ and $\delta \varepsilon_2''$, where $\varepsilon_1'' = \text{Im} \{\varepsilon_1\}$ and $\varepsilon_2'' = \text{Im} \{\varepsilon_2\}$. Note that $\beta'' = \text{Im} \{\beta(K)\}$. The resulting incremental change in the mode propagation constant $\delta \beta''$ gives the growth or decay rate of the array mode as it propagates along the longitudinal or z-direction of the laser-array structure. The sensitivity of the modal gain to local gain changes that are specific to either the low- or high-index layers can be determined by calculating the differential gain $\delta \beta''/\delta \varepsilon_j''$ which is given by

$$\frac{\delta \beta''}{\delta \varepsilon_1''} = \frac{Ak}{2k_1^2 \beta \left(A/k_1^2 + B/k_2^2\right)} \tag{4.38}$$

for the high-index layer where $j = 1$; and for the lower-index layer where $j = 2$

$$\frac{\delta \beta''}{\delta \varepsilon_2''} = \frac{Bk}{2k_2^2 \beta \left(A/k_1^2 + B/k_2^2\right)} \ . \tag{4.39}$$

The parameter A is given by

$$\begin{aligned} A = \ & k_1 d_1 \left[- \sin(k_1 d_1) \cos(k_2 d_2) - \frac{1}{2}\left(\frac{k_1}{k_2} + \frac{k_2}{k_1}\right) \cos(k_1 d_1) \sin(k_2 d_2) - \right. \\ & \left. \frac{1}{2}\left(\frac{k_1}{k_2} - \frac{k_2}{k_1}\right) \sin(k_1 d_1) \sin(k_2 d_2)\right] \ , \end{aligned} \tag{4.40}$$

and B is given by

$$\begin{aligned} B = \ & k_2 d_2 \left[- \cos(k_1 d_1) \sin(k_2 d_2) - \frac{1}{2}\left(\frac{k_1}{k_2} + \frac{k_2}{k_1}\right) \sin(k_1 d_1) \cos(k_2 d_2) - \right. \\ & \left. \frac{1}{2}\left(\frac{k_2}{k_1} - \frac{k_1}{k_2}\right) \sin(k_1 d_1) \sin(k_2 d_2)\right] \ . \end{aligned} \tag{4.41}$$

Equations (4.38 and 4.39) represent a very powerful result that can be used to analyze the sensitivity of the mode discrimination to local gain changes, changes in the layer thicknesses, and changes in the refractive indices of the layers. Furthermore, the dispersion, or K dependence, of the differential gain corresponds to the mode discrimination for different near-field phase distributions in a given mode branch.

The dispersion of the differential gain corresponding to the structure shown in Fig. 4.3 has also been calculated by *Eliseev* et al., and this result is displayed in Fig. 4.6. The solid lines correspond to the case where

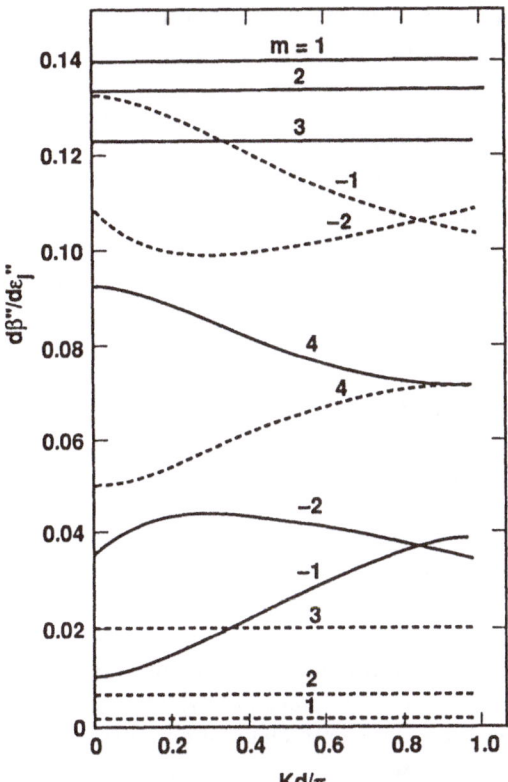

Fig. 4.6. The differential gain $\frac{\delta\beta''}{\delta\varepsilon_1''}$ (solid lines) and $\frac{\delta\beta''}{\delta\varepsilon_2''}$ (dotted lines) is plotted as a function of $K\Lambda/\pi$. The periodic array structure is that depicted in Fig. 4.3 [4.41, 4.42].

gain is added only to the high-index layers which is given by (4.38), and the dotted lines correspond to the case where gain is added only in the low-index layers which is given by (4.39). When gain is added only to the high-index layers, the $m = 1$ branch will experience the highest differential gain, with the $m = 2$ and 3 being the closest lying branches, and the $m = -1$ and -2 branches having the lowest gains. Conversely, when gain is added to only the low-index layers, the $m = -1$ branch has the highest mode discrimination at $K\Lambda/\pi = 0$. The gain of the $m = -1$ branch decreases monotonically out to about $K\Lambda/\pi = 0.8$. For $K\Lambda/\pi > 0.8$, the $m = -2$ branch then has the highest gain, and closest lying branch is the $m = -1$, with the $m = 1 - 3$ branches having the lowest-gain values.

The $m = 1$, 2 and 3 mode branches, which are evanescently-coupled modes, are seen to have no dispersion regardless of where the local gain is added in this infinite array model. Therefore, these mode branches offer no mode discrimination against different near-field phase distributions, and those regions where gain is present will tend to oscillate independently of each other. Such modes of operation will produce a low-coherence output beam with wide beam divergence that is characteristic of a single-array element. Hence, the benefit of increased brightness is not possible in an array that operates on these mode branches.

In order for the array to operate in a mode with a unique near-field phase distribution, the differential gain vs K must have a maximum; and for optimum beam quality, the highest gain should occur only at $K = 0$, which corresponds to a constant phase front across the near field of the array. For the example under consideration, this is realized at $K = 0$ on the $m = -1$ branch only when gain is added to the low-index layer. Interestingly, when the array is uniformly pumped the total differential gain is obtained by summing (4.38 and 4.39); and for all branches depicted in Fig. 4.6, this results in a total differential gain that is independent of K. This is an important result because it illustrates that single-lateral mode operation of a very large laser-array comprised of a periodic variation of the dielectric permittivity cannot be obtained unless the periodic variation also includes the imaginary part of the dielectric permittivity.

Recall that a similar result was derived in Chap. 2, by calculating the lateral mode confinement factor of an infinite array and using translational invariance. The more detailed Bloch-function analysis gives us the additional condition that *the gain should be placed in the regions where the refractive index is lower*. In this case, each gain element supports leaky-wave modes that radiate light in the lateral plane of the array. Laser arrays that are based on leaky-wave coupling, provide strong uniform optical coupling between all elements in the array, and they are robust to carrier and thermally-induced waveguide effects [1.443]. In order to gain further insight into how to discriminate against in-phase or out-of-phase mode operation of leaky-mode coupled laser arrays from the Bloch-function analysis, it is necessary to consider laser-array structures where $k_m d_m$ for $m = 1$, 2 is an integer multiple of π [4.45]. In this case, the field solutions (4.27) are replaced with an appropriate superpositon of sines and cosines [4.45], otherwise the eigenvalue equations (4.35 and 4.36) have a trival solution. Such array structures, refered to as *resonance* or *resonant-mode* structures, are necessary if in-phase operation of a diode-laser array is desired, as discussed in Sect. 6.3. In finite-extent arrays, the end-losses, which were neglected above, can have an important influence on the modal characteristics and discrimination properties. However, Bloch-function analyses of finite-extent arrays have been recently reported [4.42, 4.43].

4.2.4 Two-Dimensional Waveguide Model of Laser Arrays

In many diode-laser arrays, the effective-index method discussed in Sect. 4.2.1 will give an accurate description of the array characteristics. There are, however, some types of arrays where this one-dimensional description is not ad-

equate. This can occur in lateral waveguides, where the index difference between the effective guide and cladding layers is sufficently large or when
the transverse structure of the effective lateral cladding layer is below cutoff
[4.46], which is less common in diode-laser arrays. Laser array structures or
waveguide structures that have lateral dielectric permitivities which contain
large index steps, $\geq 10^{-2}$ or where there is more than one transverse mode
which can be supported by the lateral cladding layers [1.235, 4.12, 4.13], violate the approximation that the electric field can be written as a product
of the lateral mode profile $E_x(x,z)$ and a transverse mode profile $F(x,y)$
which depends very weakly on the lateral coordinate x. The antiguide array
which operates on the principle of leaky-wave coupling is an example of one
such structure [1.229, 1.235, 4.47]. For this type of array, the effective-index
approximation generally does not give correct predictions for either the near-
field intensity patterns or the resonance characteristics of the array structure.
This reasons for this are described in more detail in Sect. 6.3.

Two-dimensional arrays of phase-locked *Vertical-Cavity Surface Emitting
Laser* (VCSEL) arrays cannot be analyzed properly using the effective-index
method, because these structures are distributed over a plane, as opposed
to the linear-geometrical arrangement of most edge-emitting diode-laser arrays. Fortunately, for most two-dimensional VCSEL structures the modes
are well approximated by a product of eigenfunctions, each of which is the
solution to a one-dimensional wave equation [4.48]. In general, models that
separate the two-dimensional array problem into two one-dimensional problems are computationally very efficient. In [4.49, 4.50], *Amantea* et al. used a
network-model approach to derive analytic expressions that can be used for
calculating all of the eigenvalues of a two-dimensional array structure that
can be separated into a pair of effective one-dimensional structures.

Basically, in laser-array structures where the optical coupling between array elements occurs in two dimensions, the transverse mode profile will be
strongly-coupled to the lateral position. It is then necessary to use the two-
dimensional Helmholtz equation to analyze the modal properties of these
array structures. If we look for solutions of the form $E(x,y)[\exp i(\beta z - \omega t)]$
in solving the three-dimensional wave equation (4.6), then we find that the
mode profile $E(x,y)$ in the plane perpendicular to the direction of propagation satisfies the two-dimensional Helmholtz equation

$$\frac{\partial^2 E}{\partial x^2} + \frac{\partial^2 E}{\partial y^2} = [\beta - \varepsilon(x,y)]\,E \ , \tag{4.42}$$

where β is the complex mode propagation constant and $\varepsilon(x,y)$ is the dielectric permittivity in the $x - y$ plane. *Hadley* [4.47] has done a numerical
analysis of the periodic infinte array of buried-ridge waveguide lasers that is
displayed in Fig. 4.7. This structure is representative of the type of transverse
structure that could correspond to the effective one-dimensional lateral structure in Fig. 4.3. In the following discussion, we will see how for this example,
the two-dimensional calculation of the array in Fig. 4.7 reveals details of the

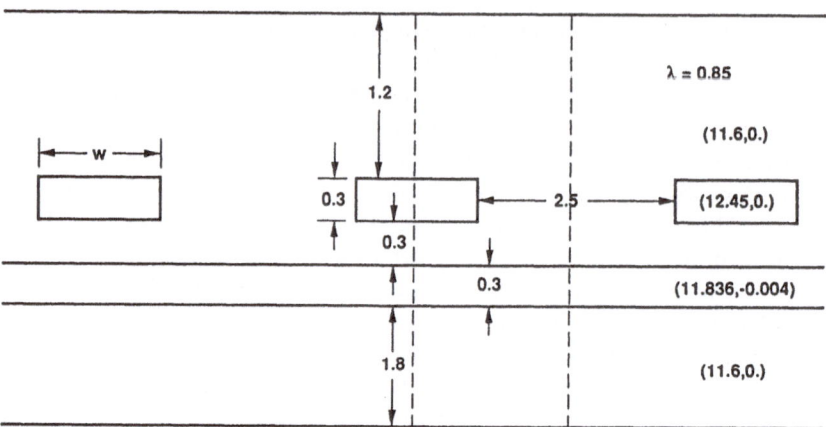

Fig. 4.7. Cross-sectional drawing of the infinite periodic buried-ridge waveguide array giving the dimensions, and in parentheses, the complex dielectric constants that were used in the analysis of *Hadley* [4.47].

near field that cannot be modeled using the sort of effective one-dimensional calculation that was discussed in Sect. 4.2.3.

Since the array is assumed to have infinite extent, symmetric boundary conditions are applied. Then it is only necessary to calculate the electric field in the region indicated between the dashed lines in Fig. 4.7. *Hadley* calculated some of the even and odd modes of the structure in Fig. 4.7 for several different widths w of the rectangular guide regions. The results of the near-field calculation are presented in Fig. 4.8. Note that the near-field of the leaky-wave even mode with $w = 2.4\,\mu m$ in Fig. 4.8 bears a qualitative similarity to the $m = -1, K = 0$ mode in the Bloch analysis of the effective one-dimensional array, that appears in Fig. 4.5. Both modes exhibit an intensity maximum in the lower-index region, and three secondary maxima in the higher-index region. Though in the two-dimensional structure, the secondary maxima are displaced in the transverse direction, and contained within the high-index regions. The transverse translation of the intensity maxima in the higher- and lower-index regions is characteristic of all the modes in Fig. 4.8. Note that the effective-index method, being a one-dimensional model, calculates a single lateral near-field intensity profile that is assumed to vary slowly with transverse displacement. A further consequence of this limitation of the effective-index method, is that the calculated confinement factors and gains of the array modes are likely to be inaccurate. In [4.47], *Hadley* gave a quantitative example of this by comparing the confinement factors and modal gains of the array in Fig. 4.7 calculated from the effective-index method and the two-dimensional Helmholtz equation. However, for some array structures with large-index steps, where the loss is greater in the high-index region, the transverse field structure in some of the modes is much less pronounced [1.229]. In this case, the effective-index method should be more accurate, pro-

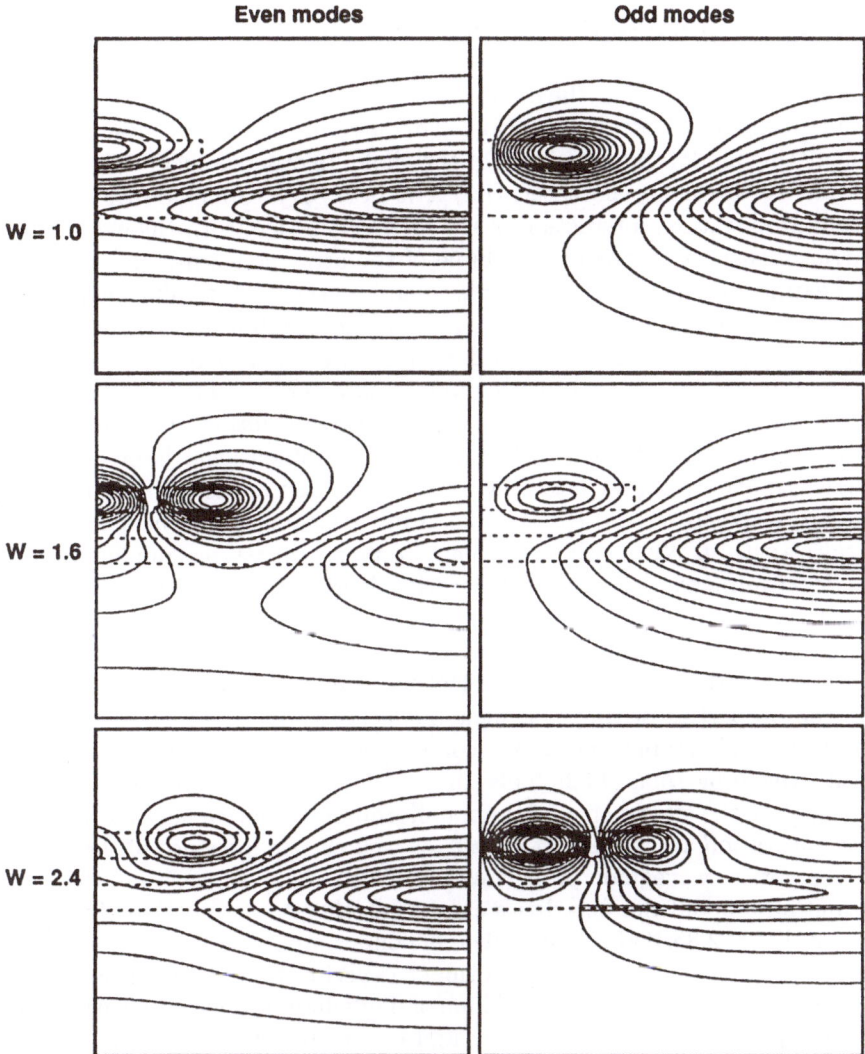

Fig. 4.8. Intensity-contour drawing of the even and odd mode near fields for the indicated guide widths of the infinite array structure appearing in Fig. 4.7 [4.47].

vided that only one-transverse mode is supported in the high-index region. In cases where the high-index region supports more than one transverse mode, the analysis must account for multiple effective-indices [4.51] in the high-index regions to correctly model the leaky-wave coupling. For this reason, it is necessary to use the full two-dimensional waveguide analysis.

4.3 Self-Consistent Models

In the previous sections, the diode-laser array was treated as a fixed dielectric structure in the lateral dimension. This sort of analysis is most useful for calculating modal characteristics at threshold. In order to accurately calculate the above-threshold characteristics, it is necessary to include the effects of beam propagation, interaction of the injected carriers with the electric field, carrier transport through the volume of the array structure, and the effects of heating on the dielectric profile. These effects can significantly modify the optical waveguide properties, and therefore, influence the performance characteristics of a laser-array under high-power operation.

A comprehensive analysis of modal characteristics of diode-laser array structures requires a simultaneous, self-consistent solution to Maxwell's wave equation for the optical field inside the three-dimensional laser cavity, the carrier-transport equation for the current flow and carrier diffusion through the array structure, and the thermal-transport equation for heat flow in the array structure. With such a rigorous treatment, the strong coupling between the local-refractive index and the carrier density, and the effects of heating on the lateral waveguide structure can be accurately modeled. Regardless of the array structure, this sort of analysis is computationally very intensive.

The importance of using numerical self-consistent modes for understanding the properties of single-element diode lasers operating above threshold has been well established, and a comprehensive review of this subject has been written by *Buus* [4.52]. This sort of self-consistent model has been used to calculate the effects of spatial nonuniformities of the carrier density due to gain saturation and transport effects, as well as, the influence of the carriers on the optical waveguide properties [2.85, 4.15, 4.53]. In addition, the power output and modal characteristics can be calculated as a function of the drive current [2.28, 2.86, 4.15, 4.54]. This is of course important for optimizing laser structures to achieve desired performance specifications for applications.

Self-consistent numerical modeling of diode-laser arrays has been shown to be indespensible for understanding and optimizing the operating characterstics of laser arrays operating at high output-power levels [4.18],[4.55]-[4.61]. The presence of injected-carriers modifies the cold-cavity lateral profile of the dielectric permittivity through the index-gain coupling and the carrier contribution to the dielectric properties. Nonuniform saturation of the lateral gain profile causes spatial-hole burning in the gain and a corresponding modification of the optical waveguide [4.62]. Nonuniform heating of the array also modifies the optical waveguide characteristics [4.18, 4.59, 4.60]. Gain saturation and heating become significant for operation at high power, where the combined effects degrade the mode discrimination and cause multi-mode operation. In reviewing the self-consistent diode-laser array models, we shall follow the approach of *Hadley* et al. [4.18], which treated both carrier and thermal transport in a self-consistent manner. We will begin by reviewing the model for the electric field propagation and then follow with the carrier transport and thermal models.

4.3.1 Electric-Field Propagation Model

Figure 4.9 is a drawing of a typical diode-laser array structure. In modeling the waveguide properties of this type of structure, the effective-index approximation, discussed in Sect. 4.2.1, is invoked to eliminate the field dependence in the y direction (perpendicular to the pn junction) at each position x along the lateral direction. This gives the effective dielectric permittivity appearing in the one-dimensional paraxial wave equation (4.10) that is used to calculate the electromagnetic propagation of the lateral modes of the array.

The effects of injected carriers and heating on the lateral modes of the laser-array are accounted for phenomenologically by including the carrier and temperature dependencies of the lateral-dielectric permittivity. Therefore, the array waveguide structure becomes a function of the operating conditions. *Hadley* has used the following form for the carrier and temperature dependent lateral-dielectric permittivity [4.18]

$$\varepsilon_{\text{eff}}\left(y, z\right) = \overline{\varepsilon_{\text{eff}}}\left(x, z\right) + \Gamma_y \Delta \varepsilon_a \left(x, z\right) + \Delta \varepsilon_T \left(x, z\right) \ , \qquad (4.43)$$

The first term in (4.43), $\overline{\varepsilon_{\text{eff}}}\left(x, z\right)$, is the effective dielectric permittivity (4.9) when no carriers are injected into the structure. The second term in (4.43) represents the contribution of the injected carriers to the lateral dielectric permittivity, where $\Delta \varepsilon_a \left(x, z\right)$ is modeled as

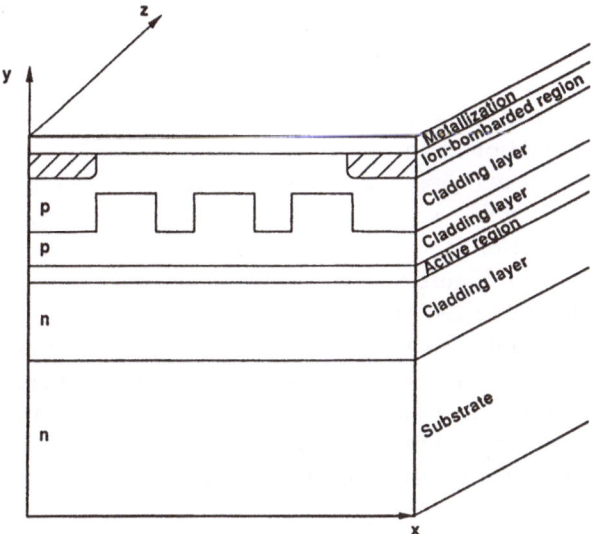

Fig. 4.9. A drawing that displays the epilayer structure of a typical diode-laser array. [4.18]

$$\Delta\varepsilon_a\left(x,z\right) = -\frac{\sigma\alpha\beta N\left(x,z\right)}{k^2} - \frac{i\beta}{k^2}\left[\sigma N\left(x,z\right) - \sigma N_{tr} - \alpha_{\mathrm{fc}}N\left(x,z\right)\right] \quad, \quad (4.44)$$

and $N\left(x,z\right)$ is the carrier density, σ and N_{tr} are the parameters used in the linear power gain equation (2.36), α is the antiguiding factor that was introduced in Sect. 2.7, and α_{fc} is the free-carrier absorption coefficient that was discussed in Sect. 2.6. The third term in (4.43) accounts for the thermal effects on the lateral profile of the dielectric permittivity, and $\Delta\varepsilon_T\left(x,z\right)$ is given by

$$\Delta\varepsilon_T\left(x,z\right) = 2\frac{\beta}{k}\frac{dn}{dT}\left[T\left(x,z\right) - T\left(0,z\right)\right] \quad, \quad\quad (4.45)$$

where dn/dT is the first derivative of the effective index of refraction with respect to temperature and $T\left(x,z\right)$ is the temperature profile, and $T\left(0,z\right)$ is the axial temperature profile. Note that in (4.45) the effects of heating on the gain have been neglected, because in most cases of practical interest this has been found to be a small effect. Equations (4.43-4.45) show how the carrier and temperature profiles in the active region are coupled to the electromagnetic field propagation in the array. The spatial dependencies of the carrier density $N\left(x,z\right)$ and the temperature profile $T\left(x,z\right)$ are determined by the carrier and thermal transport models described in Sects. 4.3.2 and 4.3.3. Although the cold-cavity dielectric permittivity has no z dependence, note that when injected carriers are present, the effective lateral dielectric permittivity can have a z dependence. It then becomes necessary to use the paraxial wave equation, to accurately model the characteristics of diode-laser arrays operating well-above threshold.

4.3.2 Carrier-Transport Model

Carrier transport is the process whereby holes injected from the p-metallization flow through the epilayer structure and undergo either radiative or nonradiative recombination with electrons in the active region. In the active region, the electron and hole transport is characterized by diffusion in the lateral direction. The first step in modeling the carrier transport is to model the flow of current from the p-electrode through the p-cladding layers and into the active region. Figure 4.10 illustrates the model of the array structure that is used to model carrier transport. The most widely-used approach is to assume that the dominant contribution to the voltage drop occurs between the p-contact and the active region, so that carrier transport can be neglected in the n-cladding layers and the substrate. Another common approximation is to assume a constant resistivity in the p-cladding layers. For a constant resistivity in the p-cladding, the two-dimensional Laplace's equation is expressed as

$$\frac{\partial^2\phi\left(x,y\right)}{\partial x^2} + \frac{\partial^2\phi\left(x,y\right)}{\partial y^2} = 0 \quad, \quad\quad (4.46)$$

where ϕ is the potential in the p-cladding layer of the array structure. In order to accurately account for axial variations in the carrier density, Laplace's

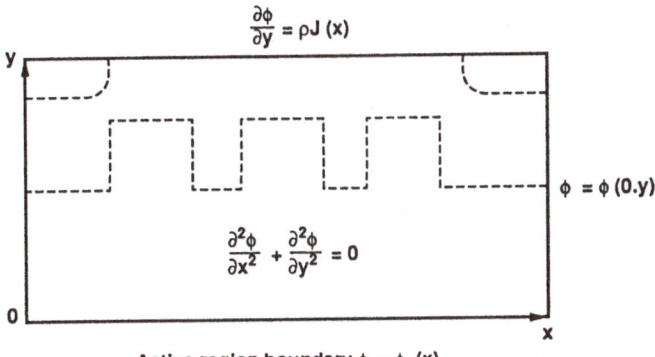

$$\frac{\partial \phi}{\partial y} = \rho J(x)$$

$$\phi = \phi (0.y)$$

$$\frac{\partial^2 \phi}{\partial x^2} + \frac{\partial^2 \phi}{\partial y^2} = 0$$

Active region boundary $\phi = \phi_f (x)$

Fig. 4.10. A cross-sectional drawing illustrating the structure used to model carrier transport in diode-laser arrays. The boundary between the p-cladding and the active region is along the x axis [4.18].

equation is solved at each z position after propagation of the optical wave according to (4.10). Note for some diode-laser structures, such as those that use a zinc diffusion in the current stripe, the assumption of a constant conductivity in the p-clad layers is not always an accurate approximation, as shown by *Papannareddy* et al. [2.85]. In these cases, it would be more accurate to use the general form for Laplace's equation, $\nabla [\sigma(x, y) \nabla \phi(x, y)] = 0$, where $\sigma(x, y)$ is the conductivity distribution in the p-clad layer.

The boundary conditions that have been employed in solving (4.46) for the potential $\phi(x, y)$ are indicated in Fig. 4.10. The current density, $J(x, y)$, is specified at the p-electrode; and at the boundary with the active region, the potential is equal to the Fermi voltage, $\phi_f(x)$, which is approximated as [2.83]

$$\phi(x, y = 0) = \phi_f(x) = \frac{k_b T}{q} \left[2 \ln \left(\frac{N}{\sqrt{N_c N_v}} \right) + A_\ell N \left(\frac{1}{N_c} + \frac{1}{N_v} \right) \right] , \tag{4.47}$$

where N_c and N_v are the density of states in the conduction and valence bands, respectively. The term that is linear in N is the first-order correction due to the Fermi-Dirac statsitics which govern the electron energy distribution in the active region. Here A_1 is a constant associated with the expansion of the Fermi-Dirac integral [2.82]. For parabolic band structure $A_1 = 1/\sqrt{8}$. The current density that is input to the active region, $J(x, y = 0)$, is calculated as

$$J(x, y = 0) = \frac{1}{\rho} \frac{\partial \phi}{\partial y} , \tag{4.48}$$

where ρ is the resistivity of p-clad layer. The current density in the active region specified by (4.48) acts as a source term in the lateral diffusion equation

$$D \frac{\partial^2 N}{\partial x^2} = -\frac{J(x, y = o)}{qd} + \frac{N}{\tau_{nr}} + BN^2 + CN^3 + \frac{g}{\hbar \omega d} \sum_n |E_n|^2 , \tag{4.49}$$

where D is the effective diffusion constant for the carriers, d is the active-layer thickness, τ_{nr} is the nonradiative recombination time, B is the spontaneous emission coefficient, C is the Auger recombination coefficient, g is the modal gain given by (2.36), and ω is the angular frequency of the photons. The first term in (4.49) is the input current density to the active region. The second term represents nonradiative carrier recombination, the third term accounts for carrier recombination resulting in spontaneous emission, the fourth term is due to Auger recombination (a nonradiative recombination process that is only important in long-wavelength lasers [1.324]), and the last term represents the stimulated recombination of carriers. The summation in the last term of (4.49) is taken over modes that are above threshold. For each mode there is a forward and backward wave when one or both ends of the cavity are terminated by partial reflectors. In single-mode operation, there will only be one forward and backward traveling wave. Note that (4.49) is also used in analyzing traveling-wave amplifiers.

Equations (4.46 and 4.49) are coupled by the boundary conditions (4.47 and 4.48) at the p-clad/active region interface. Therefore, these equations must be solved in an iterative manner, since any change in N results in a change in the Fermi voltage, ϕ_f, which inturn, produces a change in $J(x, y = 0)$, which is the source term in the active layer for the carrier diffusion equation. The analysis can be simplified if the voltage along the active layer can be set to zero. In this case, there is an analytic solution for the $J(x, y = 0)$ that can be used for multiple-stripe diode-laser geometries with homogeneous current injection across the p-electrode [4.63]. However, as shown by *Hadley* [4.18], including the carrier-dependent Fermi voltage correction, causes some of the current to be shunted away from the region of peak carrier density, and directed out towards the edges of the region under the current stripe. Intuitively, this makes sense based on the Pauli exclusion principle. In regions of high-carrier density, more of the available energy states are occupied, so injected carriers will be diverted to locations of lower-carrier density, where there are more unoccupied energy states. This effect is most pronouced at high-carrier density levels and in laser structures with broader lateral extent, where the carrier density in the lateral direction is more likely to be strongly modulated. Therefore, it could have an effect on the mode-discrimination characteristics of some diode-laser structures.

4.3.3 Thermal Model

The effects of heating on high-power diode-laser arrays were discussed in Sect. 2.8. Most self-consistent models of diode lasers which include thermal effects have accounted only for uniform heating of the active region. This choice has been motivated by the complexity of the calculation and partly because the primary cause of heating in most diode lasers is believed to be nonradiative recombination in the active region. However, other effects such as reabsorption of spontaneous emission and Joule heating due to ohmic losses give rise to sources of heat outside the active region [2.85]. As discussed by *Papannareddy* et al. [2.85], the importance of these other heat sources depends on the layer composition and the operating conditions. Though, in order to get high-power operation, high current levels are required, and Joule heating could become an important effect. To date, the effects of specific sources of heat distributed within diode-laser array structures does not appear to have been studied in detail.

Figure 4.11 depicts the cross-sectional view of the laser-array structure that is used for the thermal modeling. In this model, the source of heat is assumed to be only in the active region, as indicated. The spatial dependence of the heat source is determined by the carrier density N.

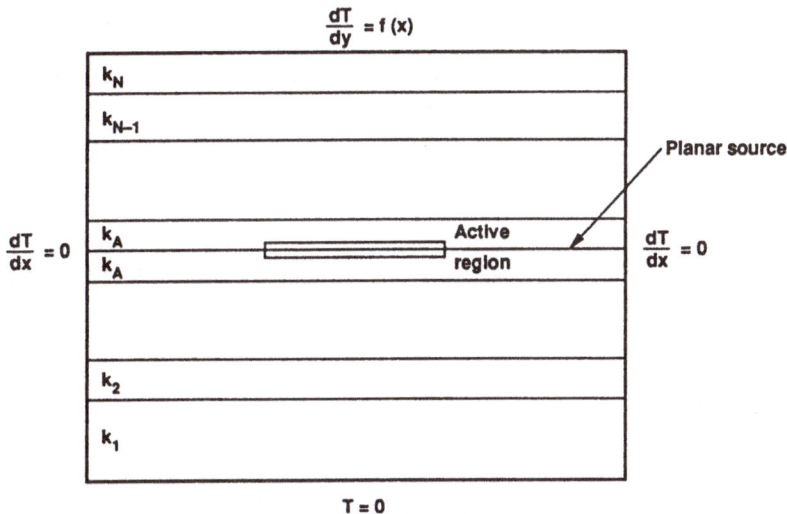

Fig. 4.11. A cross-sectional drawing illustrating the structure used to model thermal transport in diode-laser arrays [4.18].

Under steady-state conditions, thermal transport in an arbitrary two-dimensional structure, such as a diode-laser array obeys the following equation:

$$\nabla \left[K\left(x,y\right) \nabla T\left(x,y\right) \right] = -Q\left(x,y\right) \ , \tag{4.50}$$

where $K\left(x,y\right)$ is the thermal conductivity and $Q\left(x,y\right)$ is the power density of the heat source [4.64]. In general, the thermal conductivity has a spatial dependence and the heat source can be distributed continously through the structure. For the purposes of this discussion on diode-laser arrays, the heat source is restricted to the active region. In addition, within each layer of the laser stucture, the thermal conductivity is assumed to be constant. The heat sink is assumed to be fixed at $T = 0$, but provision is made for an arbitrary heat flux $f(x)$ on the p-contact. The lateral boundaries are assumed to be sufficiently far from the active region, so that there is no normal heat flux. Then the temperature profile within each layer statisfies Laplaces equation for thermal transport, which can be expressed as

$$\frac{\partial T\left(x,y\right)}{\partial x^{2}} + \frac{\partial T\left(x,y\right)}{\partial y^{2}} = 0 \ . \tag{4.51}$$

The boundary condition applied at the interfaces between the cladding layer requires continuity of the temperture and the normal heat flux. However, at the interface with the active region, which corresponds to the heat source, the following boundary condition is applied:

$$\left.\frac{\partial T\left(x,y\right)}{\partial y}\right|_{-} - \left.\frac{\partial T\left(x,y\right)}{\partial y}\right|_{+} = \frac{N\left(x,z\right)\hbar\omega d}{\tau_{nr}K_{a}} \ , \tag{4.52}$$

where K_a is the thermal conductivity of the active region. The term on the right-hand side of (4.52) relates to the heat flux that is generated in the active region due to the local carrier density. Including the spatial dependence of the heat source in the active region, will provide a more accurate simulation of thermal effects on the waveguide properties. Note that in the temperature dependent part of the lateral dielectric permittivity given by (4.45), we have $dn/dT \approx 4 \times 10^{-4}\,\mathrm{K}^{-1}$. Under high-power operation, the carrier density can exhibit a strong spatial dependence, which modifies the temperature profile and the corresponding waveguide properties.

4.3.4 Outline of Self-Consistent Numerical Procedure

In Sects. 4.3.1-4.3.3, we reviewed the optical-mode propagation model, the electronic model, and thermal model that are used in simulating the operating characteristics of diode-laser arrays above threshold. The procedure begins by selecting an appropriate trial solution for the diode-laser array mode. In many cases, the starting point is the solution to the paraxial wave equation (4.10) in the absence of carriers, and the injected current density is selected to be at or very near threshold. Starting with this trial mode, a *Fast Fourier Transform* (FFT) method is used to solve (4.10) and propagate the trial solution back and forth in the z direction, over the round-trip path in the cavity, as it evolves to a steady-state solution [4.14, 4.15, 4.18]. At each longitudinal position z_i, the carrier-transport equations (4.46)-(4.49) and thermal transport equations, (4.51 and 4.52), are solved to give $N\left(x,z_i\right)$ and $T\left(x,z_i\right)$, which are then used to correct the FFT propagator for calculating the field at the next longitudinal position, z_{i+1} (see [4.18] and references therein for details on these algorithms). In this way the quantities of interest, which are the lateral mode profile $E\left(x,z\right)$, the carrier density $N\left(x,z\right)$, and the temperature profile $T\left(x,z\right)$ are continually iterated as the mode propagates back and forth in the cavity. In order to insure accurate convergence to a steady-state, it is usually necessary to use underrelaxation techniques in the numerical evaluation, as discussed by *Hadley* [4.18, 4.65]. The power-current characteristic can be calculated by solving the self-consistent simulation for successively higher values of the injected current density. The optical power output is calculated by integrating the field

$$P_J = (1 - R_L)\, d \int E_J\,(x, z = L)\, dx \; , \tag{4.53}$$

where P_J is the power output corresponding to injected current density J, R_L is the reflectivity of the output coupler at $z = L$, and $E_J\,(x, z = L)$ is the steady-state field solution at $z = L$ of the self-consistent simulation corresponding the injected current density J. However, the self-consistent nurmerical calculation is most useful for analyzing the the effects of operating conditions on the quality of the output beam.

4.3.5 Instability of Spatial Modes

The strong tendency of filaments to form in the gain regions of diode lasers and diode-laser arrays is perhaps the primary reason that good-beam quality and mode discrimination at high-power output levels has been very difficult to achieve. Filament formation in a propagating optical beam is an inherent characteristic of laser gain media where there is a strong coupling between the local gain and referactive index [4.66], as well as passive media with a third-order nonlinearity [4.67]-[4.70]. To provide additional insight into this phenomena and its relation to the optical beam propagation and spatial-mode characteristics of diode-laser arrays, we will follow the linearized-stability analysis against filamentation of *Paxton* and *Dente* [4.66]. This analysis considers a uniform-intensity plane wave of infinite extent, as it propagates in a semiconductor gain media. A small-signal analysis of the carrier-diffusion equation (4.49) and paraxial-wave equation (4.10) for steady-state operation at the intensity of the plane wave is first done; and then, the effect of a small-sinusoidal pertubation in the lateral gain distribution is studied. Note that this method will calculate the grow rate of filaments in a sinusoidally-perturbed plane-wave. To calculate the resulting spatial mode, a full self-consistent analysis of the coupled-nonlinear equations, as described in Sect. 4.3.4, should be done. The pertubation analysis of the linearized equations, presented herein, is useful for defining the boundaries in parameter space between stable and unstable propagation of a uniform-intensity plane-wave beam in a broad-area semiconductor gain medium.

  To begin, we consider an infinte extent-plane wave, E_p, of uniform amplitude, which contains a sinusoidal pertubation in the lateral or x direction, which is expressed as,

$$E_p = (E_0 + \epsilon) \exp\,(i\delta\beta z + \gamma z) \tag{4.54}$$

where E_0 is defined as a real number such that $E_0^2 = I_0$ is the steady-state intensity of the plane wave, and ϵ is the complex pertubation that can grow or attenuate relative to the uniform steady-state field E_p. Here it is understood that the phase dependence of the steady-state field is accounted for by the $\delta\beta$ term that appears in the argument of the exponential in (4.54). Given that this is a pertubation analysis, we also have the condition that $E_0^2 \gg |\epsilon|^2$. Following [4.66], we use (2.33) to rewrite (4.49) in terms of the modal amplitude gain g as

$$s^2 \frac{d^2g}{dx^2} + g_0 - g \left(1 + \frac{|E|^2}{I_S} \right) = 0 \ . \tag{4.55}$$

where g_0 is the unsaturated gain which is expressed as

$$g_0 = \frac{\Gamma_y \sigma}{2a'(N_0)} \left[\frac{J(y=0)}{qd} + a'(N_0)(N_0 - N_{\mathrm{tr}}) - a(N_0) \right] \ , \tag{4.56}$$

s is the carrier-diffusion length given by

$$s^2 = \frac{D}{a'(N_0)} \tag{4.57}$$

and the saturation intensity I_s is represented by,

$$I_S = \frac{E_{\mathrm{photon}} a'(N_0)}{\sigma} \ , \tag{4.58}$$

where $J(y=0)$ is the source of current density input to the active region, N_0 is the carrier density corresponding to the steady-state intensity I_0, D is the carrier-diffusion coefficient, E_{photon} is the photon energy, $a(N_0)$ represents the recombination rate due to all processes except for stimulated recombination, and $a'(N_0)$ is the first derivate of $a(N_0)$ with respect to N evaluated at N_0. In general, $a(N_0)$ can be expressed as a polynomial expansion in the carrier density N_0. Specifically for (4.49), we find that

$$a(N_0) = \frac{N}{\tau_{nr}} + BN^2 + CN^3 \tag{4.59}$$

By employing (4.59) we can rewrite (4.56) as

$$g_0 = \frac{\Gamma_y \sigma}{2} \left[\eta'_i \frac{\tau_{nr} J(y=0)}{qd} + (N_0 - N_{\mathrm{tr}}) - \frac{a(N_0)}{a'(N_0)} \right] \ . \tag{4.60}$$

Upon comparing (4.60) to (2.33), we can infer an analytic expression for an internal quantum efficiency η'_i,

$$\eta'_i = \frac{1}{1 + \tau_{nr}(2BN + 3CN^2)} \ . \tag{4.61}$$

The internal quantum efficiency η' accounts for the carrier-density dependency of the spontaneous and nonradiative recombination processes that deplete the pool of carriers available for radiative recombination. At the threshold for laser operation (and above), the carrier density N is pinned. Therefore as the current is increased above the threshold value, η' will exhibit very little change, if any (note small changes in N can occur well above threshold due to nonlinear gain saturation (Sect. 2.3.4)). The pinning of the carrier density above threshold is the reason that changes in the internal efficiency are primarily caused by carrier leakage and current diffusion, as discussed in Sect. 2.3.1.

When B and C both equal 0, that is to say that spontaneous recombination and Auger recombination are neglected, $\eta' = 1$ and $a(N_0)/a'(N_0) = N_0$. In this case, we find that (4.60) and (2.34) are nearly identical. The difference being the factor of $\eta' J (y = 0)$ or product of internal quantum efficiency and current density in the active layer appears in (4.60), whereas the factor $\eta_i J$ appears in (2.34). In the case of the Rigrod analysis of Sect. 2.3.4, the current density J that appears in (2.34) is that injected at the contact, and in this context, the internal efficiency η_i is must be used as an adjustable empirical parameter to account for all parasitic processes (e.g. current leakage, carrier diffusion) that can reduce the carrier density available for stimulated recombination. In contrast, $J(y = 0)$ is given by (4.48), which is calculated from the model of Sect. 4.3.2, which accounts for carrier loss that occurs in transport from the p-contact to the active layer. The point of this brief digression is to illustrate the model dependency exhibited by some laser parameters.

To continue with the stability analysis, we subsitute (4.66) into (4.44) and (4.43), and neglect the temperature dependent term in (4.43). Then for the traveling-wave propagation represented by $E \exp(-i\beta)$ the paraxial-wave equation (4.10) becomes,

$$\frac{\partial^2 E}{\partial x^2} - 2i\beta \frac{\partial E}{\partial z} + 2i\beta g(1 + i\alpha) E - 2i\beta\alpha_{\text{loss}} E = 0 \qquad (4.62)$$

where g is governed by (4.55), α is the antiguiding factor, and α_{loss} corresponds to the field losses due to absorption and scattering. Recall that the uniform plane-wave amplitude E_0 is a solution to (4.62) with the gain g being given by (4.55) with $|E|^2 = I_0$ and the diffusion term is zero because E_0 is constant in the x coordinate. Then, substituting (4.54) into (4.62), we find that the arguments of the exponent in (4.54) can be expressed as

$$\gamma = \frac{g_0}{1 + \frac{I_0}{I_S}} - \alpha_{\text{loss}} \qquad (4.63)$$

and

$$\delta\beta = \frac{\alpha g_0}{1 + \frac{I_0}{I_S}} . \qquad (4.64)$$

Equations (4.63) and (4.64) elucidate the intensity dependence of the gain and modal propagation constant. It is seen that local intensity variations induce variations in $\delta\beta$ which correspond to local wavefront variations, and the antiguiding factor α, plays the critical role in determining the value of the local wavefront variations (4.64).

To introduce a pertubation, the gain g is expressed as the sum of a constant term g_c and a small sinusoidal perturbing term

$$g = g_c + p \sin(k_t x) \tag{4.65}$$

where $g_c \gg p$ and k_t is the spatial frequency of the gain pertubation. To begin the small-signal analysis, we need to express the perturbed gain (4.65) in terms of the perturbed plane wave (4.54). This is accomplished by substituting (4.54) and (4.65) into (4.55). Neglecting all terms higher than first order in ϵ and p one finds that

$$g = \frac{g_0}{1 + \frac{I_0}{I_S}} \tag{4.66}$$

and

$$p \sin(k_t x) = \frac{-g_0 E_0 I_S (\epsilon + \epsilon^*)}{(I_S + E_0^2)(I_S + E_0^2 + s^2 k_t^2 I_S)} . \tag{4.67}$$

The next step of the small-signal analysis is to obtain a paraxial wave equation for the pertubation ϵ experienced by the plane wave. Substituting (4.54) and (4.65- 4.67) into (4.62) and retaining terms only up to first order in ϵ and p, we find

$$\frac{\partial^2 \epsilon_1}{\partial x^2} + 2\beta \frac{\partial \epsilon_2}{\partial z} + \alpha R \epsilon_1 = 0 \tag{4.68}$$

and

$$\frac{\partial^2 \epsilon_2}{\partial x^2} - 2\beta \frac{\partial \epsilon_1}{\partial z} - R \epsilon_1 = 0 \tag{4.69}$$

where

$$R = \frac{4\beta g_0 E_0^2 I_S}{(I_S + E_0^2)(I_S + E_0^2 + s^2 k_t^2 I_S)} . \tag{4.70}$$

and ϵ has been separated into real and imaginary parts according to $\epsilon = \epsilon_1 + i\epsilon_2$. When $|\epsilon_1| \gg |\epsilon_2|$, $|\epsilon_1|$ represents the pertubation in the modulus of the plane-wave amplitude and $\epsilon_2/|\epsilon_1|$ is the pertubation in the phase. The parameter R is dependent on the steady-state intensity of the plane wave and has dimensions of cm^{-2}. We shall see below that R represents the square of a spatial frequency that characterizes the stability boundary of a sinuosidally-perturbed plane wave against filamentation.

The next step of the pertubation analysis is to calculate the growth rate of filaments due to the sinudoidal-gain pertubation (4.65). For this purpose, a trial solution for ϵ that has the same sinusodial character as the gain pertubation (4.65) is assumed. Such a solution is expressed as

$$\epsilon_1 = \epsilon_{01} \sin(k_t x) \exp(hz) \tag{4.71}$$

and

$$\epsilon_2 = \epsilon_{02} \sin(k_t x) \exp(hz) . \tag{4.72}$$

The solution that is represented by (4.71) and (4.72) corresponds to intensity and phase ripples that will grow or decay according to the sign of h. To calculate h, we substitute (4.71) and (4.72) into (4.68) and (4.69) to obtain a homogeneous pair of equations with ϵ_{01} and ϵ_{02} as the pair of unknowns. Then nontrivial solutions of ϵ_{01} and ϵ_{02} that are found for the two values of h can be expressed as

$$h_\pm = \frac{-R \pm \sqrt{R^2 + 4\alpha k_t^2 R - 4k_t^2}}{4\beta} .$$

(4.73)

The corresponding solution for ϵ_{01} and ϵ_{02} is

$$\frac{\epsilon_{02}^{(\pm)}}{\epsilon_{01}^{(\pm)}} = \frac{2\beta h_\pm + R}{k_t^2}$$

(4.74)

The solutions for $h_\pm$ are only real when the argument of the radical is positive. When the argument of the radical is negative, this term is purely imaginary so $h_\pm$ is complex, giving rise to oscillatory solutions along the axial direction. However, in this case the real part of $h_\pm$ is always negative, so that the axial oscillations decay exponentially and the plane-wave is stable against filamentation. When the argument of the radical is positive, it is seen from (4.73), that for $R > k_t^2/\alpha$ the solution corresponding to h_+ grows exponentially, while the solution corresponding to h_- is attenuated exponentially. When $R < k_t^2/\alpha$, then both h_+ and h_- correspond to exponentially-decaying solutions, and the beam will be stable against filamentation. Hence, the spatial frequency given by $\sqrt{\alpha R}$ is the low-frequency boundary where the plane wave is stable against filamentation.

A plane wave subjected to a sinusoidal perturbation with spatial frequency less than $\sqrt{\alpha R}$ will be unstable, and filamention of the beam will occur on propagation through the gain medium. In other words, when a beam of finite lateral extent is launched into the gain medium, to lowest order, the frequency components less than $\sqrt{\alpha R}$ will grow exponentially with different rates, while frequencies less than $\sqrt{\alpha R}$ will decay. Moreover, the gain is nonuniform over the 0 to $\sqrt{\alpha R}$ spatial frequency range, and this will result in a significant distortion of the lateral intensity and phase profile. Note that this analysis was done for constant E_0^2. Acutally, the steady-state intensity will grow as the field propagates through the amplifying medium, so that the growth rates $h_\pm$ of the pertubations ϵ_1 and ϵ_2 will be dependent on the axial coordinate z. Therefore, the degree of degradation in beam quality will also depend upon the length of the amplifier section.

To obtain a better insight into the intensity ranges that are stable against filamentation it is necessary to relate the steady-state intensity E_0^2 to the critical spatial frequency $\sqrt{\alpha R}$. We see from (4.70) that R is a second-order polynomial in the plane-wave intensity E_0^2, so that for a given sinusoidal pertubation there will be two intensities E_0^2 associated with each value of $R_{\mathrm{crit}} = k_t^2$. Furthermore, R has a maximum when $E_0^4 = I_S^2 \left(1 + s^2 k_t^2\right)$. As the intensity is increased beyond the value that gives maximum R, the bandwidth of spatial frequencies unstable to filamentation $\sqrt{\alpha R}$ also decreases.

Let us represent the spatial frequency of the pertubation as $k_t = 2\pi/\lambda_{\text{pert}}$, where λ_{pert} is the wavelength associated with the pertubation. Note that a gain region of equal lateral dimension will contain a single period of the pertubation. A uniform-intensity traveling wave of lateral extent w comprises spatial frequencies k that span the range from $0 \geq k \leq 2\pi/w$. When $\lambda_{\text{pert}} \gg w$, we have $k_t \ll 2\pi/w$, and the spatial frequency bandwidth that is unstable against filamentation is much less than the spatial frequency bandwidth comprising the lateral beam profile. In other words, when the lateral beam dimension is much less than the wavelength of the pertubation, filamentation will be minimized. However, when $\lambda_{\text{pert}} \approx w$ or $\lambda_{\text{pert}} \ll w$, all spatial frequency components of the lateral beam are unstable to filamentation, and therefore the lateral field profile will distort as it propagates through the gain medium. To illustrate the significant influence that α has on the filamentation process, Fig. 4.12 displays the normalized-intensity values I/I_S versus λ_{pert} that correspond to $k_t^2 = \alpha R$ for the antiguiding parameters $\alpha = 0.1, 2$, and 8. In each plot, the intensity region interior to the curve corresponds to $R > k_t^2/\alpha$, which is unstable to filamentation. For intensities outside of the curves, $R < k_t^2/\alpha$ and a plane wave is stable against filamentation. For each case plotted in Fig. 4.12, there is a value of λ_{pert} below which stability is found for all intensities.

The $\alpha = 2$ is close to the minimum antiguiding factor measured for semiconductor gain media, while $\alpha = 8$ represents more typical values. The lowest value of $\alpha = 0.1$ would not be expected for semiconductors, but it is included to underscore the critical role that the α-factor plays in determining optical beam quality. As α is decreased the range of values of λ_{pert} that is stable also increases, which means the lateral beam extent can be larger before it begins to filament. Also, if the intensity is high enough or low enough, then stability can be found. The case of interest though is the high intensity boundary. We saw in Chap. 2 that the maximum intensity that can be supported by a laser or an optical amplifier was given by (2.50). For the case considered in Fig. 4.12, this corresponds to a normalized limit intensity I/I_S of about 400, which is exceeded for $\lambda_{\text{pert}} > 35\,\mu\text{m}$ when $\alpha = 8$. A more practical limit is the catastrophic-facet damage limit, which occurs at a normalized-intensity value of ≈ 3. Thus it is seen that the intensities required to stabilize a plane wave, are in most cases above the facet damage limit. This analysis suggests that employing methods to raise the facet damage could in some cases, provide better beam quality at higher power outputs. However, this trend has not been experimentally verified. In fact, it is generally found from experiment that increasing the power output degrades the beam quality.

In Chap. 8, additional aspects of filamentation in broad-area semiconductor amplifiers and techniques for avoiding filamentation are presented. To accurately model the effects that nonuniformities can have on the lateral beam quality, it is necessary to do a fully self-consistent analysis, as discussed

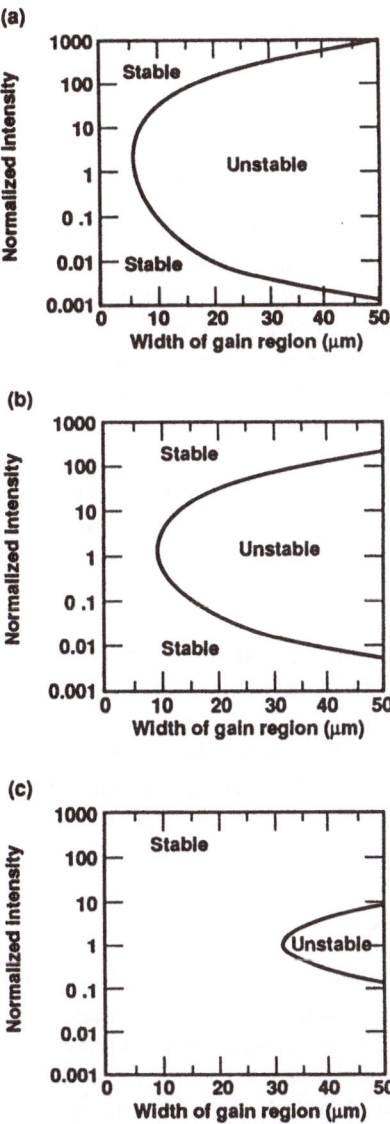

Fig. 4.12. Plot of steady-state intensity values (normalized to I_S), that satisfy $k_t^2 = \alpha R$, as a function of λ_{pert} for $\alpha = 8$ (a), $\alpha = 2$ (b), and $\alpha = 0.1$ (c). The other values for the parameters were $s = 2\,\mu\text{m}$, $\beta = 2.07 \times 10^5\,\text{cm}^{-1}$, and $g_0 = 200\,\text{cm}^{-1}$.

in Sect. 4.3.4. The pertubation analysis that has been presented here is useful because it delineates the boundaries between conditions that are stable and unstable against filamentation for specific initial values of intensity. Of course as seen in Fig. 4.12, the region of stability accessible is limited to low-intensity output levels, because of the magnitude of the α-factor and facet damage. The instability against filamentation is a formidible obstacle towards obtaining single-lateral mode operation.

4.4 Optical Coherence Theory

In Sect. 4.3, we saw how to simulate, in a self-consistent manner, the optical beam propagation, as well as the carrier and heat transport processes internal to the diode-laser array. The beam propagation model provides the near-field profile $E(x, y)$ of the light that is emitted from the output coupler of a diode-laser array. Now we will consider the diffractive effects on the laser-array beam that occur upon propagation from the output coupler, or near-field zone, to an observation point that is in the far-field zone. Having a good understanding of the characteristics of the diode-laser array output beam is very important for developing accurate interpretations of the physical processes internal to the array and their impact on the operating characteristics. It is also important for establishing the performance in optical system applications. For many applications, the most demanding performance specification for diode-laser arrays under high-power operation has been to maintain both single-frequency and single-spatial mode operation. However, many of the diode-laser arrays studied to date have shown multi-spatial mode operation at high-power output. Under these conditions, the output beam comprises partially-coherent light. Therefore, this section will review the fundamentals of optical coherence theory that are important for understanding and evaluating the properties of output beams from diode-laser arrays that may be operating multi-spectral and spatial mode.

4.4.1 Mutual Coherence

The term *coherent* is widely used to describe semiconductor lasers and laser arrays that operate in a single mode. Sometimes it is used to describe laser arrays that operate in a single-spatial mode, but multiple longitudinal modes. In other instances, it has been used to describe laser arrays that operate in both a single longitudinal and single spatial mode. In optical coherence theory, the mutual coherence gives a measure of the degree of correlation of the optical field, $E(t, \mathbf{R})$ at points within a source that have spatial coordinates $\mathbf{R}_1 = (x_1, y_1, z_1)$ and $\mathbf{R}_2 = (x_2, y_2, z_2)$, and temporal coordinates t_1 and t_2. The precise definition of mutual coherence, $\Gamma(\tau, \mathbf{R}_1, \mathbf{R}_2)$, has been given by [1.356],

$$\Gamma(\tau, \mathbf{R}_1, \mathbf{R}_2) = \left\langle E(t + \tau, \mathbf{R}_1) E^*(t, \mathbf{R}_2) \right\rangle , \qquad (4.75)$$

where $\langle \ \rangle$ represents an ensemble or time average with the averaging-time period being much greater than the longest coherence time of the source. It has been assumed that $E(t, \mathbf{r})$ is stationary, so that the mutual coherence function depends on relative time coordinate $\tau = t_1 - t_2$.

When one speaks of the *spatial coherence* of diode-laser arrays, it usually refers to the field correlation at separate points, $\mathbf{R}_1 \neq \mathbf{R}_2$, and equal times, $t_1 = t_2$ corresponding to $\Gamma(0, \mathbf{R}_1, \mathbf{R}_2)$. Spatial coherence relates to the quality of the output beam of the laser array [4.71]. Spatial coherence meaurements have been used to determine phase-locking in injection-coupled lasers [3.72]. The visibility measurement discussed in Sect. 3.4.1a is an example of the how mutual coherence can be used to study frequency locking in diode lasers. Alternatively, one can consider the field correlation at a single point, $\mathbf{R}_1 = \mathbf{R}_2$, but at different times, $t_1 \neq t_2$, which corresponds to $\Gamma(\tau, \mathbf{R}_1, \mathbf{R}_1)$. This is referred to as the *temporal coherence* or *field autocorrelation function*, and it is related to the power spectrum which provides the spectral linewidth (Sect. 3.3.4) of the laser source.

The spatial and temporal coherence properties of electromagnetic fields radiated by a source are coupled through two wave equations [1.356] so that, in general, the mutual coherence can change on propagation. It is usually assumed that the output beam of a laser can be represented as a plane wave possessing *complete spatial coherence* across its wavefront, but with a diminished temporal coherence related to the spectral width. For most conventional Fabry-Perot cavity lasers, this approximation of complete spatial coherence across the beam profile is valid provided that the output beam waist dimension is much less than the coherence length that corresponds to the spectral width of the laser output [4.72], which is indeed the case for most lasers of practical interest. Complete spatial coherence will occur even under multi-longitudinal mode operation, as long as, each longitudinal mode is associated with the same spatial mode [4.73]. Indeed, spatial coherence measurements of edge-emitting diode laser arrays are insensitive to the longitudinal mode spectrum of the array, but exhibit a strong correlation with the spatial mode spectrum of the array [1.119, 4.74, 4.75]. However, if the laser operates in multiple-spatial modes, then complete spatial coherence across the output beam is no longer possible. This is a situation that one needs to be aware of when analyzing the output beam characteristics of large-aperture diode lasers and diode-laser arrays. However, in diode-laser arrays that use dispersive elements to generate the output beam (such as grating-coupled surface-emitting lasers), complete spatial coherence also requires single-longitudinal mode operation [4.49, 4.76]. These topics are addressed in Sects. 7.1.1 and 8.2.3.

4.4.2 Cross-Power Spectrum

The specific quantities of interest for characterizing the optical properties of a partially-coherent output beam are the mutual coherence function, which is given by (4.75) and the cross-power spectrum, $S(\Omega, \mathbf{R}_1, \mathbf{R}_2)$, which is expressed as

$$S(\Omega, \mathbf{R}_1, \mathbf{R}_2) = \int_{-\infty}^{+\infty} \Gamma(\tau, \mathbf{R}_1, \mathbf{R}_2) \exp(-\mathrm{i}\Omega\tau) \, d\tau \ . \qquad (4.76)$$

It is seen from (4.76) that the mutual coherence and the cross-power spectrum are Fourier transform pairs. Note that when $\mathbf{R}_1 = \mathbf{R}_2$, the mutual coherence and the cross-power spectrum reduce to the well-known forms for the autocorrelation function and the power spectrum. In the following discussions on propagation we will be using the cross-power spectrum as opposed to the mutual coherence. The reason being that diode-laser array outputs are typically characterized using power spectrum measurements rather than autocorrelation function measurements.

4.4.3 Propagation of the Cross-Power Spectrum

Let us now consider the question of how to calculate the coherence properties as the output beam propagates diffractively through free-space. Because of the small size of the near field of most diode-laser arrays, as compared to the distances to the point of observation, the output beam characteristics of the diode-laser array, are usually measured in the far-field or Fraunhofer zone. Of course, depending on the array size and measurement geometery, situations can occur where the output beam is measured in the Fresnel zone [4.77]. In the discussion that follows, we will only consider observation points in the Fraunhofer zone. A diagram of the geometry that will be used in the following discussion is provided in Fig. 4.13. For the purposes of this discussion, the direction of propagation shall be the z direction, and we will be evaluating the coherence properties in the x — y plane which is perpendicular to the direction of propagation. Note that $\mathbf{r} = (x, y)$ will denote the coordinates in plane perpendicular to the z-direction. Since this discussion is restricted to the Fraunhofer zone, we have the following condition, $z \gg \pi x_{\max}^2 / (4\lambda)$, where $x_{\max}$ is the largest dimension in the near field [1.356]. It shall also be assumed that the resulting far-field beam divergences are narrow enough, so that Huygen's obliquity factor corrections are negligible [1.356, 1.358]. This is a good approximation for the lateral dimension of diode laser and diode-laser arrays. But for the radiated field in the transverse direction, which is $\approx \lambda$ in the near field, the obliquity factor correction is necessary to accurately calculate the far-field characteristics that are more than about 20° off the optic axis (z-axis in Fig. 4.13) [1.358, 4.78, 4.79]. The cross-power spectrum $S_F(\Omega, \mathbf{r}_1, \mathbf{r}_2)$ in the Fraunhofer zone can be expressed as [1.356, 4.72]

$$S_F(\Omega, \mathbf{r}_1, \mathbf{r}_2) = \iint H^*\left(\Omega, \mathbf{r}_1, \mathbf{r}_1'\right) H\left(\Omega, \mathbf{r}_2, \mathbf{r}_2'\right) S(\Omega, \mathbf{r}_1, \mathbf{r}_2) \, d\mathbf{r}_1' d\mathbf{r}_2' \ ,$$

(4.77)

where

$$H\left(\Omega, \mathbf{r}, \mathbf{r}'\right) = \left(\frac{1}{i\lambda z}\right) \exp\left[i\frac{\Omega}{c}\left(-z - \frac{\mathbf{r}^2}{2z} + \frac{\mathbf{r} \cdot \mathbf{r}'}{z}\right)\right] \ .$$

(4.78)

and $\mathbf{r}_n' = \left(x_n', y_n'\right)$ are the coordinates of a pair of points $n = 1, 2$ in the near-field plane, $\mathbf{r}_n = (x_n, y_n)$ are the coordinates of a pair of points $n = 1, 2$

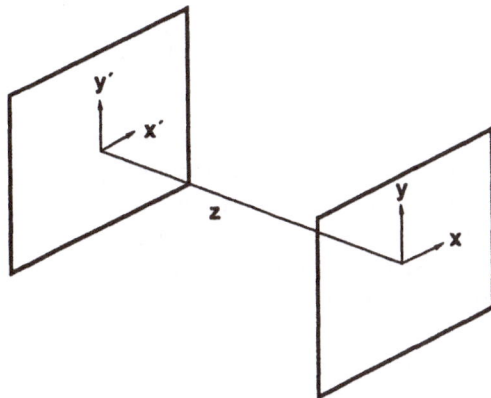

Fig. 4.13. Schematic depicting the relation of the near- and far-field planes.

in the far-field plane, and z is the distance between the near- and far-field planes. It is seen from (4.77 and 4.78) that the cross-power spectra in the plane of the far-field and near field are related by a two-dimensional Fourier transform.

The cross-power spectrum provides a complete description of the optical field. When $r_1 = r_2 = r$, (4.77) gives the power spectrum vs position in the far-field plane $S_F(\Omega, r_1, r_2)$ as

$$S_F(\Omega, r, r) = \frac{1}{(\lambda z)^2} \iint \exp\left[\frac{i\Omega\left(r_2' - r_1'\right) \cdot r}{cz}\right] S\left(\Omega, r_1', r_2'\right) dr_1' dr_2' \ .$$

(4.79)

It can readily be seen from (4.79) that the power spectrum observed at an arbitrary location in far-field plane is not simply related to the power spectrum in the near field. The possibility that the optical power spectrum can change on propagation is of interest in the field of optical coherence [4.80, 4.81]. A possible consequence of this is that a measurement of the spectral output in the far field of a diode-laser array may not necessarily give exactly the same result that one would get by doing the measurement in the near field. This effect is common in grating-coupled diode-laser arrays (Chap. 7) and laser amplifiers (Sect. 8.2), where the dispersion of the grating-output coupler introduces a wavelength-dependent tilt on the emission angle of the output beam. It occurs in multi-mode operation, and even in cases of highly-coherent, single-frequency operation, significantly different values in the spectral linewidth have been measured from grating-coupled laser amplifiers in the near- and far-field zones [4.82]. The fact that one can observe different power spectra in the near and far-fields is related to the dispersive properties of the grating and to the aperture size used in the measurement. Note that if (4.79) is integrated over all $r_\perp$, the following is obtained

$$\int S_F\left(\Omega, \mathbf{r}_\perp, \mathbf{r}_\perp\right) d\mathbf{r}_\perp = \int S\left(\Omega, \mathbf{r}'_\perp, \mathbf{r}'_\perp\right) d\mathbf{r}'_\perp \ . \qquad (4.80)$$

Equation (4.80) is Parseval's theorem, which is basically a statement of energy conservation. In diffractive propagation through a lossless medium such as free space, no light is absorbed, and therefore all light emitted from the near field propagates into the far-field zone. Although, local differences can occur between the near and far-field power spectra, Pareval's theorem tells us that if a sufficiently large aperture is used in the power spectrum measurement, then one will observe the same power spectrum in the near- and far-field zones. In practice, this is not always practical or even desirable, especially where applications are concerned. Therefore, one needs to be aware of the relation between the measurement conditions and the power spectrum.

4.4.4 Far-Field Mutual Coherence

Because of (4.76), the far-field mutual coherence is related to the far-field cross-power spectrum by

$$\Gamma_F\left(\tau, \mathbf{r}_{1\perp}, \mathbf{r}_{2\perp}\right) = \int_{-\infty}^{+\infty} S_F\left(\Omega, \mathbf{r}_{1\perp}, \mathbf{r}_{2\perp}\right) \exp\left(i\Omega\tau\right) d\Omega \ . \qquad (4.81)$$

The spatial coherence between two points in the far-field plane is obtained by setting $\tau = 0$ in (4.81) to give

$$\Gamma_F\left(0, \mathbf{r}_{1\perp}, \mathbf{r}_{2\perp}\right) = \int_{-\infty}^{+\infty} S_F\left(\Omega, \mathbf{r}_{1\perp}, \mathbf{r}_{2\perp}\right) d\Omega \ , \qquad (4.82)$$

and the intensity pattern $I_F\left(\mathbf{r}_\perp\right)$ in the far-field plane is obtained by setting $\mathbf{r}_{1\perp} = \mathbf{r}_{2\perp} = \mathbf{r}_\perp$ in (4.82) and is expressed by

$$I_F\left(\mathbf{r}_\perp\right) = \Gamma_F\left(0, \mathbf{r}_\perp, \mathbf{r}_\perp\right) = \int_{-\infty}^{+\infty} S_F\left(\Omega, \mathbf{r}_\perp, \mathbf{r}_\perp\right) d\Omega \ . \qquad (4.83)$$

Equation (4.83) states that the far-field intensity pattern is the summation of the power spectrum over all frequencies. Generally, each spatial mode of a diode-laser array will correspond to a slightly different operating frequency, as discussed in Sect. 3.3.1. Then to properly analyze the far-field output of a laser array operating multi-mode, it is necessary to measure $S_F\left(\Omega, \mathbf{r}_\perp, \mathbf{r}_\perp\right)$ as a function Ω. Experimentally, this is refered to as a spectrally-resolved far-field measurement [4.33, 4.83, 4.84]. Alternatively, a spectrally-resolved near-field measurement [1.51, 4.83] can be used to determine the near-field power spectrum, $S\left(\Omega, \mathbf{r}_\perp, \mathbf{r}_\perp\right)$ as a function Ω. The presence of more than one spatial mode in the array output beam will lead to a reduction in (4.82) the spatial coherence, since different spatial modes oscillate at different frequencies.

Equations (4.77-4.83) provide the most general formalism for calculating the optical coherence and the effects of beam propagation on the spatial and temporal coherence of the source. As is seen above, the propagation effects on the spatial and temporal coherence can be determined directly from a Fourier transform of the cross-power spectrum of the source. The next subsection will elucidate the conditions when propagation effects do not influence the power spectrum.

4.4.5 Cross-Spectral Purity

The aforementioned model of the laser output beam as a propagating plane wave possessing *complete spatial coherence*, but a reduced temporal coherence is an example of what is referred to as a *cross-spectrally pure* source [4.85, 4.86]. For a source to be cross-spectrally pure the power spectrum should be identical at any two points, which means, $S(\Omega, \mathbf{r}_{1\perp}, \mathbf{r}_{1\perp}) = S(\Omega, \mathbf{r}_{2\perp}, \mathbf{r}_{2\perp}) = S(\Omega)$. In addition, a cross-spectrally pure source will have a cross-power spectrum that is equal to the power spectrum apart from a frequency dependent phase factor. Thus, the cross-power spectrum is separable as [4.85, 4.86],

$$S(\Omega, \mathbf{r}_{1\perp}, \mathbf{r}_{2\perp}) = \Gamma(\tau_0, \mathbf{r}_{1\perp}, \mathbf{r}_{2\perp}) S(\Omega) \exp(-i\Omega\tau_0) , \qquad (4.84)$$

where τ_0 is an arbitrary temporal reference, $S(\Omega)$ is the power spectrum, and $\Gamma(\tau_0, \mathbf{r}_{1\perp}, \mathbf{r}_{2\perp})$ is the normalized mutual coherence. In a cross-spectrally pure source, the functional form of the power spectrum is independent of the position in the source. With the form (4.84) for the cross-power spectrum, it has been shown by *Saleh* [4.72, 4.73] that the power spectrum of the source will be preserved in the far-field as long, as the transverse beam dimension is much less than the coherence length of the source, which is the case for most single-spatial mode lasers, as mentioned earlier.

5. Gain-Guided Diode-Laser Oscillators

The distinction between diode-laser array oscillators, the subject of this chapter, as well as Chaps. 6 and 7, and diode-laser amplifiers, the subject of Chap. 8, is being made because of the difference in the mode-discrimination mechanisms. Laser oscillators will have both forward and backward propagating waves that provide feedback to the active region in the optical cavity; whereas laser amplifiers are injected with an external-traveling wave that determines the modal properties. Some or all elements of a laser-array oscillator are mutually coupled; however in amplifier arrays the coupling is predominantly unidirectional, with the direction being determined by the propagation direction of the dominant traveling wave, which typically originates from a single-mode oscillator section.

This chapter will review some selected examples of gain-guided diode-laser array oscillator structures and their observed performance characteristics. Gain-guided diode-laser arrays are considered to be those structures that do not contain a built in refractive-index profile within the active region. The mode-discrimination characteristics are largely determined by the injected carriers. Index-guided diode-laser array structures are presented in Chap. 6. Where possible, emphasis is placed on examples that exhibit performance characteristics that can be understood in terms of the array theorics discussed in Chap. 4. Due to the complexity of the self-consisitent model calculations and the uncertainty about the material and structural uniformity of fabricated devices, many diode-laser array oscillator designs have not yet been analyzed in detail. Therefore, some of the examples in this chapter, as well as all subsequent chapters, have been selected because they are believed to represent qualitative demonstrations of important design characteristics.

The laterally-coupled array design is perhaps what comes to mind first when one thinks of diode-laser arrays, and the first type of diode-laser array that was widely fabricated was the *gain-guided* multiple-stripe geometry array, which is discussed in Sect. 5.1. The subject of gain-tailored laser structures is considered in Sect. 5.2. Although, gain-tailored lasers typically do not comprise a coupled array of distinct laser elements, the modal characteristics of these devices are important for understanding and interpreting the output-beam properties of some gain-guided diode-laser arrays. It has been even been suggested that inadvertent gain-tailoring due to structural nonuniformities may be the explanation for those observations of single-lobed output beams from arrays that were intended to be periodic-gain structures [5.1]. With improvements in etching technologies, the unstable-resonator design which is well known in non-semiconductor lasers [2.31], has been investigated as a monolithic-semiconductor source for producing high-power coherent light, and this is the subject of Sect. 5.3.

5.1 Multiple-Stripe Gain-Guided Laser Array Structures

Figure 5.1 displays a diagram of the type of gain-guided diode-laser array structure that has been the subject of many experimental studies aimed at understanding the modal properties of these devices. Individual array elements are defined by the proton-implanted region, which acts as a current blocking layer outside the gain stripe, giving rise to a periodic gain and refractive-index variation in the lateral direction. Below the implanted layer, the current will spread and injected carriers will diffuse laterally in the active layer. Typically, the array elements are $3 - 6\,\mu m$ wide with about a $10\,\mu m$ center-to-center spacing. Although, the example in Fig. 5.1 has a multi-quantum-well active layer, single quantum-well active layers are also used in this type of array.

Early experimental studies of the modal characteristics of multiple-stripe diode-laser arrays had made the observation that these devices operated in a superposition of array modes [1.51, 4.83, 5.2]. Furthermore, to correctly model the character of the observed modes of the gain-guided arrays, it was necessary to employ an exact analysis [4.4]. In almost all cases, single-array mode operation required that the array be incorporated in an external-grating cavity [4.32, 4.84, 5.3], injection-locked using an external oscillator [3.84, 5.4, 5.5] or placed in an external cavity [4.33]. However, single-mode oscillation of mutiple-stripe diode-laser arrays can occur under certain conditions near threshold [1.57, 4.59]. Experimental observations have ascertained that the modal characteristics of mutiple-stripe gain-guided laser arrays were similar to those of broad-area lasers [4.32, 5.5]. To explain this type of behavior, it was necessary to view the coupling between the gain stripes, as occuring by radiating fields of leaky modes and not by the evanescent fields of confined modes [3.59, 5.2, 5.5]. A particularly important observation was that gain-guided laser arrays could exhibit modes that had more near-field maxima than there were gain stripes. This characteristic can be understood in terms of radiation-field coupling, where the lateral intensity profile results from the interference of the radiating modes, oriented at a slight angle with respect to the array axis, that are emitted from all gain stripes. Although certain aspects of gain-guided array behavior can be understood in terms of a coupled-mode picture (Sect. 4.2.2), to completely characterize the observed output characteristics of multiple-stripe gain-guided arrays, it has been necessary to use a fully self-consistent model, as the effects of carrier transport and heat dissipation can alter the lateral dielectric profile in a significant way.

(a)

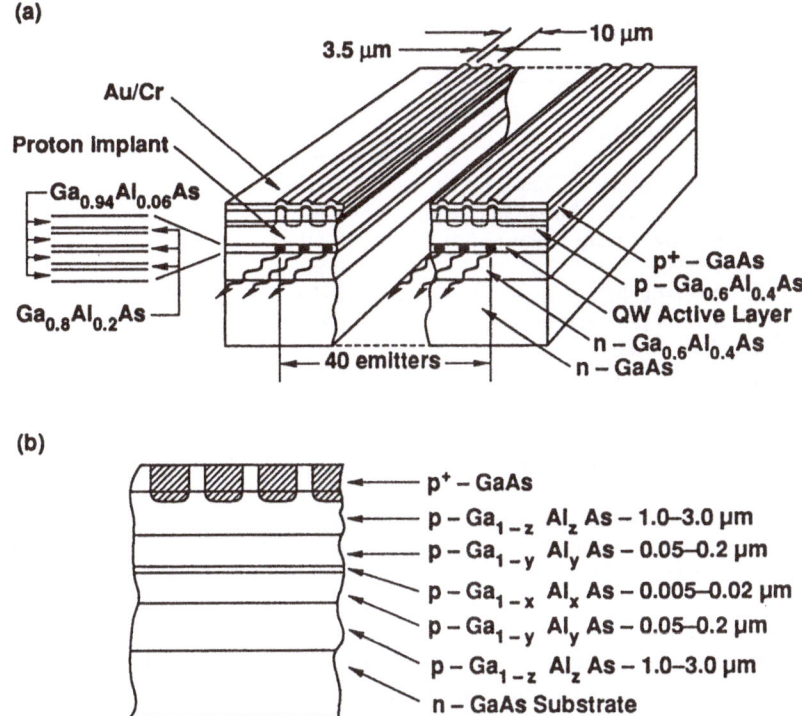

(b)

Fig. 5.1. Diagram of a multiple-stripe diode-laser array, showing the layer compositions [1.346].

The output beam of multiple-stripe gain-guided arrays is typically double-lobed, which is not desirable for most applications. This occurs because of poor mode discrimination that gives rise to multi-mode operation. Since all modes but the fundamental mode have an alternating phase distribution in the near field, arrays operating in several modes will often have a double-lobed far field. To understand and interpret the observed modal characteristics of gain-guided diode-laser arrays with periodic gain, it has been necessary to model the array using the methods discussed in Sect. 4.3. More recently, *Hadley* et al. have done a self-consistent analysis of the multiple-stripe gain-guided diode-laser array structure appearing in Fig. 5.1 and compared their calculated results with experimental observations from commercially available devices [4.60]. Table 5.1 lists the parameters that were used by *Hadley* et al. to model the gain-guided arrays. In these experiments, individual eigenmodes of a ten-stripe array were selected by injection seeding, as discussed in [3.86, 5.4], while the spectral output was monitored to insure single-array mode operation. The calculated array modes were labeled according to increasing resonance frequency. The fundamental mode, which has the desired single-lobed far field, is designated by $\nu = 1$, where ν is the mode number. The higher-order modes correspond $\nu > 1$. These higher-order modes all

Table 5.1. Parameters used in [4.59] to model the 10 stripe gain-guided array

Length	$250\,\mu m$
Active layer thickness (d)	$0.2\,\mu m$
Stripe width	$6\,\mu m$
Stripe spacing (Center-to-Center)	$10\,\mu m$
Index of cladding layer $(Al_{0.20}Ga_{0.80}As)$	3.35
Index of active layer $(Al_{0.04}Ga_{0.96}As)$	3.59
Facet reflectivities	0.99 and 0.32
Operating wavelength	815 nm
Threshold current per stripe	24 mA
Antiguiding factor (α)	2
Gain coefficient $(a = \sigma)$	$1.5 \times 10^{-16}\,cm^2$
Transparency loss $(b = \sigma N_{tr})$	$100\,cm^{-1}$
Diffusion coefficient (D)	$30\,cm^2\,s^{-1}$
Spontaneous recombination coefficient (B)	$1.4 \times 10^{-10}\,cm^3\,s^{-1}$
Nonradiative recombination time (τ_{nr})	5 ns
Auger recombination coefficient (C)	0
Index variation with temperature (dn/dT)	$4 \times 10^{-4}\,K^{-1}$

exhibit double-lobed far-field patterns, with the angular separation between lobes increases with increasing mode number ν. For higher-order modes, ν corresponds to the number of maxima in the near-field.

Figure 5.2 displays the measured and calculated near and far-field patterns when a ten-stripe array was operated at $1.01 \times I_{th}$. At this operating level, the array operated single frequency, so injection seeding was not necessary. The observed mode, which exhibits a double-lobed far field and an intensity peak at each stripe position, is very similar to the out-of-phase mode that is predicted by coupled-mode theory; and it corresponds closely to the $\nu = 10$ mode from the model calculation.

A small increase in current to $1.02 - 1.06 \times I_{th}$ produced the near- and far-field patterns in Fig. 5.3. This mode is seen to correspond to the $\nu = 11$ mode of the self-consistent calculation. Note that the calculated gain of the $\nu = 11$ modes was found to be only 3 % less than that of the $\nu = 10$, when thermal effects were taken into account.

By using injection seeding, it was possible to induce the ten-stripe array to operate in each of the low order modes $\nu = 1 - 9$. Examples of the fundamental mode, $\nu = 1$ and the $\nu = 6$ mode are shown in Fig. 5.4. For these lower-order modes, the number of maxima in the near field is less than the number of gain stripes. This characteristic can be explained using the broad-area coupled-mode model of laser arrays that is presented in Sect. 4.2.2. The periodic gain structure of a N-element array has a fundamental mode that can be approximated as a linear combination of the $m = 1$ and $m = 2N - 1$ modes, (4.20), of the broad-area laser structure. The resulting near-field intensity pattern is given by $\left| E_1^{(0)} + E_{2N-1}^{(0)} \right|^2$ [4.37]. As illustrated in Fig. 5.5, the in-

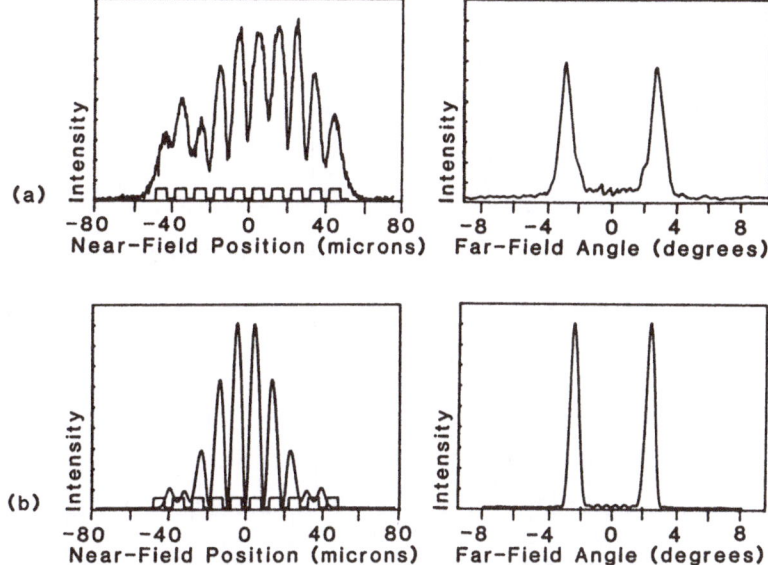

Fig. 5.2. The measured $1.01 \times I_{\text{th}}$ (a) and the calculated (b) near- and far-field patterns for the $\nu = 10$ mode of the ten-stripe array [4.60].

terference of the $m = 1$ and $m = 2N - 1$ broad-area modes for $N = 10$ results in a fundamental mode near-field pattern with a total of 9 intensity peaks.

The numerical, self-consistent treatment elucidates the influence of thermal effects on the modal characteristics of the multiple-stripe diode-laser array. Heating of the active layer results in an increase in the refractive index, which acts in opposition to the gain-induced index depression. This is illustrated in Fig. 5.6 for the lateral effective-index profile of the ten-stripe array model. A $10\,\text{K}$ increase in the active layer overcomes the index-antiguiding due to the carriers; and in effect, creates a broad-area index-guided laser with a sinusoidal pertubation due to the gain stripes.

Figure 5.7 compares the observed near- and far-field patterns of the $\nu = 15$ mode with those that were calculated both with and without thermal effects included. It is clear from Figs. 5.7a and c that better agreement is obtained with the experimental observation when the effect of heating are included in the model calculation. The importance of thermal waveguiding effects are clearly evident, as the central dip in the calculated near field without thermal effects, Fig. 5.7b, is absent in the calculated near field that includes thermal effects Fig. 5.7c.

Thermal effects were also found to have a strong influence on the round-trip losses of the modes of the ten-stripe gain-guided laser array. In the absence of thermal effects, index antiguiding causes the higher-order modes to exhibit near-field characteristics similar to those those depicted in Fig. 5.7b, with large intensity peaks at the ends of the array and a minimum in intensity in the center of the array. Array modes with this type of lateral

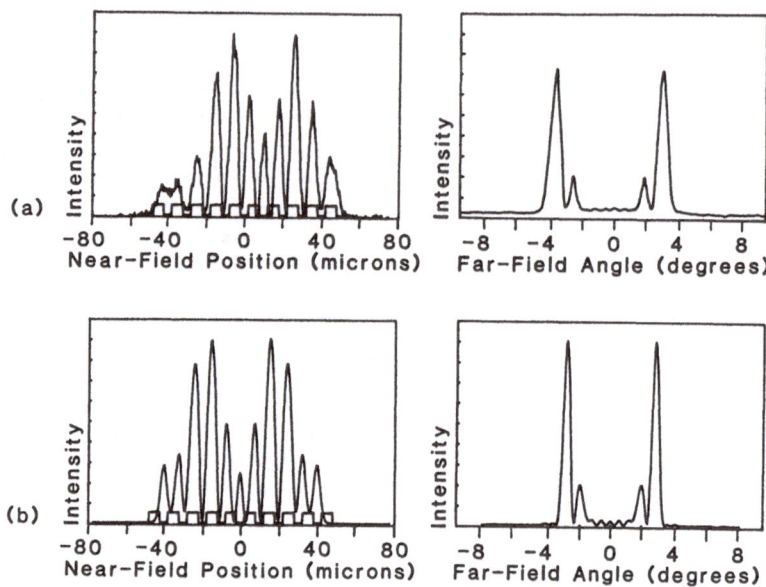

Fig. 5.3. The measured $1.02 \times I_{\text{th}}$ (**a**) and the calculated (**b**) near and far-field patterns for the $\nu = 11$ mode of the ten-stripe array [4.60].

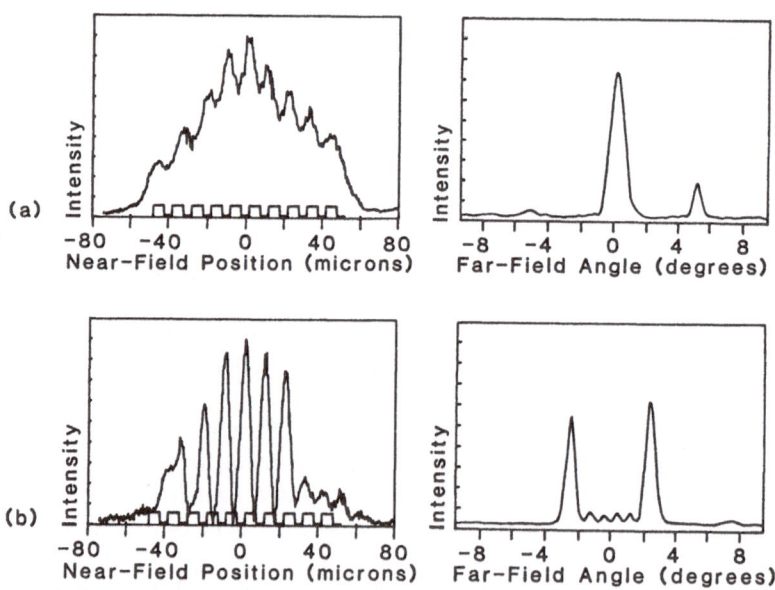

Fig. 5.4. The measured near and far-field patterns for the $\nu = 1$ (**a**) and the $\nu = 6$ (**b**) modes that were obtained under injection-seeded operation [4.60].

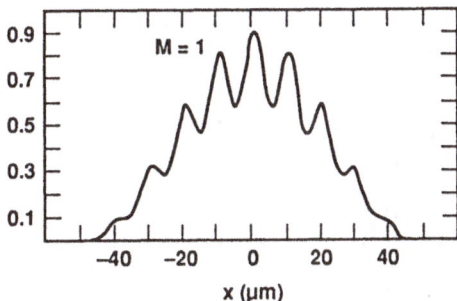

Fig. 5.5. The calculated near-field pattern for the $\nu = 1$ using the broad-area coupled mode model of [4.37].

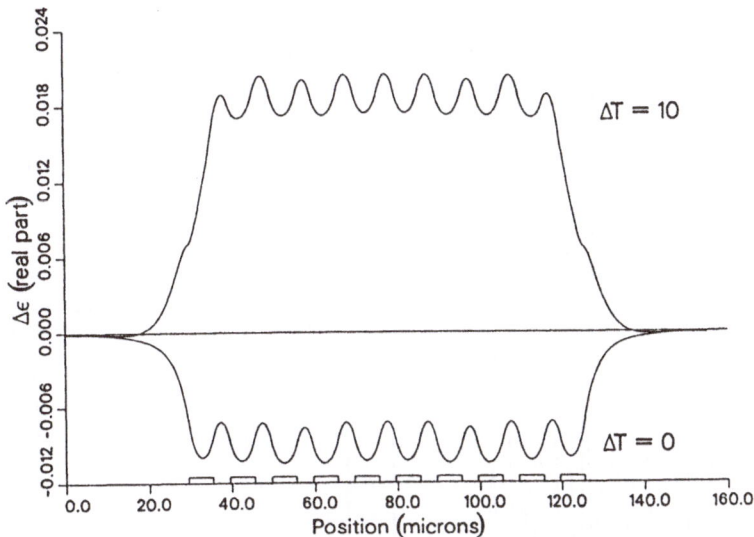

Fig. 5.6. The change in the calculated lateral effective-index profile, from (4.43) with heating of the active layer included for a $\Delta T = 10\,\mathrm{K}$, and without heating of the active layer ($\Delta T = 0\,\mathrm{K}$) [4.60].

profile, will experience higher losses because the field maxima overlap the ends of the array where the losses are higher. However, when heating occurs, the effects of index-antiguiding can be effectively nullified, as illustrated in Fig. 5.6, resulting in a nearly-uniform envelope in the near field of the array modes (Figs. 5.7a and c). With these thermal-guided lateral mode profiles, the round-trip losses will be less, as more of the mode occupies the central region of the array where most of the gain resides. The consequences of this are that more gain is available for the higher-order modes; and therefore, the modal gain will be almost independent of the array mode. This behavior is consistent with numerous experimental observations of multi-array mode operation from gain-guided diode-laser arrays [1.51, 4.32, 4.33, 4.83, 4.84, 5.2, 5.3].

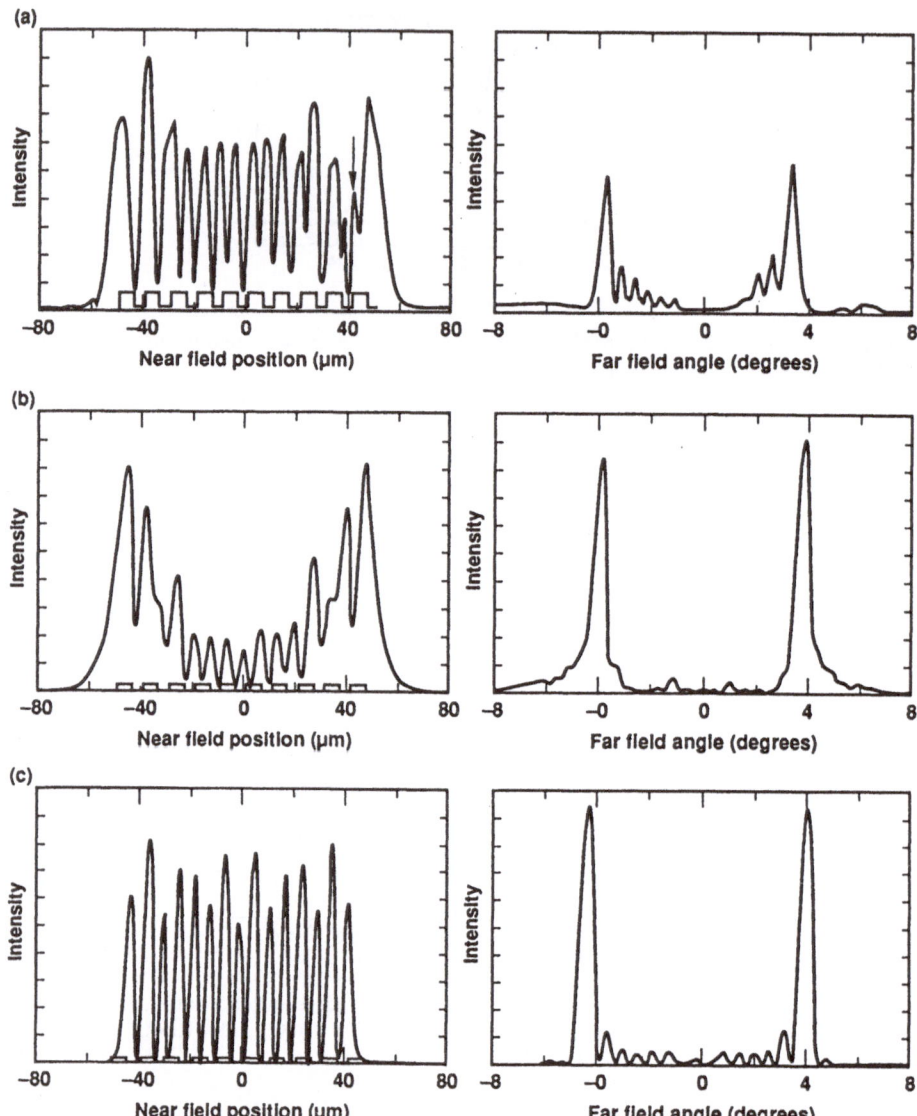

Fig. 5.7. The measured (**a**) near- and far-field patterns for the $\nu = 15$ and calculated near and far fields (**b**) without and (**c**) with thermal effects [4.59].

5.2 Gain-Tailored Structures

One approach towards generating single-lobed output beams from gain-guided lasers involved adding gain into the regions between the current stripes so that the periodic gain profile of the multiple-stripe array is effectively transformed into a broad-area gain distribution. Diode-laser array structures based on the principle of gain tailoring have exhibited single-lobed output beams

[4.20][5.6]-[5.9]. As demonstrated by *Lindsey* et al., in a periodic laser-array structure that contains a broad-area, tailored-gain distribution, the mode characteristics can be determined entirely by the broad-area gain distribution, even in the presence of a periodic array structure [4.20][5.6]-[5.8]. The structure that *Lindsey* et al. used for this purpose is illustrated in Fig. 5.8. It comprises a two-dimensional array of gain "dots", where the dot size is varied as a function of position to create a linear variation in the lateral gain profile. As discussed in [5.7], such a structure can be fabricated by using a half-tone process to produce a fractional coverage of the metal contact to the p^+GaAs cap layer. A linear variation of the density or size of the dots in the lateral direction causes the injected current density to vary in a similar manner, so that a linear gain profile is obtained.

The measured near- and far-field patterns of an asymmetric linear gain-tailored laser are shown in Fig. 5.9. This $100\,\mu m$ wide laser was fabricated using the halftone process decribed in [5.7]. As is seen from the below threshold near field in Fig. 5.9, the near-field intensity exhibits a linear asymmetry, which is consistent with the linear-gain profile. At the indicated operating

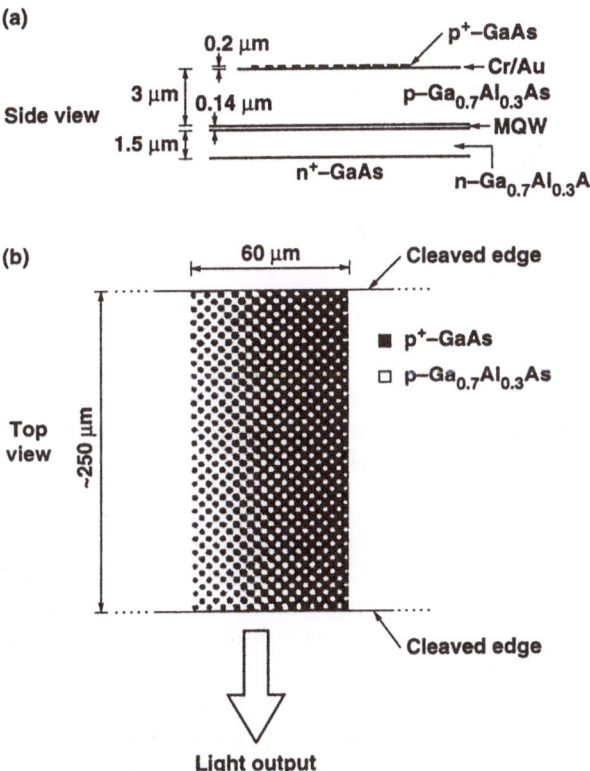

Fig. 5.8. Two-dimensional array of gain dots formed by varying the coverage of the contact to the p^+GaAs cap layer is shown in side view (a) and top view (b) [5.7, 5.8].

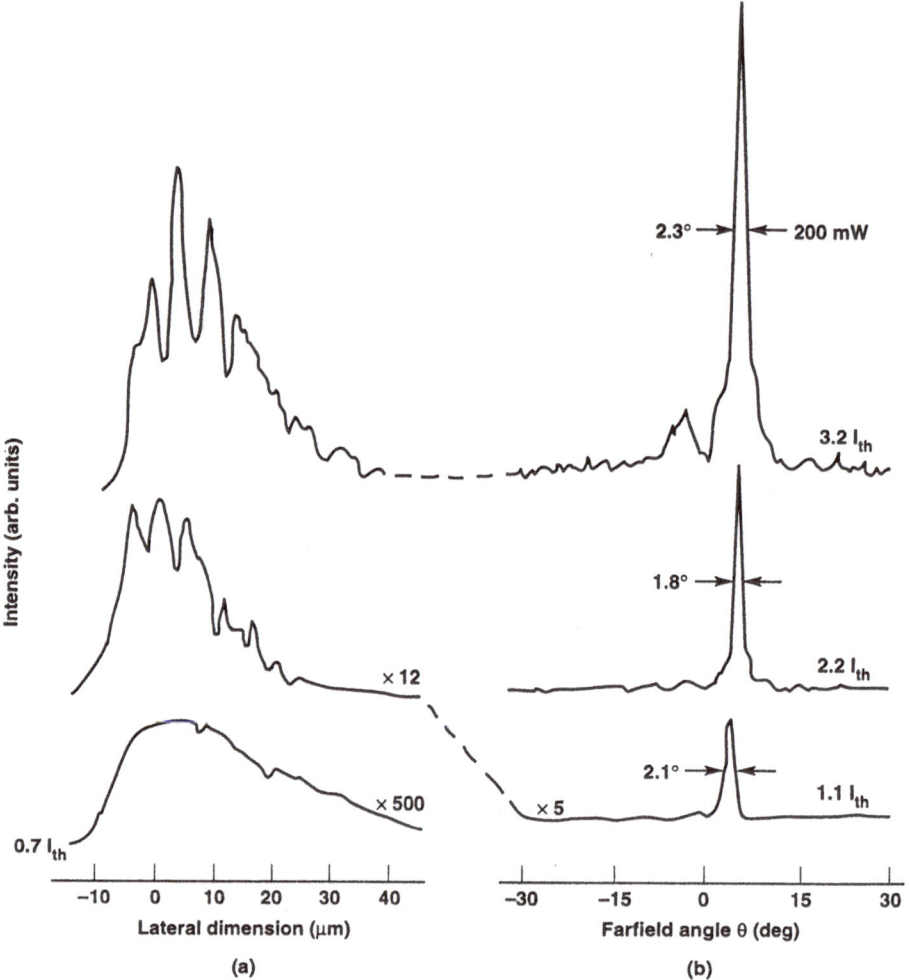

Fig. 5.9. Experimentally measured near (**a**) and far-field patterns (**b**) for the 100 μm wide gain-tailored laser of [4.20].

levels above threshold, the far-field output beams are all predominantly single lobed and emitted at an off-axis angle. However, at higher operating levels (above 200 mW), the far-field output did not remain single lobed. This was likely a consequence of gain saturation, which would tend to cause a reduction in the linear-gain profile across the laser.

The tendency of gain-tailored lasers and laser arrays to produce predominantly single-lobed output beams is an interesting contrast to the more-familiar behavior of laser arrays with a periodic-gain profile, where predominantly double-lobed output beams occur. To understand the mode characteristics of gain-tailored lasers, one needs to model a waveguide with an asymmetric dielectric profile. Our discussion of the analysis of gain-tailored structures, will follow the approach of *Lindsey* et al. [4.20]. In this analysis, analytic expressions for the unsaturated eigenmodes were found by using the effective-index approach and modeling the lateral structure as a linear asymmetric waveguide. Therefore, thermal effects, saturation, and carrier diffusion effects were not taken into account. The analysis is similar to the parabolic waveguide model discussed in Sect. 3.1.4. Instead of a parabolic permittivity, a linearly-graded, complex, effective-index profile $n(x)$, of the form [4.20]

$$n(x) = \begin{cases} n_e & x < 0 \\ n_0 - skx(\alpha + \mathrm{i}) & 0 \leq x \leq w \\ n_e & w < x \end{cases}, \tag{5.1}$$

is used. In (5.1), α is the antiguiding factor, w is the width of the gain region, and n_e is the complex index outside the gain region. The parameter s that appears in (5.1) characterizes the slope of the lateral gain gradient and it is given by,

$$s = \frac{g(0) - g(w)}{2k^2 w}, \tag{5.2}$$

where $g(0)$ and $g(w)$ are the transverse modal gains at $x = 0$ and $x = w$ respectively. The gain-tailored structure that was analyzed in [4.20] is illustrated in Fig. 5.10.

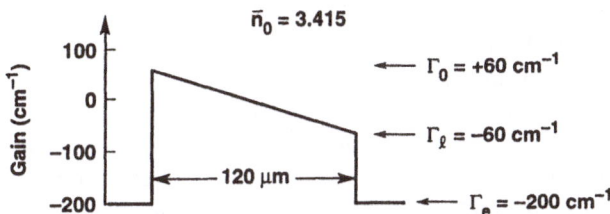

Fig. 5.10. Schematic diagram of the asymmetric linear-gain profile used in the analysis of [4.20], with a width of $w = 120\,\mu m$ and a gain gradient of $1\,\mathrm{cm}^{-1}/\mu m$. Note that $\bar{n} = \mathrm{Re}\{n_0\}$.

To calculate the modes of the asymmetric linear gain profile, *Lindsey* et al. solved (4.10) for, $\varepsilon_{\mathrm{eff}}(x) = n^2(x) \approx n_0^2 - 2n_0 k\sigma x$, under the assumption that the amplitude of the modes has no axial or z variation. In this case, the eigenfunctions are a linear combination of the Airy functions with complex argument. Applying the field continuity conditions at $x = 0$ and $x = w$, and the requirement that the field decay to zero at large distances from the gain region, yields the spectrum of eigenvalues β that is presented in Fig. 5.11. A detailed description of the actual method for calculating the eigenvalue spectrum is found in [4.20].

The complex values of β, which are indicated by the crosses in Fig. 5.11, fall along three distinct lines. The crosses denoted by $(+)_\nu$ corresponds to the branch of highest-gain modes, that undergo laser oscillation first. The

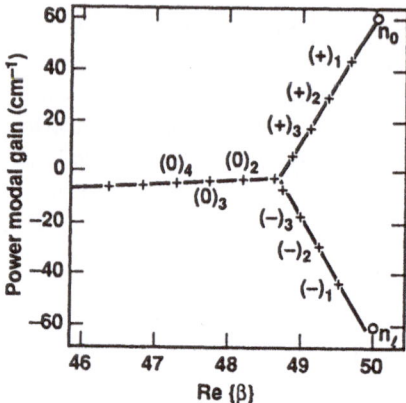

Fig. 5.11. Plot of the spectrum of complex eigenvalues β, depicting the three mode branches of the linear-asymmetric waveguide. The power modal gain, $\gamma = -2 \, \text{Im} \, \{\beta\}$ [4.20].

branches of modes denoted by $(0)_\nu$ and $(-)_\nu$ are all lower-gain modes, with the modes on the $(0)_\nu$ branch all having nearly the same gain. Note that the difference in the gain of the fundamental mode, $(+)_\nu$, and the next highest-gain modes is more than $10 \, \text{cm}^{-1}$, which considerably larger than the $\approx 1 \, \text{cm}^{-1}$ gain discrimination for multiple-stripe gain-guided arrays. In Fig. 5.12, the calculated near fields and far fields from [4.20] are shown for selected modes of the $(+)$, (0), and $(-)$ branches. The value of the mode index, ν, is indicated next to each field profile.

It is seen from Fig. 5.12 that the near field of the fundamental mode is localized at the end of the laser where the gain is largest; and as expected, it has the highest modal gain. The intensity peak of each of the higher-order modes on the $(+)$, (0), and $(-)$ branches all exhibit a shift towards the end of the array where the gain is lowest. Although with increasing mode index ν, the near fields of the (0) show only a small shift in the intensity peak, instead the width of the near field experiences a significant broadening. These trends give rise to a reduced mode overlap with the gain region for the higher-order modes of the $(+)$, (0), and $(-)$ branches which is consistent with the ordering of the power modal gains in Fig. 5.11.

A characteristic that all of the modes in Fig. 5.12 have in common is a single-lobed far-field pattern, with an off-axis tilt that depends on the mode index. *Lindsey* et al. attributed this phase tilt to a net power flow from the high-gain side to the low-gain side of the laser, which results in an off-axis angular deflection of the emitted output beam towards the low gain side. In a laser with a symmetric gain profile, the power flow in each direction is equal, resulting in on-axis emission of all modes. Therefore, under multi-lateral mode operation of a gain-tailored laser, a single-lobed output beam would still be expected to occur, but the beam divergence would be broader with the emission angle shifted further off axis due to the contributions of the

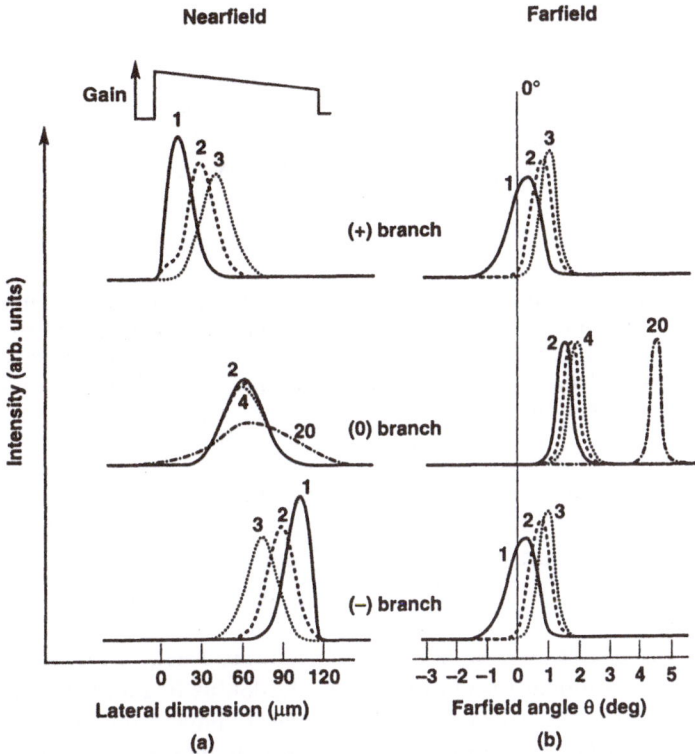

Fig. 5.12. The near-field (a) and corresponding far-field intensity patterns (b) for the indicated modes of the linear-asymmetric waveguide [4.20].

higher-order modes. As mentioned earlier, inadvertent gain tailoring due to structural nonuniformities could possibly explain some reported observations of single-lobed emission [4.83] from laser arrays with a periodic gain structure, as well as those with intentional asymmetric gain tailoring [1.80, 5.9].

5.3 Unstable-Resonator Diode Lasers

Another approach investigated for mode-control in broad-area diode lasers under high-power is the use of unstable-resonator designs. Figure 5.13 displays a schematic drawing of some of the basic unstable resonator designs. Unstable resonator are well known in non-semiconductor laser designs [2.31, 5.10]. In semiconductor lasers, unstable resonators are typically obtained by fabricating the curved facet reflectors at each end of a broad-area gain section. High-quality curved mirrors can be fabricated in semiconductor material in a well-controlled manner by using ion-beam etching techniqes [1.212, 1.472]. Since the mirror is curved in the plane of the pn junction, only the lateral mode characteristics will be effected and not the transverse mode. Unstable resonator diode lasers are considered to be

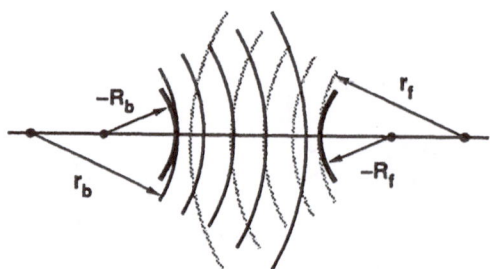

Fig. 5.13. Schematic of unstable resonator design. The mirror radii of curvature are indicated by $+R_b$ and $-R_f$, and the plus and minus signs correspond to the sense of the curvature. The radii of curvature of the oppositely propagating wavefronts are given by r_b and r_f. Note the wavefront and mirror radii of curvature are not generally coincident with each other [1.214].

gain-guided structures, because there is no built in lateral index variation in the gain section. The most common resonator design employed comprises a pair of mirrors with cylindrical curvature. Specific unstable resonator designs that have been demonstrated include a pair of concave mirrors [1.198, 1.199, 1.202, 1.208, 1.210], illustrated in Fig. 5.14a; half-symmetric cavities [1.197, 1.200, 1.205, 1.208, 1.209, 1.211, 1.212, 5.11], appearing in Fig. 5.14b; and confocal resonators [1.206, 1.208], as in Fig. 5.14c. Other designs for unstable-resonator semiconductor lasers have been considered. Theoretical analyses of laser cavities with arbitrary-shaped mirrors have been presented [1.204, 1.207] and unstable resonators comprising an axial distribution of diverging waveguide lenses have been demonstrated [1.215, 1.438]. Tapered unstable-resonator diode lasers [1.440, 1.441, 5.12], presented in Fig. 5.14d, are a design that appear to be unique to semiconductor lasers.

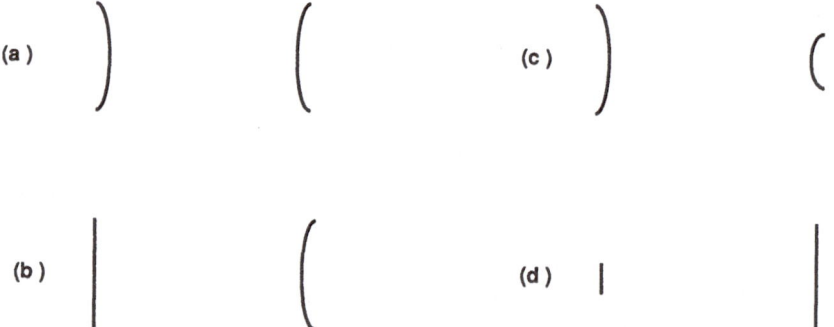

Fig. 5.14. Schematic of an unstable resonator designs that have been demonstrated in semiconductor material: (a) a pair of curved mirrors, (b) a flat high-reflector mirror and a curved output mirror (half-symmetric), (c) a large high-reflector curved mirror and small curved output mirror (confocal or self collimating); and (d) a large flat output mirror and a smaller flat high reflector at the end of single-mode waveguide section (tapered cavity).

Let us return to the basic unstable resonator design of Fig. 5.13. The curved mirrors, with radii of curvature $+R_b$ and $-R_f$, create a high-loss resonator through the introduction of lateral divergences in the oppositely propagating wave fronts with radii of curvature of r_b and r_f. The goal is to configure the mirrors so that the lowest-loss mode will produce a single-lobed output beam, and all higher-order modes are sufficiently lossy so as not to reach threshold. Another consideration in the design of semiconductor unstable resonators is that some provision should be made to suppress the tendency towards filamention and insure good beam quality.

5.3.1 Geometric Optics Analysis of Unstable Resonators

The cannonical method for analyzing the properties of unstable resonators developed by *Siegman* is to use the ABCD matrix representation to calculate the round-trip characteristics of the optical rays with respect to a specific reference plane [2.31]. It is usual to select the reference plane to be just inside the mirror, $-R_f$ in Fig. 5.13, where the output is emitted. In Fig. 5.15 an example of round-trip ray tracing is presented. To an observer situated to the right of mirror R_f in Fig. 5.15, the rays of a mode will appear to be radiating from a virtual source V, that lies a distance L_V from mirror R_f [1.199]. Note that depending on the details of the cavity design, the virtual source can be outside the resonator, as in Fig. 5.15, or it can be within the resonator. Each ray is characterized by a pair of numbers (r, r'), where r is the height of the ray relative to the optic axis and r' is the slope of the ray. A round-trip pass through the cavity, causes a ray (r, r') that originated from V to be transformed to a ray (Mr, Mr') at the output mirror R_f, where M is referred to as the *resonator magnification* or *geometric magnification*. The magnification, M, quantifies the size increase experienced by the modal wavefront upon completion of a round-trip pass through the cavity. The degree of magnification of an unstable resonator is an important parameter because it relates to the diffraction losses, as discussed below.

Using the notation in Fig. 5.15, the magnification of an unstable resonator can be expressed as [1.203]

$$M = g_1 g_2 \left[1 + \sqrt{1 - (g_1 g_2)^{-1}} \right]^2 , \tag{5.3}$$

where

$$g_1 = 1 - \frac{L}{R_b} , \tag{5.4}$$

and

$$g_2 = 1 - \frac{L}{R_f} . \tag{5.5}$$

The end losses of an unstable resonator can be represented as a distributed loss α_{el} of the form [1.203]

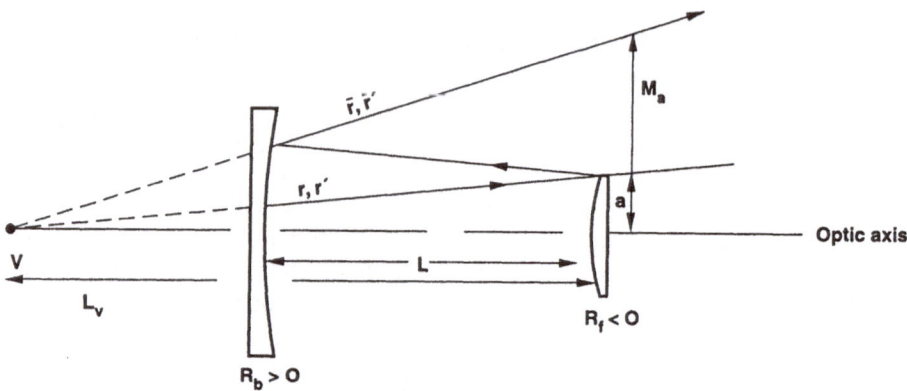

Fig. 5.15. Schematic of the round trip path traveled by a ray in an unstable resonator. L_V is the distance from the front mirror to the vitual source point of the point [5.13].

$$\alpha_{el} = \frac{1}{2L} \ln \left(\frac{M}{R_b R_f} \right) . \qquad (5.6)$$

Therefore, the threshold gain of an unstable resonator is simply given by $g_{th} = \alpha + \alpha_{el}$.

5.3.2 Unstable-Resonator Diode Lasers with Self-Collimated Output Beams

The confocal-resonator semiconductor laser displayed in Fig. 5.14c is notable because it can produce a collimated output beam [1.206, 1.208], as opposed to the divergent output beams produced by the designs in Fig. 5.14a, b, and d. This type of resonator design is also refered to as *self collimating*. In this resonator design, the confocal condition $R_b + R_f = 2L$ is satisfied, and the position of the vitual source in Fig. 5.15 is at an infinite distance from the back mirror. However, as discussed in [1.206, 1.208], the alignment tolerances on the dimensions in confocal-resonator designs dictate accuracies of better than $\pm 0.1\,\mu m$ for L and that the mirror surfaces be smooth to with $\pm 20\,nm$. These are tolerances have not yet been possible with current fabrication techniques.

It has been suggested that unstable-resonator semiconductor lasers should be designed so that no focal spot is formed between the mirrors of the resonator [1.208, 1.214]. The effects of this constraint on self-collimating resonators have been considered by *Lang* [1.214] for resonator geometries of the type that is depicted in Fig. 5.16. Note that in this geometery a collimated output beam is transmitted through the small mirror R_2, rather than being reflected off the large mirror, as in Fig. 5.14d. The high-optical powers expected at a focal spot within the resonator could cause self-focusing, resulting in beam filamentation, or even cause optical damage to the semiconductor material. It has been shown by [1.214] that the requirement of no focal spot form within the unstable resonator, dictates that the magnification M fall within the range of values

$$1 \le M \le \frac{n+1}{n-1} \ , \tag{5.7}$$

where n is the refractive index of the semiconductor material, which is typically $n \approx 3.4$ for AlGaAs. The lower-limit requirement of $M \ge 1$ is the condition that the resonator be unstable. The upper limit arises because of the condition that there be no focal spot in the resonator. For the aforementioned value of n the upper limit in (5.7) corresponds to a critical value for the magnification of $M = 1.833$. To avoid a focal spot within a semiconductor resonator with a self-collimated output beam, the magnification must be less than this critical value of 1.833. Because of the large index difference between the semiconductor and air, the size of the beam at the output mirror should not be so large that the critical angle for total internal reflection is exceeded, as this would reduce the external efficiency. To avoid internal reflection the output beam radius d should conform to

$$d < L\left(\frac{M+1}{n(M-1)} - 1\right) \ . \tag{5.8}$$

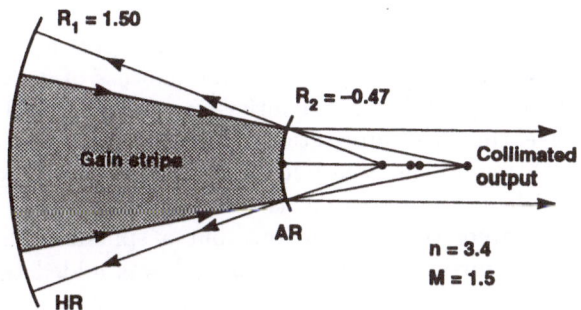

Fig. 5.16. Self-collimating unstable resonator analyzed by [1.214].

For the $M = 1.5$ and $n = 3.4$ example in Fig. 5.16, we see that (5.8) gives $d < 0.47L$. Self-collimating unstable resonators of the type analyzed by [1.214] have not yet been fabricated, though resonator designs of this type should prove to be an interesting area of future study.

5.3.3 Half-Symmetric Unstable-Resonator Diode Lasers

The half-symmetric unstable resonator, illustrated in Fig. 5.14, has perhaps been the most widely studied unstable-resonator semiconductor laser to date

[1.197, 1.200, 1.205, 1.208, 1.209, 1.211, 1.212, 5.11]. Perhaps the reason for this is that it is less demanding to fabricate a single, curved semiconductor mirror, as opposed to a pair of curved semiconductor mirrors. *Tilton* et al. have done self-consistent modeling studies of the half-symmetric unstable-resonator semiconductor laser and determined an optimized design for stable diffraction-limited operation, which has been confirmed by experimental studies [1.211, 1.213]. In their analysis, *Tilton* et al. used the effective-index approximation, where the lateral-field propagation was governed by (4.10). Thermal effects were not included, so the effective dielectric permittivity is given by (4.43) with the thermal contribution set to zero, $\Delta \varepsilon_T (x, z) = 0$. Carrier transport was modeled using a modified form of (4.49) where the effects of vertical transport on the injected current density were neglected, and a single non-stimulated recombination term was used that included both non-radiative recombination processes and spontaneous emssion. Figure 5.17 contains a schematic diagram of the external structure of the unstable resonator laser that was the subject of the theoretical and experimental studies of [1.213].

The parameters associated with the laser structure are listed in Table 5.2. For the half-symmetric resonator, the magnification M is related to the mirror radius of curvature R and the cavity length L by

$$M = \frac{\sqrt{L^2 - LR} + L}{\sqrt{L^2 - LR} - L} \ .$$

(5.9)

Note the R that appears above in (5.9), which represents the radius of curvature of the output mirror, is negative for the case of diverging waves.

As descibed in [1.213], a fast-Fourier-transform method was used to numerically solve the field propagtion. Current was injected only into the stripe of width s, and the area outside the current stripe contained within the diode width a_x, was treated as an absorbing region. The cavity was treated as if it were two flat mirrors, but a spherical phase shift (corresponding to the mirror radius of curvature) was added to the wave upon

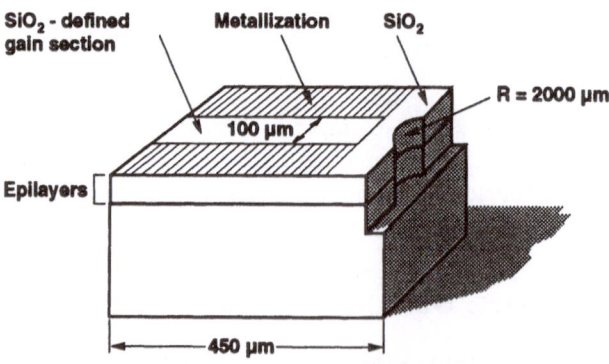

Fig. 5.17. Schematic diagram of the unstable-resonator diode laser of [1.213].

Table 5.2. Parameters used in [1.213] to model the half-symmetric unstable-resonator semiconductor laser design that appears in Fig. 5.17

Operating wavelength	837 nm
Diode width (a_x)	340 μm
Cavity length (L)	450 μm
Stripe width (s)	100 μm
Magnification range (M)	2 $-$ 3
Transparency current density (J_0)	115 A/cm^2
Confinement factor (Γ_y)	0.015
Differential gain coefficient (A)	5 A/cm
Loss (α)	5.0 cm^{-1}
Current injection efficiency (η)	0.969
Antiguiding factor (α)	2.7
Refractive index (n)	3.532
Facet reflectivities (R_b, R_f)	0.32

reflection from the curved mirror. By numerically simulating the output characteristics of various parameter sets for the half-symmetric unstable resonator design, it was found that the threshold current, power extraction efficiency, and beam quality could be optimized for magnifications in the range $2.0 \leq M \leq 3.0$ [1.211]. This restricted the mirror radius of curvature to the range $-4000\,\mu$m $\leq M \leq -1500\,\mu$m.

For the experiment of [1.213], four half-symmetric, unstable-resonator diode lasers were studied. Figure 5.18 presents the power-current characteristics of the four devices, along with that of a cleaved facet broad-area device for reference purposes. These measurements were taken under pulsed operating conditions with 100 ns long current pulses at a pulse repetition frequency of 1 kHz, so that thermal effects would not be expected to have an influence on the performance characteristics. The use of such short current pulses and low-duty cycles is desirable for minimizing the effects of heating. This is an important consideration for this study, because thermal effects were not included in the self-consistent model used in [1.213]. The devices had current stripes of width 100 μm; and from the values of magnification and radii of curvature reported in [1.213], the cavity lengths of all four devices were $\approx 450\,\mu$m.

It is seen that there is about a 67 % increase in threshold current, as well as a decrease in the differential quantum efficiency of the four unstable-resonator diode lasers relative to the broad-area cleaved facet laser. The power current charactersitics that were calculated using the self-consistent model of [1.213] are displayed in Fig. 5.19. For the calculated power-current curves, the unstable-resonator lasers exhibited only about a 36 % increase in the threshold current. This has been attributed to slight systematic deviations of the curved mirror surface from that of a cylindrical shape [5.13].

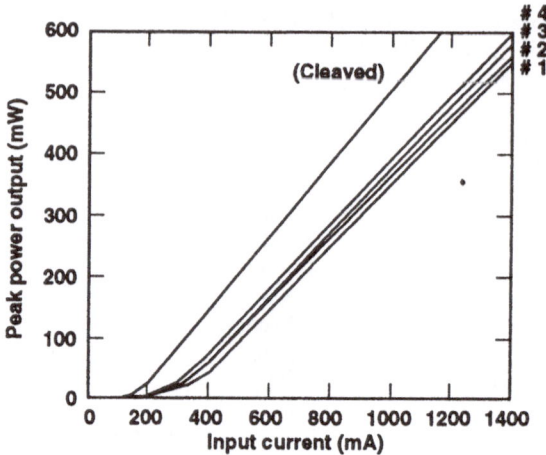

Fig. 5.18. Experimental power-current characteristics of the four unstable-resonator diode lasers: 1) $M = 2.53 \pm 0.3$, 2) $M = 2.30 \pm 0.15$, 3) $M = 2.51 \pm 0.3$, 4) $M = 2.67 \pm 0.3$, and a similar broad-area, cleaved-facet laser from [1.213].

When operating in a single lateral mode, the lateral direction of the output beam of the half-symmetric unstable-resonator laser, as is with most unstable-resonator designs, will have a diverging wave front that appears to be radiating from a virtual source that lies some distance behind the curved output mirror. In the cylindrical geometry of the semiconductor unstable resonator, a virtual line source is found. In order to obtain a quantititative measure of the beam quality, it is common to use a lens to reimage the virutal source onto a screen that is placed in the far-field zone, and measure the size of the re-imaged spot. By comparing the measured spot size with the calculated spot size (refered to as the diffraction limit), one can determine the beam quality. In addition, the power contained within the diffraction-limited bucket size was also measured. In their experiments on the half-symmetric unstable-resonator semiconductor lasers, *Tilton* et al. found

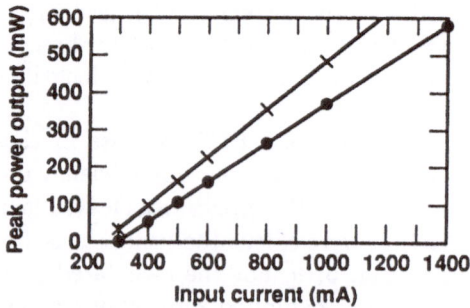

Fig. 5.19. Calculated power-current characteristics corresponding to the four unstable-resonator diode lasers and the broad-area, cleaved-facet lasers of Fig. 5.18 [1.213].

that the re-imaged near-field spot diameters remained diffraction limited up to current levels of $5I_{th}$. At current levels above $5I_{th}$, the reimaged spot size was observed to increase by about a factor of two.

In Fig. 5.20, representative output beam characteristics are displayed for one of the devices that was part of the experimental study in [1.213]. The corresponding output beam characteristics calculated from the self-consistent analysis of [1.213] appear in Fig. 5.21. Note that the flat-top profile of the calculated near field in Fig. 5.21a; whereas, the experimental measurement of the near field in Fig. 5.20 exhibits twin peaks with a large central minima. This minima in the near field is a dip in the intensity with no associated phase shift, as evidenced by the predominantly single-lobed far-field pattern observed in Fig. 5.20. However, the near-field intensity dip does increase the intensity of the two side lobes in the far field, as seen in Fig. 5.20. In comparison, the calculated far-field pattern in Fig. 5.21b that would result from the flat-top near field in Fig. 5.21a, has side lobes that have lower intensity relative to the dominant central lobe. The power-in-the-bucket measurements of the reimaged-virtual spot indicated between 59 % and 65 % of the total power output was contained within the diffraction limited spot. This should be compared with the theoretical maximum of 80 % power-in the-bucket corresponding to the diffraction-limit for the calculated near field pattern in Fig. 5.21b. The agreement between experimental and calculated results are good. A subsequent re-examination of the curved mirrors in the devices reported in the study by [1.213] revealed a slight systematic deviation of the mirror profile from that of a cylindrical shape [5.14]. This deviation manifested itself as cusp near the center portion of the mirror. When this cusp was accounted for in the self-consistent analysis, the calculated output beam characteristics reproduced the twin-peaked near-field pattern that was measured [5.14]. The cw operating characteristics, particularly the output-beam stability, of unstable-resonator diode lasers is expected to be strongly influenced by heat dissipation [5.15]. So far cw studies of the half-symmetric unstable-resonator laser have revealed good quality output beams at power levels under 100 mW [1.212, 5.15].

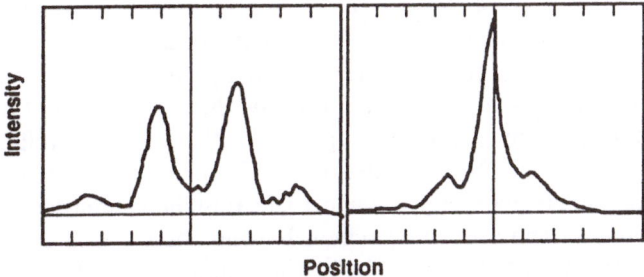

Fig. 5.20. Experimental measurements of the near- and far-field patterns (reimaged virtual source) of the device with $M = 2.51 \pm 0.3$ (number 3 in Fig. 5.18) when operated at $5I_{th}$ (one horizontal division $= 66.56\,\mu\mathrm{m}$ [1.213].

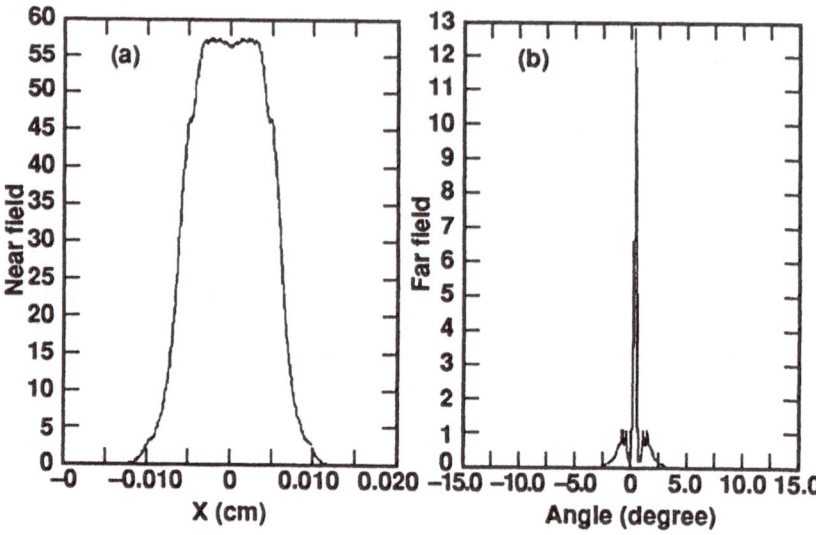

Fig. 5.21. Calculated (**a**) near- and (**b**) far-field pattern (re-imaged virtual source) of a half-symmetric unstable-resonator semiconductor laser with $M = 2.51 \pm 0.3$ operated at $5 \times I_{\text{th}}$ (other parameters are specified in Table 5.2) [1.213].

5.3.4 Tapered-Cavity Diode Lasers

The unstable-resonator semiconductor laser oscillator design consisting of a tapered-gain region terminated by flat mirrors at each end, as in Fig. 5.14d, have also been investigated [1.440, 1.441, 5.12]. Tapered-gain structures configured as single-pass power amplifiers, have been studied extensively, as discussed in Sect. 8.1.4. In the tapered-cavity unstable-resonator laser, a straight section of single-lateral mode active waveguide section is terminated at one end by a cleaved facet; and at the other end, it is integrated to a tapered cavity, as illustrated in Fig. 5.22. The tapered-gain section typically extends over a length that can range from between 2 to 3 mm, and the angle of the tapered-gain section is selected to accomodate the spread in the lateral direction due to diffraction as the mode propagates towards the large flat mirror. In this unstable-resonator design, the diverging wave front results from the diffraction in the lateral direction that occurs because there is no optical confinement in this dimension of the tapered-gain section other than the tapered-gain profile itself. The etched grooves that are indicated in Fig. 5.22 act as a mode filter by preventing optical feedback due to reflection from the regions of the facet R_1 that extend outside the single-mode active waveguide section. This is necessary to deflect the amplified spontaneous emission that propagates in the absorbing region beyond the tapered section, out of the laser cavity. As discussed in [5.12], the amplified spontaneous emission emitted from the wide end of the taper that propagates outside the taper can build up to a level where oscillation in multiple lateral modes or filamentation can occur, resulting a degradation of the beam quality.

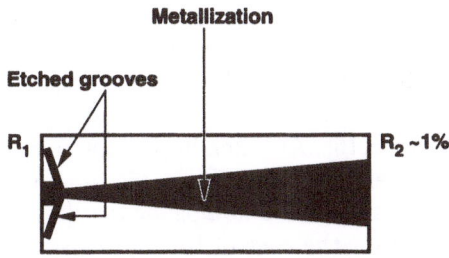

Fig. 5.22. Schematic diagram of a tapered-cavity diode-laser oscillator, where $R_1 \approx$ 30 % [1.441].

The far-field characteristics of the tapered-cavity unstable-resonator lasers that were studied by *Walpole* and *Kintzer* [1.440, 1.441] demonstrated predominantly single-lobed output beams with near diffraction-limited performance over a wide range of operating conditions. In Fig. 5.23, the measured lateral far-field characteristics are presented for a device with a 2 mm long tapered-gain section and a 215 μm wide output facet. At the $3I_{th}$ and $9I_{th}$ operating points, the measured output powers were 0.75 W and 3.2 W, respectively. Over the operating range indicated in Fig. 5.23, the far-field beam divergence (full-width at half maximum) varied between 1.4 and 1.7 times the diffraction limit. A transfer of the power from the central lobe to the pedestal, with increased injection current, is evident from Fig. 5.23. Corresponding measurements of the fraction of power contained within the central lobe spanned a range from 88 % at an operating current of 1.0 A, to 56 % at an operating current of 4.5 A. The source of the observed beam degradation was likely due to the changes in dielectric characteristics of the tapered-gain

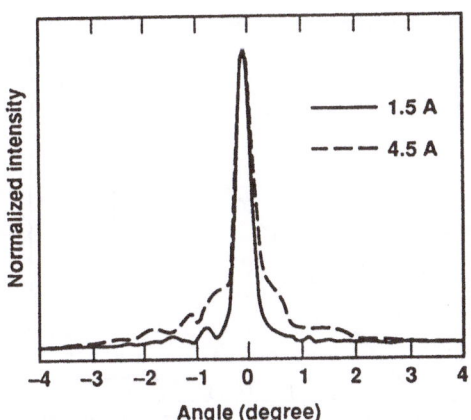

Fig. 5.23. Lateral far-field intensity pattern of the tapered laser studied by [1.441]. The solid line corresponds to cw operation at $3I_{th}$ (1.5 A) and the dotted line corresponds to cw operation at $9I_{th}$ (4.5 A) [1.441].

section with operating level. Self-consistent beam propagation model calculations of this device that accounted for local variations in the refractive index, carrier density, and temperature yielded results that gave good qualitative results with the measured power-current characteristics, as well as the observed far-field beam patterns [5.16]. Another important factor that can influence the beam quality of tapered-gain section devices, as well as broad-area devices, is material and structural inhomogeneities [5.17, 5.18]. This will be discussed in more detail for the case of broad-area semiconductor amplifiers in Chap. 8.

The maximum power characteristics were also studied by [1.441] for a tapered-cavity unstable-resonator laser that was 3 mm long with a 325 μm wide output facet. These data are displayed in Fig. 5.24. A maximum cw power of 4.2 W was obtained. At a power output of 4 W, the far-field was predominantley single-lobed (as seen in the inset to Fig. 5.24) with about 50 % of the power contained within the nearly diffraction-limited central lobe. Above the 4.2 W output level, a sudden failure occured, as indicated in Fig. 5.24 by the sharp drop in power output just above 10 A. This was attributed to catastrophic optical damage to the uncoated facet, R_2 in Fig. 5.22. Just above the 4.2 W operating level, the power density at the facet R_2 was estimated to be $\approx 1\,\mathrm{MW/cm^2}$, which closely corresponds to the facet-damage intensity for uncoated facets.

As discussed by [1.441], this result suggests that even higher-power outputs may be possible if effective solutions to the facet damage limit can be found. To achieve higher-powers, attention should also be given to heat dissipation, thermal management, and facet treatment methods to raise or eliminate the optical damage limit. The slight curvature of the power-current characterisic in Fig. 5.24 is likely due to the onset of thermal saturation. Be-

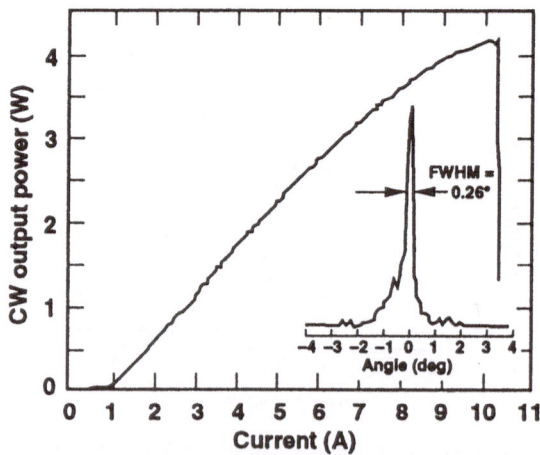

Fig. 5.24. Power-current characteristic for the 3 mm long tapered laser of [1.441]. The inset corresponds to the lateral far-field pattern measured at an operating current of 9 A.

cause of the sub-linear dependence of the thermal resistance with lateral dimension that was discussed in Sect. 2.8.2, tapered-gain section devices would be expected to exhibit an axial gradient in the active-layer temperature profile. The impact of such thermal gradients on the performance characteristics of tapered-gain lasers has not yet been ascertained. However, the impact of thermal effects has been addressed for tapered-gain laser amplifiers, as described in Chap. 8. Note that the importance of uniform heat sinking for obtaining uniformity in the local waveguide characteristics of diode-laser arrays has been analyzed by [2.88]. It may be that by appropriate tailoring of the heat sink for tapered-gain section device could also enhance the high-power performance properties.

6. Index-Guided Diode-Laser Array Oscillators

This chapter will review some selected examples of index-guided diode-laser array oscillator structures and their observed performance characteristics. Again, emphasis is placed on examples that exhibit performance characteristics that can be understood in terms of the array theories discussed in Chap. 4. Although, some of the examples in this chapter have been selected because they are believed to represent qualitative demonstrations of important characteristics.

The unstable output characteristics versus current drive exhibited by gain-guided diode lasers and diode-laser arrays made these devices unsuitable for many applications where good-beam quality, spectral purity, and stable output characteristics were requirements. This is the principal reason why only single-element index-guided lasers have been used in optical recording and compact-disk player applications. Recognition of this led to investigations of diode-laser array structures where the effective lateral dielectric permittivity was fixed by a periodic refractive index variation that was fabricated into the array structure. The idea being that if the built-in lateral variation of the refractive index was much greater than the gain-induced index depression and the index changes due to heating, then the modal characteristics could be decoupled from the operating conditions.

Laterally-coupled arrays such as the *Channeled-Substrate-Planar* (CSP) array (Sect. 6.1), *ridge-guided* array (Sect. 6.2), and the leaky-wave coupled arrays of *antiguided lasers* (Sect. 6.3), all incorporate a periodic, lateral index variation that is fabricated into the laser structure. For this reason, these types of arrays have been refered to as *index-guided arrays*. Because of the presence of carriers, all of the aforementioned array structures can exhibit characteristics of the *mixed-guides* that were discussed Sect. 3.1. Monolithic surface emitting arrays were also investigated, as they offer the promise of scalability in two-dimensions, as well as, improvements in beam quality and source brightness. These structures are dicussed in Chap. 7.

6.1 Channeled-Substrate Planar Laser Arrays

The CSP diode-laser array, which is based on the single-element CSP laser [6.1], was one of the earliest index-guided designs for diode-laser arrays that was investigated [1.71]. Note that the CSP lateral-waveguide structure is similar to the *V-channeled Substrate Inner Stripe* (VSIS) single-element laser used for optical recording and laser-audio applications [6.2]. The CSP-laser array has been fabricated using both a two-step liquid-phase epitaxy growth process [1.117, 1.120] and a single-growth process [1.71]. The basic CSP array structure is depicted in Fig 6.1. Channeled-substrate-planar arrays were fabricated by using liquid-phase epitaxy to obtain planar deposition of the epilayers of a double heterostructure laser over a wafer where a periodic array of channels had been etched [1.71, 1.117, 1.120]. The use of a double heterostructure active layer grown by liquid phase epitaxy in the CSP laser array, as opposed to the MOCVD-grown quantum-well active layers that were used in many of the gain-guided arrays discussed in Sect. 5.1, should be pointed out, as MOCVD grown material is generally recognized as being of superior quality. The periodic array of channels results in a lateral variation in the transverse waveguide structure. As discussed in Sect. 4.2.1, the effective-index approximation analysis of this type of lateral variation in the transverse-waveguide structure corresponds to a built-in variation in the effective-dielectric permittivity in the lateral direction. For this reason, the CSP array, indicated in Fig 6.1, has been viewed as an index-guided structure. However, the transverse waveguide in the regions between the channels, termed the wings, provided a passive loss in the form of a radiation leakage into the substrate [2.33, 6.3, 6.4]. This results in a modulation of the lateral gain, which will improve the mode discrimination for the 180° or out-of-phase mode.

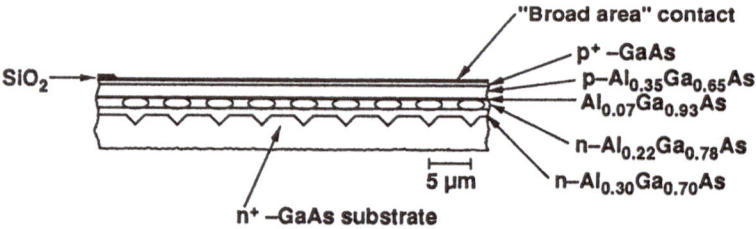

Fig. 6.1. Schematic diagram showing the epilayer structure in relation to the V-shaped channels [1.71].

Experimental studies of CSP-laser arrays have revealed diverse operating properties, depending on the proximity of the active layer to the channels and the uniformity of the structure. There are experimental observations of both index-guided array behavior [1.71, 1.117, 1.119, 1.120] and gain-guided laser behavior [3.82] in CSP-laser arrays. In structures where the built-in lateral index variation is less than the index change due to the injected carriers, the CSP array can behave as a gain-guided, broad-area laser with a small superimposed index pertubation [3.82]. Although the CSP-laser array was not intended to operate in this fashion, the technological difficulties associated with the liquid-phase epitaxy growth process can produce significant structural nonuniformities.

Let us consider first the experimental studies on index-guided CSP-laser arrays. *Matsumoto* et al. [1.117] fabricated three-element CSP diode-laser arrays, and observed single-spectral and single-spatial mode operation under cw operating conditions at power outputs up to 50 mW. Above this power level, multi-mode operation was observed. A spectrally-resolved study of the near-field pattern of the three-element array under multi-mode operation revealed array mode characteristics that were indicative of an index-guided structure. Figure 6.2 presents the experimental results of *Matsumoto* et al. [1.117] for the spectrally-resolved output of the three-element CSP array. The near-field patterns (left-hand side) and the far-field patterns (right-hand side) corresponding to the three array modes are displayed. Each mode pattern is separated by 0.2 Å.

The correspondence of the mode-pattern intensity peaks with the emitter positions is a characterisic of index-guided laser arrays, where the lateral index variation dominates any gain-guiding effects due to the injected carriers [6.5] and the optical coupling between elements can be modeled by the evanescent fields of the guided modes of the individual waveguide elements.

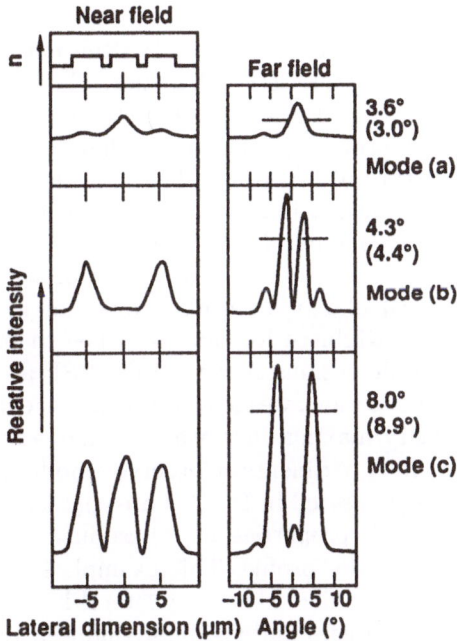

Fig. 6.2. Measured spectrally-resolved near-field (left column) and far-field (right column) characteristics of a three-element CSP diode-laser array are presented. The three modes observed are labeled as mode (a), mode (b), and mode (c); and all exhibited adjacent-separations of ≈ 0.2 Å. Note that the beam divergences calculated from the couple-mode theory of [4.21, 4.22] are indicated in parentheses [1.117].

The far-field patterns in Fig. 6.2c have beam divergences which were found to give good agreement with those calculated using the coupled-mode theory of [4.21, 4.22]. When index guiding is dominant, then some operating characteristics can be explained by treating the array modes as a linear combination of the eigenmodes associated with the isolated array elements. Recall from Sects. 4.2.3 and 5.1 that this method is not valid in gain-guided or leaky-wave coupled diode-laser arrays where radiation modes dominate the optical coupling between array elements.

The observed intensity envelope for each of the higher-order modes, modes (b) and (c) in Fig. 6.2, is uniform over the array. Similar observations of uniform near fields in eight-element CSP laser arrays were reported by *Goldstein* et al. [1.120, 6.4]. This behavior has not been predicted by the coupled-mode models of index-guide diode-laser arrays that include only nearest-neighbor coupling [4.21, 4.22]. Uniform mode intensity patterns have been predicted theoretically for 1) index-guided arrays with non-uniform channel spacings [6.6, 6.7], 2) CSP arrays with anti-meltback layers [6.4], and 3) uniform structures with small lateral-index variations (perturbed broad-area laser) and significant absorption outside the array [4.37] or when there is significant heating in the array [4.60]. However, the reason for the observed uniformity of near-field patterns in CSP arrays has not yet been experimentally identified.

As already mentioned above, there are situations when CSP-laser array structures exhibit characteristics that more closely correspond to those of broad-area lasers [3.82]. *Yu* et al. [3.82] studied a CSP array structure that exhibited modal properties that were characteristic of the Hermite-Gaussian mode structure of broad-area lasers. Figure 6.3 shows the results of a spectrally-resolved near-field measurement that were made under pulsed operating conditions. It is apparent that only in the case of mode number 8, do the intensity peaks correlate with the number of array elements.

In Fig. 6.4, the calculated intensity profiles of the Hermite-Gaussian modes from (3.27) are exhibited for the case when there is only a parabolic dielectric profile with dimensions similar to the CSP array studied by *Yu* et al.. With increasing mode index we see that the lateral extent of the mode and the number of intensity maxima increase with increasing mode index. This trend is also evident in experimentally observed mode structure in Fig. 6.3.

The lateral modes presented in Fig. 6.3 are characteristic of a broad-area laser where the waveguide properties are determined by the injected carriers and the lateral temperature profile. This example is an illustration of the necessity of fabricating a sufficiently strong lateral-index modulation in order to overcome the effects of gain and thermally-induced waveguide effects.

In Fig. 6.5, the time-averaged lateral far-field pattern of the multi-mode CSP array studied by *Yu* et al. appears. The double-lobed character of this far-field pattern indicates that the multi-mode array operation is dominated by higher-order modes with alternating phase in the near field. Note that the secondary lobe structure between the dominant lobes that is characteristic of

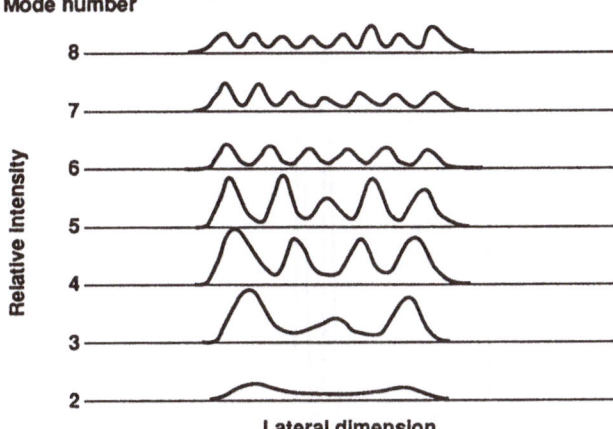

Fig. 6.3. Measured spectrally-resolved near-field characteristics of an eight-element CSP diode-laser array [3.82].

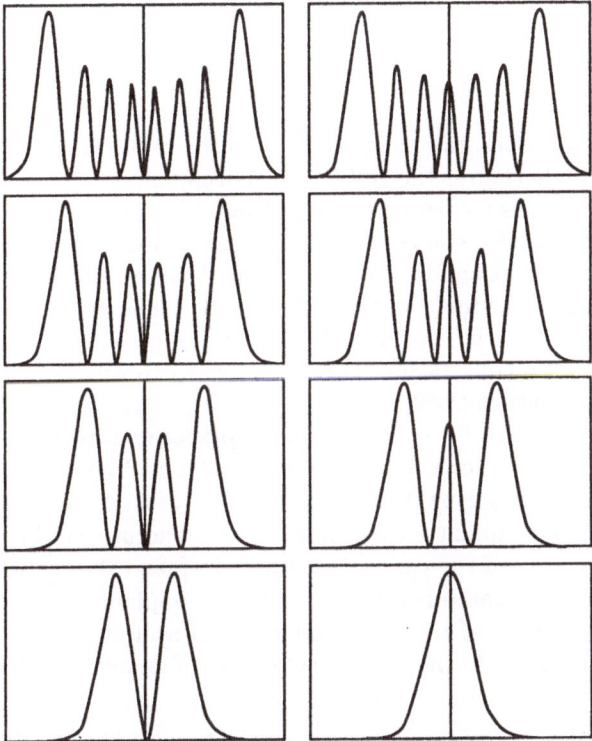

Fig. 6.4. Calculated intensity profiles for the Hermite-Gaussian modes of a parabolic-dielectric index profile.

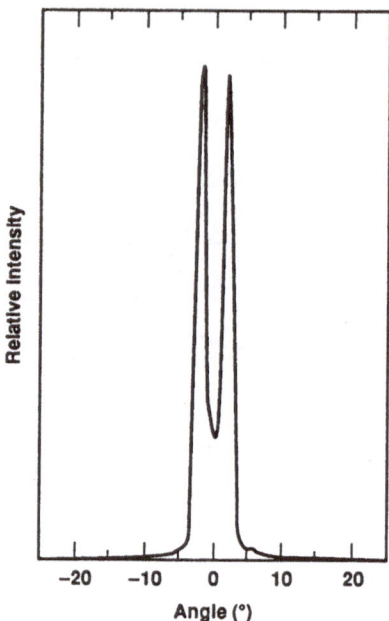

Fig. 6.5. Measured lateral far-field pattern of the multi-mode CSP array [3.82].

single-array mode operation is washed out in the multi-mode far-field pattern in Fig. 6.5. Furthermore, the full-width half-maximum beam divergence of each dominant lobe is measured to be 1.8 times broader than that expected for the diffraction-limit for single-array mode operation, which is also indicative of multi-mode operation. In multi-array mode operation, the array is not necessarily operating simultaneously in all modes. The multi-mode CSP array studied by *Yu* et al. exhibited complex dynamics in the spatial and temporal evolution of the output beam characteristics on both the picosecond and nanosecond time scales [3.82].

More recent studies have focused on nonplanar quantum-well laser arrays [6.8, 6.9], which are grown by a single-step *MetalOrganic Chemical Vapor Deposition* (MOCVD) growth over a selectively-etched corrugated substrate [6.8]. The nonplanar quantum-well laser array structure can be thought of as an improved version of the channeled substrate-planar laser. Both structures are grown over corrugated substrates, but the nonplanar quantum well laser array takes advantage of the better quality material offered by MOCVD, as well as the superior performance characteristics of quantum-well active layers. Such nonplanar-active-waveguide structures can suppress parasitic oscillation due to ASE build up along the lateral dimension. Devices as wide as 3.1 mm and 483 μm long cavity length have demonstrated pulsed power outputs as high as 8 W and threshold current densities $< 200\,\mathrm{A/cm^2}$ [1.121]. However, these devices have not demonstrated good mode discrimination characteristics. The multimode operation of these nonplanar arrays has been attributed to spatial-hole burning, as well as nonuniform current injection in the non-planar channels [6.9].

6.2 Ridge-Guided Laser Arrays

Another index-guided diode-laser array structure that has also been the sub-ject of some research is the multiple ridge-guide. This particular structure is usually less demanding to fabricate than the nonplanar laser-array structures, because the epitaxial growth of the active layer is done over a planar surface. After the active layer growth, the ridges can be fabricated by a chemical etch [1.76, 1.77][6.10]-[6.13] or dry etch [1.95]. In some cases, a second growth over the ridges is done [1.76], and this is referred to as the buried-ridge structure.

In Fig. 6.6, the geometry of a generic ridge-guided diode-laser array is dis-played. The effective index n_e and transverse mode confinement factor (2.1) Γ_y profiles are indicated. The Γ_y profile relates directly to the lateral gain profile. As discussed in [1.76, 6.12, 6.13], the details of the structure composi-tion and dimensions can also have a strong influence on the difference in gain between the laser (under the ridge) and the interelement region (outside the ridge), which we shall refer to as the *lateral gain difference*. For example, loss can be incorporated into the interelement region by selective Zn difussion, or as in the case of the CSP array structures described in Sect. 6.1, selective formation of a leaky transverse waveguide. Alternatively, the gain in the in-terelement region can be made to be higher than the gain under the ridge by selectively varing the resistivity.

Experimental studies of ridge-guided diode-laser arrays have been directed at structures with a built-in, piston-like modulation of the effective index, as seen in Fig. 6.6. Different lateral-gain profiles have been analyzed. Besides the

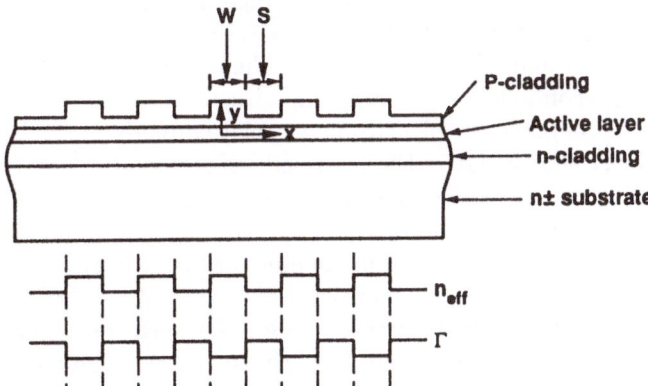

Fig. 6.6. Schematic diagram of a generic ridge-guided laser-array structure showing the resulting lateral profiles for the effective index and transverse-mode confinement factor [4.61].

uniform gain profile along the lateral direction [1.77, 1.76, 6.10, 6.11], other ridge-guided array structures have provided gain in the interelement regions by using separate contacts [1.90, 6.14]. Also by appropriate selection of the ridge-guide parameters, the gain can be made to be higher in the interelement regions than in the ridges [6.12, 6.13]. Control of the losses in the interelement regions has been investigated in structures where current is injected only into the ridges [1.279] and with distributed saturable absorbers along the lateral direction [1.95]. In many cases studied [1.76, 1.77, 1.95, 1.279][6.11]-[6.13], predominately single-lobed far-field patterns were observed over a limited range of operating conditions. These observations can be explained as being due to excess gain that can occur in the region between the ridge-guide elements and the presence of leaky-wave coupling.

    Twu et al. [4.61] have done a self-consistent analysis of ridge-guided laser array structures that is helpful for understanding the observed operating characteristics. In this analysis, the lateral modes of the array were modeled using the one-dimensional wave equation, (4.10), obtained from the effective-index method but with the axial or z dependence neglected. Also, thermal effects were not included, so the lateral dielectric permittivity (4.43) that appears in (4.10), was only influenced by the injected carriers. With these approximations, the lateral dielectric permittivity given by (4.43) reduces to, $\varepsilon_{eff}(y,0) = \overline{\varepsilon_{eff}}(x,0) + \Gamma\Delta\varepsilon_a(x,0)$, where $\Delta\varepsilon_a(x,0)$ is given by (4.44). Carrier transport was modeled in a manner similar to the treatment described in Sect. 4.3.2, but the one-dimensional approximation described in [2.82] was used to solve the two-dimensional Poisson equation. The cases of current injection only for the higher-index ridge guides were modeled, as well as uniform current injection across the entire array structure. Table 6.1 contains the device parameters that were employed in this analysis.

    In their analysis, Twu et al. calculated the mode structure of a five-element ridge-guided type diode-laser array for a range of built-in lateral effective index differences Δn_{eff} that spanned the range from $\Delta n_{eff} = 0$, which corresponds to gain-guided elements, to $\Delta n_{eff} = 1.2 \times 10^{-2}$, which corresponds to a strongly index-guided diode-laser array. The results of this analysis for the case of a stripe-contacted array are summarized in Table 6.2 and Fig. 6.7. Note that the differences in effective index relate directly to the height of the p-cladding layer that is left in the interelement region after etching.

    Table 6.2 gives the transverse mode confinement factor and the effective index in the interelement regions for each different value of Δn_{eff}. Also, given is the threshold current and mode number for the array mode that first reaches threshold. We see that increasing the strength of the built-in index guide, reduces the threshold current. This is because the regions of maximum optical field are confined more tightly to the ridge guides where most of the gain is located. The interelement transverse mode confinement factor increases as Δn_{eff} increases, because the height of the p-clad in the interelement is decreased to achieve larger values of Δn_{eff}. As the p-clad thickness is decreased between the ridges, the transverse mode in the interelement region

Table 6.1. Parameters used in [4.61] to model the 5 element ridge-guided array

Active layer thickness	$0.15\,\mu\mathrm{m}$
Ridge width	$3\,\mu\mathrm{m}$
Ridge spacing (center-to-center)	$3\,\mu\mathrm{m}$
Index of cladding layer	3.353
Index of active layer	3.59
Operating wavelength	864 nm
Antiguiding factor (α)	2
Gain coefficient ($a = \sigma$)	$3 \times 10^{-16}\,\mathrm{cm}^2$
Transparency loss ($b = \sigma N_{\mathrm{tr}}$)	$450\,\mathrm{cm}^{-1}$
Diffusion coefficient	$33\,\mathrm{cm}^2/\mathrm{s}$
Spontaneous recombination coefficient	$1.0 \times 10^{-10}\,\mathrm{cm}^3/\mathrm{s}$
Nonradiative recombination time (τ_{nr})	5 ns
Intrinsic carrier concentration	$1.79 \times 10^6\,\mathrm{cm}^{-3}$
Free carrier absorption coefficient α_{fc}	$6 \times 10^{-16}\,\mathrm{cm}^2$
Cladding layer absorption coefficient α_c	$10\,\mathrm{cm}^{-1}$
Conductance per unit area between p-contact and active layer	$4.0 \times 10^5\,\Omega^{-1}\,\mathrm{cm}^{-2}$
Sheet conductance for lateral current spreading	$13.0 \times 10^{-3}\,\Omega^{-1}$

Table 6.2. Stripe-contacted ridge-guided array parameters for different Δn_{eff}

$\Delta n_{\mathrm{eff}} =$	0	1.3×10^{-3}	3.7×10^{-3}	6.5×10^{-3}	1.2×10^{-2}
Interelement Γ_y	0.50127	0.51490	0.52689	0.53461	0.54088
Interelement n_{eff}	3.426091	3.424789	3.422327	3.419589	3.414692
I_{th} [mA]	148.9	145.8	141.1	140.2	139.6
Threshold mode number	1	1	5	4	4

is pushed more toward the active layer and n-clad, resulting in an increase in the transverse-mode confinement factor in the interelement region.

Figure 6.7 presents the near-field profiles of some of the array modes at threshold for the different values of Δn_{eff} considered. The mode numbers were assigned according to the decreasing propagation constant, which is equivalent to decreasing effective index. Therefore, the modes labeled by a 1 always correspond to the fundamental or in-phase mode, and successively higher modes are indicated by the larger integers. Fundamental mode operation occurs at threshold only for the weakest index guide where $\Delta n_{\mathrm{eff}} = 1.3 \times 10^{-3}$. *Twu* et al. also found that the array mode discrimination decreased with increasing Δn_{eff}. Additional modes also appear for larger Δn_{eff} as each isolated ridge waveguide can support more than one mode in this situation.

The mode ordering found at threshold is not preserved as the injection current to the array is increased. For the case of the $\Delta n_{\mathrm{eff}} = 1.3 \times 10^{-3}$, higher-order modes begin to oscillate at higher injection currents along with the fundamental mode, and stable in-phase mode operation is no longer possi-

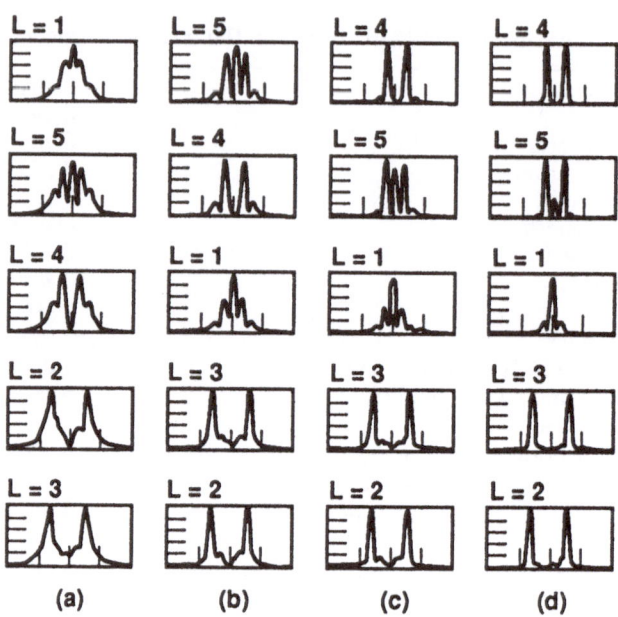

Fig. 6.7. Calculated near field of array modes for (a) $\Delta n_{\text{eff}} = 1.3 \times 10^{-3}$ (b) $\Delta n_{\text{eff}} = 3.7 \times 10^{-3}$, (c) $\Delta n_{\text{eff}} = 6.5 \times 10^{-3}$, and (d) $\Delta n_{\text{eff}} = 1.2 \times 10^{-2}$. The modes are ordered from top to bottom according to largest modal gain, with the modes at the top being the ones that reach threshold first [4.61].

ble. A principal reason for the higher-order modes being supported at higher operating currents is spatial-hole buring in the lateral carrier distribution, as discussed in [4.62]. Note that in Fig. 6.7 the intensity distribution of the fundamental mode is peaked at the center of the array. Therefore, carriers will be preferentially depleted at the center of the array. The carriers at the ends of the array are not depleted as much by the fundamental mode, and therefore, they can provide gain for the higher-order modes, since these modes have intensity maxima near the ends of the array. The mode stability of index-guided laser arrays had also been analyzed with a self-consistent model by *Whiteaway* et al. [6.15]. *Whiteaway* et al. concluded that the modification of the built-in dielectric profile by injected carriers and interaction of the carriers with the optical field would make stable in-phase mode operation of real index-guided lasers, such as the ridge guided array, very difficult to achieve. However, these same effects were shown to have a stabilizing effect for an index-guided array designed to operate in the out-of-phase mode [6.16].

In the analysis of [4.61], the dependence of stable fundamental-mode operation of ridge-guided laser arrays on the contact geometry was also investigated. The results for stripe and broad-area contact geometries for the case of the $\Delta n_{\text{eff}} = 1.3 \times 10^{-3}$ index-guided array are summarized in Table 6.3. The analysis of [4.61] indicates that the array with the broad-area contact geometry operates in the fundamental mode to a slightly higher power out-

Table 6.3. Comparisons of mode thresholds in ridge-guided array with $\Delta n_{\text{eff}} = 1.3 \times 10^{-3}$ for different contact geometries [4.61]

Contact Geometry	Stripe	Broad Area
Threshold current	145.8 mA	140.0 mA
Current/power when 2nd mode oscillates	164.0 mA /17.1 mW	159.3 mA /20.3 mW
Current/power when 3rd mode oscillates	233.9 mA /93.4 mW	299.9 mA /176.0 mW
Power ratio for 1st to 2nd mode when 3rd mode oscillates	1 : 0.854	1 : 0.565

put. We saw in Table 6.2 that the transverse-mode confinement factor can be larger than the mode confinement under the ridge guide. Therefore, when uniform current injection is applied across the array, one can have a situation where the gain can actually be larger in the lower-refractive index regions outside the ridge guides. Earlier cold-cavity analyses of ridge-guided arrays [3.23, 4.3, 4.30] considered the case of placing gain in the interelement regions between the ridges as a means of obtaining in-phase mode operation. It was revealed by *Fujii* et al. [4.3] that a direct analysis (that accounted for leaky-wave coupling) was necessary to find all modes of these type of index-guided arrays, as the coupled-mode analyses [3.23, 4.30] did not calculate all of the modes of the array. Recall from Sect. 4.2.3, that the Bloch analysis tells us that in order to obtain good mode discrimination for either the in-phase or out-of-phase mode, the lateral gain profile should be modulated with the gain maxima being located in the regions of lower refractive. This condition supports leaky-wave modes in the array. The weakly-index guided ridge-guided array with uniform pumping is an example of a structure where this can occur [6.12, 6.13]. This is a possible explanation for the predominantly single-lobed operation that has been observed from many ridge-guided laser arrays over limited ranges of operation.

Both experimental observations [1.76, 1.77, 1.95, 1.279][6.11]-[6.14] and theoretical analysis [4.61, 4.62, 6.15] show that stable single-spatial-mode operation in ridge-guided laser arrays was not obtained at high-power outputs. The projection of the maximum power output of 20.3 mW at a current level of $1.14 I_{\text{th}}$ for single-array-mode operation [4.61] is quite low in relation to the single-mode capabilities of single-element diode lasers. This is also the case for a more recent experiment where the lateral loss profile was tailored to enhance the gain discrimination for the in-phase mode [6.14]. For this device, stable in-phase mode operation up to about $1.5 I_{\text{th}}$, and a gradual broadening of the far-field beam was observed up to $4.0 I_{\text{th}}$.

However, as discussed in Chap. 2, the maximum power conversion efficiency typically occurs at or above $5 I_{\text{th}}$ for good-quality laser material. To take full advantage of the material capabilites, an array should be able to operate in a single mode at the current levels where the power conversion efficiency is at or near its maximum value. In order to overcome the modal instabilities present in the ridge-guided laser array, the built-in dielectric profile needs to be robust against the carrier-induced, optical-field induced, and thermal induced changes in the lateral waveguide structure.

6.3 Leaky-Wave-Coupled Arrays

The inability of the multiple-stripe gain-guided laser array (Sect. 5.1) and the index- guided laser arrays (Sects. 6.1 and 6.2) to operate single mode at cw power outputs above what was possible from single-element diode lasers motivated research into alternative structures for high-power diode-laser arrays [1.443]. As we have seen in Sects. 5.1, 6.1, and 6.2, the modification of the built-in dielectric profile of the laser array due to combined effects of injected carriers, interaction of the optical field with the carriers, and non-uniform heating causes a severe modal instability, even in perfectly uniform structures, so that multi-array mode operation results. This results in a poor-quality output beam. When structural non-uniformities are present, we saw in Sect. 3.4.2 that the frequency locking between array elements can break down, and this is yet another mechanism for destroying coherent operation in a diode-laser array.

An array structure that was developed specifically to be robust against the aforementioned mechanisms of coherence reduction was the leaky-wave-coupled antiguided diode-laser array. As its name implies, this structure operates on the principle of leaky-wave coupling between gain elements that are configured as negative-waveguide structures. In the nomenclature of the three-layer waveguide model of Sect. 3.1, a negative-waveguide structure is where $\Delta n < 0$, and optical waveguiding can occur when the condition expressed by (3.21) is satisfied. The original motivation of *Ackley* et al. for employing gain stripes that operated in a leaky mode was to produce a strong coupling between all array elements [1.74, 6.17]. This was indeed demonstrated in the earliest leaky-wave-coupled arrays that were developed [1.74]. As a consequence of the leaky-mode operation, these lasers had high internal losses which reduced the power conversion efficiency. In [1.74], it was proposed and demonstrated that the impact of the leaky-mode losses could be reduced by increasing the number of gain stripes in the array.

Botez et al. [1.218]-[1.221] proposed using leaky-wave-coupled arrays of antiguided lasers with a large built-in lateral index step (large meaning $\Delta n_{eff} \geq 5 \times 10^{-2}$) to provide immunity to the carrier and field induced changes in the lateral dielectric profile, which had been the source of instability in positive index-guided arrays. To further reduce the radiation losses associated with the leaky waveguides, the structures demonstrated in [1.218]-[1.221] used closely-spaced antiguided elements ($\approx 5\,\mu m$ center-to-center spacing) as opposed to the more-widely spaced ($\approx 15\,\mu m$ center-to-center spacing) antiguided arrays that had been previously demonstrated

[1.74, 6.17]. In addition, *Botez* et al. recognized the resonant character of leaky-wave-coupled diode-laser arrays and quantified the relation between resonant coupling and the mode discrimination properties by introducing the resonant optical-wave coupling model. In this model, the array of antiguides is treated as an effective lateral grating, so that an analytic expression for the resonance condition for some of the lateral modes could be obtained [1.219, 1.220, 4.43]. The concept of the resonant-coupling model has been generalized to include structures that support multiple transverse modes [4.51] To accurately model the modes of arrays of leaky-wave-coupled antiguided lasers, the numerical methods discussed in Chap. 4 should be used. However, the resonant-coupling picture of leaky-wave coupling is simpler and provides a more intuitive explanation of the basic modal characteristics of antiguided arrays.

The single-mode-performance characteristics demonstrated by antiguided-laser arrays have been superior to those of previous laterally-coupled laser-array structures. Reports of stable, single-spatial-mode operation to cw power outputs up to 1 W have been demonstrated [2.41]. Under pulsed operating conditions, single-lobed output beams (with nearly-diffraction limited angular divergences) have been observed at power outputs up to 2 W [1.233, 6.18] from resonant structures; and from nonresonant structures, single-lobed output beams (with angular divergences several times the diffraction limit) have been observed at power outputs as high as 32 W [1.235, 6.19]. In Sect. 6.3.1 we review the resonant-optical-wave coupling model, Sect. 6.3.2 presents structures of antiguided arrays, Sect. 6.3.3 dicusses the modal characteristics, Sect. 6.3.4 presents more rigorous modeling of modal properties, and Sect. 6.3.5 describes experimental and theoretical studies of the above-threshold mode stability of antiguide arrays.

6.3.1 Resonant Optical-Wave Coupling Model

The following discussion on resonant leaky-wave coupling follows the approach of [1.219, 1.220, 4.38, 6.20]. Consider the schematic diagram of an array of antiguides in Fig. 6.8a. The low-index regions, often referred to as *antiguide cores*, have an effective index n_0. The high-index regions, often termed the *interelement regions*, have an effective index n_1. The optical ray trajectories in the antiguide core and interelement regions, of the array structure are indicated, along with the reflection or bounce angles ϕ_i at the interface. As the light rays propagate in the antiguide core, at each reflection off the interface with the interelement region, some of the light is refracted into the surrounding higher index interelement region, as illustrated in Fig. 6.8a. The propagation constant for the lateral modes is indicated by β where $\beta = n_{\text{eff}} k$. As we saw in Sect. 4.2.3, when the gain is located in the regions of lower index, as in an antiguided-laser array, the leaky-wave-coupled array modes (which are the lowest-loss modes) have effective indices that are slightly less than the index of the waveguide layer, so $n_{\text{eff}} \leq n_0$.

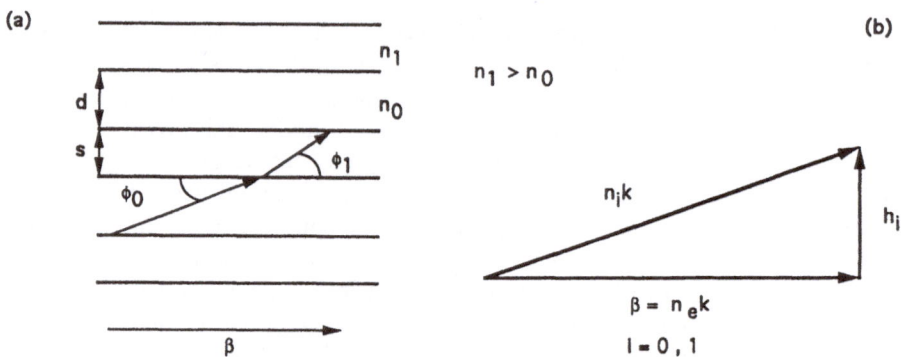

Fig. 6.8. Schematic diagram depicting the geometric optics representation of a leaky-mode coupled array of waveguides (**a**), and the vector relationship between β, h_i and the bounce angles ϕ_i (**b**).

In both the interelement region and antiguided core, there is a real lateral-mode propagation vector h_i (in the antiguide core $i = 0$, and in the interelement region $i = 1$), that is related to β and the bounce angles ϕ_i, as seen in Fig. 6.8b. Mathematically, the lateral propagation vector is given by

$$h_i = \sqrt{(n_i k)^2 - \beta^2} \ . \tag{6.1}$$

Associated with each lateral propagation vector h_i, is a lateral wavelength $\lambda_i = 2\pi/h_i$ that is given by

$$\lambda_i = \frac{\lambda}{\sqrt{n_i^2 - n_{eff}^2}} \ . \tag{6.2}$$

When the interelement and antiguide core regions support a single transverse mode, resonant-wave coupling occurs for those modes where propagation of the lateral wave through a single period $\Lambda = d + s$ of the array results in a phase change by an integer multiple of π (see [4.51] for analysis of resonant condition when more than one transverse mode is involved). Figure 6.9a illustrates an example of the relation between the lateral wavelengths and the near-field intensity pattern for an in-phase resonant mode of an antiguided array. The resulting near-field pattern arises from the interference of the two oppositely-propagating lateral waves. Hence, the relative phase of the adjacent near-field maxima is shifted by π radians, giving rise to an alternating phase sequence in Figs. 6.9b and c. When the phase change is an even multiple of π over a period Λ, the modal near-field pattern also has a period of Λ, as seen in Fig. 6.9b, and all of the antiguide cores radiate in phase and constructively interfere with each other. These modes are strongly coupled because light radiated from any given element propagates across the lateral extent of the array with negligible losses. For odd multiples of π, the modal near-field patterns will have a period of 2Λ, resulting in a 180° phase difference between adjacent antiguide cores, as displayed in Fig. 6.9c. At the

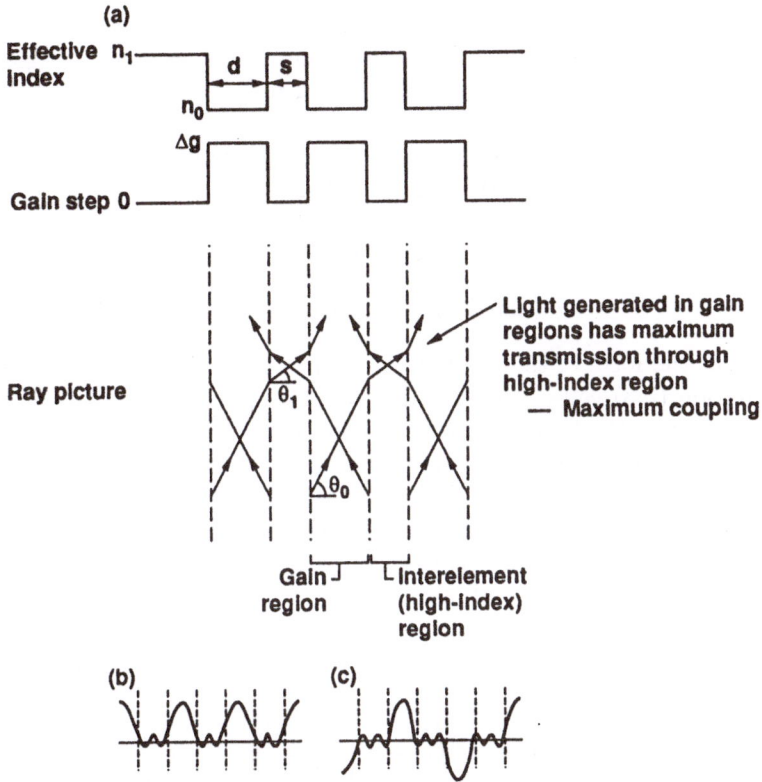

Fig. 6.9. Drawing of a antiguide array illustrating the geometric relation of the lateral propagation vectors at resonance (**a**) [6.19], the near-field pattern of the in-phase resonant mode for the case $q_1 = 3$ and $q_0 = 1$ (**b**) [1.235], and the near-field pattern of the out-of-phase resonant mode for the case $q_1 = 4$ and $q_0 = 1$ (**c**) [1.235].

adjacent interelement-core interface, the radiated light is strongly reflected back towards the core that it was radiated from, so that it couples in-phase. Therefore, optical coupling between the antiguide cores across the array is weak for these modes, and adjacent-core elements will emit light with a 180° relative phase shift.

Using the lateral propagation vectors h_i, the resonant-wave coupling condition can be expressed as

$$h_0 d + h_1 s = q\pi \; , \tag{6.3}$$

where $q = 1, 2, 3, \ldots$. Equation (6.3) is the basic equation for analyzing the resonant modes of the leaky-wave-coupled antiguided array. As seen by the example in Fig. 6.9, a resonant mode will have an integer number of half-lateral wavelengths in the core and interelement regions. Then the lateral propagation vectors satisfy $h_0 = q_0 \pi / d$ and $h_1 = q_1 \pi / s$, respectively, in the antiguide core and interelement regions, with $q_0 + q_1 = q$. In this case, the eigenvalue equations (4.35 and 4.36), that were derived using the Bloch analysis, reduce to a trivial form that will not yield the eigenvalues. At resonance, $K\Lambda = q\pi$ and the eigenfunctions correspond to linear combinations of sines and cosines [4.38].

The analysis used for the resonance-coupling condition is analogous to the model that is applied in analyzing the diffraction orders of gratings. In fact, by using the vector relation displayed in Fig. 6.8b, the resonant wave-coupling condition can be expressed in terms of the ray bounce angles ϕ_i

$$dn_0 \sin(\phi_0) + sn_1 \sin(\phi_1) = q\lambda , \qquad (6.4)$$

which is of the form of the grating equation.

Equation (6.3) can be simplified to yield a more useful form of the resonance condition. For isolated antiguide cores with $d \geq 2\,\mu m$, the bounce angle ϕ_0 in the antiguide core is at most $3°$, as shown by *Engelmann* et al. [3.13]. This occurs because the width of the core is much larger than the wavelength of light in the core, so the diffractive spread in the propagation vectors is small. In this situation, $h_0 = q_0\pi/d$ is a good approximation, which means that $\lambda_0 = 2d/q_0$ is also a good approximation. Here $q_0 = 1, 2, \ldots$ is the mode number for a single antiguide element, with $q_0 = 1$ being the fundamental mode. Applying this approximation to (6.2), we see that n_{eff} for the array mode is given by

$$n_{\text{eff}} = \sqrt{n_0{}^2 - \left(\frac{\lambda(q_0)}{2d}\right)^2} . \qquad (6.5)$$

For typical values of $n_0, \lambda,$ and d, we see from (6.5) that $n_0 - n_{\text{eff}} \leq 10^{-2}$. In analyzing most arrays of antiguides, only the $q_0 = 1$ (the fundamental mode of the core) and $q_0 = 2$ (the first-order mode of the core) cases have been found to be important. Substituting (6.5 and 6.2) into (6.3) and using the approximations $\lambda_0 = 2d/(q_0)$ and $n_0 \approx n_1$, the resonance condition is simplified to

$$\Delta n_{\text{res}} = \frac{\lambda^2}{8n_0} \left[\left(\frac{q_1}{s}\right)^2 - \left(\frac{q_0}{d}\right)^2\right] . \qquad (6.6)$$

For fixed antiguide core width d and interelement width s, (6.6) can be used to calculate the index difference $\Delta n = n_1 - n_0$, necessary for the mode corresponding to a specific q_0 and q_1 to be at resonance. Recall that the integer q_1 corresponds to the number of near-field intensity maxima in the interelement region and the integer q_0 corresponds to the number of near-field intensity maxima in the antiguide core. The modes that satisfy the resonance condition have either in-phase or out-of-phase character. As is illustrated in Fig. 6.9, modes with $q_0 = 1$ and q_1 being an odd integer, or $q_0 = 2$ and q_1 being an even integer, will exhibit in-phase character; whereas, modes with $q_0 = 1$ and q_1 being an even integer, or $q_0 = 2$ and q_1 being an odd integer, will exhibit out-of-phase character.

The usual convention for numbering the modes of the antiguide array is the integer pair (ℓ, m), where $\ell = q_0 - 1$ and $m = q_1$ [6.20]. In addition, the array-mode number L for an N-element array is given by $L = (m + 1)(N - 1)$ for $\ell = 0$; and when $\ell = 1$, the array-mode number is $L = (m + 1)(N - 1) + N$. The array-mode number L corresponds to the total number of nulls, or intensity minima, in the near-field intensity profile.

We have seen how the resonant optical-wave condition has established the phase characteristics of the resonant modes of the antiguided arrays. But it is extremely important to elucidate the discrimination properties of these resonant modes. This will be treated in the Sect. 6.3.4, using an exact analysis. However, by extrapolating the radiative loss of a single antiguide core, we can obtain a more physical understanding of the radiation loss mechanism in antiguided arrays.

Using a numerical calculation, *Engelmann* et al. [3.13] have shown that the radiation loss $\alpha_R(1)$ experienced by a single-antiguide core of large index step is

$$\alpha_R(1) = \frac{(q_0 \lambda)^2}{d^3 n_0^2 \sqrt{2 \Delta n / n_0}} \ . \tag{6.7}$$

The radiative loss for the resonance modes of an N-element array of antiguides can be expressed as

$$\alpha_{RR}(N) \approx \frac{\alpha_R(1)}{N} \ . \tag{6.8}$$

Note that (6.8) has actually not been derived analytically. In [6.20], a comparison of the exact calculation of radiation losses versus the number of array elements showed excellent agreement with (6.8). Justification of the form given by (6.8) has also been presented with intuitive physical arguments. As discussed in [6.20], at resonance the interelement regions effectively become transparent. Therefore, the antiguide array can be modeled as being equivalent to a single element of width Nd that is operating in the higher-order mode with $q_0 = N$. In this case (6.7), has the form $\alpha_R(1) \propto 1/N$. A recent derivation of an analytic theory of finite-arrays of antiguides [4.43] has revealed the exact expression for α_{RR} to be

$$\alpha_{RR}(N) = \frac{\alpha_R(1)}{N} \left\{ 1 + \frac{1}{m^2} \left[\left(\frac{s}{d} \right)^2 + \left(\frac{s}{d} \right)^3 \right] \right\}^{-1} \ . \tag{6.9}$$

The exact expression (6.9) exhibits the $1/N$ dependence; and for most cases of practical interest the approximate expression (6.8) is accurate to 5% or better [4.43].

To gain further insight into the characteristics of resonant optical-wave coupling, consider a ten-element array with $n_0 = 3.40$, $\lambda = 0.86\,\mu\text{m}$, $d = 3\,\mu\text{m}$, and $s = 2\,\mu\text{m}$. Figure 6.10 represents a plot of the mode radiation losses versus Δn, the lateral-index differential, for modes with $L = 27, 36$, and

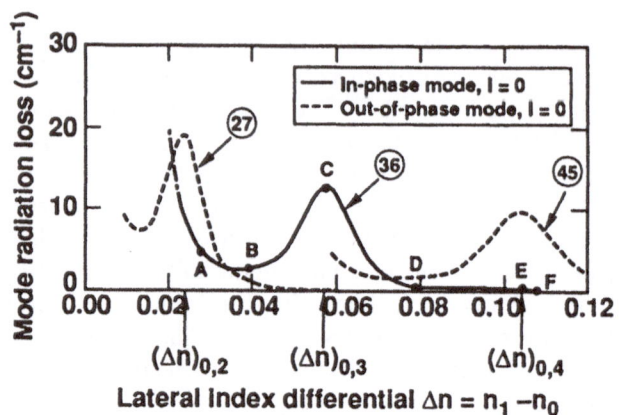

Fig. 6.10. Mode radiation loss versus Δn for a 10 element array. The values of the lateral-index differential that correspond to the resonance are indicated by $(\Delta n)_{\ell,m}$ [6.20].

45. The points along the lateral differential-index axis where the resonance condition (6.6) is satisfied are designated, as described in the caption.

At resonance, each mode experiences a maximum in the radiation loss. This makes sense because at resonance light is most strongly radiated into the surrounding array structure from the antiguided cores. For the $L = 27$ and $L = 36$ modes, the lowest losses are obtained on the large Δn side of the resonance maximum. For the in-phase $L = 36$ mode, the lowest-radiation loss occurs for $\Delta n \geq 0.08$. Note that radiation losses are not the only factor that influence the mode discrimination. The vertical epilayer structure impacts the gain discrimination through the selective introduction of loss in the interelement regions and the two-dimensional mode confinement factor, $\Gamma_{\Delta y:n}$ that was introduced in Sect. 3.2. Also, the resonant-optical-wave model does not account for the near-resonant array modes, termed adjacent modes (described on page 216 and in Sect. 6.3.3) that can effectively compete for the gain. To accurately account for all of the aforementioned loss mechanisms, a complete numerical analysis is required.

Mehuys et al. have presented an experimental demonstration of the principle of resonant optical-wave coupling in antiguided arrays [6.19]. In this experiment, a group of 40-element antiguided arrays were fabricated all having core widths of $d = 2.0\,\mu m$, but the width of the interelement region s spanned the range from $1.0\,\mu m$ to $2.6\,\mu m$. The measured far-field outputs as a function of s that were made are displayed in Fig. 6.11.

The observed far-field patterns change from being predominantly single-lobed (indicating in-phase mode operation) to a double-lobed pattern (indicating operation an out-of-phase mode), as s is changed by $0.4\,\mu m$. Good agreement was found by [6.19] between the observed mode cycling period of $0.4\,\mu m$ for s and that calculated using (6.6) with the expected value for the lateral-index differential of $\Delta n = 0.175$. With increasing values of s, the

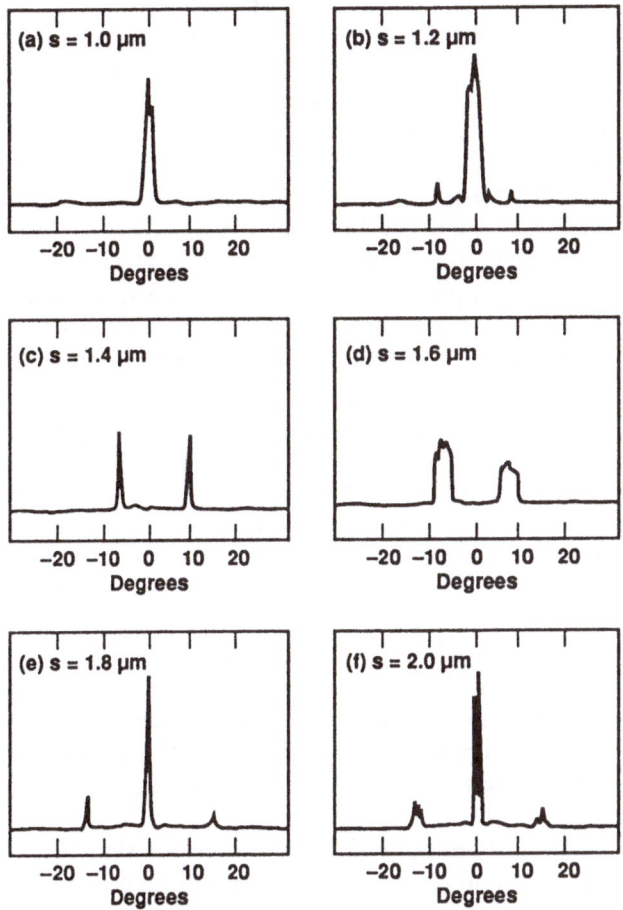

Fig. 6.11. Measured far-field patterns versus interelement width s from the experiment of [6.19]. The values of interelement widths used are indicated [6.19].

intensity of the secondary-lobe structure tends to increase. This happens because the number of lateral wavelengths, λ_1, contained within the interlement region must increase with s, giving rise to an increasing number of near-field maxima in the interlement region, of the sort illustrated in the example of Fig. 6.9. These 40-element antiguided arrays were found to operate in a stable single lateral array mode at cw power outputs of several hundred mW.

The resonant optical-wave model provides an intuitive understanding of radiation coupling in antiguided arrays and the characteristics of the lateral modes at resonance. Besides, the modes that display resonance behavior, there are other allowed modes that do not exhibit the phase periodicty of resonance modes. These non-resonant modes have near-field patterns with periodicities that are non-commensurate with the dimensions of the core and the interlement regions across the array. As a result, most of these modes have very high losses, and they do not compete effectively for the gain. However, those non-resonant modes with array mode number L that differs only by ± 1 with that of a resonance mode can effectively compete for the gain. These are refered to as *adjacent modes*. Recall that L corresponds to the total number of nulls in the lateral-mode profile; therefore, the near-field patterns of adjacent modes are nearly-periodic with the array structure, and over the entire extent of the array a half-wave phase shift accrues. This results in a 180° relative phase shift between both halves of the array. In Fig. 6.11, the observed mode for $s = 2.0\,\mu\mathrm{m}$ with a twin-lobed sub-structure of the dominant and secondary far-field peaks is an example of an adjacent mode. As discussed in Sects. 6.3.3 and 6.3.4, other means such as tailoring the interelement loss and confinement factor are employed to discriminate against adjacent modes.

6.3.2 Transverse and Lateral Structures of Antiguided Arrays

Figure 6.12a presents a schematic diagram of a generic antiguide array, indicating the functions that are associated with the different waveguide regions. The active layer is continuous across the array, and the interelement sections contain additional passive waveguide and loss regions that are not present in the core elements. In the interelement region, the passive guide confines light most strongly as illustrated in Fig. 6.12a, so that the transverse-confinement factor in the active layer is reduced. The composition and thickness of the passive guide are selected to optimize the two-dimensional confinement factor for the desired mode of operation. The addition of the loss region will discriminate against those modes that have a larger overlap with the interelement regions.

There are two basic antiguided array structures that have been investigated experimentally, and they are diagramed in Fig. 6.12b and c. The structure in Fig. 6.12b is refered to as the *Complimentary-Self Aligned* (CSA) stripe array [1.235]. The CSA array typically contains a GaAs quantum-well active layer. In the interelement region, the p − GaAs acts as the loss region and the p − $Al_{0.3}Ga_{0.7}As$ functions as the passive guide. There is at least one example of a CSA array containing an InGaAs quantum-well active layer having been fabricated [1.233]. In this case, both the p − GaAs and the p − $Al_{0.3}Ga_{0.7}As$ layer act as the passive guide region and there is no loss region. Similarly, the *Self-Aligned-Stripe* (SSA) array has been investigated only with both GaAs [6.18] and InGaAs quantum-well active layers [1.225, 1.231, 1.232, 6.21]. Here, the n − GaAs layer will act as the passive guide region, because of its higher index; and because it is n-type, it is intended to function as a current-blocking layer, effectively reducing the gain in the interelement active layer relative to the gain in the active layer core.

6.3.3 Basic Modal Characteristics of Antiguided Laser Arrays

In many cases the modal characteristics of antiguided arrays can be adequately modeled by a numerical analysis of the passive structure. This is

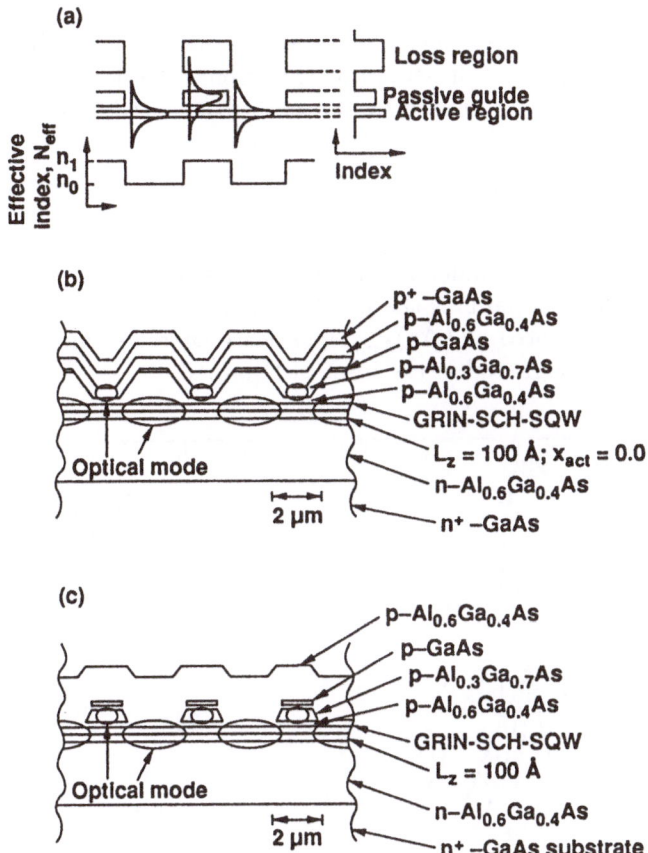

Fig. 6.12. Generic diagram of antiguide array structure depicting the functions of the layers in (**a**), diagram of a CSA array structure in (**b**), and diagram of SAS array structure in (**c**) [1.235].

possible because the large lateral-index differential makes antiguides largely immune to the effects of injected carriers and heating. The effective-index method, presented in Sect. 4.2.1, has been used most often to analyze antiguided arrays. Being a one-dimensional model, the effective-index method is computationally less demanding than the two-dimensional model. For this reason the effective-index method, is well suited for finding the allowed modes of a particular structure. Therefore, it is preferred in design studies that are aimed at determining the structural parameters necessary for optimizing the mode discrimination. As discussed in Sect. 4.2.4, the effective-index method is not always accurate for modeling arrays with the lateral index differentials that are sometime used in antiguide arrays. Some examples of this are presented in Sect. 6.3.4.

Numerical analyses of antiguided-laser arrays using the one-dimensional effective-index approximation of the two-dimensional structure have shown that these devices have a spectrum of allowed modes which is considerably more complex than that presented by the resonant optical-wave model. This situation is illustrated in Fig. 6.13, where the calculated radiation loss versus lateral index differential is plotted for the lowest-loss modes of a 10-element lossless leaky-wave-coupled array of antiguides with $d = 3\,\mu m$, $s = 1\,\mu m$, $n_0 = 3.40$, and $\lambda = 0.86\,\mu m$. For this calculation, the variation in lateral-index differential was obtained by varying n_1, the effective-index of the interelement region. The array-mode index L is indicated next to each curve, and along the lateral-index differential axis the points where the resonance condition (6.6) is satisfied are denoted as $(\Delta n)_{\ell,m}$. The phase character and the core-mode number, ℓ, are designated in the inset of Fig. 6.13.

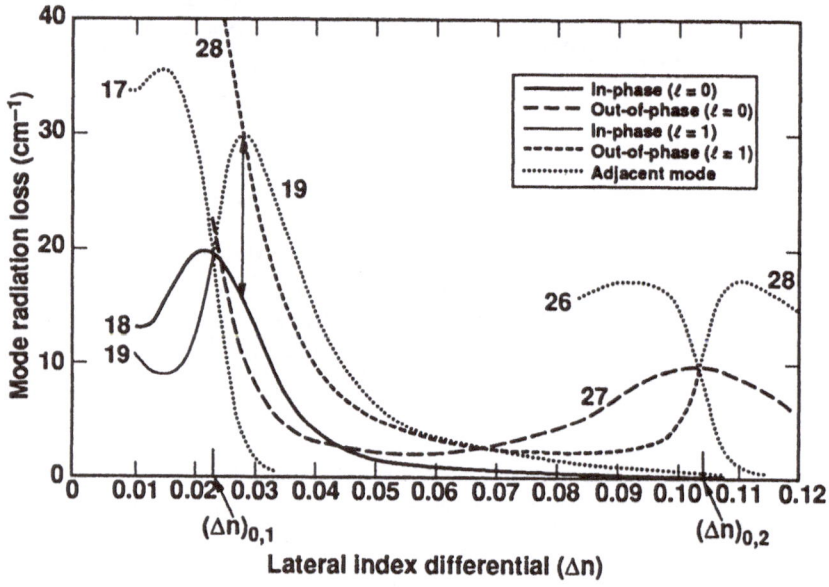

Fig. 6.13. Calculated lateral radiation losses versus lateral-index differential are plotted for a 10-element antiguide array for the structure analyzed in [6.20].

The mode discrimination is seen to depend critically on the lateral-index differential, as the character of the individual modes can be considerably altered. For example, only modes 18 and 27 reach points where the resonance condition occurs over the range of Δn values considered. The number 17 and 26 modes, which are adjacent modes, approach cutoff, respectively, in the neighborhood of the resonances of the 18 and 27 modes. For a leaky-wave mode, cut off means that the mode becomes evanescent in the core regions, so that the mode is confined to the interelement regions, where the local gain is lowest. Although, the mode radiation loss goes to zero for the modes at cutoff, the two-dimensional confinement factor (3.38) decreases significantly so that cutoff modes have almost no influence on the gain discrimination properties.

Also, indicated in Fig. 6.13 is a transformation of both the $L = 19$ in-phase mode and the $L = 28$ out-of-phase mode into adjacent modes. To understand the nature of this transformation, we need to study the evolution of the near-field intensity pattern as Δn is increased. Figure 6.14 shows the near-field intensity patterns for modes 17, 18, and 19 for selected values of Δn that correspond to the case modeled in Fig. 6.13. At the $\Delta n = 1.6 \times 10^{-2}$ point, mode 19 is identified as $\ell = 1$ because the field pattern in the antiguide cores has a null, and it is in-phase because there are no interelement maxima, so $m = 0$. The field in the core regions in the center of the array is symmmeric with the null falling at the center of the core. However, at the ends of the array the null is shited away from the core centers towards the center of the array. At $\Delta n = 2.4 \times 10^{-2}$ it is seen that in the end-core regions, the null is shifted towards the boundary with the adjacent interelement region; while in the center of the array, the core elements have nulls at or near their centers. This is the point where mode 19 begins to transform into an adjacent mode. When $\Delta n = 2.8 \times 10^{-2}$, the transformation to an adjacent mode is complete. Most of the core elements exhibit predominantly single-peaked near fields. Only the central core region has a null at the center. This null at the center of the array indicates that the relative phase between the two halves of the array is shifted by 180°. Since the core regions of the adjacent modes do not all exhibit the same single-core mode characteristic, there is not a unique assignment for ℓ and m. However, the mode number L, the total number of nulls in the arrays near field, is preserved, so L provides a more general means of identifying the array modes. Also, the near-field intensity envelope of mode 19 has changed dramatically. The maximum intensity moves from the center of the array to the ends of the array, which is consisitent with the increase in radiation loss seen in Fig. 6.13.

Let us consider the evolution of the two other modes in Fig. 6.14. At the lowest value of Δn, mode 17 has an adjacent mode character not unlike that of mode 19 at $\Delta n = 2.8 \times 10^{-2}$. With increasing Δn, nulls move into the outer core regions from the interelement boundary that is closest to the array center, resulting in a larger portion of the near field being confined to the interelement regions. The intensity-envelope maxima moves from the ends of the array to the center of the array, because the mode becomes more tightly confined resulting in a decrease in radiation losses. As indicated in Fig. 6.13, mode 17 reaches cutoff just above $\Delta n = 0.03$, where it becomes evanescent. By comparison, mode 18, with $\ell = 0$ in-phase character, shows small changes in the near-field phase characteristics as Δn is varied, and only a moderate variation in the intensity envelope. The differences in the spatial charcteristics of the antiguide modes can be used to improve mode discrimination either through the introduction of interlement losses or by incorporating a intracavity spatial filter into the array structure [1.443].

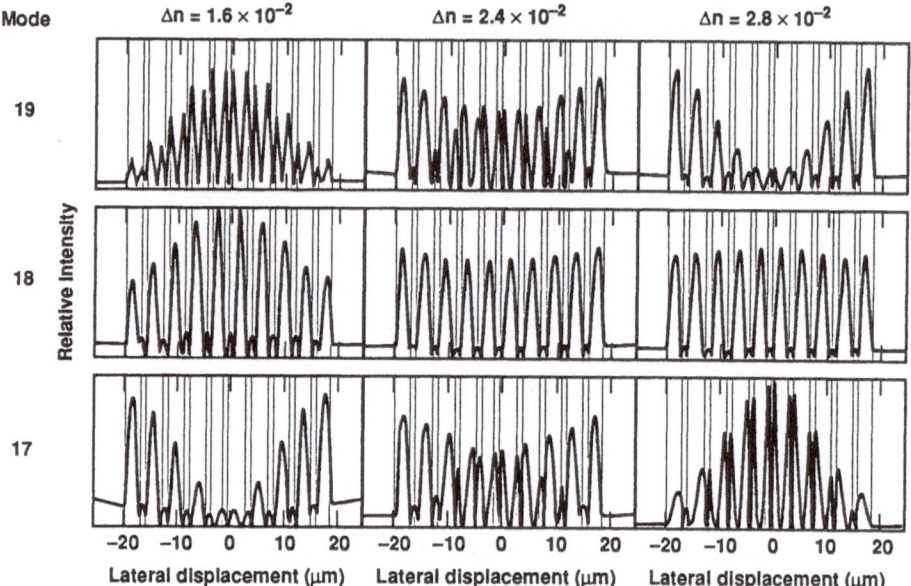

Fig. 6.14. Calculated near-field intensity patterns for modes 17, 18, and 19 for the structure analyzed in [6.20] at the following selected values of Δn: $\Delta n = 1.6 \times 10^{-2}$ the point of maximum loss for mode 17, $\Delta n = 2.4 \times 10^{-2}$ the resonance point for mode 18, and $\Delta n = 2.8 \times 10^{-2}$ the point of maximum loss for mode 19 [6.20].

The aforementioned models of the modal characteristics of antiguided-laser arrays have been corroborated by experimental studies [1.231, 1.443]. In spectrally-resolved studies of the far-field output as a function of power, *Major* et al. have identified a correlation between far-field broadening and multi-lateral mode operation in structures similar to that displayed in Fig. 6.12c that were intentionally designed to operate at a point away from the resonant-coupling condition [1.231]. The power current characteristic of the device that was studied is displayed in Fig. 6.15. A predominately single-lobed far-field pattern was observed. Although, as illustrated by the two examples in the inset of Fig. 6.15, the beam-divergence was observed to increase with increasing drive current [1.443].

The spectral output of this array, shown in Fig. 6.16, exhibited a rich structure, which is highlighted in Fig. 6.16c. At the 200 mA operating point predominantly single-wavelength operation is seen at 982 nm; and at the 400 mA operating point, two principal wavelengths, $\approx$ 982 nm and 978 nm are evident. The spectral region spanned by the principal wavelengths is displayed on a higher-resolution scale in Fig. 6.16c. *Major et al.* [1.231] have identified the fine structure in Fig. 6.16c as two distinct sets of longitudinal modes, each corresponding to a different lateral mode of the antiguided array. Each longitudinal mode sequence has a wavelength separation of 2.5 Å, as noted in Fig. 6.16c.

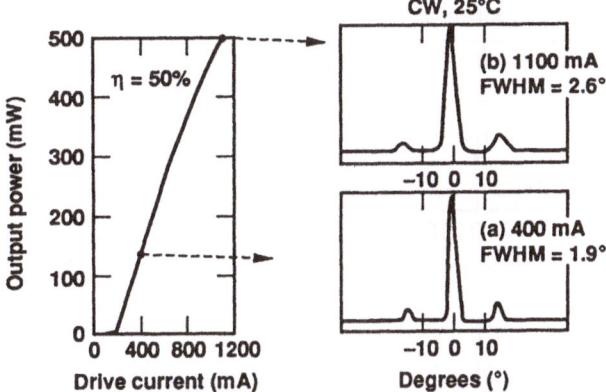

Fig. 6.15. CW power-current characteristic of the array that was used in the spectrally-resolved far-field experiment of [1.231]. The far-field output at operating currents of 400 mA and 1100 mA are shown in the insets (**a**) and (**b**), respectively [1.231].

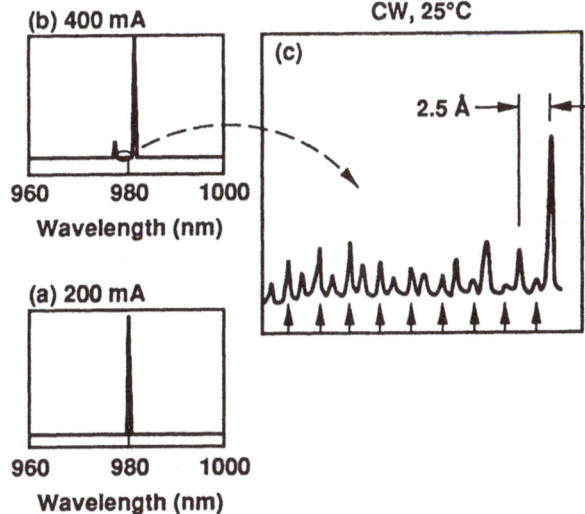

Fig. 6.16. Spectral output of the array operating at 200 mA (**a**) and 400 mA (**b**) [1.231]. In (**c**) the indicated spectral range is shown under higher resolution. The arrows denote the longitudinal mode sequence associated with the principal mode at 978 nm [1.231].

The spectrally-resolved far-field patterns that were determined experimentally appear in Fig. 6.17c and d, and the corresponding calculated far-field patterns appear in Fig. 6.17a and b. These calculated far-field patterns were generated using a one-dimensional effective-index model that did not include the effects of carrier injection or heating. The mode at 982 nm corresponds to an in-phase mode, while the mode at 978 nm, corresponds to

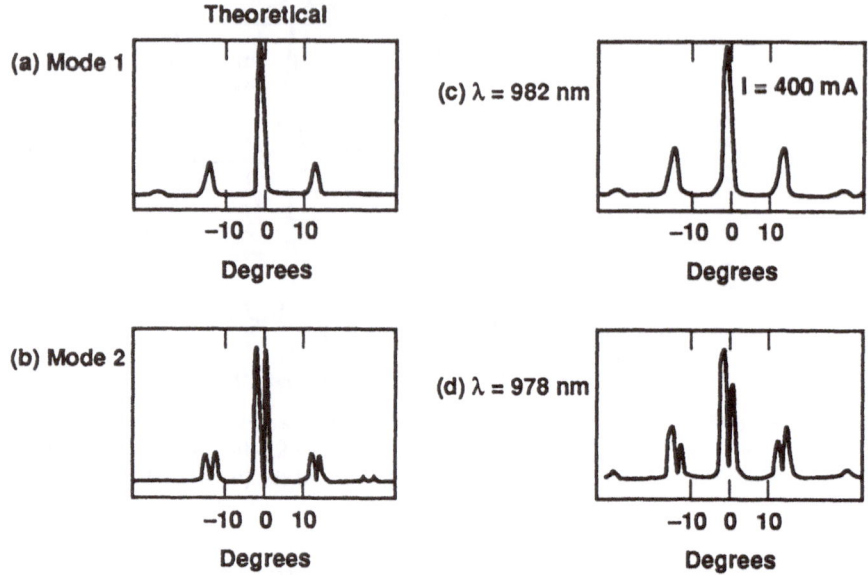

Fig. 6.17. Spectrally-resolved far-field study of the antiguided array operating at 400 mA; (**a**) theoretical model of far-field attributed to the 982 nm mode, (**b**) theoretical model of far-field attributed to the 978 nm mode, (**c**) spectrally-resolved far-field pattern measured for 982 nm wavelength, and (**d**) spectrally-resolved far-field pattern measured for 978 nm wavelength [1.231].

the adjacent mode. The 978 nm mode is identified as the adjacent mode by the doublet-structure of each far-field lobe. This is caused by a 180° relative phase shift between the two halfs of the array, as is illustrated in Fig. 6.14.

6.3.4 Rigorous Modeling of Structural Tolerances for Antiguided Laser Arrays

So far we have only used the effective-index method to describe the modal characteristics of the antiguided arrays. As discussed in Sect. 4.2.4, the effective-index method is not always accurate for structures such as the antiguided array where the transverse-mode profile can exhibit a strong dependence on the lateral position and the interelement regions can support more than one transverse mode. In optimizing the design of an array it is more convenient to use the effective-index method, but this approach has some limitations, which we will now consider.

Given the resonant nature of the modal losses of the antiguide, an important question that needs to be addressed is, *how robust is the mode-discrimination mechanism in real resonant-optical-wave laser arrays to the type of structural nonuniformities that can occur in the material growth and device fabrication process?* This question has been addressed by *Hadley* et al. in [1.229], who used a numerical approach to model CSA-type antiguide arrays as two-dimensional passive waveguide structures. The most realistic representation of a CSA-type antiguide structure that was treated is presented in Fig. 6.18. As was seen in Fig. 6.12, antiguided-laser array structures are generally non-planar. To accurately account for the nonplanar properties of the structure in a full two-dimensional waveguide model, the structure was discretized using an irregular triangular mesh, as shown in Fig. 6.18, so that the dielectric permittivity was constant within each region. Then solutions of the two-dimensional Helmholtz equation (4.42) were matched (using the appropriate boundary conditions) at the interfaces of the triangular regions. A more detailed description of the computational aspects of this analysis can be found in [1.229]. At each end of the array, labeled as radiation boundary in Fig. 6.18, the field is represented as a linear combination of those transverse eigenmodes that correspond to radiation fields that propagate away from the structure.

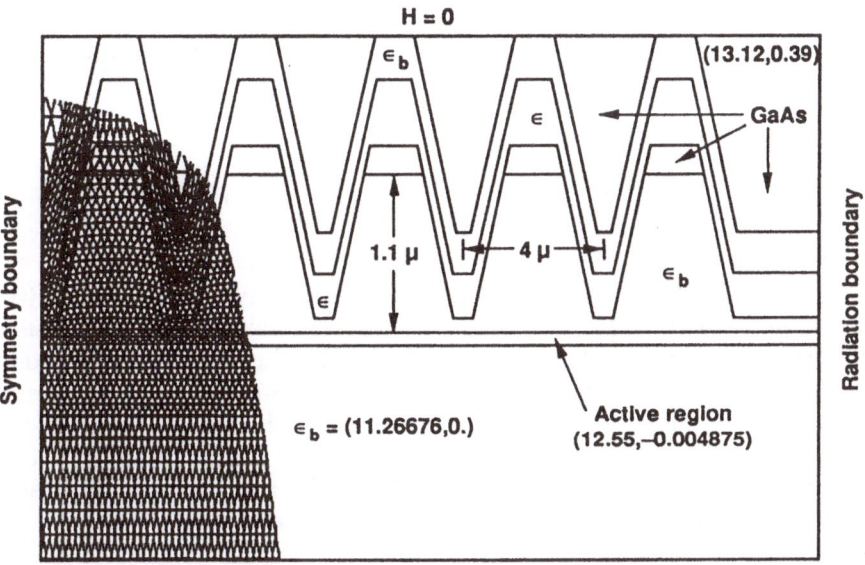

Fig. 6.18. Diagram depicting the geometry used to model the effects of nonplanar-antiguided-laser array structures. Only half (five elements) of the symmetric ten-element array are shown. On the left-hand side, the irregular-triangular mesh that was used to approximate the nonplanar geometry is illustrated [1.229].

In their analysis, *Hadley* et al. considered the calculated-modal gain as a function of the layer with permittivity ϵ for the structure in Fig. 6.18. The results of this numerical calculation are presented in Fig. 6.19. Note that modes are designated using both the mode index ν, where modes are labeled according to increasing resonance frequency; and the mode index L, where modes are classified by the number of nulls in the near field. According to

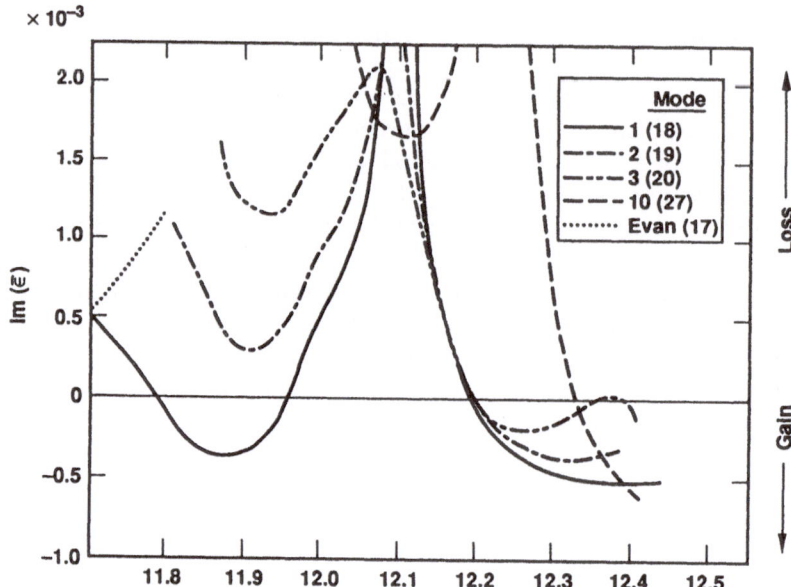

Fig. 6.19. Modal gain as a function of the layer with permittivity ϵ [1.229].

Fig. 6.19, there are two regions for ϵ where the in-phase mode, designated by $\nu = 1$ and $L = 18$, is most favored.

The regions where in-phase mode operation is favored corresponds to a range of only 0.04 in the Al fraction of the AlGaAs layer with permittivity ϵ, that appears in Fig. 6.18. This is a demanding constraint on the compositional uniformity, and it could be an explanation for the reported low-yields in fabricating CSA-type antiguide arrays that operate in the in-phase mode [1.229]. As an alternative, [1.229] suggested that the higher-yield SAS-type antiguide structure be used. Generally speaking, it would appear that there is still a need to investigate designs of antiguided arrays that are less sensitive to small variations in the structural compositions and geometries.

6.3.5 Above-Threshold Mode Stability of an Antiguided Array

For the purposes of calculating effects of spatial-hole burning on the mode stability of resonant antiguide laser arrays, the one-dimensional self-consistent model has been used to simplify the calculation [4.44, 6.22]. Here we will consider the more detailed analysis of *Nabiev* et al. reported in [6.22] that included carrier diffusion effects (Note that the self-consistent analysis by the same group of researchers in [4.44] neglected carrier diffusion). In their analysis, *Nabiev* et al. [6.22] ignored the axial dependence of the dielectric permittivity. This means that all the z dependence in the paraxial equation (4.10) was neglected, resulting in a one-dimensional wave equation involving only the lateral variable x. Also, thermal effects were not included, so that

$\varepsilon_{\text{eff}}(x)$ only included effects of the injected carriers. This analysis pointed out the influence that spatial-hole burning and carrier diffusion can have on the operating characteristics of resonant and noresonant antiguide laser arrays.

A nonresonant eleven-element array was modeled with the following set of structural parameters: antiguide core width $d = 3\,\mu$m, interelement region width $s = 2.8\,\mu$m, antiguide dielectric constant of 12.25, and lateral permittivity differential $\Delta\varepsilon = 0.2$. Note that the lateral index differential is related to the lateral permittivity differential $\Delta\varepsilon$ by $\Delta\varepsilon = \Delta n\,(n_1 + n_0)$. In Fig. 6.20, the calculated power-current characteristic of the eleven-element array operating in the $L = 40$ in-phase mode is presented for electron diffusion length values of $L_d = 0$ and $3\,\mu$m and interelement loss values of $\alpha_{\text{abs}} = 20$ and $100\,\text{cm}^{-1}$. Note that the power output is given in units of $\lambda/(2\pi\sigma\tau)$, and it is displayed as a function of the dimensionless pumping parameter $\rho = [J\tau/(ed) - N_{\text{tr}}]/N_{\text{tr}}$. The $L_d = 3\,\mu$m case is close to the value expected for typical devices, and is therefore more physically meaningful than the $L_d = 0\,\mu$m case. It is clear from Fig. 6.20, that carrier diffusion improves performance, since it increases the slope efficiency and decreases the threshold. This occurs because the carriers in the interelement regions, which see less optical field, will move into the antiguide cores, where the higher optical field causes a rapid depletion of the carriers due to stimulated recombination. For the same diffusion length $L_d = 3\,\mu$m, increasing the interelement loss results in an increase in threshold and decrease in slope efficiency. However, threshold for the $L = 41$ exhibits only a slight increase when L_d is increased from $0\,\mu$m to $3\,\mu$m. Although the $100\,\text{cm}^{-1}$ interelement loss in-

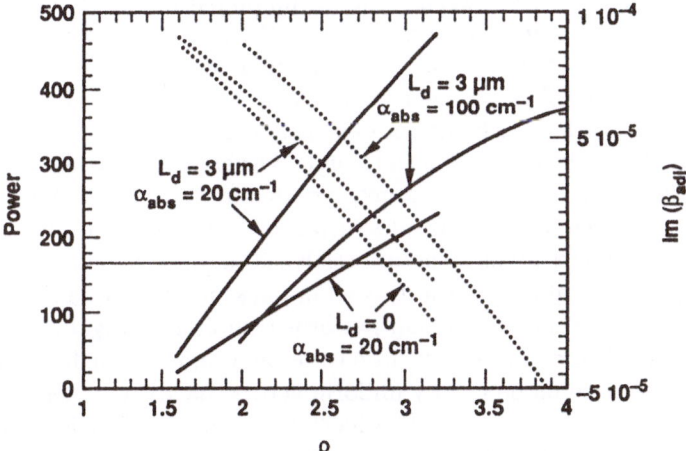

Fig. 6.20. Calculated power output versus pump parameter (solid line) for the $L = 40$ nonresonant eleven-element array modeled by [6.22] for the indicated values of the carrier diffusion length L_d and the interelement absorption α_{abs}. The dotted lines correspond the imaginary part of the propagation constant Im $\{\beta_{\text{adj}}\}$ for the $L = 41$ adjacent mode. The horizontal line that intesects the vertical axis at Im $\{\beta_{\text{adj}}\} = 0$ indicates the threshold for the $L = 41$ adjacent mode [6.22].

creases the threshold of oscillation for the $L = 41$ adjacent mode, the power current characteristics of the $20\,\mathrm{cm}^{-1}$ interelement loss case are, as expected, superior over the range of single array mode operation.

The sublinear power versus pump parameter characteristic displayed in Fig. 6.20 for the case where $L_d = 3\,\mu\mathrm{m}$ and $\alpha_{\mathrm{abs}} = 1000\,\mathrm{cm}^{-1}$ is due to a self-focusing effect. A nonresonant array operating in the in-phase mode has an intensity maximum at the center of the array, when the lateral-index differential is less than the resonant value, as illustrated by the example in Fig. 6.13. Therefore at higher optical powers, the carrier density in the center of the array will be preferentially depleted, resulting in an increase in the local effective index at the array center relative to the ends of the array. This results in a self-focusing effect that confines the mode more tightly to the center of the array. There is less field at the ends of the array where there is more available gain, so that the overlap of the optical mode and the available gain decreases with drive. The example presented in Fig. 6.20 illustrates the influence of carrier diffusion and spatial-hole burning on the stability of the in-phase mode operation of a nonresonant array above threshold.

Nabiev et al. [4.44, 6.22] investigated a number of antiguide array structures by varying the values for the core and interelement widths, as well as the lateral permittivity differential. Unlike the nonresonant structures, it was found for the case of resonant structures that carrier diffusion and spatial-hole burning were significant factors that gave much-improved mode discrimination for above-threshold operation. An optimum near-resonant structure geometry was identifed as having a core width of $d = 3\,\mu\mathrm{m}$, an interelement region width $s = 1\,\mu\mathrm{m}$, an antiguide dielectric constant of 12.25, and a lateral index differential of $\Delta\varepsilon = 1.11$. The power-pump parameter characteristic for this optimized structure appears in Fig. 6.21. The power-pump parameter curves for the three different values of the interelement loss α_{abs} are almost identical, indicating that the near-field pattern of the $L = 40$ in-phase mode has very little overlap with the interelement regions. The sensitivity of the adjacent modes is clearly evident, although neither of the two adjacent modes reach threshold for the range of pump parameters considered. Note that the experimental device of [2.41] (results displayed in Fig. 6.25) is an example of the optimum near-resonant structure analyzed in [4.44, 6.22].

The improved stability of the $L = 40$ in-phase mode can be better understood by inspecting the modal field patterns exhibited in Fig. 6.22. Since the array is designed to operate very close to resonance, the near-field pattern of the $L = 40$ in-phase mode is very uniform across the array so that very little spatial-hole burning can occur. The increased lateral dielectric permittivity differential, along with the narrower interelement region, causes an increase in the power radiated into the central lobe of the far-field pattern [2.41, 4.42]. For the pattern in Fig. 6.22c, about 65 % of the total power was calculated to be in the central lobe.

As mentioned earlier, having a large lateral-index difference between the core and the interelement regions is expected to improve the robustness

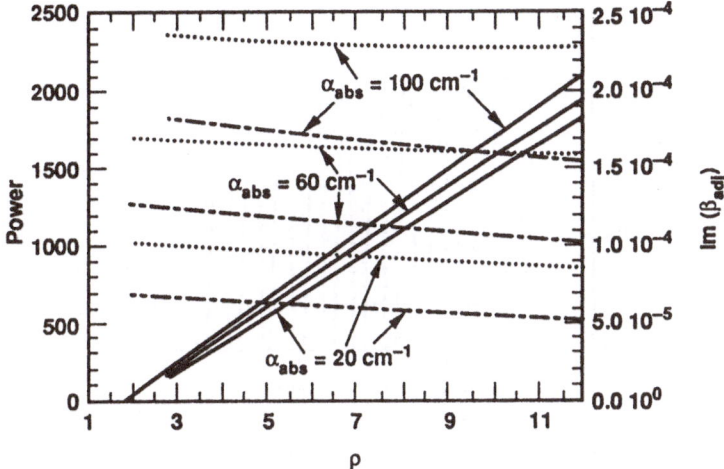

Fig. 6.21. Calculated power output versus pump parameter (solid line) for the $L = 40$ in-phase mode of the near-resonant eleven-element array modeled by [6.22] for a carrier diffusion length $L_d = 3\,\mu$m and the indicated interelement absorption α_{abs}. The dotted lines correspond to the imaginary part of the propagation constant $\mathrm{Im}\,\{\beta_{adj}\}$ for the $L = 41$ adjacent mode, and the dotted-dashed line corresponds to the $L = 39$ adjacent mode [6.22].

against thermal induced changes in the lateral dielectric profile. *Hohimer et al.* [1.232] have done an experimental and theoretical study on a non-resonant antiguided laser array structure that had a lateral-index differential of $\Delta n < 10^{-2}$ [1.225]. Under cw operation in the in-phase mode, *Hohimer et al.* observed a broadening of the central lobe, along with an increase in the relative intensity of the side lobes. Using the two-dimensional model of [1.229], which is described in Sect. 6.3.4, *Hohimer et al.* explained the observed changes in the far-field pattern as being caused by a thermal-induced decoupling of the elements at the ends of the array. As described in [1.229], thermal effects where incorporated into the two-dimensional way in a phenomenological manner. Assuming a $dn/dT = 4 \times 10^{-4}\,\mathrm{K}^{-1}$ for all material layers, the refractive index of each material layer was adjusted in the core and interelement regions according to a lateral temperture profile that corresponded to an $8\,\mathrm{K}$ rise at the center of the array relative to the ends of the array.

Figure 6.23a-c present the cw near- and far-field patterns at progressively higher drive levels, Fig. 6.23d shows the calculated near- and far-field patterns in the absence of thermal effects, and Fig. 6.23e displays the calculated near- and far-field patterns (close to threshold) for a quadratic-like lateral temperature profile with an $8\,\mathrm{K}$ temperature rise at the array center. The effect that heating exhibits in Fig. 6.23e is to confine the mode more strongly to the center of the array. A possible explanation for this behavior that was posed by [1.232] is a mismatch of propagation constants across the array. In

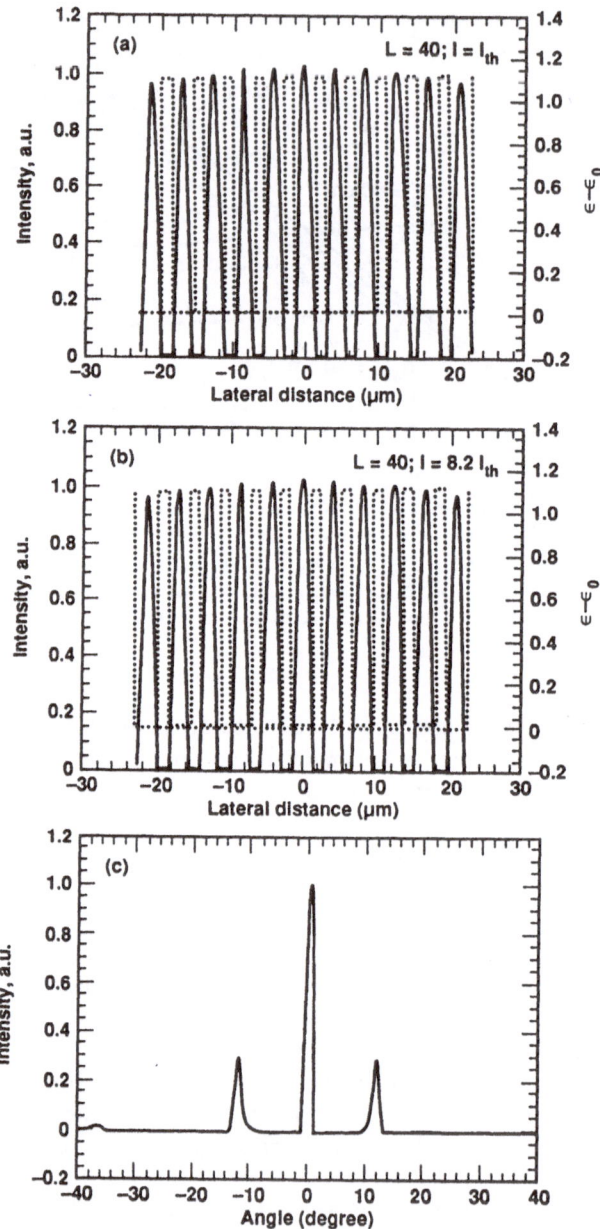

Fig. 6.22. Calculated near-field pattern for the $L = 40$ in-phase mode of the near-resonant eleven-element array (**a**) at threshold, (**b**) at $8.2 \times$ threshold, and (**c**) the far-field pattern for $\alpha_{abs} = 20\,cm^{-1}$. The dotted lines denote the lateral-dielectric permittivity profile, with the vertical axis on the right [6.22].

other words, the lateral temperature profile gives rise to a lateral variation in the lateral index differential across the array, so that there may not be a unique resonant condition, as expressed by (6.6), for the entire array. For a suffciently large lateral variation, smaller sub-arrays within the larger array may be characterized with respect to a local resonance condition. This allows for the possibility of decoupling of the subarrays with operating conditions. As seen in Fig. 6.23a-c, the near field becomes increasingly confined to the center of the array with increasing drive current. The far-field beam divergences of 1.54° in Fig. 6.23a, 1.94° in Fig. 6.23b, and 2.65° in Fig. 6.23c are consistent with the diffraction limit expected for the narrower near-field patterns. At higher drive levels where the effects of nonuniform lateral heating would be expected to be more pronounced, there is less optical field observed at the ends of the array. This can have deleterious effects on the stability of the in-phase mode operation. Recall that the adjacent modes typically have near-field patterns that are peaked at the ends of the array with a minimum in the center of the array, so that under certain conditions oscillation can occur in the adjacent mode [1.232].

Antiguided-diode laser arrays with lateral-index differentials of $\Delta n \approx 4 \times 10^{-2}$ and with geometries selected to optimize the in-phase mode stability [4.44, 6.22] have been studied experimentally [2.41]. This particular

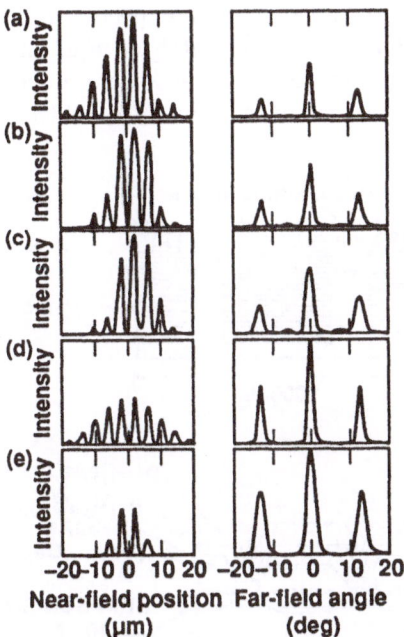

Fig. 6.23. Near- and far-field patterns for the leaky-mode array under cw operation for (**a**) 1.4 × threshold, (**b**) 2.4 × threshold, (**c**) 3.4 × threshold, (**d**) calculated far-field pattern in the absence of thermal effects, and (**e**) calculated far-field pattern with an 8 °C temperature rise at the center of the array [1.232].

array was of the type pictured in Fig. 6.12b [1.226]. It comprised twenty gain elements with core widths of $d = 5\,\mu m$ and interelement widths of $d = 1\,\mu m$, and exhibited stable predominantly single-lobed far-field operation. Because of the relatively larger core width used in this structure, the contribution of edge-radiation losses, as expressed by (??, to the mode discrimination would be expected to be diminished. Therefore, it was anticipated by *Zmudzinski* et al. [2.41] that monolithically integrated Talbot filters would be needed to discriminate against the adjacent modes and the higher-order out-of-phase mode. Figure 6.24 contains a schematic diagram of the cavity geometry that indicates the placement of the Talbot spatial filters. The Talbot filters operate on the principle of Fresnel self-imaging [1.466, 1.470]. Each Talbot filter is a region with no built in lateral waveguide structure, so that the diffraction can occur in the lateral direction. By precise selection of the length (within the Fresnel zone), the adjacent and higher-order out-of-phase in an antiguided array can be made to experience a higher round-trip loss than the desired in-phase mode [6.20, 6.23].

The measured lateral far-field patterns at different power outputs for the array in [2.41] are displayed in Fig. 6.25. At the lowest power output of $0.1\,W$, $75\,\%$ of the power output was measured to be within the dominant central far-field lobe. In the $0.5 - 1.0\,W$ range, a stable far-field pattern was observed with a beam divergence that was found to be $1.7\times$ the diffraction limit of $0.4°$ full-width at half-maximum expected for the $120\,\mu m$ wide near field array aperture. It is clearly evident from Fig. 6.25b and c that in the $0.5 - 1.0\,W$ range less of the total power is contained within the dominant central lobe, as the side lobes are more intense and there is a broad pedestal

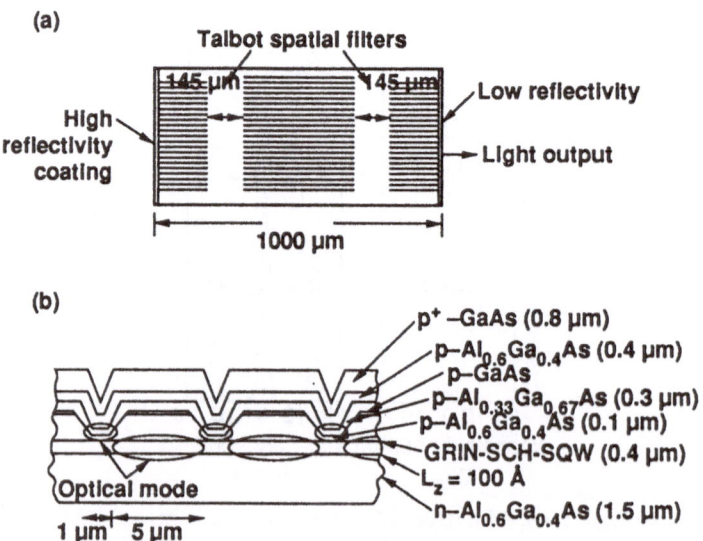

Fig. 6.24. Array cavity geometry with a pair of integrated Talbot filters (a) and diagram of the antiguide array structure (b) [2.41].

of intensity about the base of the central lobe. A possible explanation for this suggested in [2.41], is the presence of lateral thermal gradients, of the type described by [1.232], that provide self-focusing for the adjacent array mode so that it suffers less diffraction loss in propagating through the Talbot filter. This could allow the adjacent mode to reach threshold at higher drive levels, without appreciably modifying the characteristics of the in-phase mode. No thermal-induced narrowing of the near fields versus drive was observed [2.41] for the case of a resonant array with large index step, as was found [1.232] in the near fields of Fig. 6.23a-c, for a nonresonant array of lesser index step. It is interesting to note that the broadening of the diffraction-limited far-field caused by the thermal-induced narrowing of the near field between Fig. 6.23a and c for the smaller Δn array of [1.232] corresponds to a factor of 1.72, which is about the same as the factor of 1.7 broadening observed for the larger-index step array of [2.41]. The magnitude of the far-field broadening observed in [2.41] is similar to that observed in [1.231] (Fig. 6.15) which was shown by spectrally-resolved analysis of the far fields (Fig. 6.17) to be caused by oscillation of the array in a adjacent mode, as well as the fundamental mode.

Although antiguide arrays have demonstrated significant performance improvements in terms of realizing a stable, predominantly single-lobed far-field output beam over a wide-range of operating conidtions, the nature of the far-field broadening and loss of power in the dominant central lobe is not yet fully understood in all cases. It has been suggested that dynamic instabilities (initiated by saturable losses in the cavity) could be a source of the apparent degradation in beam quality observed in antiguided-diode laser arrays [6.24]. Theoretical modeling of the operating frequencies of antiguided arrays, obtained from a numerical analysis of (3.63), indicate that near-degenerate frequecies can occur for the in-phase mode, adjcacent modes, and out-of-phase mode even when there is good mode discrimination [6.25]. The presence of nonlinearities or structural non-uniformities in laser arrays with closely separated operating frequencies may result in spectral instabilities [4.35]. For example, self-pulsations have been observed in diode-laser arrays that have been attributed to the presence of an intracavity saturable loss [6.26]. It has been experimentally confirmed that the presence of saturable loss in the interelement regions of antiguided arrays can cause sustained self pulsations of the light output [6.27]. However in the same study [6.27], it was reavealed that in antiguided arrays with sufficiently narrow interelement regions, self-pulsations in the light output could be avoided. The subjects of dynamic and spectral stability and their relation to output beam quality could well be a interesting area for future studies in antiguided-laser arrays.

Another recent area of interest in antiguide diode laser array research is the possibility of two-dimensional arrays. As discussed in [6.28], two-dimensional arrays can be configured by suitable arrangement of one-dimensional arrays so that the light radiated at the ends of each one-dimensional array is coupled into another one-dimensional array. It has been suggested

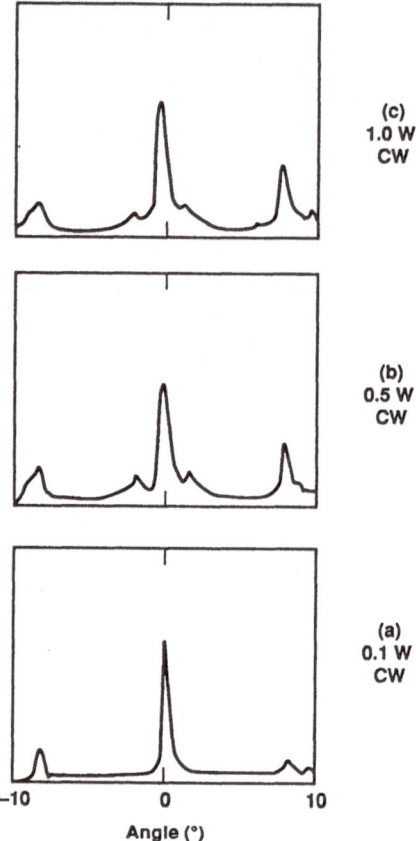

Fig. 6.25. Antiguided array far-field patterns for cw power outputs of (a) 0.1 W, (b) 0.5 W, and (c) 1.0 W [2.41].

[6.28] that this sort of array architecture, which should provide more global coupling between all laser elements, may have a broader frequency-locking bandwidth than laser array structures that rely on nearest-neighbor coupling (Sect 3.4.2).

7. Surface-Emitting Diode-Laser Arrays

Surface-emitting diode-laser arrays have been investigated as an approach for making scalable two-dimensional laser arrays with performance characteristics suitable for optical systems applications (Sect. 1.3.4). The compelling benefit offered by two-dimensional laser arrays is that when all elements are locked in frequency and emitting in-phase, the brightness will scale as the square of the total number of elements in the array. As already mentioned in Sect. 1.3.3, the brightness of a one-dimensional array scales linearly with the number of elements in the array. The waveguide properties of a surface-emitting diode-laser array can either be gain or index guided, depending upon the structure of the gain elements comprising the array. Moreover, the edge-emitting gain-guided structures presented in Chap. 5 and the index-guided structures of Chap. 6 can be used as the basic gain element for a surface emitting array, as the surface-emitted light output is generated over the plane of the array by the process of distributed-output coupling.

In this chapter we shall focus our attention on Grating-Coupled Surface-Emitting Diode Laser Arrays in Sects. 7.1 and Sect. 7.2, while Vertical Cavity Surface Emitting Laser Arrays are discussed in Sect. 7.3. Note that the etched facet surface emitting laser array is not included, as this structure has been used primarily for incoherent two-dimensional arrays [1.170], an exception to this being the structure of [1.475], which appears in Fig. 1.15.

7.1 Fundamentals of GSE Diode-Laser Arrays

The *Grating-Coupled Surface-Emitting* or *Grating-Surface-Emitting* (GSE) diode laser has been investigated as a potentially scalable approach for fabricating monolithic one- and two-dimensional arrays of diode lasers [1.194]. Grating-coupled surface-emitting laser arrays operate on the principle of distributed feedback in the same manner as single-element *Distributed Feedback Lasers* (DFB) and *Distributed Bragg Reflector* (DBR) edge-emitting diode lasers [7.1], as represented in Fig. 7.1. However, the GSE laser array uses distributed output coupling in the form of a segmented or continuous second order grating for generating the output beam along the length of the array cavity. Cleaved facets are not necessary for providing optical feedback to the gain medium. The absence of facets is a potential advantage in that

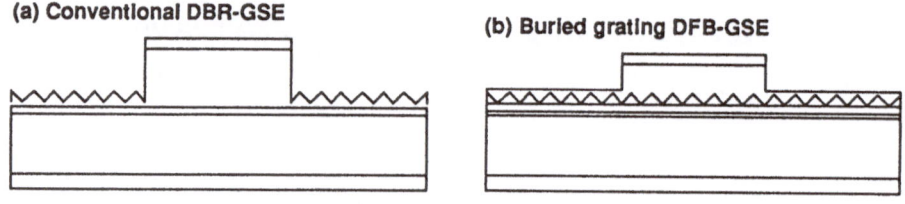

Fig. 7.1. Diagrams of a DBR-GSE (a) and DFB-GSE (b) laser structure.

catastrophic facet damage is not a failure mechanism in a GSE diode laser device.

The principles of distributed-feedback and grating-output coupled wave-guide elements that are central to our understanding of the operation of GSE diode-laser arrays are reviewed in Sect. 7.1.1. In Sect. 7.1.2, some of the important design considerations and principles of GSE laser array operation are presented. In Sect. 7.2, the mode discrimination properties and observed operating characteristics are analyzed and discussed.

7.1.1 Grating Feedback and Output-Coupling Elements

The fundamentals of distributed-feedback and grating-output coupling can be understood by modifying the three-layer waveguide model of Sect. 3.1 to include a grating at the interface of two of the layers as shown in Fig. 7.2. The path of a ray, designated as k_i, propagating in the waveguide mode, with propagation constant $+\beta$, is shown along with the expected rays for the diffracted orders k_0, $k_{\pm 1}$, and k_{-2} from a second order grating. To analyze the interaction between the waveguide mode and the grating, in the simplest form, we can apply the grating equation to the waveguide-mode ray and the diffracted ray [7.2], and find that,

$$n_g \sin\left(\theta_g\right) + n_c \sin\left(\theta_c\right) = \frac{m\lambda}{\Lambda} \ , \qquad (7.1)$$

where n_g is the refractive index of the guide layer, n_c is the refractive index of the cladding layers, θ_g is the bounce angle of a ray in the waveguide mode, θ_c is the angle of a diffracted light ray in the $k_{\pm m}$ order, λ is the wavelength of light in air, Λ is the grating period, and $m = 0, \pm 1, \pm 2, \ldots$ is the order number of the diffracted ray. Recall that the real part of the modal propagation constant, β, is related to the ray angle by (3.11), so that $\beta = 2\pi n_{\text{eff}}/\lambda = n_g \sin\left(\theta_g\right)$. This fact can be used to express (7.1) as

$$n_{\text{eff}} + n_c \sin\left(\theta_c\right) = \frac{m\lambda}{\Lambda} \ . \qquad (7.2)$$

The utility of (7.2) is that for a given waveguide of effective index n_{eff} and grating of period Λ, we can calculate the diffraction angle θ_c associated with each order m of the grating. The condition for distributed feedback

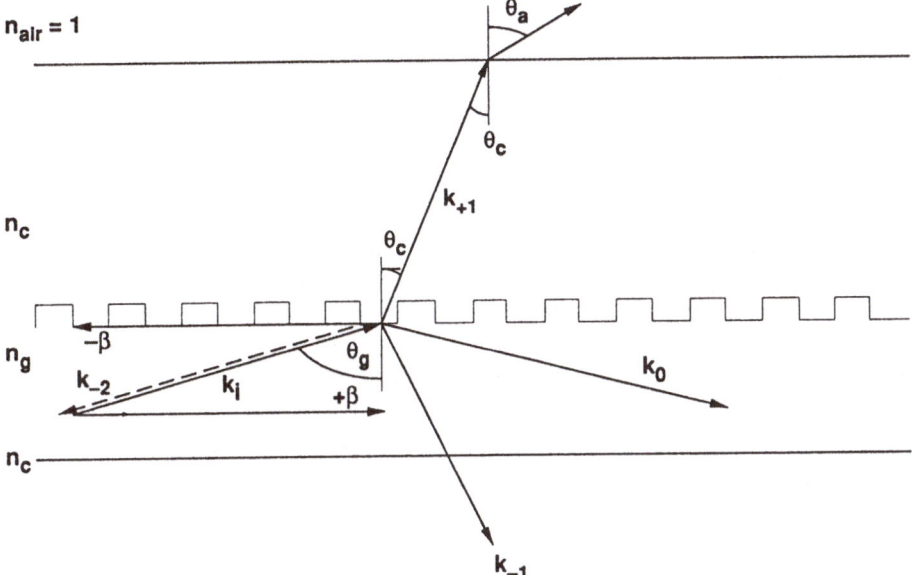

Fig. 7.2. Diagram of a three-layer waveguide containing a grating. The parameters that are used in the grating equation (7.1) are indicated.

will correspond to $\theta_c = 90°$, so that the incident ray k_i is diffracted into the ray $k_{-2} = -k_i$. Therefore, at every point in the waveguide where there is grating, light propagating in the waveguide with propagation constant $+\beta$, is diffracted into the waveguide mode that propagates in the opposite direction (with propagation constant $-\beta$) to the incident mode, hence the term distributed feedback. We find then, that (7.1) can be expressed in the simpler form,

$$n_{eff} + n_c = \frac{m\lambda}{\Lambda} . \tag{7.3}$$

when the distributed feedback condition, $\theta_c = 90°$, is satisfied. The more familiar form for the Bragg condition (7.3) is,

$$\Lambda = \frac{m\lambda}{2n_{eff}} , \tag{7.4}$$

where it is assumed that $n_{eff} \approx n_c$, which is a good approximation for semiconductor diode-laser waveguides. It is seen from the (7.4), that when the grating period corresponds to an integral number of half-wavelengths of the waveguide mode (in the semiconductor material), there will be distributed feedback.

Now let us consider the conditions for both distributed feedback and output coupling to occur simultaneously. Output coupling normal to the wafer surface will occur for diffraction orders where $\theta_c = 0°$. However, gratings can produce diffraction orders that will give light emitted at angle with respect to the surface normal (i.e., $\theta_a \neq 0$ in Fig. 7.2). For this reason, when analyzing the output-coupled grating orders it is more convienent to characterize the surface emission in terms of θ_a. Referring to Fig. 7.2 and using Snell's law at the interface between the cladding and air $(\sin(\theta_a) = n_c \sin(\theta_c))$, we find that,

$$\sin(\theta_a) = \frac{m\lambda}{\Lambda} - n_{\text{eff}} , \qquad (7.5)$$

When the condition for distributed feedback is satisfied, the grating period for Bragg scattering resonances is given by (7.4), and the expression for the output coupling angle (7.5) becomes,

$$\sin(\theta_a) = n_{\text{eff}} \left(\frac{2m}{p} - 1 \right) \qquad (7.6)$$

where $p = 1, 2, 3, \ldots$ is the order of the Bragg grating and m is the diffracted order. Since $n_{\text{eff}} \approx 3.3$ for semiconductor laser waveguides, a first-order Bragg grating, which corresponds to $p = 1$, will not have a solution to (7.6). Hence, there are no diffraction orders that correspond to surface emission for a first-order grating. From (7.2), it is seen that for a first-order grating, the only allowed diffraction order is the $m = 1$ with $\theta_a = 90°$, which corresponds to distributed feedback. When $p = 2$, which is the case that is illustrated in Fig. 7.2, we have a second-order Bragg grating and (7.6) is satisfied for $m = 1$ with $\theta_a = 0°$. Therefore, light diffracted in the k_{+1} order is radiated normal to the surface, while light diffracted in the k_{-1} order is radiated into the substrate. The $m = 2$, k_{-2} order corresponds to the distributed feedback, as seen from (7.2) which for $p = 2$ and $m = 2$ gives $\theta_c = 90°$. For a third-order Bragg grating $p = 3$ in a semiconductor waveguide, the $m = 1$ and 2 diffraction orders are emitted at angles that are totally internally reflected at the air-substrate interface, and the $m = 3$ order provides the distributed feedback, hence there is no surface emitted light. Note that those orders that do not have a solution to (7.6), suffer total internal reflection, and the internal diffraction angle, θ_c, of these orders can be found from (7.1) or (7.2). In general, for a Bragg grating of order p, normal surface emission will occur only for the diffraction order where $m = p/2$. Light emitted at an angle with respect to surface normal ($\theta_a \neq 0°$) can occur in higher-order gratings for those diffraction orders that satisfy

$$p > \frac{2mn_{\text{eff}}}{n_{\text{eff}} + 1} . \qquad (7.7)$$

The light diffracted into radiation modes is also emitted into the substrate, as illustrated by the k_{-1} order in Fig. 7.2, and lost unless steps are taken to recover this light. Approaches that have been to used to recover this light are metallized gratings [1.174, 1.184, 1.484], multi-layer substrate reflectors [1.195, 7.3, 7.4], and a superstrate reflector with an anti-reflection coating on the surface of the substrate [1.194, 1.243, 1.474, 7.5]. Finally, distributed feedback can also be obtained by using a pair of nonresonant gratings that are oriented perpendicular with respect to each other in a ring-laser cavity [7.6, 7.7]. This grating geometry has recently been investigated as a possible means to obtain single-mode operation in ring-configured broad-area edge-emitting semiconductor lasers [7.6, 7.7].

When the Bragg condition (7.4) is not satisified, the grating will act primarily as an output-coupling element, with the off-normal emission angle of the output-coupled light being given by (7.5). This is often referred to as a *nonresonant* grating-output coupler. The nonresonant situation occurs by tuning either the wavelength λ or the grating period Λ so that (7.4) is no longer satisfied. In this case, the order that would normally provide distributed feedback, is not diffracted into the waveguide mode; but instead, it is diffracted into an out-going radiation mode that propagates into the cladding regions, but is not emitted into air due to total internal reflection. This non-resonant mode of operation of grating-output couplers in laser waveguide structures was the subject of investigations of leaky-wave distributed-output couplers in optically-pumped, thin-film dye lasers [7.8], optically-pumped semiconductor lasers [1.35, 1.483], and grating-output-coupled semiconductor diode lasers [1.171, 1.183, 1.484, 1.485][7.9]-[7.13].

The ray-optics model has been used, in some cases, to obtain good approximations of the coefficients for distributed feedback and distributed output coupling [7.2, 7.14] from symmetric, as well as asymmetric slab waveguides. *Zory* [7.2] has developed a phenomenological method for applying the ray-optics model to analyze the grating coupling coefficients for an asymmetric slab waveguide that contains a grating, in the form of a corrugated boundary, at one of the guide/cladding layer interfaces. Referring to Fig. 3.2, one first solves for the mode bounce angle θ_m (3.11) and the optical mode width w_m (3.15) in the absence of a grating for the fundamental or $m = 0$ mode. It is then possible to define a bounce rate B_0 for the fundamental transverse mode of the waveguide, which can be expressed as

$$B_0 = \frac{1}{2w_0 \tan(\phi_0)} \ . \tag{7.8}$$

The optical power in the fundamental mode is then represent as plane wave that is incident on the grating at an angle $\phi_0 = (\pi/2) - \theta_m$. Referring to Fig. 7.2 and setting $\phi_0 = \theta_g$, we can identify the various diffraction orders that result in power loss for the fundamental mode. A diffraction efficiency $\eta_0^{(m)}$ can be associated with each diffracted order k_m of the grating. These diffraction efficiencies $\eta_0^{(m)}$ corresponds to the fractional amount of incident power that is radiated into each of the diffraction orders of the grating. The coupling coefficient or power loss coefficient from the fundamental waveguide mode into a grating order is simply given as the product of the diffraction efficiency and the bounce rate to obtain $\alpha_0^{(m)} = \eta_0^{(m)}B_0$. To apply this method, one needs to to calculate or measure the diffraction efficiencies $\eta_0^{(m)}$ of the grating. In the example presented by *Zory* in [7.2], the grating corrugation was assumed to be sinusoidal and the corresponding diffraction efficiencies were calculated from general expressions that were derived by *Marcuse* [7.15]. Comparing values of the grating-output coupling coefficient $\alpha_0^{(-1)}$ calculated with the ray-optics model to a more accurate numerical analysis based on a phyiscial optics model by [7.16], revealed good agreement for grating depths $\leq (3\lambda)/(8\pi)$ [7.2]. Note that the grating-output coefficient $\alpha_0^{(-1)}$ that we speak of here is for a single traveling wave in the guide. It was pointed out by *Zory* [7.2, 7.8] that in a grating-output coupled DFB laser, the laser would likely select a mode where the output coupled light from the counter-propagating traveling waveguide modes would destructively interfere to minimize losses. For a symmetric three-layer waveguide, *Zory* [7.2] also found that the ray-optic calculation of the grating-coupling coefficient for distributed feedback gave good agreement with a numerical calculation by *Streifer* et al. [7.17] that used coupled-wave analysis.

The grating-output coupling and distributed-feedback functions are considerably more complicated than the aforementioned geometric ray-optics analysis. Accurate modeling of grating-coupled waveguides will usually require a physical optics approach, of the type presented in the next section. Even in the case of the three-layer waveguide, the thickness of the waveguide layer plays a critical role in the output coupling as this layer acts as a Fabry-Perot etalon for the light emitted normal to the waveguide layer [7.18, 7.19]. Depending upon the guide-layer thickness, the resulting interference effects can cause most of the light to be emitted either into the substrate or superstrate. This effect is especially important in grating-coupled semiconductor waveguides, where there are typically more than three layers. Then the thickness of layers other than the guide layer can have a significant influence on the output coupling properties [1.195, 1.243][7.3]-[7.5].

Based on theoretical analysis [7.21]-[7.24], another factor that will have a strong influence on the grating-coupling coefficients is the shape or blaze of the grating. The use of an asymmetric or blazed-grating shape is well know in the field of discrete diffractive optics [1.356] and grating coupled dielectric waveguides [7.25] as a means for enhancing the diffraction efficiency of one order at the expense of the other orders. Although the art of fabricating blazed-grating structures in semiconductors has been around for more than a decade [7.26, 7.27], blazed gratings have typically been not been used in semiconductor diode lasers. With the projected benefits of using blazed grating-output couplers in active-grating surface-emitting MOPAs [6.25], the fabrication of blazed gratings in semiconductors has been revisited [6.25, 7.28]. Further research and development in this area is needed to ascertain whether blazed grating couplers in semiconductor waveguides will provide high-output coupling efficiencies and to perfect the processing techniques required to control the grating shape.

7.1.2 Principles of GSE Laser Array Design and Operation

During the five-year period immediately following the initial demonstration of the DFB laser [7.29], there followed a flurry of research on semiconductor-diode lasers with grating structures that operated on the principle of distributed feedback and distributed-output-coupling [1.35, 1.171][1.483]-[1.489] [7.9][7.30]-[7.32]. These early devices contained nonresonant grating-output couplers or Bragg gratings that were either of fourth or sixth order in the surface emitters. *Luk'yanov* et al. in [1.171] describe what appears to be the first operation of a linear array of GSE diode lasers. This array which is displayed in Fig. 7.3, comprised homojunction DFB diode-laser elements with sixth order Bragg gratings. This two element homojunction laser array had a threshold current density of $12\,\mathrm{kA/cm^2}$, output beam divergence of $0.5°$, and a spectral bandwidth of $4\,\text{Å}$. These performance characteristics suggest that it was unlikely that the array elements operated in a mutually coherent manner.

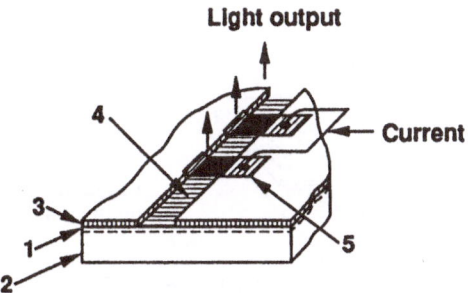

Fig. 7.3. Diagram of a DFB homojunction laser array with segmented contacts so that output coupling occurs through the windows in the contact [1.171].

Improvements in material quality, processing technology, and grating fabrication techniques led to a renewed interest about ten years latter [1.194] in grating-coupled surface emitting lasers and laser arrays. Beginning in 1986, double heterostructure GSE laser array structures with second order gratings, of the type shown in Fig. 7.4, were fabricated [1.173, 7.33], as well as DFB surface-emitting lasers and laser arrays [1.174, 1.184] with metallized gratings. These arrays exhibited sub-Ångstrom spectral bandwidths and showed evidence of mutual coherence over the entire extent of the array, as evidenced by the narrowing of the far-field divergence when the number of array elements were increased. However, the threshold current density of $\approx 10\,\mathrm{kA/cm^2}$ indicates that these double heterstructure laser arrays suffered from high-losses associated with the unpumped DBR sections.

The DBR architecture displayed in Fig. 7.4a was initially the preferred structure for fabricating GSE diode-laser arrays. The GaAs substrate is not transparent to the operating wavelengths of GaAs and AlGaAs active layers.

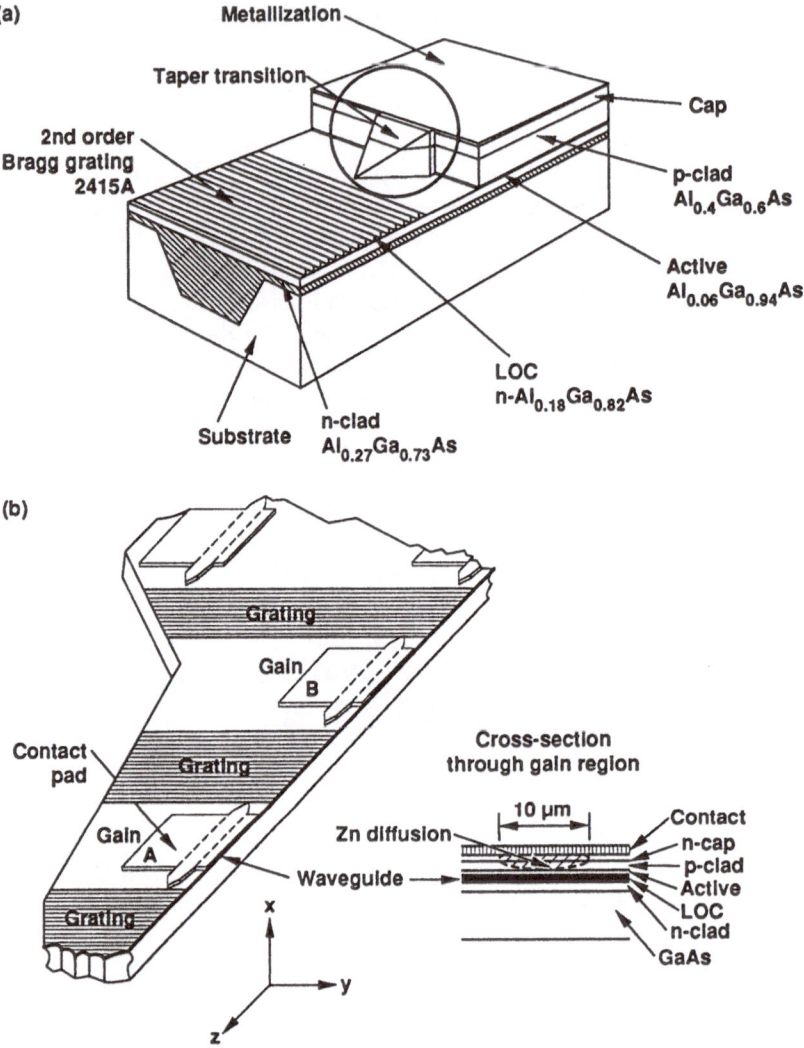

Fig. 7.4. Diagram of a double heterostructure DBR-LOC laser element (a) and a wafer section containing a monolithic array of DBR-LOC lasers (b) [7.11].

Therefore, output-coupled light would suffer significant absorption losses if transmitted through the substrate, unless a window were etched into the substrate [1.174, 1.184] or the operating wavelength of the laser were shifted to be longer than the wavelength corresponding to the bandgap energy of the substrate [1.181, 1.474]. Placing the second-order grating at the air-cladding interface provides direct surface emission of light into air. The design issues and criteria of selecting the remaining p-clad thickness above the waveguide layer are discussed in detail by [1.194].

An alternative means for generating surface-emitted light from a grating structure, is to use a metallized grating in a DFB laser and etch a window in the GaAs substrate as indicated in Fig. 7.5. The Au grating serves also as the electrical contact on the superstrate side of the structure. In a DFB surface-emitting laser with a metal grating, it is important that the optical losses of the metal deposited on the grating be very low. For this reason, Au, which has a high-reflectivity ($> 90\%$) for wavelengths in the $0.8 - 1.0\,\mu$m range [7.34], has been the metal of choice in structures that contain metallized gratings in the active waveguide section. The details of surface-emitting DFB laser designs with metal gratings have been analyzed by [7.35, 7.36].

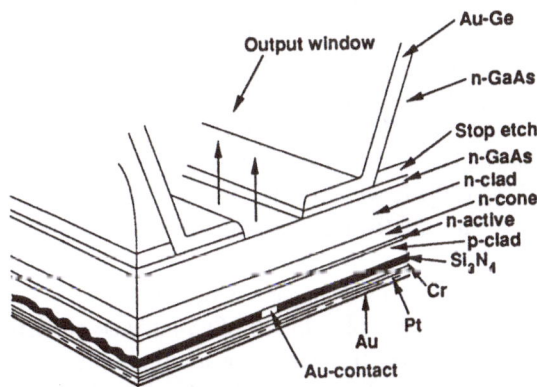

Fig. 7.5. Cross-sectional view of a surface-emitting DFB laser that contains a second order grating for providing optical feedback and output coupling [1.174, 1.184].

An important consideration in the design and fabrication of the DBR-GSE laser element is the interface between the active gain section and the passive DBR waveguide section, as well as the absorption losses resulting from the unpumped active layer in the DBR waveguide section. To minimize these sources of loss *Uematsu* et al. [7.32] showed that the use of a separate confinement heterostructure or *Large Optical Cavity* (LOC), as shown schematically in Fig. 7.4a, offered some improvement. The LOC layer reduces the active layer confinement factor in the DBR section, so modal absorption losses are lower and higher efficiencies are obtained. To reduce losses at the interface, *Evans* et al. [7.11] fabricated a tapered waveguide, illustrated in Fig. 7.4a, at the interface of the gain and DBR waveguide sections.

The DBR laser appears to be well suited as the fundamental element or unit cell for a scalable monolithic diode-laser array. As shown in Fig. 1.14 and Fig. 7.4b, the active laser and passive DBR sections can be fabricated along a common waveguide section with alternating gain elements and passive second-order DBR waveguide sections. Facet reflectors are not required because the DBR waveguides provide frequency selective feedback to the gain sections, in second order, as well as the output light normal to the wafer surface in first order. Optical coupling between adjacent array elements along the waveguide is supplied by the light transmitted through the common DBR waveguide section.

The introduction of quantum-well active-layer structures in surface-emitting DBR arrays had a profound effect on improving the performance characteristics. *Kojima* et al. [7.37] fabricated a single element DBR laser, similar to that shown in Fig. 7.4a, but the LOC and active layers were replaced with by a multi-quantum-well active layer. Besides having a lower confinement factor, it was demonstrated by *Kojima* et al. that the optical absorption loss in the unpumped multi-quantum well DBR waveguide layer could be saturated with only a few mW of optical power. This meant that a several hundred micrometer length of unbiased quantum-well diode laser could act as a low-loss waveguide for transmitting light between active elements on the same wafer. Similar behavior and improved performance characteristics were observed in surface-emitting DBR lasers that incorporated single-quantum well GRINSCH structures [7.38, 7.39], as well as linear arrays of surface-emitting DBR lasers [1.175, 1.176]. It was also found by *Evans* et al. [7.38]-[7.40], that the transmission at the DBR-gain section interface ranged between 0.92 to 0.99, depending on the structure design and thickness of p-clad layer that was left above the guide layer. This is to be compared with the 0.46 interface transmission for the case of a DBR-LOC surface-emitting laser without tapered coupler, and just over 0.9 with a tapered coupler [7.11]. Tapered couplers, as depicted in Fig. 7.4a have proved difficult to fabricate, resulting in a low yield of acceptable devices. The demonstration that unbiased quantum-well laser/waveguides could act as low-loss passive-waveguide sections was a significant development for all of integrated optics, in particular, monolithic semiconductor MOPAs, which are the subject of Chap. 8.

Because of their superior performance characteristics, high-interface transmission and low absorption losses, the DBR-GSE laser with a quantum-well active layer, similar to the structure illustrated in Fig. 7.8 became the element of choice for GSE-laser array structures. In principle, the GSE array can be scaled in the lateral direction (in the plane of the pn-junction) by using any of the laterally-coupled laser array designs presented in Chap. 5 and 6, as well as others. Two examples of two-dimensional GSE array structures that incorporate lateral and longitudinal optical coupling are illustrated in Fig. 7.6 and Fig. 7.7. The type of two-dimensional arrays appearing in Fig. 7.6 demonstrated characteristics indicative of coherent operation of a two-dimensional laser array as evidenced by significant narrowing of the output beam in both the longitudinal and lateral directions [1.177, 1.182, 1.464], as well as, electronic-beam steering [7.41, 7.42]. However, as discussed in [1.474, 4.76], these types of arrays, in general, do not provide good mode discrimination, being very sensitive to structural and material nonuniformities. As mentioned in Chap. 6, the laterally-coupled ridge-guided laser array has poor mode discrimination properties, so a GSE laser array with such gain elements comprising multiple-coupled ridge-guided laser elements would

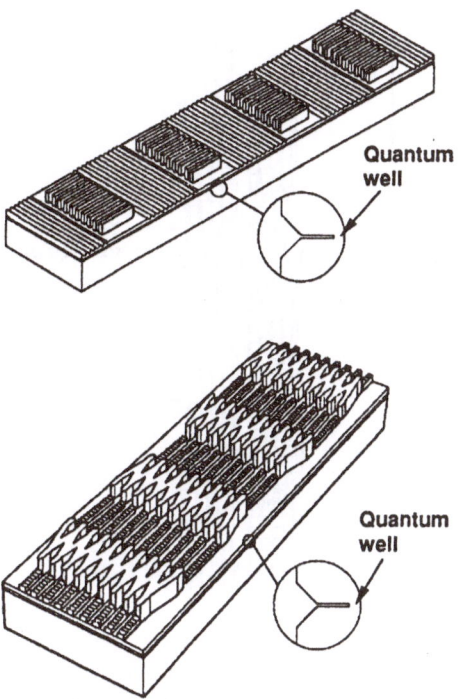

Fig. 7.6. Top view of a two-dimensional (10×10) GSE laser array that uses a ridge-guided array structure in the gain section (**a**), and a detailed sketch of a two-dimensional array with Y-branched waveguides in the gain sections connected by passive ridge-guide DBR waveguides (**b**) [1.464].

generally not be expected to operate in a single-array mode. However, as discussed below, with improvements in material uniformity, the independently-addressable contacts associated with each gain section could be adjusted to induce single-frequency operation at selected operating points [7.43].

Further improvements in material quality and in the laser design, produced single-quantum well DBR lasers with threshold current densities that were the same as an equivalent edge-emitting laser [7.40, 7.44]. This was a significant development because it showed that the optical losses due to scattering at the interface and absorption in the passive DBR section that had limited the performance characteristics of these lasers, could be almost eliminated. To accomplish this, gain sections of 300 μm in length (longer that the 200 μm lengths previously used) were used to reduce the end losses. As discussed in Sect. 2.2, cavity lengths of $> 200 \, \mu m$ are necessary to minimize the threshold-current density of quantum well lasers. In addition, the grating period was selected so that the DBR laser would operate $\approx 100 \, \text{Å}$ to the long wavelength side relative to where an equivalent Fabry-Perot laser would operate. At this longer operating wavelength, as described in [7.40, 7.44], the absoprtion in the DBR waveguide section would be reduced further and band

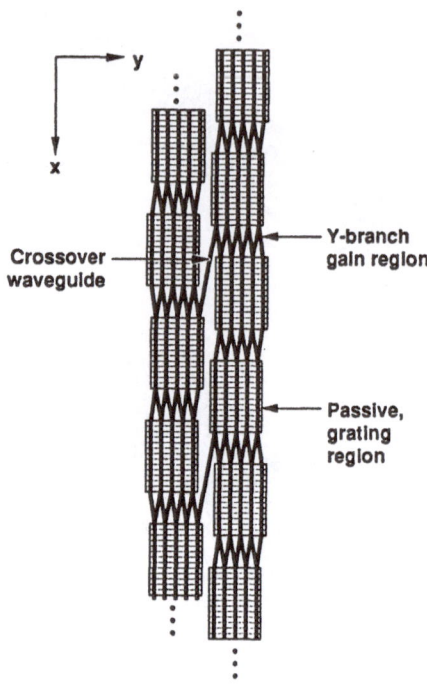

Fig. 7.7. Rendition of a two-dimensional GSE array with Y-branched gain sections that incorporates mutiple crossover waveguide sections [7.42].

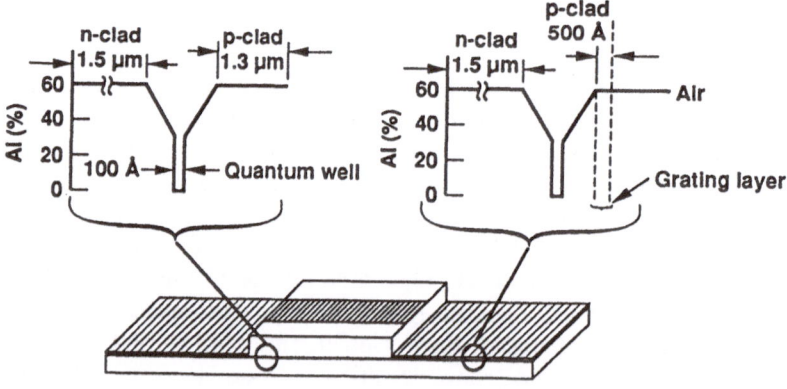

Fig. 7.8. Diagram of a surface-emitting DBR laser with a quantum-well active layer [7.40, 7.44].

gap shrinkage, described in Sect. 2.6, in the active section would provide sufficient gain at the selected operating wavelength. The reduction of losses in the quantum-well surface-emitting lasers is necessary to avoid the gain flattening effects, discussed in Sect. 2.2, which can limit the power output of the device when losses are not minimized [7.45].

The choice of grating period is important because it sharply constrains the operating wavelength of the laser, and this can strongly influence the operating characteristics of the laser. This is illustrated in Fig. 7.9 where the photoluminescence spectrum of a GaAs quantum-well laser structure is presented along with selected operating wavelengths that correspond to different operating characteristics. The low-threshold DBR laser and the Fabry-Perot laser operation were discussed in the previous paragraph. When the grating period is selected so that the operating wavelength is on the short wavelength side of the gain peak, the DBR waveguide section can be made to be highly absorbing, so that greater optical powers are required to saturate the loss in the DBR waveguide. As demonstrated in [7.46, 7.47], in this case a DBR waveguide section in a GSE array can act as an intra-cavity saturable absorber and the array can function as a high-speed optical switch with sub-nanosecond rise time. Another benefit that is achieved by selecting the operating wavelength on the short wavelength side of the gain peak is that the antiguiding factor or linewidth broadening factor, presented in Sect. 2.7, can be made to be smaller [2.4, 2.62, 7.48] because of the increase in differential gain with photon energy in quantum wells. A reduction in the antiguiding factor is desirable because it will reduce the tendency of filamentation in the gain section, and in the case of single frequency operation provide a narrower spectral linewidth.

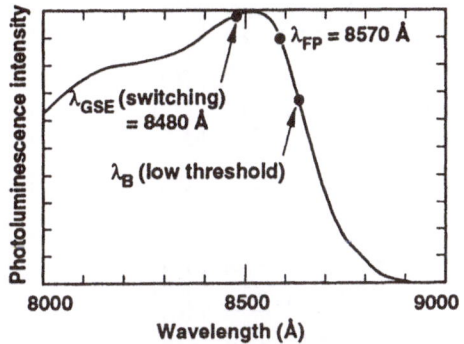

Fig. 7.9. Photoluminescence spectrum of a GaAs quantum well laser structure. The operating wavelength of an edge-emitting Fabry-Perot laser is designated by λ_{FP}, the operating wavelength of the low-threshold surface-emitting DBR laser of [7.40, 7.44] is designated by λ_B, and the operating wavelength of a GSE laser array that exhibited high-speed switching is designated by λ_{GSE} [7.44].

Another technological development that impacted GSE laser structures, as well as semiconductor diode lasers in general [1.296], was the development of strained-layer quantum well structures. This was briefly alluded to in Sect. 1.3.1. Through band gap engineering, psuedomorphic strained-layer quantum well structures with improved performance characteristics such as lower threshold and better reliability were realized. The ability to shift the operating wavelength to wavelengths longer than that of the band gap wavelength of the substrate made it possible to take the output-coupled light directly through the substrate [1.181, 1.474]. An additional benefit afforded by this capability was that junction-down mounting could be done, for more efficient heat removal, without interfering with the output beam [1.181, 1.184, 1.474]. Additional developments in materials growth technology [1.316, 1.332] and the aforementioned design advances, contributed to improvements in the performance characteristics of GSE laser arrays.

Material and structural nonuniformities in GSE arrays can give rise to a spread in operating wavelengths of the DBR laser elements of several Å over distances of several mm to a cm over a wafer. This well exceeds the locking bandwidth of ≈ 1 Å estimated in the strong-coupling limit for Fabry-Perot lasers that was discussed in Sect. 3.4.2. Because of the independently-addressable contacts associated with each gain section in a GSE array, it was possible to adjust the current to each gain section so as to induce single-frequency operation at selected operating points [1.194, 4.76, 7.43] up to power outputs of several hundred mW. From a ring-configured GSE array, depicted in Fig. 7.10, *Liew* et al. [7.43] have demonstrated single-frequency

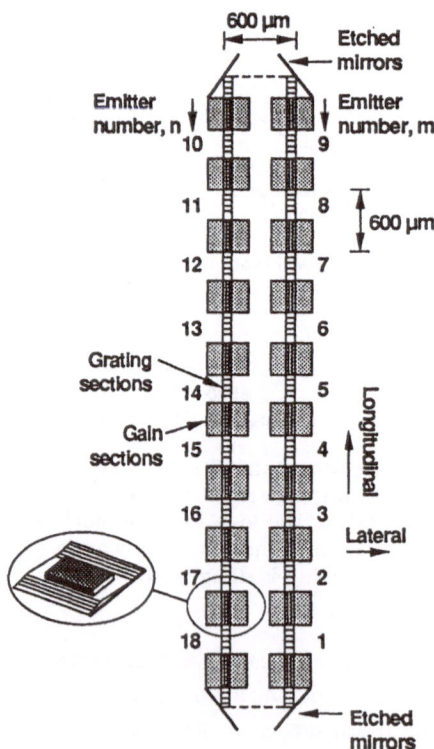

Fig. 7.10. Diagram of a ring-configured GSE laser array where each gain section comprises ten $2\,\mu$m wide ridge waveguides on $4\,\mu$m centers [7.43].

operation (at an isolated operating point at $\approx 2 \times$ threshold) with a narrow bandwidth characteristic of single mode operation.

The cw operating characteristics of this array at $1.9 \times$ threshold are presented in Fig. 7.11. Both the low-resolution spectrum, as measured with a grating spectrometer, and a high-resolution spectrum measured with a scanning Fabry-Perot interferometer are displayed. Single-frequency operation with a linewidth of $\approx 28\,\text{MHz}$ is evident. As discussed in [4.76], the frequency separation of adjacent lateral modes of the ridge-guided gain elements was estimated to be in the 3 to $10\,\text{GHz}$ range, which is significantly larger than the observed linewidth of $\approx 28\,\text{MHz}$. Although the spectral data indicates single-array mode operation, the observed far-field pattern does not exhibit good beam quality.

The random structural nonuniformities, and in some cases stresses induced by the mounting, combined with the current settings that give single-frequency do not, in general, produce a well-defined phase relationship of the light emitted from the DBR sections [1.464, 4.71]. Instead, what usually occurs is a random relative phase shift between the light emitted from the DBR sections [1.194, 1.464, 4.71] that gives rise to a multi-lobed far-field pattern, of the sort that appears in Fig. 7.11. Mutual coherence measurements of this GSE ring array, appearing in Fig. 7.12 are all $\geq 70\,\%$ indicating a high degree of temporal coherence (as expected for frequency-locked operation) between all the gain sections of the array. This is consistent with the degradation of the far-field pattern being due to random time-independent phase shifts

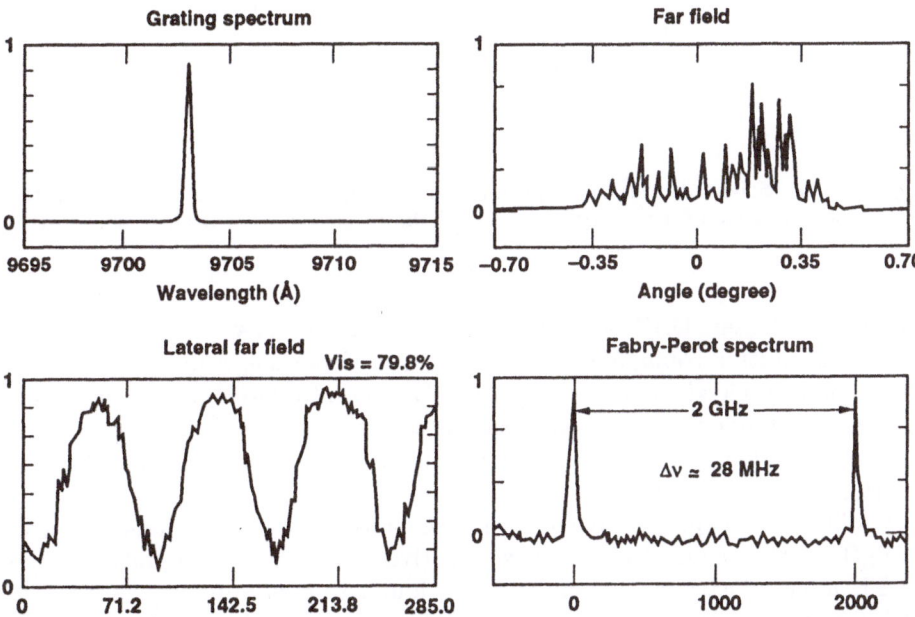

Fig. 7.11. Performance characteristics of a ring-configured GSE laser array [7.43].

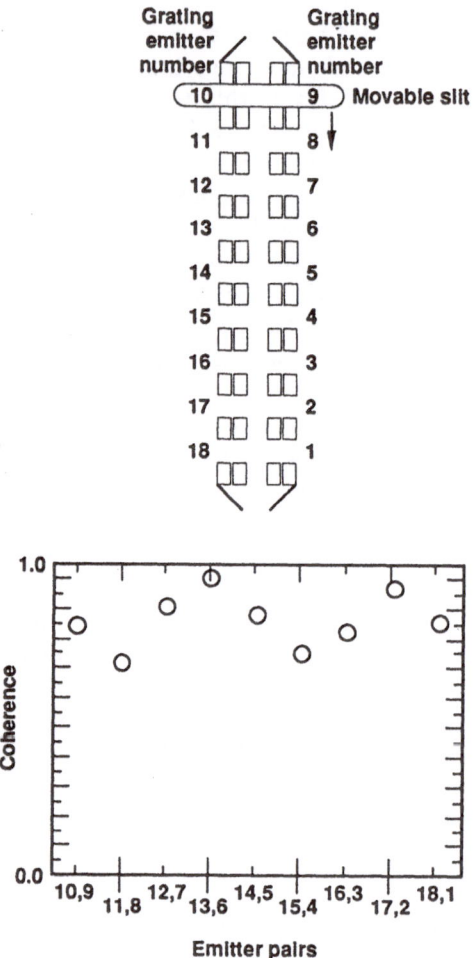

Fig. 7.12. Schematic of the experimental arrangement used for measuring the mutual coherence of a ring GSE array and the measured coherence versus emitter pair [7.43].

between different DBR emitters. The poor beam quality observed from the multiple-emitter configured GSE array, even under single-frequency operation, had provided some additional motivation into developing devices with gratings extended into the gain sections to produce a single large-area emitter. The GSE-DFB laser [1.174, 1.184] is one such example of this type of surface-emitting laser.

With the all epilayers of the structure transparent to the operating wavelength of the strained quantum well active layer, the buried-grating GSE-DFB laser, illustrated in Fig. 7.13 was found to offer superior characteristics in terms of mode discrimination, narrower spectral linewidths and lower threshold current densities than that of the GSE-DBR laser with the grating

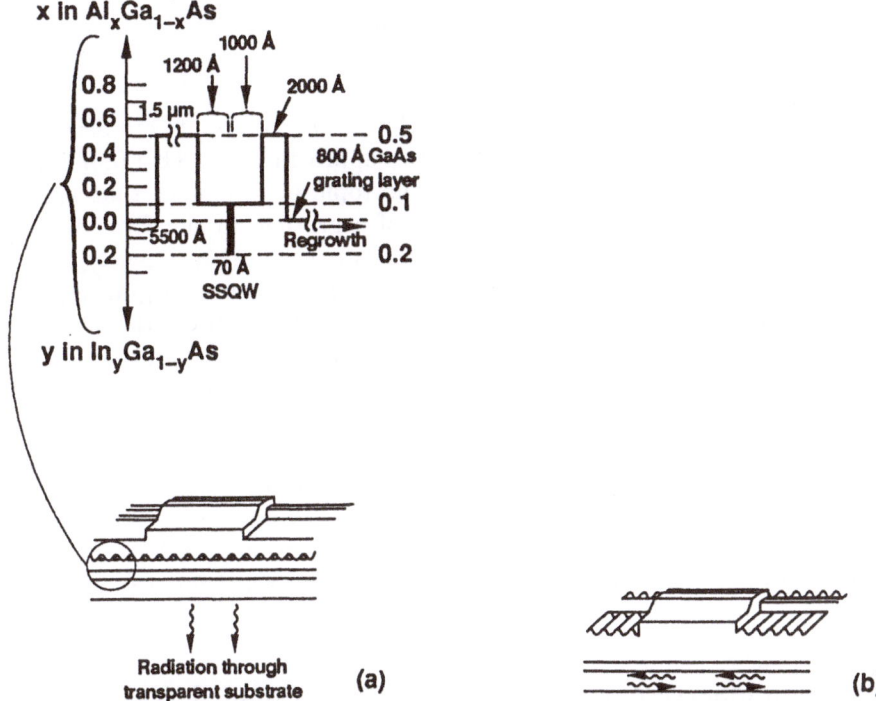

Fig. 7.13. Illustration contrasting the buried-grating GSE-DFB laser structure with a *Strained Single Quantum Well* (SSQW) active layer (**a**) to that of the GSE-DBR laser structure with a relief grating at the p-clad/air interface (**b**) [7.50].

at the p-clad/air interface [1.295, 7.49, 7.50]. As discussed in [7.50], there is essentially no difference in the effective indices for the waveguide mode in the pumped and unpumped sections of the DFB-GSE; and therefore, the dielectric discontinuity at the active/passive interface is virtually eliminated. Furthermore, the strength of a buried grating can be more accurately controlled, since it is determined predominantely by the epilayer thickness. In contrast, the thickness of the relief grating is determined primarily by the uniformity of the wet or dry etching process used to fabricate the grating, which can introduce thickness nonuniformities in the grating. *Liew* et al. first demonstrated thresholds of 9 mA from a 5 μm wide single-element ridge-guided GSE-DFB lasers, and more recently thresholds as low as 4 mA [7.51]. The ability to fabricate low-threshold DFB (or DBR) lasers along with integrated low-loss passive waveguide sections on the same wafer, was an important factor that led to the development of the active-grating surface-emitting amplifier, as well as, some of the other monolithic MOPA semiconductor structures discussed in Chap. 8.

7.2 Mode Discrimination Properties of GSE Laser Arrays

The basic functions of a grating in a GSE laser were elucidated in Sect. 7.1.1 within the context of the geometric optics approximation. This familiar description is useful because it presents the topic in a manner that appeals to our intuition. However, the presence of a grating, in the form of a periodic corrugation of a boundary between two layers (i.e. a relief grating) or as a periodic variation in layer composition (i.e. a buried grating) adds a significant complication to the analysis of the modal properties of a laser waveguide structure. Moreover, the dispersive nature of the grating induced distributed-feedback and grating-output coupling have a significant influence on the mode discrimination properties of a laser array containing an intracavity grating. As an illustration of the sort of complexity that is encountered in a general analysis, the numerical approach of *Butler* et al. [7.5, 7.52] for modeling the dispersion characteristics of grating-coupled waveguides is presented in Sect. 7.2.1. In Sect. 7.2.2, the coupled-wave approach for modeling GSE lasers and laser arrays is introduced, and in Sect. 7.2.4 an analysis of the ring-configured GSE array is presented.

7.2.1 Boundary-Element Model of Grating-Coupled Waveguides

Consider the multi-layer laser waveguide structure that is depicted in Fig. 7.14, wherein each layer is of homogeneous composition, and the boundary between the layers of permittivities ε_1 and ε_3 corresponds to a relief grating. Since the periods of gratings that are of interest for GSE lasers are comparable to the wavelength of light in the material, the approximations discussed in Sects. 3.1, 4.1, and 4.2 for the axial dependence of the mode need to be modified. Then following [7.5, 7.52], the electric fields of the TE modes, $E_x(y, z)$ within a homogeneous layer can be modeled as satisfying the scalar Helmholtz equation,

$$\left(\frac{\partial^2}{\partial y^2} + \frac{\partial^2}{\partial z^2}\right) E_x(y, z) + k^2 \varepsilon E_x(y, z) = 0 \tag{7.9}$$

where k is the free-space propagation constant and ε is the dielectric permitivity within a given homogeneous layer. Because of the periodicity of the grating layer the solution of (7.9) in the mth layer can be expressed as a Floquet-Bloch expansion,

$$E_{m,x}(y, z) = \sum_n \psi_{mn}(y) \exp(-\gamma_n) \tag{7.10}$$

where

$$\gamma_n = i\beta + n\frac{2\pi}{\Lambda}, \tag{7.11}$$

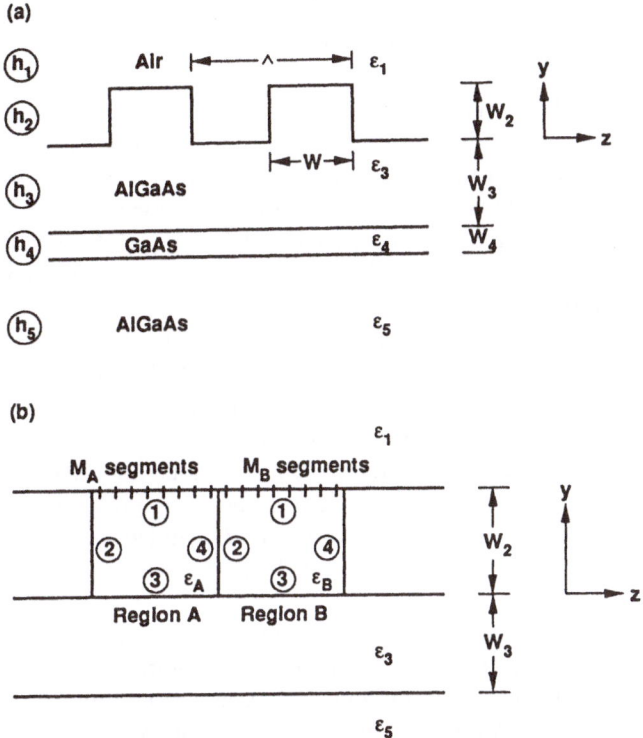

Fig. 7.14. Layer compositions and geometry of a laser waveguide structure that contains a relief grating [4.76, 7.5].

γ_n is the complex modal-propagation constant, and $n = 0, \pm 1, \pm 2, \ldots$. For the multi-layer structure in Fig. 7.14 in those layers other than the grating layer, the transverse field amplitudes $\psi_{mn}(y)$ that appear in (7.10) are of the form,

$$\psi_{1,n} = A_{1n} \exp\left[h_{1n}(w_2 - y)\right]; \text{ where } y > w_2 , \tag{7.12}$$

$$\psi_{3,n} = A_{3n} \cos(h_{3n}y) + B_{3n} \sin(h_{3n}y); \text{ where } -w_3 > y > 0 , \tag{7.13}$$

$$\psi_{4,n} = A_{4n} \cos(h_{4n}y) + B_{4n} \sin(h_{4n}y); \text{ where } -w_3 - w_4 > y > -w_3 , \tag{7.14}$$

and

$$\psi_{5,n} = A_{5n} \exp\left[h_{4n}(w_3 + w_4 + y)\right]; \text{ where } y < -w_3 - w_4 . \tag{7.15}$$

The complex-valued transverse wavenumbers satisfy $h_{mn}^2 = \pm\left(k^2 \varepsilon_m + \gamma_n^2\right)$ where the $-$ sign is taken in layers $m = 1, 5$ and the $+$ sign is taken in layers $m = 3, 4$. Note that in layers of finite thickness $m = 3, 4$, the transverse field amplitudes correspond to waves that propagate in both the $+y$ and $-y$ directions; whereas, in the semi-infinite layers $m = 1, 5$ only the transverse field components that decay exponentially with distance y from

the laser waveguide are retained, because of the requirement that there are
no sources of radiation at infinite distances from the waveguide. By matching the tangential field components $E_{m,x}$ and $dE_{m,x}/dy$ at the interfaces of
the homogeneous layers (where $E_{mn,x}$ is given by (7.10)), a relation between
the transverse-wave amplitudes in the homogeneous layers is obtained. Recall
that this procedure was followed in Sect. 3.1 to derive the secular equation
(3.9) for the symmetric three-layer waveguide. However, the presence of the
relief grating at the boundary of layers 1 and 2 in Fig. 7.14 has, in effect, introduced an inhomogenous layer of periodic composition along the z direction
into the waveguide structure.

In the next step of the analysis, the electric field solution, E_x, within the
grating must be found, and through continuity of the tangential-field components, related to the fields in the homogeneous layers $m = 1, 3$ that are
adjacent to the grating layer. The approach used in [7.5] is to express (7.9)
as an integral equation, by invoking Green's second identity. This method
provides a simplification in that it allows one to express the electric field
in the grating layer in terms of the values of E_x and it's normal derivative,
$\partial E_x/\partial n$, only on the boundary enclosing the homogeneous regions within the
grating layer. The boundary element method is used to then discretize the
integral representation of E_x into N segments. This yields a set of N homogeneous equations where the are $2N$ unknowns: E_x and it's normal derivative
$\partial/E_x \partial n$ on the segments comprising the discretized boundary. Continuity
of the field at the boundaries with the $m = 1$ and 3 layers, as well as the
boundaries of the homogeneous regions within the grating layer, complete the
relation between the grating-layer fields and the transverse-field amplitudes
in the adjacent layers, obtaining a system of $2N$ equations with the $2N$ unknowns being the boundary fields and their normal derivatives. As discussed
in [7.5], the condition that the determinant of the coefficient matrix of the $2N$
equations vanish will provide the modal propagation constants β in (7.11).
Once β is determined, the values of the field and their normal derivatives
can be calculated and the Floquet-Bloch expansion of the field (7.10) can
be evaluated in the homogeneous layers of the structure. This approach is
computationally very demanding, but it provides details on the character of
the radiation fields emitted into both the substrate and superstrate, and it
should give accurate results for the grating-coupling coefficients that can be
used to model distributed feedback and grating-output coupling in GSE laser
arrays and laser amplifiers.

The dispersion characteristics of the grating-coupled waveguide in
Fig. 7.14 that were calculated in [7.5] are exhibited in Fig. 7.15. In this calculation, the waveguide was assumed to be passive and the absorption losses were
set to zero. The normalized optical frequency is designated by $k\Lambda = 2\pi\Lambda/\lambda$.
It is plotted as a function of both the normalized mode wavevector $\mathrm{Re}\{\beta\}\Lambda$
and the normalized mode attenuation constant $\mathrm{Im}\{\beta\}\Lambda$. In Fig. 7.15, the dispersion is displayed over a range that includes the cases where the first and
second Bragg resonances are satisfied. Recall that the first Bragg resonance

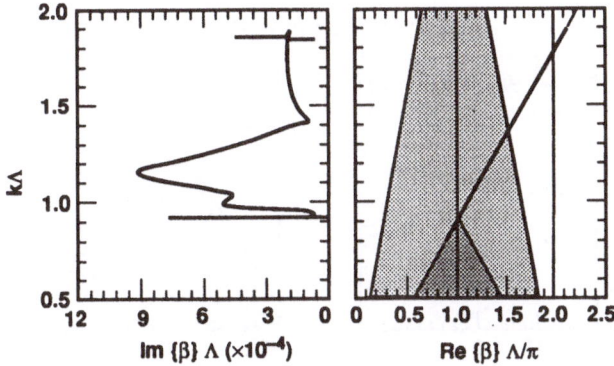

Fig. 7.15. Dispersion characteristics of the complex mode propagation constant corresponding to the passive grating-coupled waveguide structure exhibited in Fig. 7.14 [7.5].

occurs when (7.4) is satisified for $m = 1$; and the second Bragg resonance occurs when (7.4) is satisified for $m = 2$. If Λ is kept constant, then the plots in Fig. 7.15 give the wavelength dependence of the complex mode propagation constant.

The mode-attenuation constant shows a rapid increase as the wavelength approaches the first Bragg resonance where $\mathrm{Re}\{\beta\}\Lambda/\pi = 1$. Figure 7.16 presents a magnification of the dispersion at the first Bragg resonance. Since the absorption losses were set to zero, the contributions to mode attenuation constant come about entirely from the diffractive radiation losses associated with the grating. The stop-band, where the mode propagation constant becomes purely imaginary, is indicated by the apparent discontinuity in the $k\Lambda$ versus $\mathrm{Re}\{\beta\}\Lambda/\pi$ plot. Within the stopband the mode attenuation constant increases significantly. Both the spectral width of the stopband and the magnitude of the mode-attenuation constant is observed to increase with grating depth, as the coupling between the guided mode and the grating layer is strengthened. These are characteristics that are well known in DFB lasers that contain a first-order grating [7.1, 7.53, 7.54], where the two lowest loss modes occur at the wavelengths on the edges of the stopband.

Above the first Bragg resonance, the mode attenuation increases, due to radiation into the substrate (layer 5 in Fig. 7.14), and goes through a maximum. Then it drops to a roughly constant level. The cusp that appears at $k\Lambda \approx 1.4$ is where radiation into air (layer 1 in Fig. 7.14) begins to occur. Below $k\Lambda \approx 1.4$, the light diffracted by the grating into radiation modes suffers total internal reflection. As $k\Lambda$ is increased further there is a sharp feature in the mode-attenuation constant that appears at $\mathrm{Re}\{\beta\}\Lambda/\pi = 2$ the second Bragg resonance. In Fig. 7.16, a magnification of the second Bragg resonance and it's immediate neighborhood is depicted. In contrast to the first Bragg resonance, there is no stopband in the $k\Lambda$ versus $\mathrm{Re}\{\beta\}\Lambda/\pi$ plot. Only a slight dip on the right at $\mathrm{Re}\{\beta\}\Lambda/\pi = 2$ is evident. The mode-attenuation

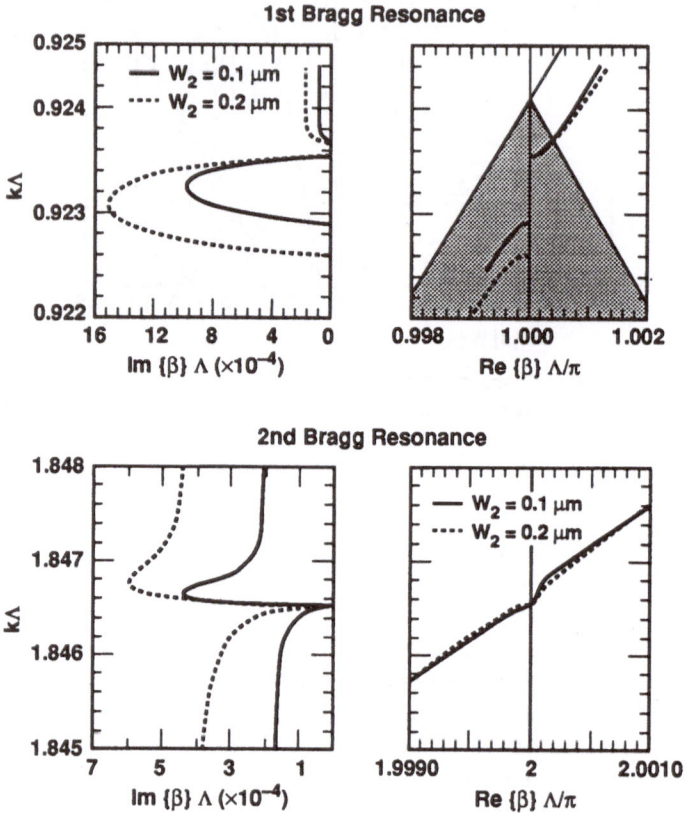

Fig. 7.16. Magnification of the dispersion in the vicinities of the first and second Bragg resonances for the two of grating depths indicated [7.5].

constant is found to drop to a very low value very near $Re\{\beta\}\Lambda/\pi = 2$, and then rises again sharply. The origin of this feature can be understood in terms of the interference that occurs between the radiation waves that are generated by the guided modes that propagate in the $+z$ and $-z$ directions in the waveguide. An intuitive description of this phenomena, using the coupled-wave model, is presented in Sect. 7.2.2.

7.2.2 Coupled-Wave Model of GSE Laser Arrays

A simpler approach used for modeling grating-coupled surface-emitting laser arrays is the coupled-mode approximation that was discussed in Sect. 4.2.2, where the corrugated grating layer is treated as a pertubation that couples the forward and backward-traveling waves in the cavity, as well as generating the radiating wave that corresponds to the output coupled beam. Coupled-mode analysis has been widely used to calculate the modal characteristics of single-element DBR and DFB lasers [7.1, 7.53, 7.54]. Depending upon the basis set of modes used in the Floquet-Bloch expansion, the coupled-mode method can become less accurate when the grating depth is comparable to the wavelength of light in the material. As mentioned in Sect. 4.2.2, the accuracy of the coupled-mode method depends on the number and character of the set of modes that comprise the coupled-mode expansion. The more general numerical approach of [7.5, 7.52] is expected to be more accurate, since the degree of accuracy is determined by the size of the boundary elements used to discretize the grating layer. In addition to being computationally less demanding, the coupled-mode analysis of grating-coupled waveguides yields a simple set of equations that will improve our understanding of the physics governing the interaction of the guided modes and it's relation to the mode discrimination properties of grating-coupled surface-emitting laser arrays.

The coupled-mode formalism of *Kazarinov* et al. [7.54, 7.55] and *Streifer* et al. [7.17, 7.19, 7.21, 7.22, 7.56] has served as the foundation for much of the modeling of grating-coupled surface-emitting lasers and laser arrays that contain second-order gratings [1.195, 4.49, 4.50, 7.20, 7.35, 7.36, 7.57, 7.58]. In this model, the electric field for a single transverse mode, $E(x, z)$, is expressed as a Floquet-Bloch expansion similar to (7.10) and the effect of the periodic grating layer is modeled as a periodic dielectric perimittivity and expressed using a Foruier series as,

$$\varepsilon(y, z) = \varepsilon_0(y, z) + \sum_{n \neq 0} \xi_n(y) \exp\left(-i\frac{2\pi m}{\Lambda}\right) \tag{7.16}$$

where $\varepsilon_0(y, z)$ is the permittivity in the absence of the grating plus a contribtution of the $m = 0$ term of the Fourier expansion. The coupled-mode equations are obtained from (7.9) by using a Floquet-Bloch representation of the field solution and the Fourier representation of the dielectric permittivity (7.16). The coupled-wave equations for a single-transverse and lateral mode propagating in a waveguide with a periodic grating that acts both as a feedback element and an output coupler are derived in [7.56]. A form of the coupled-mode equations that is more useful for treating arrays includes the field dependence on the lateral coordinate [7.36],

$$\frac{1}{2i\beta_0}\frac{\partial^2 R}{\partial x^2} + \frac{\partial R}{\partial z} = (-\alpha + i\Delta\beta - \zeta_1) R + i(\kappa^* + \zeta_2) S \tag{7.17}$$

$$\frac{1}{2i\beta_0}\frac{\partial^2 S}{\partial x^2} - \frac{\partial S}{\partial z} = (-\alpha + i\Delta\beta - \zeta_1) S + i(\kappa^* + \zeta_4) R \tag{7.18}$$

where

$$E_x(x, y, z) = \phi(y) [R(x, z) \exp(i\beta_0 z) + S(x, z) \exp(-i\beta_0 z)] , \tag{7.19}$$

$\beta_0 = 2\pi/\Lambda$, $R(x, z)$, and $S(x, z)$ represent the slowly-varying axial dependence of the lateral-mode propagating in the $+z$ and $-z$ directions and $\phi(y)$ represents the transverse mode profile. The coefficients that appear in (7.17) and (7.18) are interperted as follows: α is the amplitude absorption coeffcient, $\Delta\beta = (\beta - \beta_0)$ is the deviation of the mode propagation constant from the Bragg condition, ζ_1 is represents the contribution of the radiation losses due to the grating, κ is the grating coupling coefficient for distributed feedback, and ζ_2, ζ_4 are corrections to κ due to the radiation waves that are generated by the grating. The calculation of the coefficients ζ_1, ζ_2, ζ_4, and κ is quited involved and more details can be found in [7.17, 7.19, 7.21, 7.22, 7.56]. The values of these coefficients are dependent on the details of the waveguide structure, as well as, the depth and profile of the grating corrugations. For example, when the grating has a rectangular profile and the depth that is small compared to the wavelength of light, the approximation $\zeta_1 = \zeta$ and $\zeta_2 = \zeta_4 = \zeta$ can be used [7.57, 7.58]. In the absence of a grating, the radiation coefficients ζ_1, ζ_2, and ζ_4 are all equal to zero, (7.17) and (7.18) reduce to the more familiar form of the coupled-mode equations [7.1, 7.53, 7.54] used to model DFB and DBR laser structures, where radiation coupling does not occur.

The diffracted electric field at the surface, E_S, is given by the sum of the field amplitudes R and S in (7.19),

$$E_S(x, z) \sim R(x, z) + S(x, z) \tag{7.20}$$

The total optical power that is radiated from a grating, P_{out} is expressed by integrating the Poynting vector over the area of the grating emitter,

$$P_{out} = 2\zeta_1 \int |R(x, z) + S(x, z)|^2 dz dx . \tag{7.21}$$

Upon inspection of (7.21), the loss contribution of the radiation output coupling is found to be of a different nature than the absorption losses. Each of the oppositely travelling waves, designated by the amplitudes R and S in (7.19), will produce a radiation wave. The power output coupled by the grating is proportional to the interference that occurs between these two radiation waves, and this can have a significant effect on the amount of power diffracted away by the grating. Since R and S are dependent on the mode propagation constant β, we see that the mode discrimination properties will be influenced by the radiation losses of a laser or laser array that incorporates a grating-output coupler. In the case of DFB lasers with a second-order grating and non-reflecting ends, it is often the case that the mode with the lowest losses is the mode where the two grating-coupled radiation waves experience a destructive interference, so that P_{out} is minimized [7.2, 7.8, 7.36, 7.57, 7.58]. However, this characteristic has not been experimentally observed for in GSE-laser arrays, as discussed on page 262.

The interference properties of the R and S waves are illustrated in Fig. 7.17. For a DFB laser, we have the requirement that each wave must build up from zero as it propagates from it's respective end towards the center of the device. With this boundary condition, the intensities each of the R

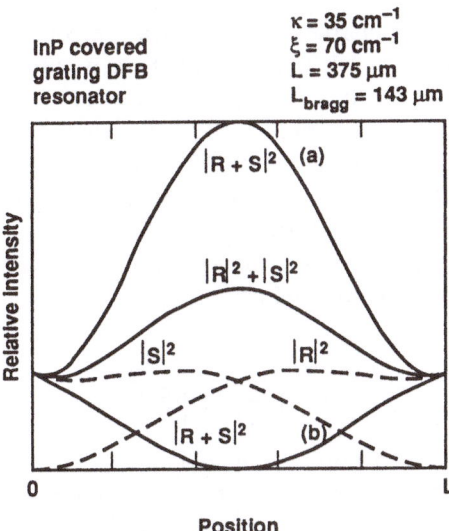

Fig. 7.17. Calculated intensity distributions versus position of the R and S waves for a grating-coupled DFB diode laser. The curve (**a**) is the near-field intensity pattern of the mode with the lowest losses, while curve (**b**) corresponds to the mode with higher losses, where the radiated waves interfere constructively. The length of the grating-coupled waveguide section is L = 375 μm [7.59].

and S waves are indicated by the dotted lines that appear in Fig. 7.17. Then interference between the R and S waves is absent at the ends where one of the waves is zero. Maximum interference occurs at the center where the R and S wave amplitudes become nearly equal. As evidenced by Fig. 7.17, the mode that exhibits destructive inteference, (b) curve, will have the lowest radiation losses.

For a laser that operates on the principle of grating-output coupling, this is undesirable, as the lowest loss mode will also have the lowest differential quantum efficiency. There are exceptions to this which have been demonstrated in monolithic, buried-grating surface-emitting DFB lasers terminated with passive low-loss unpumped grating sections [7.50], where the Bragg resonance of the pumped grating section is shifted slightly relative to the Bragg resonance of the unpumped grating sections, due to the gain induced index depression that occurs in the pumped section. Note that modal characteristics of edge-emitting DFB lasers can be significantly modified when terminated with external reflectors [7.60, 7.61]. Monolithic DFB-GSE lasers terminated with passive unpumped grating sections have been analyzed by [7.50]. In the monolithic DFB-GSE, the passive unpumped grating sections can function as high-resolution dichroic filters that have a high reflectivity for the mode that undergoes constructive interference and a low reflectivity (or high transmissivity) for the mode that suffers a destructive interference. In this way it is possible to improve the efficiency for radiative output coupling.

In modeling the mode characteristics of grating-coupled laser arrays such as those presented in Sect. 7.1.2, the coupled-wave approach has been most widely used [1.195, 4.49, 4.50, 7.20, 7.58], because it is computationally simpler than the more exact approaches such as that discussed in Sect. 7.2.1. For modeling GSE laser arrays, (7.17) and (7.18) are used to model the propagation characteristics of active or passive waveguide sections containing gratings. In the case of an active section the $-\alpha$ term appearing in (7.17) and (7.18) must be replaced by $g - \alpha$, where g represents the complex, saturable modal gain which can be expressed as [7.58],

$$g = \frac{g_0 \left(1 - ia\right)}{1 + \left|\frac{R}{I_S}\right|^2 + \left|\frac{S}{I_S}\right|^2} \; , \tag{7.22}$$

where g_0 is the unsaturated gain for the field and I_S is the saturation intensity introduced in Sect. 2.3.4 and a represents the antiguiding factor discussed in Sect. 2.7. Note that although the notation is identical, the unsaturated field gain g_0 used in this coupled-mode analysis is equal to half of the unsaturated power gain used Sect. 2.3.4. The antiguiding factor a has been included in (7.22) to account for gain-induced changes in the optical path lengths of the gain sections. This effect can have an important influence on the relative phasing of the light emitted from the DBR sections. Waveguide sections that do not contain grating sections, are also governed by (7.17) and (7.18), but with $\kappa = 0$ and $\zeta_1 = \zeta_2 = \zeta_4 = 0$. In the case of the single-spatial mode active waveguide section commonly used in linear GSE laser arrays, (7.17) and (7.18) reduces to,

$$\frac{\partial R\left(x, z\right)}{\partial z} = \left(g - \alpha + i\Delta\beta\right) R \tag{7.23}$$

and

$$\frac{\partial S\left(z\right)}{\partial z} = -\left(g - \alpha + i\Delta\beta\right) S \; . \tag{7.24}$$

An example of the above-threshold coupled-mode model calculation of a linear GSE laser array comprising 10 gain sections and 11 DBR section output couplers is found in [7.58]. A diagram of the array structure that was modeled is presented in Fig. 7.18. The waveguide and grating parameters used for this analysis were $\kappa = 15\,\text{cm}^{-1}$, $\zeta = 1.5\,\text{cm}^{-1}$, $\alpha = 3\,\text{cm}^{-1}$ (field loss), $a = 0$, and $g_0 = 20\,\text{cm}^{-1}$ (field gain). In Table 7.1 the calculated characteristics of the four lowest loss modes are listed for the case when each end of the array is terminated by an infinitely long passive waveguide section that does not contain a grating. The columns labeled by $\Delta\beta$ and g_{th} correspond to the results of a linear analysis of the structure in the absence of gain, meaning that $g_0 = 0$ in (7.22). The other columns represent the result of the nonlinear analysis for $g_0 = 20\,\text{cm}^{-1}$ (field gain) and $a = 0$. The total optical power (noramlized to I_S) that is generated in the array is denoted by P_{tot}. The remaining columns, expressed as percenatges of the total optical power are

Table 7.1. Calculated modal properties for a ten-element GSE array

$\Delta\beta$ (cm^{-1})	$\Delta\beta_{sat}$ (cm^{-1})	g_{th} (cm^{-1})	P_{tot} $(/I_S)$	P_{abs} (%)	P_{out} (%)	P_{end} (%)
11.1	11.0	11.0	2.6	55.3	4.6	40.1
−11.6	−11.5	13.2	1.9	46.2	19.6	34.2
19.4	19.5	15.9	1.2	40.6	6.8	52.6
−20.1	−20.2	17.1	0.8	37.6	13.5	48.9

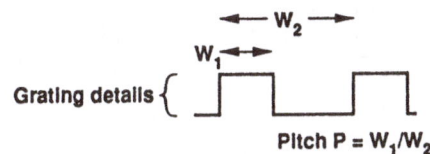

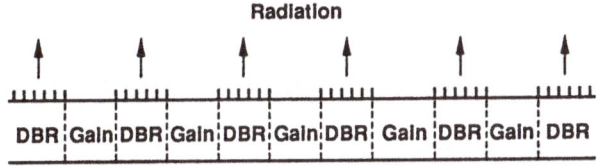

Fig. 7.18. Geometry of a linear ten-element GSE-array indicating the parameter values used in the analysis of [7.58]. The length of the gain and DBR sections is 150 μm.

P_{abs}, the power loss due to absorption; P_{out}, the power radiated away by the grating output coupling; and P_{end}, the power that is lost at the ends due to the aforementioned termination condition. It is important to realize that P_{out} should be interpreted as the maximum available power for output coupling, as this coupled-wave model calculation does not explicility incorporate the radiation field that is coupled into the air. Therefore, the distribution of radiated light emitted into the substrate and superstrate is not accounted for and the effects of absorption and interference from other layers in the structure are also not included. Note that coupled-wave analyses have been developed for the purpose of analyzing substrate reflectors on grating-coupled waveguides [1.195], and an example of the results of such an analysis are presented in Fig. 8.23.

Upon inspection of the P_{out} column, we see that there is a factor of four difference in the maximum power available for output coupling for the two lowest loss modes. This disparity is caused by the same interference effect illustrated in Fig. 7.17 for a single-element DFB laser. The calculated near-field intensity and phase distributions for the two lowest loss modes of the ten gain section GSE laser array are displayed in Fig. 7.19. In Fig. 7.19a, we see the near-field intensity of the lowest-loss mode with $g_0 = 11.0\,cm^{-1}$ is maximized at each end of the array; while at the center of the array, the

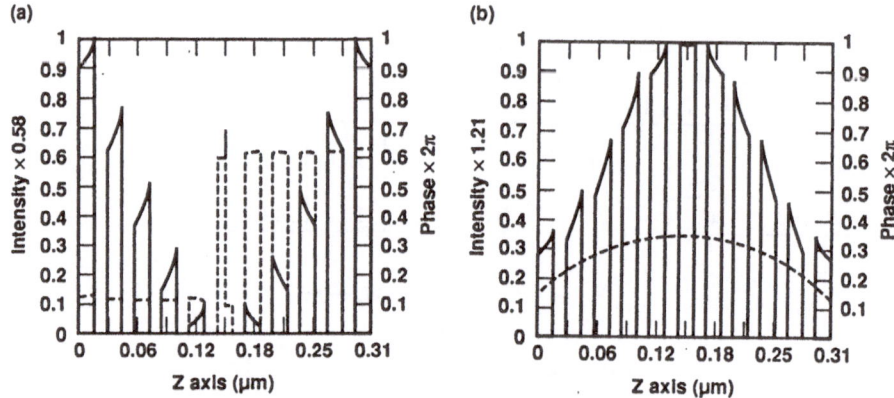

Fig. 7.19. The near-field intensity (solid lines) and phase (dotted lines) for the eleven-element DBR-emitting sections of a ten-element GSE laser array are plotted for the modes with $g_{th} = 11.0\,cm^{-1}$ (a) and $g_{th} = 13.2\,cm^{-1}$ (b) [7.58].

intensity is nearly zero. There is also an abrupt 180° relative phase shift that occurs in the middle of the center DBR section, so that the right and left halves of the array emit out-of-phase. This type of near field will give rise to a double-lobed far-field pattern. Considering Fig. 7.19b, the mode with $g_0 = 13.2\,cm^{-1}$ has a near-field intensity pattern that is peaked at the center of the array and decreases to minimum values at each end of the array. The intensity profile of this mode is accompanied by a nearly-uniform phase distribution across the array, which corresponds to a predominately single-lobed far-field pattern. The difference in interference of the radiated light for these two modes occurs because of the wavenumber detunings $\Delta\beta$ relative to the Bragg condition. The $g_{th} = 11.0\,cm^{-1}$ corresponds to $\Delta\beta < 0$, and exhibits a destructive interference at the center of the near-field pattern. In contrast, the $g_{th} = 13.2\,cm^{-1}$, that occurs for $\Delta\beta > 0$, has a constructive interference at the center of the near-field pattern. This characteristic is consistent with the more accurate result depicted in Fig. 7.16, where $\Delta\beta < 0$ also corresponds to minimum radiation losses and $\Delta\beta > 0$ to maximum radiation losses.

One can think of the GSE laser array with passive DBR waveguide sections as a type of long-cavity DFB laser with the feedback and gain functions are represented by $N + 1$ discrete passive-DBR waveguide sections and N active waveguide sections. In this picture, the GSE near-field patterns that appear in Fig. 7.19 correspond to a sampled version of the DFB near fields of Fig. 7.17. However, the GSE array will generally have a much longer cavity length; and therefore, the wavelength separation between the two lowest loss modes is considerably less than in the case of a conventional single-element DFB laser. For a single-element DFB laser, the wavenumber detuning, $\Delta\beta_s$, of the two lowest loss modes, which are typically on either side of the stop band, is dependent on the magnitude of the grating-coupling coefficient, κ [7.53, 7.62]. An example of this dependence is found in Fig. 7.20, where the

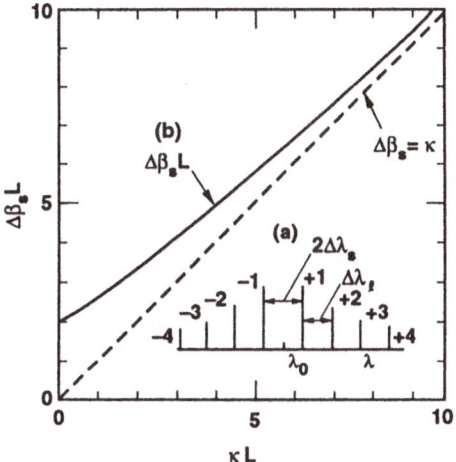

Fig. 7.20. Schematic representation of the longitudinal mode spectrum of a DFB laser (a), and plot of $\Delta\beta_S L$ versus κL, where L is the cavity length of the DFB laser (b) [7.62].

calculated $\Delta\beta_s$ versus κ from [7.62] is displayed for the case of a DFB laser containing a first order grating. Without loss of generality, we can apply the result of Fig. 7.20 to our discussion relating to GSE-laser arrays with second-order gratings.

Note that the wavelength separation between adjacent modes that appear on either side of the stop band, centered on λ_0 in Fig. 7.20a, is given by (2.6), and the wavenumber detuning vs κL for the two lowest-loss modes is indicated by the solid line in Fig. 7.20(b). To find the wavelength separation, $\Delta\lambda_s$, from $\Delta\beta_s$, one uses the expression [7.62]

$$\Delta\lambda_s = \Delta\beta_s \frac{\lambda^2}{2\pi n_g} \qquad (7.25)$$

where n_g is the group index given by (2.7). It is evident from Fig. 7.20b that $\Delta\beta_s \sim 1/L$, and thus the wavelength separation of the lowest-loss modes in a DFB laser decreases with increasing cavity length for a fixed κ. Considering the GSE array modeled by [7.58] as the type of long-cavity DFB laser already mentioned, we find $\kappa L = 4.73$ for the GSE array. From Fig. 7.20b, it is found that $\kappa L = 4.73$ gives $\Delta\beta_s L \approx 5.5$, which corresponds to $\Delta\beta_s \approx 17.5\,\mathrm{cm}^{-1}$, or $\Delta\lambda_s \approx 0.6\,\text{Å}$ from (7.25), for the 0.315 cm long GSE-laser array. The value of $\Delta\beta_s$ for the GSE laser array estimated from Fig. 7.20b gives fair agreement to the value of $\Delta\beta_s = 22.7\,\mathrm{cm}^{-1}$ from Table 7.1, or $\Delta\lambda_s \approx 0.8\,\text{Å}$ from (7.25), found by the more accurate coupled-wave model calculation of the GSE-laser array of [7.58]. One should not expect perfect agreement however, because the grating in the GSE laser array is not distributed along the entire cavity length.

In practice, the type of modal near fields portrayed in Fig. 7.19 have not yet been clearly identified. Inhomogenities in the material and epilayer structure can cause an effective randomization of the optical-path lengths in the gain and DBR waveguide sections. Of course, this makes it difficult to get all array elements to frequency lock [1.195]; and when it is possible to induce all the array elements to become frequency locked, the resulting modal near-field patterns do not usually resemble those calculated for a uniform array. Here, the use of external phase shift plates has been suggested as a remedy [1.194, 4.71]. The prediction that the lowest-loss mode of the type illustrated in Fig. 7.19b has not been verified by experiment on a GSE array that was known to be operating single frequency. This is probably due to the inhomogenities present in actual arrays which destroy the symmetry of the structure. Nevertheless, single-frequency operation at cw powers in the $\approx 10 - 100$ mW range has been observed.

In an experimental study carried out by *Liew* et al. [4.76, 7.63], the near- and far-field output patterns were measured from a linear GSE array with six gain and six DBR sections. This array operated single frequency with an observed spectral linewidth of 24 MHz [4.76]. This observed linewidth is much less than the ≈ 12 GHz frequency separation between adjcant modes expected for the 3.3 mm long cavity. Figure 7.21 presents the measured near-

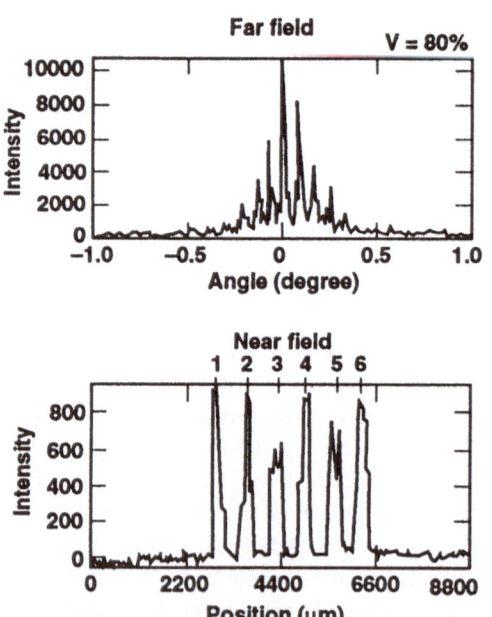

Fig. 7.21. The measured near- and far-field patterns of the GSE laser array of [4.76, 7.63] under single-frequency operation. The number 1 indicates the position of the first gain section and the number 6 indicates the position of the last gain section. The numbers along the upper-horizontal axis denote the number of the DBR section.

and far-field patterns for the GSE laser of [4.76, 7.63] that corresponded to single-frequency operation. To obtain single-frequency operation, it was necessary to tune the currents to each of the gain sections. At these operating currents the cw output power was measured to be about $\approx 50\,$mW and the differential quantum efficiency $\approx 10\,\%$. The far-field intensity has a dominant single lobe, but there is clearly a significant level of intensity in the side lobes and each lobe appears to be atop a broad pedestal. The near-field intensity pattern appears to be fairly uniform, and it does not appear to exhibit the any nulls of the type predicted in the coupled-wave model calculations.

To explore this further, individual measurements were made of the near- and far-field pattern emitted from each of the six DBR section, and these data are presented in Fig. 7.22. None of the individual DBR section near-field patterns display a clear null in intensity; however, the near field of DBR sections 3 and 5 both exhibit a central dip in intensity but the corresponding far-field patterns are predominatly single-lobed. Basically the observed far-field patterns in Fig. 7.22 are all single lobed, which indicates that there are no 180° phase shifts present in any of the DBR section near fields. The far-field patterns from DBR sections 4, 5, and 6 appear to deviate several tenths of a degree off the optic axis, which is an indication of a phase tilt present in the near field of the corresponding DBR sections. The origin of this phase tilt in this particular experiment was not determined, although the occurance of such nonuniform phase tilts in the DBR sections of GSE arrays can be caused by bowing of the chip that may occur in mounting [1.194], as well as inhomogenities in the optical path length differences of the gain and DBR sections of the array [1.464, 7.41].

Let us turn our attention to the output-coupling efficiency of the GSE laser array. Both the structure modeled in Table 7.1, as well as the experimental structure discussed in [4.76, 7.63], exhibited power-output efficiencies that are quite low compared to the power extraction efficiencies given in Sect. 2.3 for edge-emitting lasers optimized for high efficiency. For the experimental device, approximately half the radiated light is emitted into the substrate, where it is lost. Making this correction, would at most raise the observed external quantum efficiency of the single-frequency array in [4.76, 7.63] to $\approx 20\,\%$, which is still low when compared to edge-emitting laser arrays. Note that GSE laser arrays have demonstrated external quantum efficiencies greater than $40\,\%$ per surface [1.194], but in these cases the gain sections were not frequency locked, as evidenced by measured output spectral widths of several Å. In this mode of operation, the gain sections oscillate independently of each other or as groups of subarrays. As already mentioned, a different situation is encoutered for single-frequency operation of a GSE laser array. The principal reason for this is that for fixed absorption and end losses, the lowest-loss array modes must seek to minimize the radiation losses due to the grating-output couplers. This applies even when inhomogenities are present in the array.

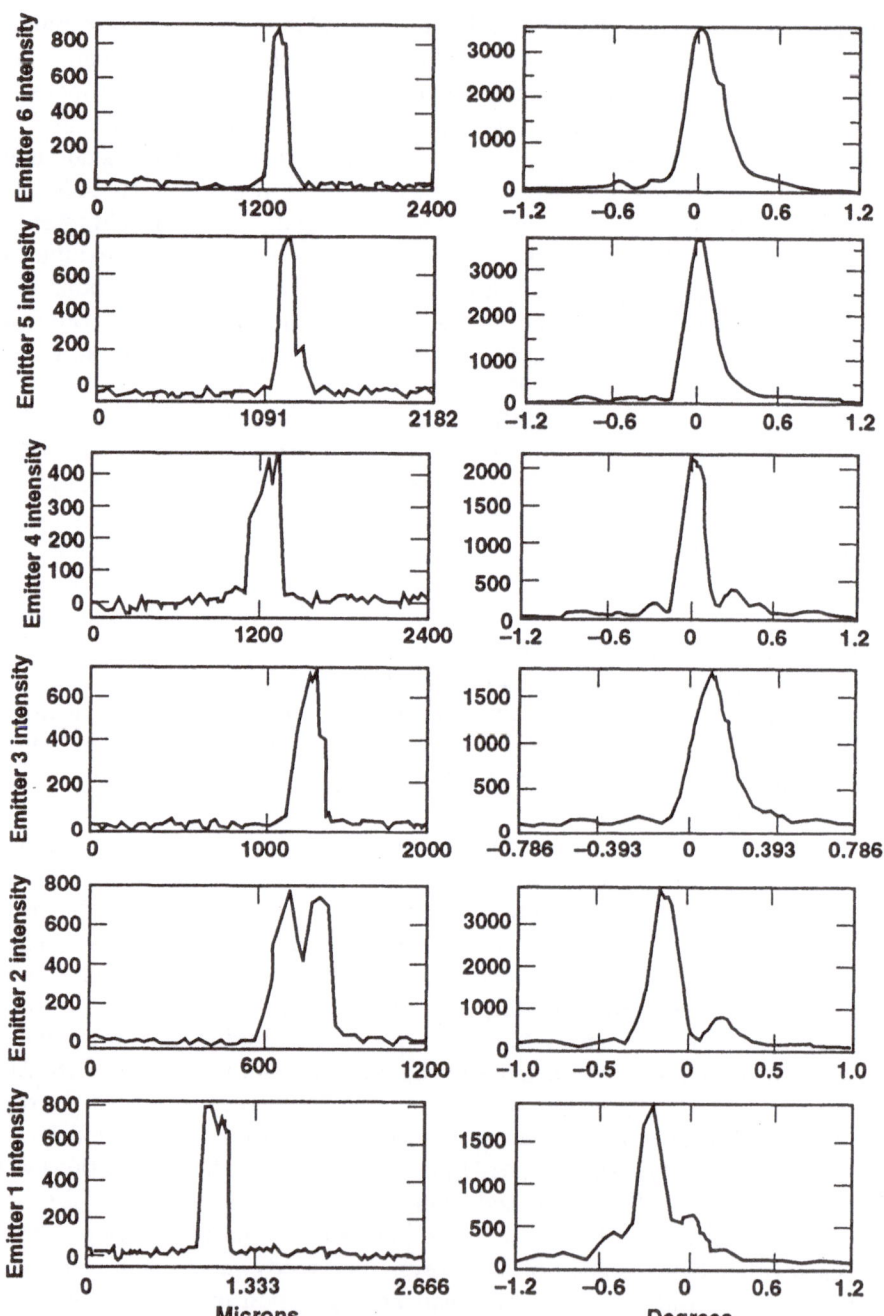

Fig. 7.22. Near- and far-field patterns of the individual DBR sections of the GSE laser array corresponding to the operating conditions in Fig. 7.21. The DBR section number appears on the lrft hand side [4.76, 7.63].

In spite of the tendency of the lowest-loss mode of a uniform GSE array to radiate least efficiently, there are structural modifications that can be applied to optimize the power output coupling efficiency. For example, in the model GSE laser-array structure presented in Fig. 7.18 and Table 7.1, it was found that by increasing the length of the two end DBR sections, P_{out} for the $g_{th} = 13.2\,cm^{-1}$ could be significantly increased. Note that the $g_{th} = 13.2\,cm^{-1}$ mode corresponds to the near-field in Fig. 7.19, which has the most efficient output coupling. The calculated result of [7.58] for the dependence of P_{out} on the end DBR section length is displayed in Fig. 7.23. By increasing the length of the end DBR sections, some of the power lost due to the end losses, P_{end}, in Table 7.1, can be converted into useful surface-emitted power for the $g_{th} = 13.2\,cm^{-1}$ mode; however, the modes that radiate less efficiently, which are the $g_{th} = 11.0\,cm^{-1}$ lowest-loss mode and the $g_{th} = 15.9\,cm^{-1}$ mode, exhibit virtually no improvement by increasing the length of the end DBR sections. Another approach that has resulted in improved external efficiency under single-frequency operation is the ring-geometry array that is depicted in Fig. 7.10. The mode discrimination characteristics of the ring-configured GSE-laser array are presented in Sect. 7.2.4.

Another approach for optimizing the performance characteristis of GSE laser arrays is to design the structure so that the power flowing into each gain section of the array is as uniform as possible over the extent of the array [4.49, 4.50]. When this condition is met, maximum utilization of the gain is obtained so that the conversion efficiency is maximized. In addition, the mode that has optimum overlap with the gain regions will be most robust to the destabilizing effect of nonlinear gain saturation that can occur as the array is operated at power levels well above threshold. A figure of merit that can be used to quantify the degree of power uniformity in an array is the

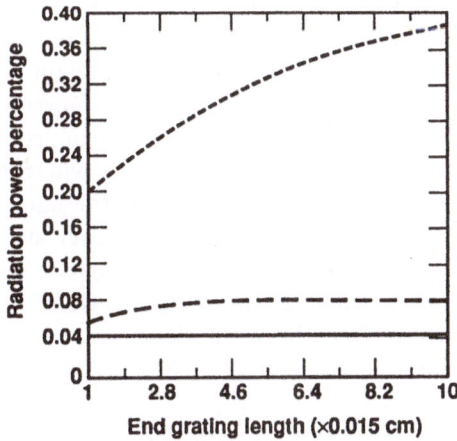

Fig. 7.23. Calculation of P_{out} versus length of the end DBR sections is plotted for the $g_{th} = 11.0\,cm^{-1}$ (solid line), the $g_{th} = 13.2\,cm^{-1}$ (dotted line), and the $g_{th} = 15.9\,cm^{-1}$ [7.58].

root-mean-square power deviation, ΔP_{rms}^2, which can be expressed as

$$\Delta P_{\text{rms}}^2 = \sum_{m=1}^{N} (P_m - P_{\text{avg}})^2 \; , \qquad (7.26)$$

where N is the number of gain sections in the array, $P_m = |R_m(0)|^2 + |S_m(L)|^2$ is the power flowing into the mth gain section, and P_{avg} is the average power which can be expressed as

$$P_{\text{avg}} = \frac{1}{N} \sum_{m=1}^{N} P_m \; . \qquad (7.27)$$

Optimization of a GSE laser array structure can be accomplished by varying the design parameters so that ΔP_{rms}^2 minimized. Note that the condition expressed by (7.26) can be used in above-threshold models so that, in principle, the performance can even be optimized for a specific operating range. In general, this sort of optimization is critically dependent on the values of the grating-coupling coefficients, as the dispersive nature of the DBR sections represent a coherent loss within the cavity of the GSE array. Here again we can draw an analogy to the single-element DFB laser. As discussed in [7.53], for an optimized DFB laser structure the modal intensity distribution in the cavity should be as uniform as possible; and for this to occur, the condition $\kappa L \approx 1$ must be satisified. When $\kappa L \ll 1$, the DFB laser is said to be undercoupled and the optical intensity in the gain region is peaked at the ends of the laser with the minimum intensity being found at the center of the gain region. For $\kappa L \gg 1$, the DFB laser is said to be overcoupled and the optical intensity is peaked at the center of the gain region and a minimum at the ends of the laser. Similar trends have been found and analyzed for GSE laser arrays using a network model for the GSE array [4.49, 4.50]. However, it is important to realize that for the GSE laser array with passive DBR sections, the observed near-field pattern does not generally provide an accurate representation of the power that is flowing into the gain sections. As already mentioned, the near-field pattern of a GSE laser array represents the local interference pattern $|R(z) + S(z)|^2$ of the grating-coupled light from the counter-propagating traveling waves in the cavity, whereas the power flowing into a gain section of length L corresponds to the sum $|R(0)|^2 + |S(L)|^2$. To optimize a GSE laser-array structure of arbitrary size so that the power into each gain section is as uniform as possible across the entire array, it is most convenient to use a network approach described in Sect. 7.2.3, where all the elements and interfaces between elements in the GSE array can be represented as a matrix.

7.2.3 Network Models of GSE Laser Arrays

The network model of laser arrays treats the complete array as a single large-laser cavity comprising gain sections and DBR sections that are connected together in a common waveguide network [4.49, 4.50, 4.76]. As discussed by *Amantea* and *Carlson* [4.50], the advantage in applying network theory to analyze diode-laser arrays is that it provides a formalism for handling very-large one- and two-dimensional arrays of coupled-laser cavities. Here an analogy can be made to integrated-circuit technology, where large numbers of transistors are connected together. We see in Fig. 7.24 a schematic diagram of the basic network element for a DBR-GSE laser array. As seen in Sect. 7.2.2, a set of first-order equations governs the propagation of the two traveling waves, R and S, in the gain and DBR waveguide sections in Fig. 7.24. For a single-spatial mode waveguide section containing a grating, (7.17 and 7.18) are used and the terms involving $\partial/\partial x$ are neglected. The general solution of the coupled differential equations (7.17 and 7.18) gives a pair of linear equations

$$S_k(L) = r_k R_k(L) + t_{k+1} S_{k+1}(0) \tag{7.28}$$

and

$$R_k(0) = r_k S_k(0) + t_k R_{k-1}(L) \tag{7.29}$$

in the traveling wave amplitudes $R_k(L)$ and $S_{k+1}(0)$, which are incident on the grating scattering network (passive DBR section), and the traveling wave amplitudes $S_k(L)$ and $R_{k+1}(0)$ reflected by the grating scattering network. In (7.28 and 7.29), r and t are, respectively, the frequency depedent field relectivity and transmissivity of the grating-coupled waveguide section, which also depend on the grating coupling coefficients κ and ζ. Note that (7.28 and 7.29) can be used for active or passive grating-waveguide sections. For a gain or passive waveguide section that does not contain a grating, (7.23) and (7.24) are used. A linear analysis is obtained by letting the gain g that appears in (7.23 and 7.24) and/or (7.28 and 7.29) be a constant. Alternatively, a nonlinear analysis would use the expression for the saturable gain given by (7.22) in place of a constant g. As seen from (7.28 and 7.29), the grating scattering network in Fig. 7.24 is represented mathematically as a 2×2 scattering matrix, to account for distributed feedback, and the gain sections are characterized by a transmission function $\exp(i\beta z)$, where β is the complex modal-propagation constant. An advantage of the network theory is that one need not be restricted to a particular model of the grating-coupled waveguide. All that is required in the network theory is that the grating sections be represented by a frequency dependent reflectivitiy and transmissivity at each

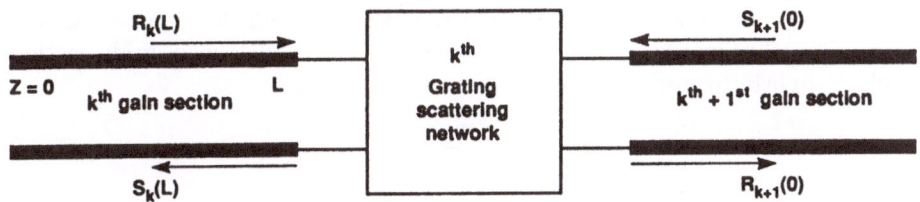

Fig. 7.24. The basic network element for a DBR-GSE laser array [4.50].

interface. Therefore, one is not restricted to using the coupled-mode model of the grating (Sect. 7.2.2), but other models such as that of the boundary-element method of Sect. 7.2.1 could as well be employed.

The matrix representations for the passsive DBR sections, gain sections, DFB or active-grating sections, as well as the boundary conditions encountered at the interfaces and ends of the arrays can be derived and combined, as detailed in [4.49, 4.50], to obtain a matrix set of homogeneous equations. The eigenvalues of the matrix correspond to the modal-propagation constants and the eigenvectors correspond to the traveling-wave field amplitudes at the interfaces between the elements of the array, which are indicated in Fig. 7.24. A central result of the network approach, as applied to a uniform DBR-GSE laser array with N gain elements, is that the modal-oscillation condition can be reduced to the form of that for a single Fabry-Perot type gain element with a multi-branch effective reflectivity. The effective round-trip gain condition is

$$\exp\left[-2L\left(g_{mn} + i\beta_{mn}\right)\right] = R_m^2\left(\beta_{mn}\right) \qquad (7.30)$$

where L is the length of a gain section, $m = 1, \ldots, N$ is the index for the branch associated with the gain sections in the array, n is the number of wavelengths in the gain section, g_{mn} is the threshold field modal gain, β_{mn} is the modal propagation constant, and $R_m^2\left(\beta_{mn}\right)$ is the effective multi-branch reflectivity. The effective reflectivity can be expressed as the product of the DBR waveguide reflectivity and the eigenvalues of a longitudinal-coupling matrix. This calculation is somewhat involved, and details on the derivation of the longitudinal coupling matrix and it's eigenvalues can be found in [4.49, 4.50]. The amplitude part of (7.30) gives the modal gains g_{mn} and the phase part of (7.30) gives the propagation constants β_{mn} associated with the different branches. Besides treating uniform arrays, the network model can also be used to treat GSE arrays with nonuniformities in both the gain and DBR sections. In addition, the linear network model can be extended to model above-threshold modal characteristics by incorporating a saturable gain, as already mentioned, and this is referred to as the nonlinear-network model [4.76, 7.64]. These two features have been employed to model the effects of material nonuniformities and to optimize above-threshold characteristics.

In an effort to quantify the deleterious effects of material and structural inhomogenities on the mode discrimination properties of GSE laser arrays, Amantea [4.76, 7.64] has used the nonlinear-work model to calculate the modal characteristics for a three-element GSE array that contains a stepped-linear grade in the effective index of the DBR waveguide sections across the array. In other words, the effective index of a given DBR waveguide section differs slightly from a nominal value of n_e according to

$$\delta n_e\left(p\right) = \frac{2n_e\Delta(p-1)}{(N-1)} , \qquad (7.31)$$

where p is the number of the DBR section when counting from left to right in the array, and Δ is the total relative change in the effective index di-vided by the number of gain sections, and N is the number of gain sections. The effect of the step-like grade on the effective index is to cause a shift in the Bragg condition in the DBR sections across the array. In Fig. 7.25, the multibranch-effective reflectivity and mode detunings are presented for a three-element array ($N = 3$ in (7.31)) with a constant grade in effective index across the array. Note that the larger the magnitude of the effective reflectivity, the lower the losses of the mode. For the case of a unifrom array, we see from Fig. 7.25 that there is a dominant mode near $\Delta\beta = 0$ with an effective reflectivity magnitude that is well above all other modes, hence some measure of mode discrimination is provided. However, as the magnitude of the index grade across the array is increased, the branch with the largest effective reflectivity is diminished, as the bandwidth of the effective reflectivity broadens with increasing index grade. The reduction in effective reflectivity means that the threshold gain of the dominant mode is increased at the lower values of the index grade. For the largest-index grades, the dominant branch has essentially merged with the other branches with lower-effective reflectiv-

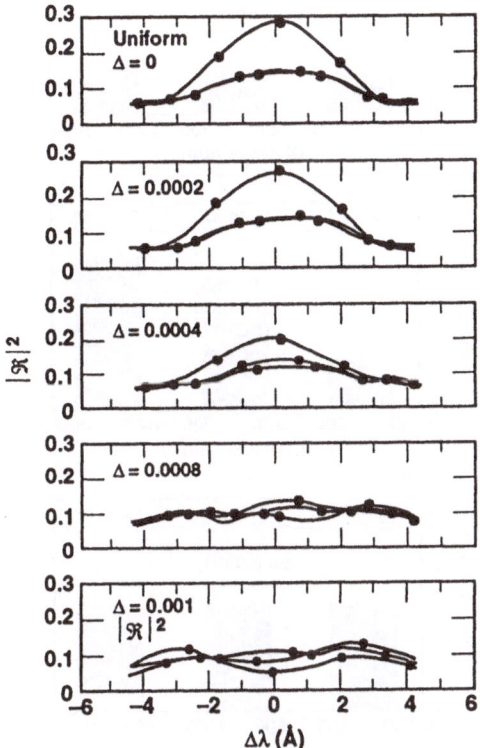

Fig. 7.25. The solid lines correspond to the three branches of the effective reflectivity for the three element GSE array. The points indicate the actual operating points, reletive to the Bragg condition, where the phase condition is satisifed and laser oscillation can occur [1.474, 4.76].

ities, and the mode discrimination is effectively eliminated. In this situation, multimode operation would be expected, since the spectral selectivity of the DBR sections appears to have been almost completely negated by the graded effective index.

As discussed in [1.474, 4.76], there are current tuning algorithms that can be implemented to restore, to some degree, the mode discrimination associated with the dominant branch of the effective reflectivity. The nonlinear network theory can be used to simulate the effects of current tuning by making independent adjustments to the phase and unsaturated gain for each gain section. Proceeding in this manner, it was found that by minimizing ΔP^2_{rms}, as given by (7.26), the gain discrimination could be recovered [7.64]. To demonstrate feasibility of this idea, the three-element GSE array in Fig. 7.25 with

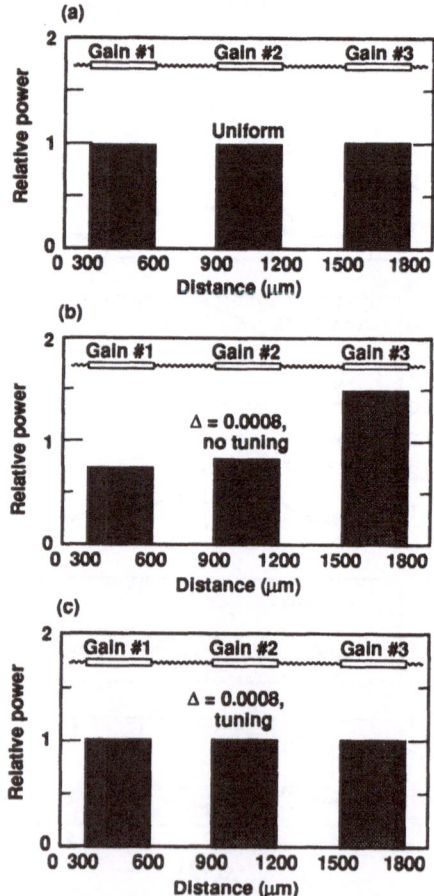

Fig. 7.26. The power flowing into each gain section of a three element GSE array are shown for effective-index grades of $\Delta = 0$ (a), $\Delta = 0.0008$ (b), and $\Delta = 0.0008$ (c) with the phases and unsaturated gains tuned to minimize ΔP^2_{rms} [1.474].

effective-index grade of $\Delta = 0.0008$ was simulated. The results of minimizing ΔP_{rms}^2 are depicted in Fig. 7.26. For reference, Fig. 7.26a displays P_m,
the power flowing into each gain section for the uniform three element GSE
array where $\Delta = 0$. We see from Fig. 7.26b that the grade in the effective
index produces a large nonuniformity in the power flowing into the gain sections. As seen in the corresponding effective reflectivity plot in Fig. 7.25,
when $\Delta = 0.0008$ there is no longer any mode discrimination. Comparing
Fig. 7.26b and c, it is evident that the optimized distribution of power flow
into the gain sections can be completely recovered by the simulated effect of
current tuning.

The multi-branch reflectivity that corresponds to Fig. 7.26c is displayed
in Fig. 7.27. The form of the dominant branch has been largely recovered,
although the maxiumum of the reconstructed reflectivity is ≈ 0.2, whereas the
uniform array exhibited a maximum reflectivity of nearly 0.3. Furthermore,
the mode discrimination is not quite as good as it was for the case of the
uniform array. Current tuning algorithms based on this simulation have met
with some success in experimental devices [1.474]. However, the process of
having to independently adjust the current to each gain section of a multiple
gain section array is quite tedious and the resulting single-frequency operating
condition is not always stable. These are some of the reasons why diode-laser
arrays with independently adjustable gain sections may not be practical in
applications that require stable, single-frequency operation.

The nonlinear network model has also been applied to simulate above-
threshold, single-frequency operation of GSE-laser arrays in an effort to estimate the maximum differential quantum efficiency expected from a uniform
array [7.64]. The results of this analysis on a uniform ten-element GSE-DBR
array analysis gave a value of 15 %, which is a little less than the value
of 19.6 % calculated in the analysis of [7.58]. An alternative structure that

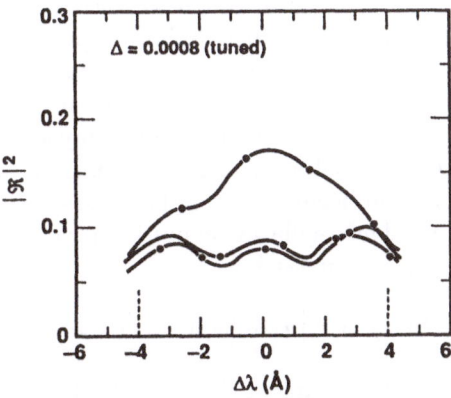

Fig. 7.27. Calculated effective reflectivities versus wavelength detuning for the
three element GSE array with $\Delta = 0.0008$ with the phases and unsaturated gains
tuned to minimize ΔP_{rms}^2 [1.474].

was also investigated was a GSE-DFB array of the type reported in [1.295], where the second-order grating is continued into the gain sections. This sort of structure might be expected to have a higher differential quantum efficiency, because output coupling will also occur in the gain sections where the light is being generated, and the intensity is maximized for the lowest-loss modes. Simulations of single mode GSE-DFB laser arrays have exhibited external efficiencies of about 35 %, which is a considerable improvement, though still less than the differential quantum efficienies demonstrated by edge-emitting diode lasers [7.64]. The ring-configured GSE array exhibited in Fig. 7.10 is yet another example of a structual modification that can improve the differential quantum efficiency of GSE laser arrays.

7.2.4 Network Analysis of Ring-Configured GSE Laser Array

The ring-configured GSE laser array such as that illustrated in Fig. 7.10 has been investigated both experimentally [4.76, 7.43] and theoretically [4.76, 7.64], as the ring geometry had been suggested by *Uematsu* et al. [7.32] as a means for reducing the end losses in diode lasers that rely on grating-output coupling. As seen in the Table 7.1, as much as 40 % of the total power generated is lost through the ends, and increasing the length of the end DBR sections did not improve the power output of the lowest-loss mode. The benefit of the ring-configured laser is that end losses are effectively eliminated and the possibility exists for realizing true uniform power flow into all gain sections if the ring-laser structure itself is uniform. Note that in the ring-configured GSE laser array, the DBR waveguide sections provide strong coupling bewteen the waves that travel in opposite directions. These modes shall be refered to as *coupled-ring modes*. This is an important feature that distinguishes ring-configured GSE laser arrays from the conventional ring-laser design [3.51], where there is little or no coupling between the oppositely-propagating traveling waves. These traveling-wave modes shall be refered to as *pure-ring modes*. In principle, both types of modes can occur in a GSE array, though for reasons given below, the coupled-ring modes usually have the lowest losses. Both of these types of modes must satisfy the periodic boundary condition of the N-element ring laser array structure, which is the mode must exactly repeat itself after N gain sections.

A network analysis of uniform GSE ring-laser arrays has been been developed by *Amantea* [7.64]. The oscillation condition for the coupled-ring modes an N-gain section array is expressed as

$$\exp\left[-L\left(g_{\nu n} - i\beta_{\nu n}\right)\right] = r\left[q\cos\left(\phi_\nu\right) \pm \sqrt{1 - q^2\sin^2\left(\phi_\nu\right)}\right] \qquad (7.32)$$

where r is the wavelength-dependent field reflectivity of the DBR sections, $q = t/r$ is the ratio of the DBR section wavelength-dependent field transmissivity t to the DBR section field reflectivity r, n is the number of wavelengths in the ring cavity, and ϕ_ν is expressed as

$$\phi_\nu = \frac{2\pi\nu}{N} \quad \text{where} \quad \nu = 0, 1, \ldots, N - 1 \ . \tag{7.33}$$

The index ν identifies the reflectivity branches that are associated with the number of gain sections. For modes on the $\nu = 0$ branch, the traveling waves have equal amplitudes as they enter each gain section. Therefore, these modes would be expected to be the most stable, as identical power flows into each gain section and the gain saturates uniformly across the array. The oscillation condition for pure ring modes is

$$\exp\left[-L\left(g_{\nu n} + i\beta_{\nu n}\right)\right] = t \ . \tag{7.34}$$

It can be shown that $\max\left(|t \pm r|\right) \geq |t|$. This means that near the Bragg condition, where $|r|$ is near it's maximum value, there will be at least one coupled-ring mode with losses lower than the pure-ring mode losses. Far from the Bragg condition, where the DBR reflectivity tends towards zero, the coupled-ring oscillation condition (7.32) and the pure-ring mode oscillation condition (7.34) will have nearly identical form, so there will be virtually no discrimination between these modes. Of course, coupled-ring modes with frequencies well outside the Bragg reflectivity will have higher losses than coupled-ring modes near the Bragg condition. Therefore, pure-ring modes far from the Bragg condition will have higher losses than coupled-ring modes near the Bragg condition, provided that the Bragg condition lies near the peak of the gain profile.

Let us turn our attention to the $\nu = 0$ lowest-loss coupled-ring modes of the DBR-GSE array. An interesting feature of the $\nu = 0$ mode branch is that the oscillation condition (7.32), and hence the effective reflectivity, is independent of the size N of the array because $\phi_\nu = 0$. This characteristic is unique to ring-geometry GSE laser arrays, and it has positive implications for the projected mode discrimination properties of these devices, as the array is scaled to larger sizes. Figure 7.28 illustrates the scaling properties of the $\nu = 0$ coupled-ring mode branch as a function of the number of gain sections in the ring array. The effective reflectivity and threshold gain of the lowest-loss mode are independent of the number of gain elements, but the nearest mode characeristics do depend on N, as indicated in Fig. 7.28. In the limit of $N \to \infty$, the spectrum becomes continuous as it converges on the lowest-threshold mode and the mode discrimination will vanish. The absence of end losses in the uniform ring GSE laser array is what sets the lowest threshold mode apart from the other modes. In contrast, to achieve as low a modal loss in a linear GSE laser array, the array would have to be scaled to element numbers of $N \approx 20$, where the mode disrimination is diminished.

As a further illustration of the distinction between the scaling characteristics of ring and linear GSE arrays, let us consider Fig. 7.29, which depicts threshold gain and effective reflectivity as $1/N$. Although, the threshold gain decreases with increasing N, the mode discrimination between the modes of an N element and $N - 1$ element array decreases rapidly, as expected. The

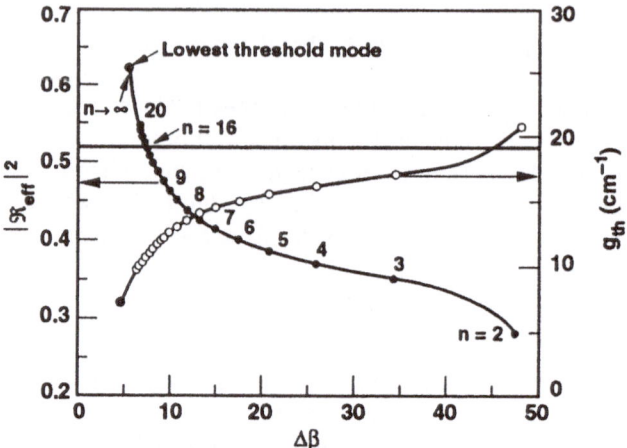

Fig. 7.28. Calculated effective reflectivity $|R_{eff}|^2$ (along left vertical axis) and threshold gain g_{th} (along right vertical axis) of a GSE ring laser as a function of detuning from the Bragg condition. The left-most dots correspond to the values of $|R_{eff}|^2$ and g_{th}, for the lowest-loss mode of the $\nu = 0$ branch, which are independent of N. The solid dots give $|R_{eff}|^2$ and the open dots indicate the corresponding values of g_{th} of the nearest mode of the $\nu = 0$ branch for the indicated array sizes [7.64].

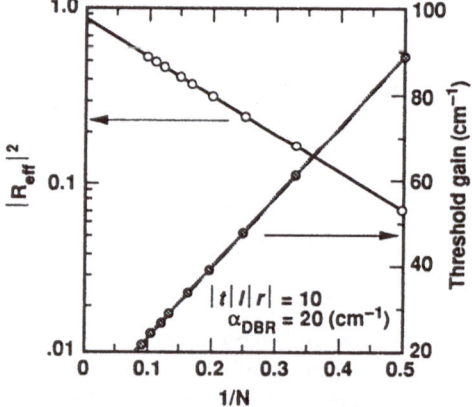

Fig. 7.29. Calculated effective reflectivity $|R_{eff}|^2$ (along left vertical axis) and threshold gain g_{th} (along right vertical axis) of a linear GSE ring laser as a function of $1/N$. The dots correspond to the result of the network calculation, while the lines represent the results of an approximate model based on energy conservation [7.64].

close proximity of the threshold gains of sub-arrays implies that there may be a tendency for the array to oscillate as subarrays when operated further above threshold. In contrast, the lowest-low mode of a ring GSE is the same for arrays of all size N. In a ring geometry, there are no smaller-size sub-ring GSE arrays. The smaller sub-arrays of a ring GSE array must be linear arrays, which will generally have higher losses than the ring modes. For these

reasons, the mode discrimination of a ring GSE laser array would be expected
to scale to larger array sizes than linear GSE laser arrays.

The issue of the differential quantum efficiency of the ring GSE laser array
also needs to be analyzed. The total differential-quantum efficiency of the ring
GSE laser array can be expressed as the product of the single gain section
quantum efficiency and the DBR section radiation efficiency. For a uniform
ring GSE array operating in the lowest-loss mode, the electric field will be
asymmetric with a zero crossing in the center of the DBR section to minimize
the coherent loss due to output coupling. With equal length gain and DBR
sections, this means that the quantum efficiency asscociated with a single
gain section is approximately 50 % regardless of the value of the incoherent
optical losses. However, the radiation efficiency of the DBR sections has a
strong dependence on the dissipative optical losses. Figure 7.30 displays a
representative graph of the ring GSE quantum efficiency versus dissipative
optical losses. The single-gain section and DBR radiation efficiency are also
indicated. At the lowest loss values, the differential-quantum efficiency is in
the range of 40 % to 45 %, which is an improvement over the differential-
quantum efficiencies expected for the linear GSE laser arrays presented in
Sect.7.2.3. However, even with this improvement the differential-quantum
efficiency of the ring GSE array is still about a factor of two lower than the
differential quantum efficiency of optimized edge-emitting diode laser arrays.

The experimental ring GSE studies of [7.43], some of which are presented
in Fig. 7.11 are in qualitative agreeement with the predictions of the network
analysis. As seen in Fig. 7.10, these experimental ring laser arrays were not
uniform because of the pair of etched corner turning mirrors at either end

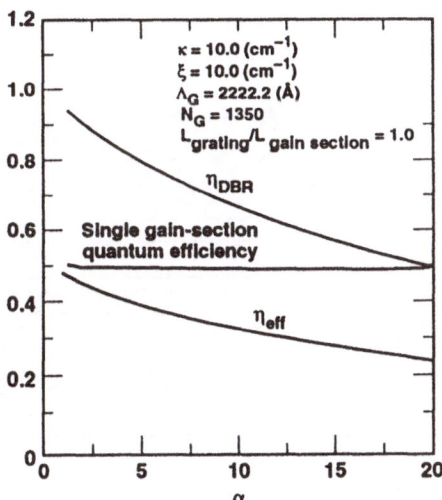

Fig. 7.30. The ring GSE array differential-quantum efficiency η_{eff}, single-gain section quantum efficiency, and DBR radiation efficiency η_{DBR} are plotted as a function
of the optical losses α [7.64].

of the array. A differential quantum efficiency of 22 % per side was measured for this nonuniform ring GSE laser array.

The experimentally measured performance characteristics of GSE laser arrays under single-frequency operation [1.474, 7.43] have exceeded some of the more pessimistic theoretical performance predictions [1.195]. However, the limited range of observed single-frequency operation and the difficulties associated with obtaining good-beam quality led researchers to investigate master-oscillator power-amplifier GSE lasers and laser arrays [1.236]-[1.245][1.490]. This class of surface-emitting device is presented in Sect. 8.2.

7.3 Vertical-Cavity Surface-Emitting Laser Arrays

Two-dimensional arrays of *vertical-cavity surface-emittting laser* VCSEL structures [1.136]-[1.138][1.158][7.65]-[7.72] have been investigated as an approach to obtain a high-brightness diode-laser source [7.70, 7.71]. Vertical-cavity, surface-emitting, diode-laser array structures, such as that illustrated in Fig. 7.31, are an interesting contrast to the conventional edge-emitting Fabry-Perot diode-laser arrays and the GSE laser arrays that we have considered so far. In the vertical cavity laser approach, the length of the gain region is typically several μm. The resulting separation between adjacent longitudinal modes from (2.6) is a few hundred Å. For this large a mode separation, we see from (3.61) that, at most, one longitudinal mode is contained within the gain profile. To insure that a mode falls within the gain profile, distributed-Bragg reflector stacks are used for both reflectors with sufficiently large reflectivities > 95 % to keep the end losses (2.15) at an acceptable level [1.158, 7.72]. The lateral optical waveguide in most vertical cavity lasers is determined by the current injection into the active volume, as well as thermal

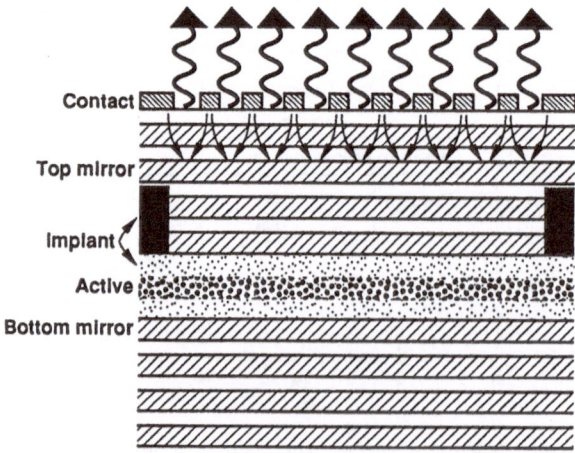

Fig. 7.31. Schematic diagram of a VCSEL array that uses a grid contact to define the array elements [7.70].

effects. For diameters of the gain element between 5 and 10 μm, filamentation does not typically occur and single-lateral mode operation is realized under most operating conditions [1.158]. Filamentation can occur in larger diameter VCSELs, due to carrier and thermal induced waveguide effects, and spatial nonuniformities in the mirror reflectivities [7.67]. Index-guided VCSEL arrays have also been fabricated and investigated [7.65]. Leaky-wave coupled VCSEL arrays have been analyzed theoretically, and found to have better mode discrimination properties as compared to both gain- and index-guided VCSEL arrays. In terms of ease of fabrication, gain-guided VCSEL arrays use a simpler process that does not usually require a regrowth, as does index-guided or leaky-wave coupled VCSEL arrays. The elements of gain-guided VCSEL arrays are typically defined using either reflectivity modulation [7.67, 7.68] or a grid-contact geometry [7.66, 7.70]. Besides the conventional square or rectangular structure, there are a variety of geometries that are possible with two-dimensional VCSEL arrays. For example, circular, hexagonal, triangular have been theoretically analyzed [1.136] with the hexagonal structures being demonstrated [1.137]. A two-dimensional VCSEL structure with an electrode configuration corresponding to a Fresnel lens has been recently demonstrated [7.73].

Vertical-cavity surface-emitting lasers have very small active volumes, in the $10^{-14} - 10^{-13}$ cm^{-3} range, which is orders of magnitude smaller than the $10^{-11} - 10^{-10}$ cm^{-3} size active volume of a conventional single-mode Fabry-Perot edge-emitting laser. Due to the small active volumes, VSCELs have threshold currents in the mA range, and high-series resistances, typically $\geq 20\,\Omega$, that increases the operating voltage and electrical power dissipated within the laser. For these reasons, the maximum room temperature cw power output from a VCSEL has been so far limited to the several mW range.

7.3.1 Scaling of Two-Dimensional VCSEL Arrays

The small size of VCSELs and their vertical-cavity geometry allow for monolithic integration into array structures with areal densities of $\geq 10^6$ lasers/cm^2. Even at power outputs of a ≈ 1 mW per laser, the potential power output from a 1 cm^2 VCSEL array of ≈ 1 kW is considerable. Individual vertical-cavity surface-emitting lasers have good output beam characteristics, typically emitting a moderate-angular divergence ($\approx 10°$), near-unity aspect ratio output beam. The VCSEL array is attractive as an approach for a coherent, beam-quality source where the source brightness will scale with the size of both dimensions of two-dimensional array, and the unity-aspect ratio of the output beam can be preserved.

As already mentioned, each element of the VCSEL array can be designed so that it operates in a single longitudinal and lateral mode. Given that VCSEL cavity lengths are several μm in length and Fabry-Perot lasers are several hundreds of μm in length, inspection of (3.87) tells us that the locking-bandwidth for synchronous operation of a pair of VCSELs is about two orders of magnitude larger than that of Fabry-Perot lasers. Therefore, the VC-SEL arrays should be more immune to some of the phenomena discussed in Sect. 3.4, that have severely limited the frequency-locking bandwidth of the longer length diode lasers with the cavities in the plane of the pn junction. In the strong-coupling limit for Fabry-Perot lasers discussed in Sect. 3.4.2, the locking bandwidth was estimated to be $\approx 1\,\text{Å}$. A strongly-coupled VCSEL array would then be expected to have a locking bandwidth of tens of Ångstoms to perhaps as large as $100\,\text{Å}$. This range of magnitudes in the locking bandwidth is comparable to or larger than the wavelength shifts that could result from material or structural nonuniformities, injected carriers, and thermal effects, so that frequency locking in VCSEL laser arrays should be much more robust than in conventional edge-emitting Fabry-Perot lasers.

To obtain a comparative estimate of the number of elements that can be effectively frequency locked in VCSEL arrays and conventional edge-emitting arrays in the presence of small random wavelength shifts between array elements, we can apply the phase-diffusion model for very-large arrays that was presented in Sect. 3.4.2. For sufficiently large arrays (> 100 elements), the maximum number of elements in a one-dimensional array is given by (3.105); and for a two-dimensional array, it is expressed by (3.106). Figure 7.32, displays results of a calculation of the maximum number of array elements using (3.105) and (3.106) for parameters representative of both VCSEL arrays and conventional edge-emitting arrays. The assumptions used in the phase diffusion model correspond more closely to the VCSEL array structures. Note that the case of the one dimensional conventional edge-emitting laser array, is not shown because $N_{1-D} < 100$ over the range of wavelength spreads shown. The RMS wavelength spreads encountered in actual devices would be expected to be in the range of $5 - 20\,\text{Å}$. This corresponds to an array size of between 10^4

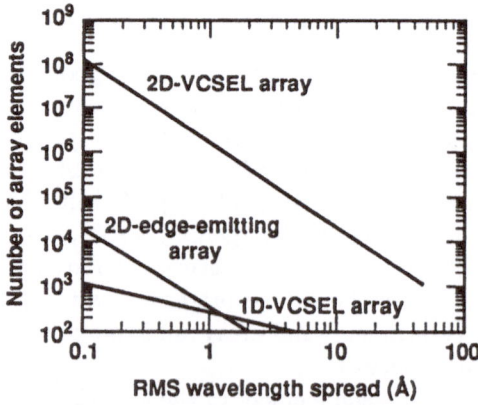

Fig. 7.32. Log-Log plot of the number of array elements (in the strong-coupling limit) as a function of the rms wavelength spread, $\sqrt{\langle \delta \lambda^2 \rangle}$ using (3.105) for one-dimensional and (3.106) for two-dimensional arrays, and VCSEL elements with $5\,\mu m$ cavity length and conventional edge-emitting laser of cavity length $500\,\mu m$.

and 10^5, which is significantly less than $\geq 10^6$ element estimate based on the capacity of a cm^2 wafer.

According to the phase-diffusion model result in Fig. 7.32, the two-dimensional VCSEL array can phase lock orders of magnitude more elements than either the one-dimensional VCSEL or a two-dimensional conventional edge-emitting array. Besides the benefit derived by using shorter cavity lengths, it is clear from comparing the cases of the one and two-dimensional arrays, that there may be a significant advantage to using two-dimensional arrays, when random deviations of the array element wavelengths are present. A two-dimensional array will have multiple nearest neighbors and an additional spatial dimension, as compared to a one-dimensional array where each element has only two nearest neighbors. At any point, in the one-dimensional array where adjacent elements operate outside the locking bandwidth, frequency locking in the array would be terminated. One the other hand, for a two-dimensional array there are more nearest neighbors so there is a greater chance of having an adjacent pair of array elements and there are many more possible pathways through a two dimensional array geometry, so that it is less likely that the formation of sub arrays (contained within the larger two-dimensional array) that are not frequency locked can occur.

Although, the VCSEL arrays would seem to be robust against frequency unlocking due to structural nonuniformities, the level of lateral-mode discrimination in these two-dimensional structures is expected to be small [4.48], making single-lateral mode operation difficult to achieve. In addition, a major challenge that needs to be met in the development of large-area coherent VCSEL arrays is the high-level of heat dissipation. Thermal effects are expected to be the dominant factor that limits the performance characteristics of VCSEL and VCSEL array devices [7.74].

7.3.2 Modal Properties of Two-Dimensional VCSEL Arrays

The above-threshold mode discrimination mechanisms have not yet been fully investigated. Analyses of passive VCSEL array structures have been accomplished using a coupled-mode approximation for finite index-guided structures [1.136, 1.138], as well as a seperable-solution approximation to the two-dimensional Helmholtz equation for the case of arrays of infinite extent [4.48]. The observed modes of two-dimensional gain-guided VCSEL arrays have been elucidated using a phenomenological approach [7.69] that models the lateral modes as solutions to the two-dimensional Helmholtz equation, which is given by (4.42). In the rectangular coordinate system, the general solution can be expressed as,

$$\psi_{\mu,\nu}(x,y) = \sum_{n,m} a_n a_m \exp\left(i2\pi\xi_{n,\mu}x\right) \exp\left(i2\pi\chi_{m,\nu}y\right) , \qquad (7.35)$$

where μ and ν are the mode indices, a_n and a_m are the Fourier coefficients, and n and m are the summation indices. The form for the modal solution given by (7.35) is separable in x and y so that the two-dimensional lateral modes can be represented as a product of the modes of two, one-dimensional array structures, which leads to a considerable simplification in computing the modes of the two-dimensional array. As has been shown by *Hadley* [4.48], the modes of a two-dimensional array can be represented by a separable form when the two-dimensional dielectric profile can be well approximated as a sum of two one-dimensional periodic dielectric profiles $u_1(x)$ and $u_2(y)$ respectively for each of the x and y dimensions. In other words, $\epsilon(x, y)$ that appears in (4.42) is represented as,

$$\epsilon(x, y) \approx \epsilon_b + \delta\epsilon\left[u_1(x) + u_2(y)\right] \tag{7.36}$$

where ϵ_b is the background dielectric constant. Along with the form for the dielectric profile given by (7.36), one needs to approximate the boundary conditions of the two-dimensional array, as described in [4.48], in order to obtain a separable solution to the two-dimensional Helmholtz equation (4.42) of the form, $E_{\mu\nu}(x, y) = E_\mu(x,) E_\nu(y)$. Hence, the separable form for the two-dimensional lateral modes, as expressed by (7.35), can be readily used to analyze the modal content of a VCSEL array using the far-field output beam. This can be accomplished by expressing the spatial frequencies, $\xi_{n,\mu}$ and $\chi_{m,\nu}$ that appear in (7.35) as,

$$\xi_{n,\mu} = \left(n - \frac{1}{2}\frac{\mu - 1}{N - 1}\right)\left(\frac{1}{d_x}\right) = \frac{\sin\left(\theta_{n,\mu}^x\right)}{\lambda}, \tag{7.37}$$

and

$$\chi_{m,\mu} = \left(m - \frac{1}{2}\frac{\nu - 1}{M - 1}\right)\left(\frac{1}{d_y}\right) = \frac{\sin\left(\theta_{m,\mu}^y\right)}{\lambda}, \tag{7.38}$$

where N and M are the number of elements in the x and y directions, d_x and d_y are the periods of the array in the x and y directions, and $\theta_{n,\mu}^x$ and $\theta_{m,\mu}^y$ are the modal emission angles in air, as measured in the $x - z$ and the $y - z$ planes, with the z-direction corresponding to the surface normal. From (7.35), we see that the eigenmmodes of the array comprise a coherent superposition of plane waves with a discrete spectrum of spatial frequencies given by (7.37) and (7.38). The fundamental mode will have a dominant on-axis single-lobed far-field pattern. For this to occur, the terms of the near-field expansion (7.35) must have identical phases (modulo 2π). This means that the spatial frequencies (7.37) and (7.38) that comprise the fundamental mode can only differ by even multiples of $2\pi/d_x$ or $2\pi/d_y$, respectively. It follows, then that the fundamental mode corresponds to the set of frequencies $\xi_{0,1}, \xi_{\pm1,1}, \xi_{\pm2,1}, \cdots$, and $\chi_{0,1}, \chi_{\pm1,1}, \chi_{\pm2,1}, \ldots$, with the dominant contributions coming from the terms containing $\xi_{0,1}$ and $\chi_{0,1}$. In Fig. 7.33a, a one-dimensional illustration of the plane-wave expansion of the fundamental or in-phase mode is presented. The corresponding far-field pattern has a dominant on-axis single lobe due to the dominance of $\xi_{0,1}$, with weaker symmetric side lobes resulting from $\xi_{\pm1,1}$.

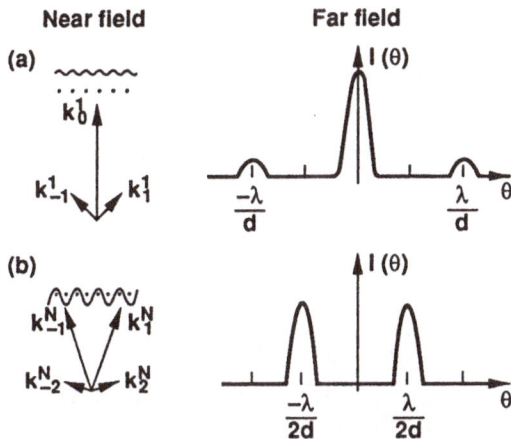

Fig. 7.33. Pictorial representation of the plane-wave expansion of (a) the fundamental or in-phase mode, $(\mu = 1, \nu = 1)$ and (b) the out-of-phase mode, $(\mu = N, \nu = N)$. Note that the wave vectors k_n^μ are related to the spatial frequencies by $k_n^\mu = 2\pi \xi_{n,\mu}$ [7.71].

The out-of-phase mode is identified as that mode where there is a 180° relative-phase shift between the light emitted by adjacent elements of the array. The corresponding plane-wave expansion as expressed by (7.35) can only contain terms where adjacent elements in the x and y directions have phase differences of π/d_x or π/d_y, respectively. By inspection of (7.37) and (7.38), it is seen that the out-of-phase mode corresponds to the set of frequencies $\xi_{\pm 1,N}, \xi_{\pm 2,N}, \cdots$, and $\chi_{\pm 1,N}, \chi_{\pm 2,N}, \cdots$, with the dominant contributions coming from the terms containing $\xi_{\pm 1,N}$ and $\chi_{\pm 1,N}$. The plane-wave expansion of the out-of-phase mode is depicted using a one-dimensional representation in Fig. 7.33b. A double-lobed far-field pattern results because of the dominance of the $\xi_{\pm 1,N}$ spatial frequencies. Smaller side-lobes (not shown) will occur due to the weaker spatial frequencies at $\xi_{\pm 2,N}$.

The character of the two-dimensional far-field patterns for some of the more commonly occuring modes of VCSEL arrays is illustrated in Fig.7.34. The dominant-lobe structure in the far field of each of the modes comes about because of the interference that results from the specific combination of plane waves that occur in (7.35). For example, the dominant terms in the plane-wave expansion of the out-of-phase mode are those with spatial frequencies, $\xi_{\pm 1,N}$ and $\chi_{\pm 1,N}$. These four plane waves have emission angles that satisfy $\theta_{1,N}^x = \theta_{1,N}^y$, giving rise to a quartet of plane waves, with each plane wave being oriented at 45° degrees with respect to the x and y axes. When propagated into the far-field zone, this plane-wave quartet forms the predominantly four-lobed structure illustrated in Fig. 7.34a.

The question of mode discrimination can also be investigated using the plane-wave expansion model. Because of the assumption of separability, the modal gain can be found by using a confinement-factor approach analogous

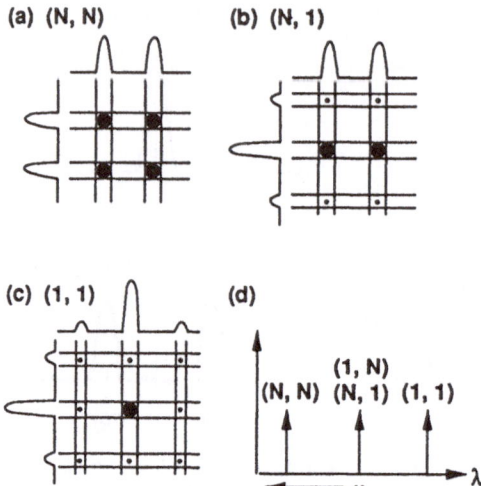

Fig. 7.34. Illustration of the two-dimensional far-field structure for (a) the $(\mu = N, \nu = N)$ or out-of-phase mode, (b) $(\mu = N, \nu = 1)$ mode, (c) the $(\mu = 1, \nu = 1)$ or in-phase more, and (d) the relative separations of the corresponding modal frequencies. Note that the $(\mu = 1, \nu = N)$ is obtained from $(\mu = N, \nu = 1)$ by performing a 90° rotation [7.71].

to that expressed by (3.38) and (3.39). For the two-dimensional VCSEL array the normalized product, $|\psi_{\mu,\nu}(x,y)|^2 g(x,y)$, is integrated over the entire exent of the two-dimensional array, where $g(x,y)$ represents the distribution of available gain in the two-dimensional array. Note that since $g(x,y)$ is the available gain, it depends on the distribution of injected carriers, as well as the distribtuion of the total losses (including output coupling losses). Using this method to analyze an $N \times N$ grid-contact 2-D VCSEL array structure of the type illustrated in Fig. 7.31, *Morgan* et al. found that the maximum modal gain occured for the $(\mu = N, \nu = N)$ or out-of-phase mode [7.69]. However, further investigation may yet yield modified designs that would preferentially support the $(\mu = 1, \nu = 1)$.

A qualitative agreement of the predicted modal gain and the far-field character with experimentally observed modal properties of two-dimensional VCSEL arrays has been found by *Morgan* et al. [7.69]. The arrays studied were 8×8 grid-contact structures with $d_x = d_y = 10\,\mu\text{m}$ and operated at a wavelength of $0.84\,\mu\text{m}$. Stable single $(\mu = N, \nu = N)$ array mode operation was observed in some cases, as evidenced by a symmetric four-lobed far-field pattern with angular separation of $\approx 5°$ between the lobes. For this structure geometry, a lobe separation of 4.8° is calculated by using (7.37 and 7.38) of the plane-wave expansion.

Such single array-mode operation has not been found to occur consistently in this type of VCSEL array structure, as described in [7.69]. Multiarray mode operation has also been observed and explained in terms of the modes illustrated in Fig. 7.34. By spectrally resolving the far-field output, it was found that at threshold single ($\mu = N, \nu = N$) array mode operation occured; and as the injection current was increased, the ($\mu = N, \nu = 1$) and ($\mu = 1, \nu = N$) array modes reached the threshold for oscillation, followed by the ($\mu = 1, \nu = 1$) mode at higher current levels. It has been suggested that spatial hole burning may be the cause of the observed multi-mode operation. Because of the high level of heat dissipation, these VCSEL arrays were operated under pulsed conditions, with the 8×8 grid-contact structures attaining a maximum peak power output of about 0.5 W. Larger 10×10 have produced peak power outputs as high as 1 W, however, single mode operation was not obtained at these higher power levels. The mode discrimination properties of VCSEL arrays above threshold is not yet well understood. A complete analysis should include both the effects of injected carriers and thermal effects on the modal properties, since heat disspation is a dominant effect. As already mentioned, the leaky-mode coupled VCSEL array has been suggested by [4.48] as a candidate structure that may offer better mode discrimination. Antiguide VCSEL lasers have recently been demonstrated [7.75, 7.76], opening the possibility of future investigations on antiguided VCSEL laser arrays.

8. Master-Oscillator Power-Amplifier Diode Lasers

The inability of edge-emitting gain-guided (Chap. 5) and most laser-array oscillators (Chap. 6), as well as surface-emitting laser arrays to operate single frequency, and at brightness and power level above single-element diode lasers, provided an impetus to researchers to investigate alternative structures. In recent years, the *Master-Oscillator Power-Amplifier* (MOPA) architecture, which was already well known in the fields of large high-power gas, dye, and solid-state laser systems [1.272, 2.32, 2.37, 2.38, 4.16][8.1]-[8.3], as well as optical telecommunications [8.4]-[8.7], has emerged as a preferred approach for obtaining high-power, single-frequency, beam-quality, diode-laser sources. The mode discrimination can be considerably simpler in a MOPA diode-laser array, as compared to a diode-laser array oscillator. In a MOPA architecture, the output of a low-power, single-frequency laser oscillator is injected unidirectionally into an optical amplifier of greater output-power capacity. Here, what is meant by unidirectional coupling is that a negligible light level from the amplifier is injected back into the oscillator. This is also referred to as nonreciprocal coupling. In principle, the unidirectional coupling means that the frequency characteristics and power output levels can be set independently of each other. Of course, the advantage here is that the spectral- and spatial-mode instability with drive current, which is pervasive in diode-laser array oscillators, can be avoided. A critical requirement for all MOPA designs is that the light levels injected back into the oscillator from the amplifier be sufficiently low that the frequency stability of the oscillator is maintained and self oscillation of the power amplifier is avoided.

The primary focus of this chapter will be on monolithic MOPA diode-laser sources for high-power single-frequency operation. Reference will be made to some of the work on high-power discrete broad-area MOPA-diode lasers, as these experiments have contributed significantly to improving the understanding of the nature of lateral-beam instabilities and filamentation in broad-area semiconductor-diode amplifiers and diode-laser arrays. In Sect. 8.1, edge-emitting MOPA structures, such as amplifier arrays, broad-area, and tapered amplifiers, will be presented, while Sect. 8.2 will deal with grating-coupled, surface-emitting MOPAs.

8.1 Edge-Emitting MOPAs

Efforts to use the semiconductor-diode laser as an optical amplifier date back
to the initial discovery of laser oscillation in GaAs [1.15][8.8]-[8.16]. In the earliest of these experimental investigations, homojunction semiconductor-laser
amplifiers were operated at cryogenic temperatures [1.15][8.8]-[8.12]. It was
demonstrated that 1/4-wave anti-reflect coatings of $CaWO_4$ [8.8] or SiO_2
[8.9, 8.11] appropriately deposited onto the cleaved facets of a diode-gain
section could reduce facet reflectivities to a level where small-signal gains in
excess of 1000 were measured by injecting light from an external diode-laser
oscillator. Similar increases in small-signal gain were realized by angling of the
end facets of the diode amplifier at an angle of 10 to 20° degrees with respect
to the optic axis of the amplifier [8.12, 8.16]. Later investigations revealed that
semiconductor amplifier devices could be fabricated to be closely coupled to
a laser oscillator with separations of 0.2 to 2.0 μm between the oscillator and
amplifier [8.12, 8.15], or even monolithically integrated on the same wafer
[8.14]. In some cases, the monolithic design was believed to be a benefit over
the discrete optics approach of using a lens system to couple the signal from
an external oscillator into the amplifier [1.15, 8.8], since mechanical adjustments were eliminated [8.12]. However, it was noted that the the discrete
MOPA architecture could offer superior optical isolation between the oscillator and amplifier elements, thus providing a high degree of unidirectional
coupling [1.15, 8.11]. In addition, coherent amplification was investigated by
Vuilleumier et al. [1.15] using a laser system where a single diode-laser oscillator was used to drive a monolithic array of diode-laser amplifiers. This
phased-array diode-laser system, which also provided independent-phase cor-
rectors for each element of the amplifier array, demonstrated the feasibility
of coherent amplification using phased-arrays of diode-laser amplifiers [1.15].
With the development of the double heterostructure, more detailed studies of
the room-temperature cw operating characteristics of single-element diode-
laser amplifiers were done [2.29][8.13]-[8.15][8.17].

One of the earliest examples of a monolithic MOPA was reported by
Kishino et al. [8.14]. A schematic diagram displaying a side-on view of the
integrated laser-amplifier/detector structure appears in Fig. 8.1. The optical
coupling between the laser and amplifier was accomplished by using an integrated twin-guide structure to provide a continuous passive-waveguide section
(labeled output guide in Fig. 8.1) that extended from the laser section output
to the input of the amplifier section. Both the laser and amplifier/detector
sections had independent electrical contacts, and the lateral dimension of the
MOPA was 100 μm. Under low-duty cycle short-pulsed operating conditions,
this monolithic MOPA provided 130 mW peak-power output. The critical element necessary for realizing monolithic-semiconductor MOPAs, as well as
other integrated optical devices, was the ability to fabricate integrate active
and passive waveguide networks at the wafer level. An important development in this area was the observation, discussed on page 242, that unbiased

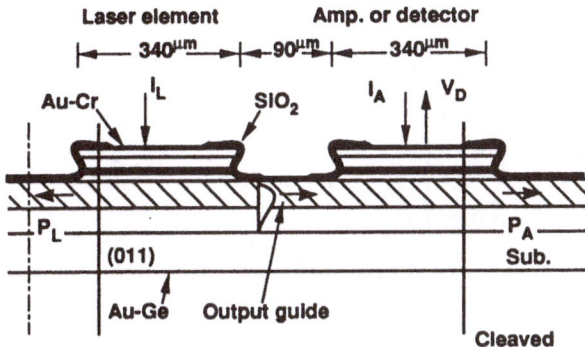

Fig. 8.1. Schematic diagram of the monolithic MOPA based on an integrated twin-guide structure. With no electrical bias to the amplifier section, the same device functioned as an integrated laser/detector [8.14].

quantum-well gain sections could act as low-loss integrated waveguides for transmitting light from the laser oscillator to the power amplifier of a monolithic MOPA.

Although, the mode discrimination is more straightforward in MOPAs, the theoretical maximum-power output from an optical amplifier is the same as that for laser oscillator of the same size and epilayer structure. The power-density limitation (2.52) due to loss-limited saturation, presented in Sect. 2.4, essentially determines the scaling properties of all laser oscillators and laser amplifiers. As illustrated by the example in Fig. 2.18, the optimum reflectivity of the output-coupling facet for maximum-energy extraction decreases exponentially with cavity length, and approaches values $< 0.01\%$ as the cavity length approaches the $1/\alpha$ absorption length, which characterizes the onset of loss-limited saturation. At these long cavity lengths, the laser oscillator is similiar to a double-pass amplifier.

In order to obtain power outputs in the 1 W or more range from a MOPA, it is necessary to consider semiconductor amplifiers with larger-active volumes than the semiconductor amplifiers that support only a single-spatial mode. In this respect, the degree of scaling required for a specific power output is the same for laser oscillators and laser amplifiers. As an example, Fig. 8.2 displays the scaling characteristics for a single-spatial mode semiconductor laser amplifier of width $3\,\mu m$, as calculated using the Rigrod analysis of Sect. 2.3.4. The parameters used for this calculation [1.244] were $g_0 = 100\,\mathrm{cm}^{-1}$, $P_S = 40\,\mathrm{mW}$, and $\alpha = 15\,\mathrm{cm}^{-1}$ which are indicative of single-spatial mode InGaAsP/InGaAs multi-quantum well amplifiers that operate at wavelengths in $1.5\,\mu m$ range [8.18, 8.19]. The power output versus length saturates beyond the $1/\alpha$ length regardless of the input power. To maximize the power output from the amplifier at any length, it is seen from Fig. 8.2 that the input power should be sufficiently large to saturate the local gain. However, the amplifier gain $G = P_{out}/P_{in}$ is diminished as the input power is increased; and for input powers that exceed the limit power $P_{Lim} = A_m I_{Lim}$

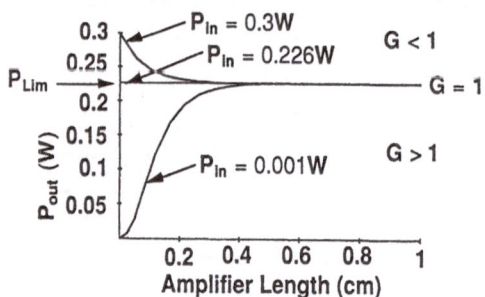

Fig. 8.2. Calculated power output versus length for a semiconductor optical amplifier of width $3\,\mu m$ for the indicated input powers. [1.244]

where I_{Lim} is given by (2.50), the amplifier gain $G < 1$. Note that the data presented in Fig. 8.2 are representative of typical performance. As discussed later of page 297, monolithic InGaAsP/InGaAs MOPAs comprising single-mode buried heterostructure with a 6 quantum-well active layer have operated up to 370 mW cw [8.20].

As gain-guided diode-laser arrays became more readily available in the mid 1980s, experimental investigations were carried out on the use of these sources as high-power optical amplifiers. Control of the spatial- and spectral-mode properties of gain-guided laser arrays was demonstrated by using optical injection for both regenerative-amplifier structures [1.216, 3.84, 3.85, 3.87, 5.4, 5.5, 8.21, 8.22] and traveling-wave amplifier architectures [8.23]-[8.25]. In addition, experiments on broad-area diode-laser optical amplifiers revealed conditions where predominately single-lobed, far-field output beams could be obtained [3.63][8.26]-[8.28].

8.1.1 Semiconductor-Diode Amplifier Arrays

Improvements in semiconductor materials and processing technology have made it possible to scale the type of amplifier-array architecture suggested by [1.15] to very large sizes. As already discussed in Sect. 1.4.1, the monolithic-amplifier array of *Krebs* et al. [1.460], which is presented in Fig. 1.10 comprised an elaborate integrated active-waveguide splitting network that used the output of a single external oscillator to drive, in parallel, a monolithic array of 400 semiconductor-diode amplifiers that was 12 mm wide. This amplifier array demonstrated 22 W of total peak-power output under pulsed operating conditions. This device did not contain any means for correcting the relative phase shifts, but fringe visibility measurements of amplifiers separated by 5 mm exhibited a mutual coherence of 0.8. Spectral measurements of the light output revealed that 87 % of the output was at, or very near, the oscillator frequency, with the remainder of the light being emitted as amplified spontaneous emission spread over a bandwidth on many nanometers. Such large-scale amplifiers arrays have been considered as a potential means for transmitting power over large distances in space [8.29]. However for efficient-power transmission, the limits on random-phase and random-pointing variations between the amplifier elements, require active-phase correction for each array element.

8.1.2 Discrete Broad-Area MOPAs

In considering the broad-area diode laser or gain-guided diode-laser array as a high-power source, we are still confronted with the issues of loss-limited saturation and catastrophic-optical-facet damage, presented in Sect. 2.3.5, that limit the power output capabilities of single-element diode-laser oscillators. Recall that the maximum-theoretical power density possible is given by (2.50). For the values of small-signal gain g_0 and α listed for GaAs in Table 2.1, we find a theoretical-maximum power density of $20\,\mathrm{MW/cm^2}$ (higher values can be obtained for larger values of g_o), which is about a factor of 20 greater than the catastrophic-optical-facet damage intensity limit of $\approx 1\,\mathrm{MW/cm^2}$. The catastrophic-optical-facet damage intensity limit corresponds to a power output-per-unit width of $\approx 0.025\,\mathrm{W/\mu m}$, which means that the minimum width expected to give reliable 1 W power output will be about $40\,\mu\mathrm{m}$. Applying the stability analysis of Sect. 4.3.5, specifically Fig. 4.12b, we see that even for an antiguiding factor of 2, a $40\,\mu\mathrm{m}$ wide traveling-wave amplifier should be unstable to low-spatial-frequency pertubations at power levels from about 2 mW to 20 W, which spans the practical operating range.

Most of the experimental studies on broad-area MOPAs have employed a discrete architecture with an external-laser oscillator because this architecture has the following benefits: 1) superior optical isolation between the oscillator and amplifier elements, thus providing a high degree of unidirectional coupling as noted by [1.15, 8.11], 2) independent control of the laser oscillator operating wavelength so that it can be tuned to the maximum gain of the amplifier, and 3) the ability to modify the intensity- and/or phase-profile of the input light. Experimental studies of discrete broad-area semiconductor MOPA systems have shown that predominantly single-lobed operation is possible at power outputs of 1 W or more under cw operation [2.42, 8.27, 8.30] and to over 10 W under pulsed-operating conditions [2.42, 8.31, 8.32]. In order to avoid the tendency towards filamentation at higher-power outputs in both broad-area semiconductor amplifiers, as well as gain-guided diode-laser array amplifiers, experimentalists found it necessary to angle the input beam from the oscillator with respect to the facet normal of the amplifier [3.63, 3.84, 3.87, 5.4, 5.5][8.30]-[8.34]. Apparently, this has been the case for both the configurations depicted in Fig. 8.3, that have been investigated. In Fig. 8.3a, when the input facet has an antireflection coating, the configuration is termed a *single-pass amplifier*. Similarly, in Fig. 8.3b when the input facet has an antireflection coating, the device is said to be a *double-pass amplifier*. If the input facet in either Fig. 8.3a or b has sufficient reflectivity to form an

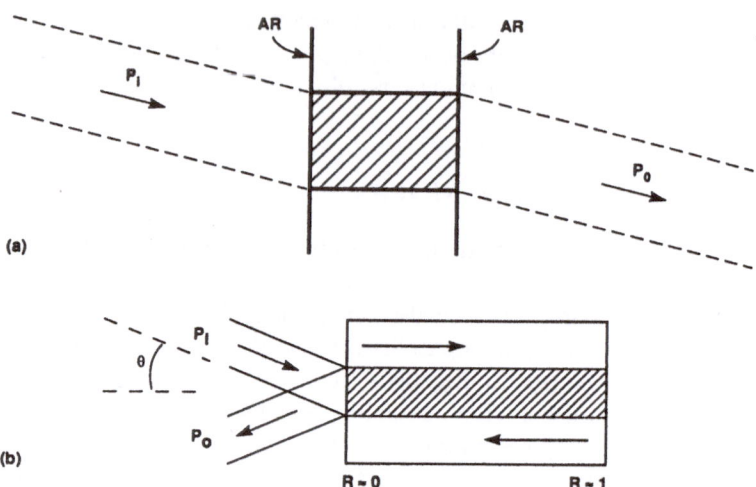

Fig. 8.3. Schematic of the single-pass (**a**) and double-pass (**b**) broad-area MOPA configurations. [8.34]

effective Fabry-Perot cavity, then the amplification process is regenerative. The general observation that angling the injected beam with respect to the input facet can produce single-lobed far-field beams has been explained using several models.

Experimentally observed output-beam characteristics of regenerative broad-area, diode-laser amplifiers (configured as illustrated in Fig. 8.3b) have been modeled [3.63] as active Fabry-Perot cavities wherein an amplified-diffracting wave propagates in a zig-zag fashion through amplifier, as it is reflected from each facet. After each pass, a portion of the wave is emitted into air, then the resulting far-field pattern comprises a coherent superposition of these amplified waves. A pictorial representation of this description appears in Fig. 8.4. Functionally, this is similar to the quasidistributed-output-coupling architecture [1.38] that was presented in Fig. 1.4. However, to obtain good beam quality from this sort of amplifier it is necessary that the profile of the beam inside the amplifier be sufficiently wide so that successive passes overlap [3.63]. In this situation, the injected light interacts with most of the gain volume and suppresses the free-running modes of the amplifier.

It was revealed by *Abbas* et al. [3.63] that single-lobed off-axis output beams could be obtained over a narrow range of external-injection angles ($3 - 5°$) and explained within the context of a Gaussian beam propagating in a Fabry-Perot amplifier. Furthermore, beam steering (in the far field) was observed by varying the frequency of the injected light or the bias current to the amplifier. At larger angles of injection, the zig-zag wave did not fill the amplifier cavity sufficiently and free-running modes of the broad-area laser were observed to oscillate. With smaller-injection angles, the beam profile of adjacent passes overlap so that most, if not all, of gain volume was used by

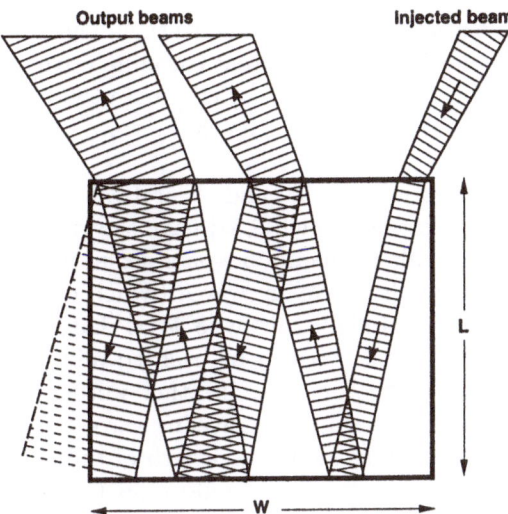

Fig. 8.4. Schematic representation of a broad-area semiconductor amplifier as a Fabry-Perot cavity with multiple-output beams [3.63].

the zig-zag propagating amplified wave. In the absence of any injected light from an external oscillator, the free-running, broad-area semiconductor amplifier was observed to oscillate in multiple longitudinal and lateral modes. Within this model, single-lobed far-field operation of the regenerative broad-area amplifier can be viewed as injection locking to one of those modes of the broad-area laser that produces the desired output beam characteristics. Generally speaking, the Gaussian character (or any other beam-profile characteristic, for that matter) of injected light is not necessarily preserved on propagation through a semiconductor diode-laser amplifier (Sect. 4.3.5); and to accurately model broad-area semiconductor amplifiers, a self-consistent beam-propagtion model is required (Sect. 4.3). A specific example of such a calculation for broad-area semiconductor amplifiers is presented in this section on page 292.

Angling the orientation of the input beam from the oscillator with respect to the amplifier axis has the effect of adding a phase tilt to the traveling wave that is injected into the amplifier. In the spatial-frequency domain, this means that there is a uniform translation of all spatial-frequency components that comprise the beam profile of the injected signal. Recall that the amplifier-stability analysis of the paraxial beam-propagation equation (Sect. 4.3.5) tells us that for a traveling wave of finite-spatial extent in the lateral dimension, the gain pertubations at spatial frequencies below a critical value $k_t = \sqrt{\alpha R}$, where R is given by (4.70), are unstable and experience an exponential growth dependent on spatial frequency; while those pertubations at spatial frequencies above the critical value of k_t are stable and will decay exponentially. The nonuniform amplification of the spatial-frequency components of the finite-extent beam causes filamention. Therefore, a uniform shift of the spatial frequencies of a finite-extent traveling wave towards frequencies higher than $k_t = \sqrt{\alpha R}$ should reduce or eliminate the beam instability that contributes to the formation of filaments as the wave propagates through the amplifier.

Both experimental studies [8.33] and theoretical simulations using a self consistent beam propagation model [8.34] have shown that beam filamentation is an inevitable result in the double-pass geometry in the case of exact opposite propagation of the traveling waves (which occurs when $\theta \approx 0°$ in Fig. 8.3b). A reason suggested for this by [8.34] is that, when the input beam is angled, the beam filaments formed by each traveling wave in the amplifier will tend to wash each other out. In other words, the peaks and nulls of the filaments associated with each traveling wave would not generally be expected to be commensurate along the length of the amplifier, so that the effects of nonuniform-gain saturation would tend to be homogenized in an angled double-pass geometry. In comparing the single-pass and double-pass geometry amplifiers, it is found that for the identical amplifier structures at the same current level, the single-pass amplifier will produce about 15 % less power output [8.33]. The reasons for this is that the gain saturation in the double-pass amplifier is more uniform along the length, which improves the energy-extraction efficiency, and the effective gain-length product is larger in a double-pass amplifier than for a single-pass amplifier of the same length.

However, comparative experiments of similar structure single-pass and double-pass broad-area semiconductor amplifiers have found that for longer-pulse operation (i.e., μs range), that the single-pass amplifier configuration (with angled input beam) produces a higher-power diffraction-limited output beam [8.30, 8.33]. On the other hand, the double-pass amplifier exhibited higher-peak powers for pulse durations less than 300 ns, which is consistent with the higher-extraction efficiency present in the double-pass geometry [8.33]. The fact that different time scales were required in each case suggests that thermal effects play an important role in determining the beam quality. The influence of heat dissipation on beam quality is explored in more detail beginning on page 305 in Sect. 8.1.4. A proposed explanation for the observed tendency of the double-pass amplifier towards filamentation is that in the double-pass amplifier there is an interaction, due to Fresnel reimaging or Talbot self-imaging, between the oppositely propagating beams, when the input beam is near normal incidence to the input facet (Fig. 8.3b) [8.30, 8.35]. Some experimental observations of filamentation in the near field of double-pass amplifiers have been attributed, in part, to Talbot self imaging [8.30, 8.35]. Moreover, it has been suggested that a periodic-lateral pertubation imposed on the forward-traveling wave will, upon reflection for the second pass, be imaged in the backward-traveling wave to create yet another periodic pertubation for the forward-traveling wave in the vicinity of the input facet [8.33]. This explanation does not address the reason for the pulse duration dependence that was observed in the comparative experimental studies of [8.33].

In real semiconductor-diode amplifiers random nonuniformities in the amplifier structure due to slight compositional or dimensional fluctuations within the epilayers, thermal nonuniformities, or random fluctuations across the intensity profile of the oscillator beam act as pertubations that provide the seeds for beam filamentation. An illustration of this effect is found in [5.18], where a numerical simulation of a $100\,\mu$m wide by $2200\,\mu$m long single-pass broad-area semiconductor amplifier was done using the beam-propagation method (Sect. 4.3) and accounting for local temperature-induced changes in the dielectric profile due to heat dissipation. Note that in this simulation the input beam was not tilted with respect to the amplifier axis. The model calculation included a nonresonant grating-output coupler within the amplifier to provide a continuous picture of the evolution of the filaments, as the traveling wave propagated along the amplifier. Figure 8.5 displays input beams both with and without random fluctuations and the resulting edge-emitted output beams from the amplifier output. The intensity profiles of the input beams without random-intensity fluctuations (Fig. 8.5a) and with random-intensity fluctuations (Fig. 8.5d) exhibit very-slight differences. In the absence of any random-intensity fluctuations, the Gaussian-intensity profile (Fig. 8.5b) is largely preserved at the output of the amplifier and a predominantly single-lobed far-field pattern (Fig. 8.5c) is found, with most of the power contained within the dominant lobe. In the case of the random fluctuations, the amplifier-output beam profile (Fig. 8.5e) displays well-defined filaments and a corresponding far-field pattern (Fig. 8.5f) with multiple side-lobe structure, indicating that a significant amount of the power is not contained within the dominant-central lobe. By running simulations for different random-intensity profiles, it was found by [5.18] that the pattern of the filaments would change, but the period of the filaments was approximately the same for a fixed set of operating conditions.

8.1.3 Monolithic Edge-Emitting MOPAs

The experimental work on discrete broad-area semiconductor MOPAs would seem to indicate that the angled single-pass MOPA architecture is preferable for providing good-beam quality at high-power cw operating conditions. A similar conclusion had been found also, as already mentioned, for gain-guided laser arrays used as regnerative amplifiers. This was taken into consideration in the monolithic MOPA structures of [1.216, 1.217], displayed in Fig. 8.6. The single-element gain-guided laser oscillator is fabricated in close proximity to facilitate optical coupling, and at a 1.4° tilt with respect to the optic axis of a gain-guided diode-laser array. *Hohimer* et al. made this selection of the 1.4° tilt angle, as it corresponded to the internal-emission angle (under free-running operation) for optimum injection locking that had been found in discrete injection-locking experiments [3.84, 3.87, 5.4, 5.5]. As indicated in Fig. 8.6a, the basic structure had a length of $250\,\mu$m. However, as discussed in [1.216, 1.217] longer devices were fabricated and studied, but such devices

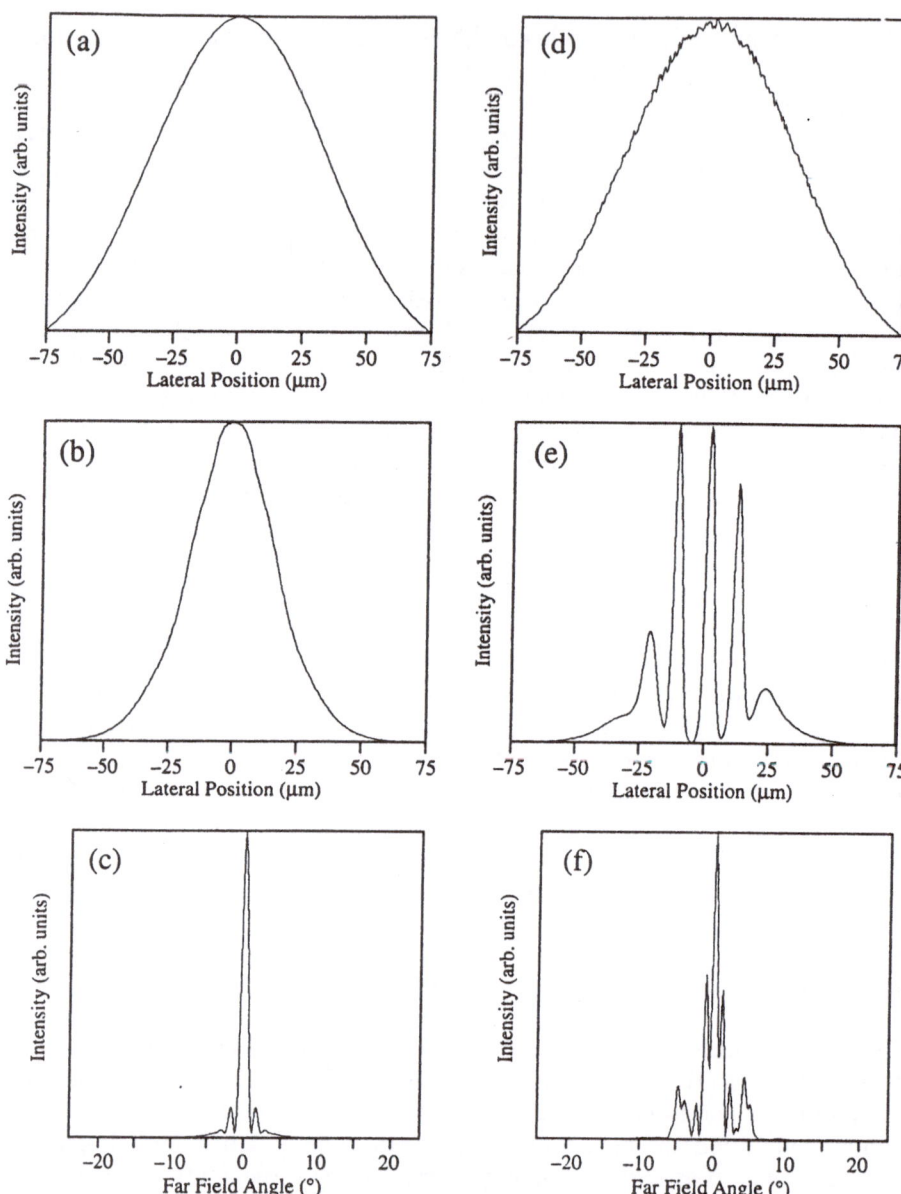

Fig. 8.5. The intensity profiles in the absence of random fluctuations for the Gaussian input beam to the amplifier (**a**), the corresponding output beam of the amplifier (**b**), and the resulting far-field pattern (**c**). Intensitiy profiles in the presence of a 2 % peak-to-peak random-intensity fluctuation on the Gaussian input beam (**d**), the corresponding output beam of the amplifier (**e**), and the resulting far-field pattern (**f**) [5.18].

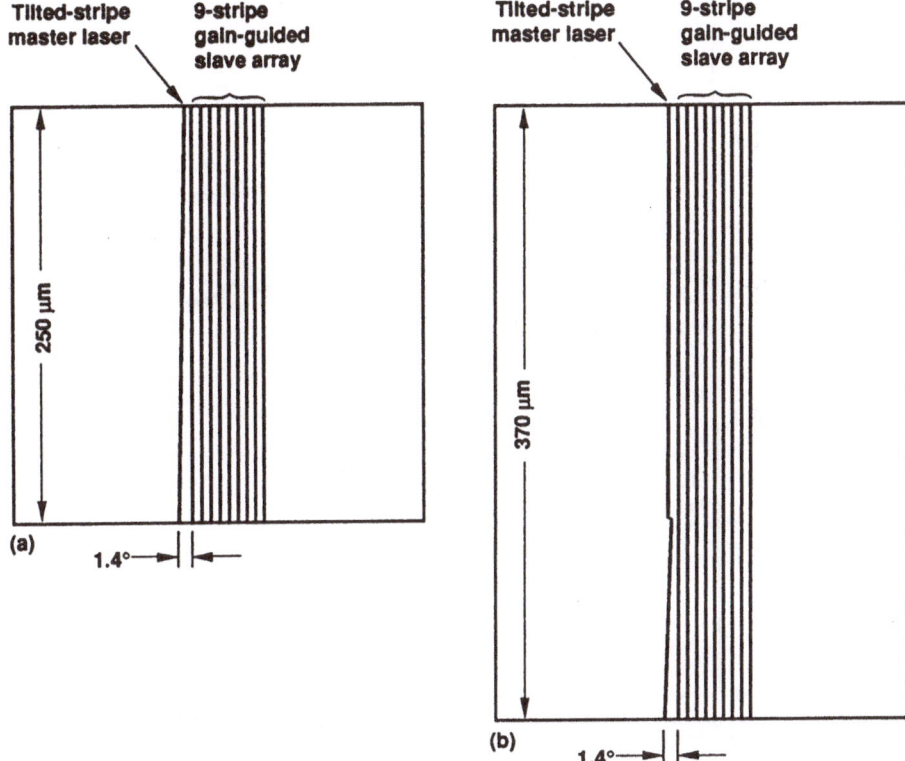

Fig. 8.6. Schematic diagram presenting the basic geometry of the monolithic injection-locked gain-guided diode-laser array (**a**). Example of a longer device with lateral discontinuity in the oscillator section (**b**) [1.216].

contained a lateral discontinuity, as illustrated in Fig. 8.6b. Independent electrical contacts were made to the oscillator and amplifier sections.

The monolithic injection locked gain-guided arrays of [1.216, 1.217] exhibited similar properties (e.g., off-axis single-lobed far-field patterns that could be steered by varying the frequency of the master oscillator) as had been observed in the discrete injection-locking experiments [3.84, 3.87, 5.4, 5.5]. Figure 8.7 presents the observed power-current characteristics and far-field patterns of the monolithic laser-array amplifier at selected cw power-output levels for both injection-locked and free-running operation. The effects of injection locking of the output characteristics are clear from Fig. 8.7. Under injection-locked operation, a predominantly single-lobed far-field pattern is exhibited, however, progressively more power falls outside the dominant lobe as the power output is increased. This particular device had a cavity length of 500 μm, so it comprised two repetitions of the basic structure that appears in Fig. 8.6a. It should be pointed out that the lateral discontinuity may be of dubious benefit, as it is unlikely that such a discontinuity could improve the spectral purity of the laser oscillator. In fact, *Hohimer* et al. [1.217] found

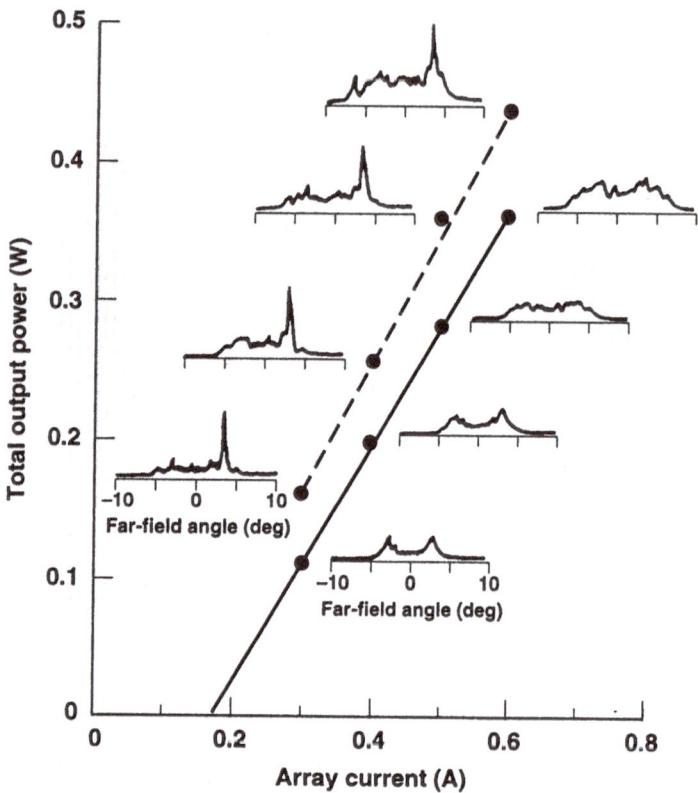

Fig. 8.7. The measured power-current characteristics of the gain-guided laser-array amplifier under free-running (solid line) and injection-locked (dashed line) operating conditions. The insets display the measured far-field patterns at the operating levels indicated by the data points [1.217].

it necessary to injection-lock the monolithic gain-guided master oscillator in order to make it operate in a single frequency. Moreover, the power emitted into the single-lobe of the far-field pattern was considerably enhanced over that when the master oscillator was not operating single frequency. It was also discovered that at higher currents to the amplifier (400 mA), it was necessary to tune the laser oscillator current so that the oscillator wavelength could be shifted as necessary to track the shift in the gain profile of the amplifier array.

It should be pointed out that angling of the input beam has been found not to be necessary when an external-laser oscillator is used to injection lock arrays of resonantly-coupled arrays of antiguided lasers [8.36, 8.37]. Recall from Sect. 6.3.1, that arrays of leaky-mode waveguides emit a radiation mode at an angle with respect to the array axis that is determined by the large-index step between the antiguide cores and the interelement regions. When the condition for resonant coupling is satisified, the radiated light will constructively interfere in the antiguide cores, as depicted in Fig. 6.9, which can result in single-frequency operation of the array. For this reason, injection-locking of antiguided arrays can be accomplished by injecting light into just one of the antiguide cores. In contrast to the injection-locked leaky-wave-coupled gain-guided arrays, the injection-locked resonant antiguide array did not exhibit beam steering when the frequency of the external laser oscillator was tuned [8.37]. This illustrates the importance of how employing a strong built-in lateral index profile in a diode-laser amplifier can improve the stability of the output beam.

Single-mode index-guided monolithic MOPAs have been investigated as a source for pumping Er-doped fiber amplifiers. An example of a structure designed for this purpose by *Koren* et al. [8.20] is depicted in Fig. 8.8. A six well strtained-layer multi-quantum-well active layer was used to raise the saturation level of the output power. To improve the power performance of the single-mode index-guided diode-laser amplifiers, the passive waveguide, that connected a $2.5\,\mu$m wide single-mode DBR laser to a $5\,\mu$m power amplifier, was tapered in a gradual manner to expand the size of the optical mode to match the larger width of the amplifier. In effect, the mode area A_m has been increased so that the maximum-power level possible from the amplifier is scaled accordingly. As indicated in Fig. 8.8, a *high-reflectivity* (HR) coating was applied to the rear facet of the DBR-laser oscillator, and an *anti-reflection* (AR) coating was applied to the output facet of the amplifier. An adiabatic-mode expander for the transverse mode was employed at the amplifier output to reduce the aspect ratio of the near field. The monolithic MOPA of *Koren* et al. [8.20] demonstrated $370\,$mW of cw power output at a wavelength $1.48\,\mu$m. The near-field spot size was $3.7\,\mu$m in the lateral dimension, and $1.9\,\mu$m in the transverse dimension. In terms of spectral performance, this MOPA exhibited two peaks, with relative peak intensities of about $10\,$dB that were separate by $8\,$nm in wavelength [8.20]. These two wavelengths were attributed to the

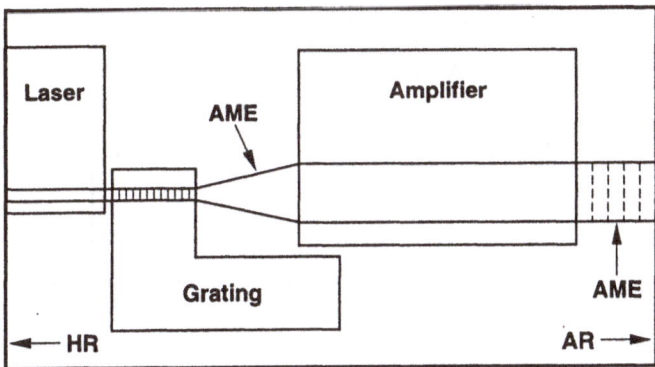

Fig. 8.8. Schematic drawing of a monolithic-MOPA structure that employs an *adiabatic-mode expander* (AME) between the single-mode DBR-laser oscillator and a wider amplifier to expand the lateral size of the optical mode injected into the amplifier. [8.20].

fundamental and first-order transverse modes that were supported in the Bragg section (labeled as grating in Fig. 8.8) of the monolithic MOPA.

Another type of monolithic MOPA that has been investigated is a single-pass broad-area MOPA, that provides for angled injection of the light to the amplifier input. A monolithic version of the angled single-pass MOPA configuration has been realized by using a leaky-waveguide structure to generate an angled-input beam to a broad-area gain section [8.38, 8.39]. Figure 8.9 presents a schematic diagram of this device. The DBR laser and preamplifier are fabricated along a common $4\,\mu$m wide index-guided lateral-waveguide structure that is angled with respect to the cleavage plane of the wafer. Note details on the nature and fabrication of the index guide in this structure has,

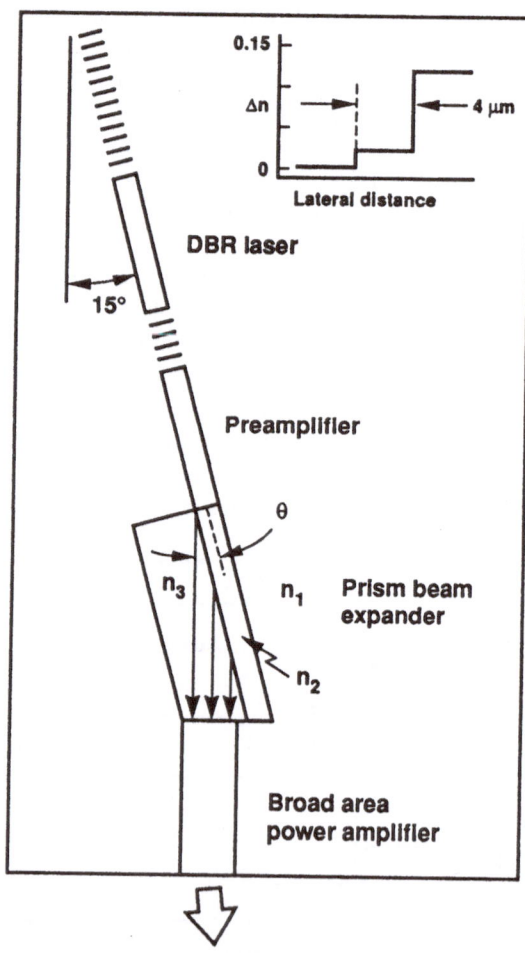

Fig. 8.9. Diagram of a monolithic single-pass broad-area amplifier. The inset at the top indicates the relative effective-index profile across the broad-area amplifier and waveguide section that generates the leaky-wave input to the amplifier [8.39].

so far, not been disclosed in any technical publication. A leaky-waveguide or *prism-beam expander* follows the preamplifier section. The effective index of the transverse-waveguide structure in the broad-area amplifier n_3 in Fig. 8.9 is made to be greater than the index n_2 of the guide layer of the index so that radiation is emitted into the broad-area amplifier at an angle $\theta = \cos^{-1}(n_2/n_3)$ relative to the waveguide axis. For the device fabricated by [8.39], the injection angle was $\theta = 12.3°$, meaning that the beam was tilted at about a 3° angle with respect to the output facet. This translates to an output beam that is emitted into air at an angle of about 9° off the facet normal.

The performance characteristics of this monolithic single-pass broad-area semiconductor MOPA were evaluated under low-duty cycle pulsed operating conditions, where single-frequency operation was observed along with a predominantly single-lobed far-field output beam up to a peak power of 800 mW, and a measured differential-quantum efficiency of 48 %. A primary consideration for optimizing the energy extraction in this device is that the power of the light injected into the broad-area amplifier be sufficiently high so that the amplifier gain is saturated and amplified spontaneous emission is suppressed. For the device studied by [8.39], it was estimated that a power of ≈ 120 mW was injected into the broad-area amplifier by the leaky-wave beam expander. Part of the basis for this estimate was the observed performance characteristics of the monolithic oscillator-preamplifier device alone [8.40].

8.1.4 Tapered MOPAs

A limitation that occurs in scaling the lateral dimension of the broad-area MOPA is that the single-mode power required at the input to saturate the gain also increases with the width of the amplifier. Given that single-frequency diode-laser oscillators presently operate at ≈ 100 mW, scalable MOPA architectures were considered that were not limited by the power output limits of single-element oscillators. Both the tapered-MOPA, which is treated in this section, and the grating-coupled surface-emitting MOPAs in Sect. 8.2 are examples of scalable MOPA designs that are not necessarily constrained by the power ouput available from the single-mode master oscillator.

The tapered-amplifier design was originally proposed as a means for overcoming the decrease in extraction efficiency on scaling excimer laser-amplifier lengths to the point where the product of the cavity length and absorption loss exceeds unity [8.41]. Similar analyses have been done of tapered laser amplifiers in semiconductor-gain media [8.42]-[8.44] and short-wavelength metal-vapor lasers [8.45]. Initial experiments in monolithic-semiconductor MOPAs employed passive-tapered waveguides to provide adiabatic mode expansion of the transverse mode at the amplifier input [8.20]. This was done to increase the mode size in the ampliifer so that a higher power output could be obtained. The basic tapered amplifier structure is displayed in Fig. 8.10.

Although a uniform-linear taper is displayed in Fig. 8.10, nonlinear [8.43], as well as nonuniform linear tapers have also been considered [8.44]. It should be pointed out that for efficient-energy extraction to occur the expanding-lateral taper in the width of the gain region should match the expansion rate of the beam, as it propagates through the amplifier.

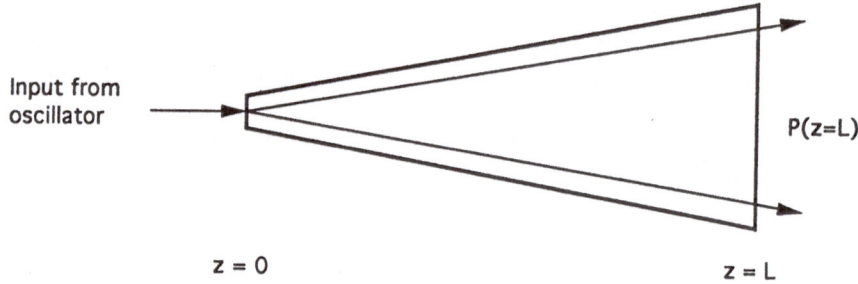

Input from oscillator

P(z=L)

z = 0

z = L

Fig. 8.10. Schematic of a linearly-tapered optical amplifier.

An adiabatic increase in the lateral dimension of the gain medium along the direction of propagation of the traveling-wave input has the effect of producing an axial dependence in the saturation intensity, which can maintain the intensity in the amplifier at a level that is near optimum for realizing maxiumum energy extraction at absorption length products greater than unity. This can be elucidated by using the approach of [8.41] that treats the amplifier using the Rigrod model (Sect. 2.3.4) where only a single-traveling wave is considered, but with an intensity in the tapered amplifier that is dependent on the cross-sectional area of the mode according to $I(z) = P(z)/A(z)$. The total derivative of $I(z)$ with respect to z is expressed as

$$\frac{dI}{dz} = \left(\frac{\partial I}{\partial A}\right)\frac{\partial A}{\partial z} + \frac{\partial I}{\partial z} \ . \tag{8.1}$$

The term $\partial I/\partial z$ in (8.1) corresponds to the variation in intensity with z when the area A is a constant, which is given by (2.31). Substituting (2.31) into (8.1) and using $I(z) = P(z)/A(z)$, we find that the axial dependence of the power in the tapered amplifier is governed by

$$\frac{dP}{dz} = \left(\frac{g_0}{1 + \frac{P}{A(z)I_S}} - \alpha\right)P \tag{8.2}$$

where the effective-saturation power $A(z)I_S$ will increase with increasing z. As a further illustration of the effect that tapering the gain region of an amplifier has on the power output, Fig. 8.11 presents the results of using (8.2) to calculate the power output when the tapered amplifier cross-sectional area is $A(z) = (d/\Gamma_y)[w + 2z\tan(\delta)]$ where δ is the half angle, in radians, of the linear taper, $w = 3\,\mu$m is the lateral dimension at $z = 0$, $\Gamma_y = 0.02$ is the

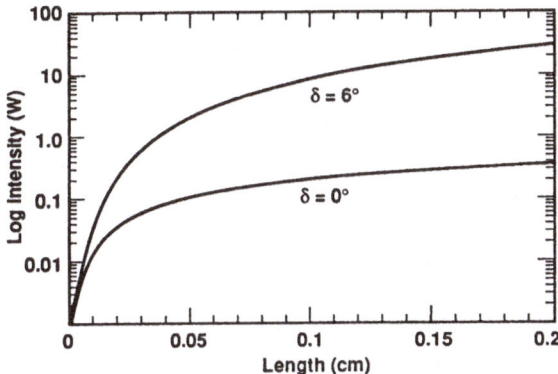

Fig. 8.11. Log plot of the power outputs of a linearly-tapered semiconductor amplifier and a semiconductor amplifier without a taper.

transverse-mode confinement factor, and $d = 80$ Å is the transverse dimension of the single quantum-well active region. The two curves correspond to a half-taper angle of 6° and 0°, which is not tapered. For this model calculation the parameters used were $g_0 = 400\,\mathrm{cm}^{-1}$, $\alpha = 5\,\mathrm{cm}^{-1}$, and $I_S = 0.5\,\mathrm{MW/cm}^2$. One absorption length corresponds to a length of 0.2 cm, and the amplifier without a taper is saturated at a power of ≈ 250 mW from the 3 μm wide emitting aperture, whereas the 6° tapered amplifier exhibits about 25 W of power from a 423 μm wide emitting aperture, for the same 0.2 cm cavity length. Of course, the areal extent and gain capacity of the 6° tapered-gain region in Fig. 8.11 is more than two orders of magnitude larger than the same length of untapered amplifier, so a proportionately larger-current density would be required for the same small-signal gain. At a length of 0.2 cm, the power density in the case of the 6° tapered-gain region is about 5.9 MW/cm², while that of the untapared amplifier is about 8.3 MW/cm². This illustrates another characteristic of the tapered amplifier, which is that the power density at the facet is kept slighty lower than an untapered amplifier as the length is scaled.

The effects of saturation in power output at one absorption length are still apparent for the tapered amplifier in Fig. 8.11, however, there is still an increasing slope on the log-power plot as the length is extended beyond one absorption length. This occurs because the gain in the center of the tapered region becomes saturated down to the level of the losses, while at the edges of the lateral taper, where more gain is added as the length is increased, the gain is not completely saturated so amplification still occurs in this region. Therefore, the amplified power is added at the wings of the input beam, as the gain becomes fully saturated in the center of the amplifier. This can lead to a distortion in the near field of the tapered-amplifier output beam. As will be discused below, proper treatment of this phenomena requires a beam-propagation model calculation.

The feasibility of tapered-semiconductor amplifiers for producing $> 1\,\mathrm{W}$ cw power outputs with near-diffraction limited single-lobed output beams was demonstrated by *Walpole* et al. [1.439, 8.46]. In this experiment a discrete MOPA arangement was employed with a Ti:sapphire laser serving as the master oscillator. The tapered amplifier is very similar to the tapered-cavity unstable resonator appearing in Fig. 5.22, that was also developed by *Walpole* and *Kintzer* [1.440, 1.441]. However, in the case of the tapered amplifier the input width was $10\,\mu\mathrm{m}$, the length was $2\,\mathrm{mm}$, the output width was $200\,\mu\mathrm{m}$, and both the input and output facets had anti-reflection coatings. When injected with $80\,\mathrm{mW}$ from the Ti:sapphire master oscillator, the amplifier exhibited the power-current characteristic displayed in Fig. 8.12. Both the total-power output, as well as the power contained within the first nulls in the lateral far-field pattern is presented in Fig. 8.13. The fact that almost $80\,\%$ of the total power is contained within the dominant lobe of the far-field is a good indication of the near-diffraction limited performance of this tapered MOPA. Note that the far-field beam in the tapered MOPA, as with the unstable-resonator tapered-cavity diode-laser design of Sect. 5.3.4, is highly astigmatic because of the diffraction of the traveling wave in the laterally-tapered gain region. The wavefront in the transverse direction, or perpendicular to the pn junction, will appear to diverge from the output facet because of the single-mode confinement provided by the GRINSCH waveguide structure. However, the wavefront in the lateral direction appears to eminate from within the amplifier from a virtual source that is placed a distance L/n_{eff} from the output facet, where L is the amplifier length and n_{eff} is the effective index of the transverse-waveguide structure. The astigmatism was removed in the lateral far-field measurement of Fig. 8.13 by using the technique described in Sect. 5.3.3, wherein a spherical lens is used to reimage the virtual source

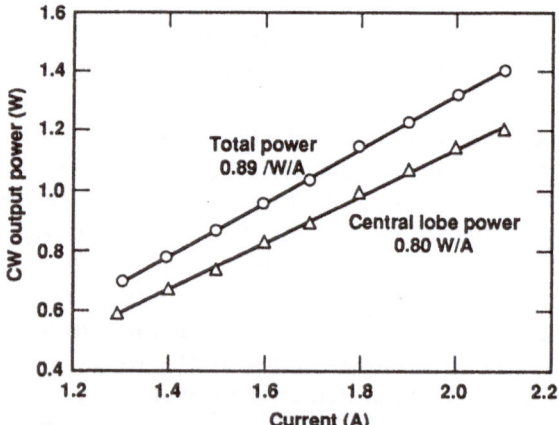

Fig. 8.12. The cw power-current characteristic of tapered semiconductor amplifier when injected with $80\,\mathrm{mW}$ of power at $977\,\mathrm{nm}$. [1.439]

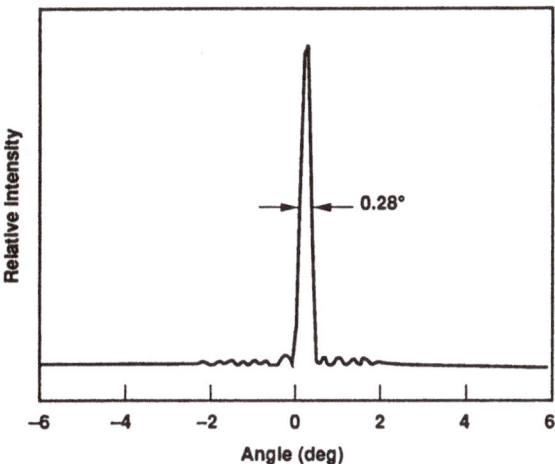

Fig. 8.13. The far-field intensity pattern at 1.44 W with the quadratic-phase curvature removed. [1.439]

onto a screen or detector and the resulting profile is scaled according to the focal length of the lens.

As the current to the amplifier is increased, gradients in the lateral-thermal profile and lateral carrier-density profile give rise to lensing effects [8.25] which cause slight changes in the distance that the virtual source appears behind the output facet. Results of beam-propagation model calculations of near-field characteristics of the tapered-amplifier structure that have been done by [5.16] are in agreement with the 5 % variation in lateral-wavefront cuvature that was measured by [1.439]. This model calculation included effects that could cause spatial nonuniformities in the gain and index within the amplifier such as gain saturation, heat dissipation, and nonuniform strain such as might result from the patterned-metal contact that was used to define the tapered-gain region. The principal factors found to influence the beam quality were the gain saturation through the gain-induced index depression, which is characterized by the α or antiguiding factor, and the lateral-thermal gradient that produces a quadratic-phase curvature causing an apparent shift in the position of the virtual source [1.439]. Improved near-diffraction limited performance of the tapered-semiconductor amplifier with injection from an external oscillator has been recently exhibited at cw power-output levels of several Watts [1.440, 1.441, 2.40, 8.47].

Following the successful demonstration of the discrete-tapered MOPA by [1.440, 1.441], monolithic versions of the tapered-MOPA were investigated [1.246, 8.48], and in some cases, exhibited promising performance characteristics [1.246]. A schematic diagram of a monolithic-tapered amplifier appears in Fig. 8.14. This monolithic MOPA comprised an index-guided DBR single-mode laser oscillator, index-guided preamplifier, and a $\approx 6°$ tapered power

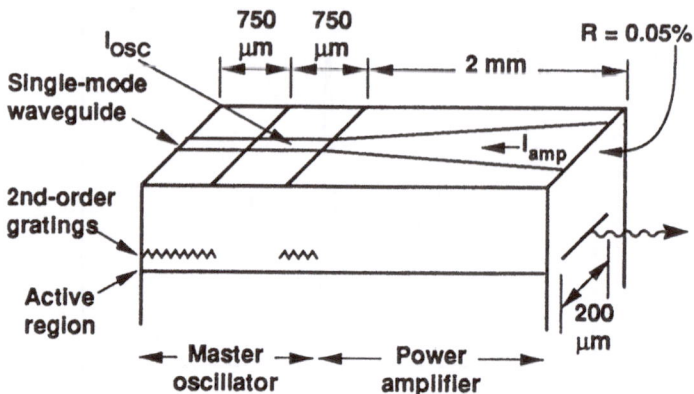

Fig. 8.14. Schematic diagram of a monolithic-tapered MOPA structure. The width of the index guide in the DBR laser/preamplifier master oscillator section was $3.5\,\mu m$. [1.246]

amplifier that were all fabricated on the same wafer. The axial dimensions of the integrated components are displayed in Fig. 8.14.

As discussed in [1.246], monolithic-tapered MOPAs of the type shown in Fig. 8.14 have demonstrated single-wavelength operation at power outputs in excess of 1 W, while maintaining predominantely single-lobed, near-diffraction-limited far-field patterns. These measured performance characteristics are presented in Figs. 8.15 and 8.16.

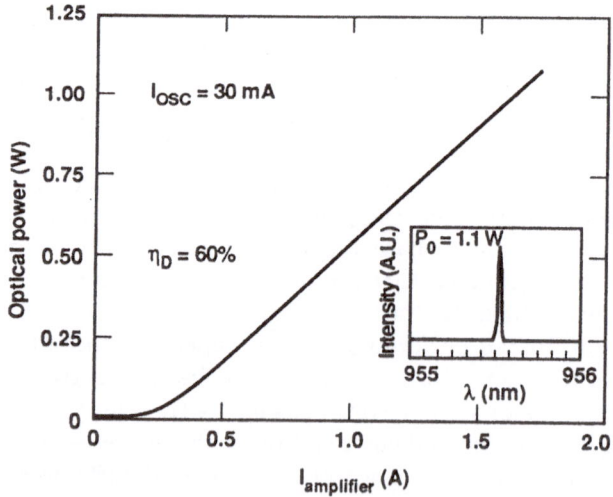

Fig. 8.15. The power output of a monolithic-tapered MOPA as a function of the current to the tapered amplifier. A representative spectrum illustrating single-wavelength operation is displayed in the inset. [1.246]

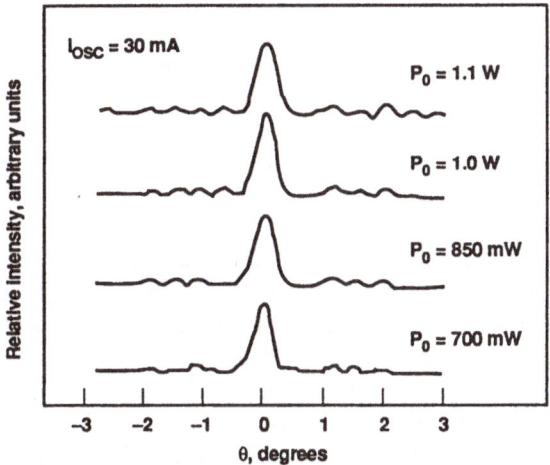

Fig. 8.16. Measured far-field intensity patterns of a monolithic-tapered MOPA at the indicated power-output levels. Note that these measurements were corrected using the method described in Sect. 5.3.3 to remove astigmatism. [1.246]

Although the monolithic-tapered MOPA in [1.246] has demonstrated the desirable operating characteristics up to the 1 W level, these characteristics are sensitive to spatial nonuniformities in the saturated-index and lateral-temperature profile, as well as amplified-spontaneous emission and internal reflections that are directed back into the master oscillator. As was the case with the broad-area semiconductor amplifier in Sect. 8.1.2, random spatial nonuniformities in the index can cause a significant degradation in the beam quality of tapered MOPAs [5.17, 8.48, 8.49]. A drawback of the monolithic approach is that it is difficult to prevent reflections from the output facet, as well as amplified spontaneous emission, from being injected back into the master oscillator. Light injected back into the master oscillator can desta-bilize the single-mode operation of the oscillator and cause self-oscillation of the amplifier. In the cases where an external oscillator is used to drive the semiconductor-power amplifier, it is possible to reduce light injected back into the oscillator either by using an optical isolator or spatial filters. This is perhaps the principal reason that monolithic MOPAs have so far not yet exhibited the level of performance or reproducibility in performance attained by their discrete MOPA counterparts. The present lack of an inte-grable semiconductor-waveguide isolator or waveguide spatial-filter analogs, that are available using discrete optical elements, is an obstacle that needs to be overcome in order to improve the performance characteristics of monolithic MOPAs.

The performance characteristics and beam quality of the tapered-MOPA structures similar to the experimental devices of [1.246] have been modeled numerically by *Lang* et al. [8.49] using a self-consistent beam-propagation model similar to that described in Sect. 4.3. Effects of carrier diffusion were not included, and the small-signal modal gain was expressed using the form given by (2.34) where the lateral taper of the gain region was accounted for in the lateral dependence of current density J. In the expression for the complex-effective index (4.43) employed by [8.49], an antiguiding factor of 2.5 was used, and a detailed accounting of some of the heat-dissipation mechanisms within the laser structure was incorporated into the model of the thermal effects on the local index.

The local heat dissipation was modeled as being due to the electrical-power dissipation Q_{elec}, the power absorbed by unsaturable losses Q_{abs}, and the stimulated emission power minus stimulated absorption Q_{stim}, which is also refered to as *photon cooling*. Hence the total local-heat dissipation $Q(x, z)$ was expressed as $Q = Q_{elec} + Q_{abs} + Q_{stim}$. As discussed in [8.49], this model does not include energy transport due to spontaneous emission, amplified-spontaneous emission, or optical scattering. Neglecting transport effects due to spontaneous emission is valid when the input power to the amplifier is sufficiently close to the saturation level, so that the level of the stimulated emission far exceeds that of the spontaneous emission. Note that in semiconductor-power amplifiers, where the input intensity is sufficient to saturate the gain, the local-carrier density and corresponding electrical energy is decreased as the optical-energy extraction is much more efficient.

To calculate, the temperature $T(x, z)$ as a function of position within a tapered amplifier, [8.49] employed the convolution $T(x, z) = f_T(x) \circ Q(x, z)$ which incorporates a 1-dimensional thermal-response function $f_T(x)$ that is dependent on the lateral position x. This approximation does not account for longitudinal-temperature gradients, such as might be expected to occur in a tapered amplifier because of the sublinear dependence of the thermal resistance on the lateral dimension of the gain region (see Sect. 2.8.2). Evidently, these appear to be a satisfactory assumption because the lateral gradients in the structure have been found to play the dominant role in determining the phase and amplitude of the traveling wave as it propagates through the amplifier.

The results of the numerical calculation of [8.49] are consistent with the experimental measurements of [1.246] for the expected range of input powers from the oscillator. Figure 8.17 illustrates the relation between output power and input power for a fixed amplifier current of 1.5 A. It is seen that efficient-energy extraction from the tapered-semiconductor MOPA is critically dependent on having the input power in the 10 − 100 mW range, which is well into saturation.

An especially interesting result of the calculation of [8.49] was the evolution of the output beam with operating currents. Examples of the calculated output beam as a function of drive current are presented in Fig. 8.18. The input conditions corresponding to these near-field patterns was a Gaussian-profile beam of width $2\,\mu m$ and a power level of 10 mW focused onto the $10\,\mu m$ wide-input end of the tapered MOPA. At the lowest-current value, or smallest small-signal gain, in Fig. 8.18a the near-field pattern at the output

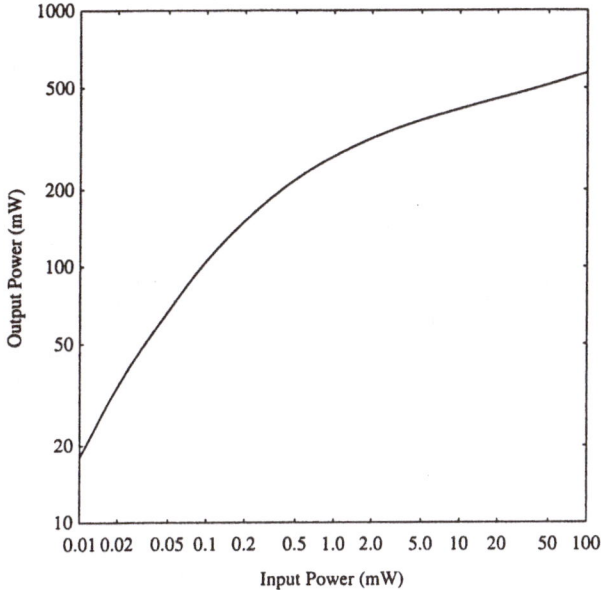

Fig. 8.17. Log-log plot of the calculated power output of a tapered- semiconductor amplifier, operated at a constant current of 1.5 A, versus input power with a $2\,\mu m$ wide Gaussian input beam [8.49].

end exhibits a Gaussian character. However, at the larger-current levels, or larger values of the small-signal gain, the near-field pattern has evolved to a flat-top intensity profile in Fig. 8.17b; and at the highest-value of the small-signal gain in Figs. 8.17c, the flat-top intensity-profile beam begins to develop peaks at each lateral edge of the gain region. The origin of these peaks has not been completely investigated, but they may be related to the lateral-gradient in the effective index that occurs at the edge of the tapered-gain region. The flat-top intensity profile on the output beam present at higher currents develops because the gain has been saturated down close to the level of the losses in the central portion of the beam, where the intensity and the gain-length product is largest. Note that in this calculation a material loss of $\alpha = 5\,cm^{-1}$ was assumed, which corresponds to an absorption length of 2 mm, that is equal to the length of the tapered amplifier.

At still higher-current levels > 3 A, filamentation was evident in the model calculations of [8.49], with the spatial distribution of the filaments varying with drive current. Recall from Sect. 2.6, that filaments evolve because in regions of higher-optical intensity the gain will be reduced due to saturation. Because of the coupling between the local gain and refractive index, the local index is increased in regions of higher-gain saturation. This results in a localized self-focusing of the optical beam, which creates a filament. Generally speaking, a filament will not be sustained along the entire length of the amplifier. Eventually the local gain that a filament sees, becomes completely

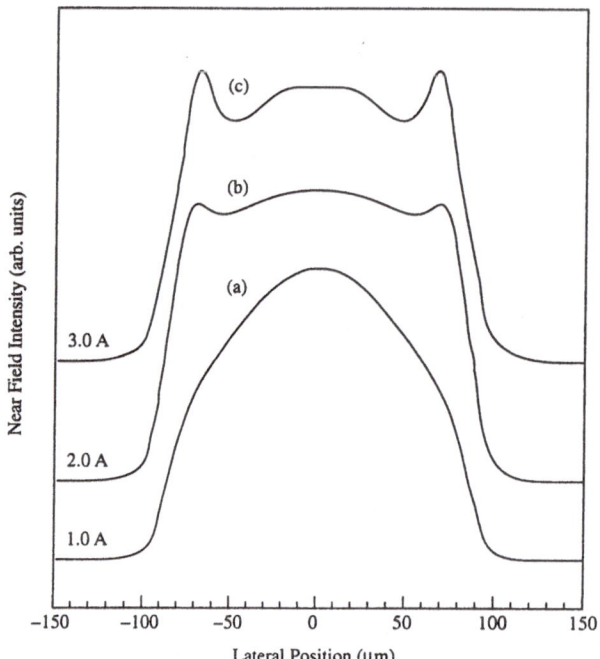

Fig. 8.18. Near-field intensity patterns at the output facet of a tapered-semiconductor amplifier (not to the same relative scale) for amplifier currents of 1.0 A (**a**), 2.0 A (**b**), and 3.0 A (**c**). [8.49]

saturated, at which point the filament dissapears and generates a secondary set of filaments. The near-field sequence with current depicted in Fig. 8.18a-c illustrates how a single Gaussian input beam can begin to generate higher-intensity filaments at the edges on the gain region when the gain at the center of the beam is saturated.

The problem of filamentation in tapered-semiconductors can be critically dependent on random-spatial nonuniformities [5.17, 8.48, 8.49]. Beam-filamentation effects in tapered amplifiers have also been modeled by *Amantea* [5.17, 8.48], who found that random-spatial nonuniformities that give rise to a root-mean-square deviations in the effective index of $\Delta n_{eff} \geq 0.001$ causes strong filamentation resulting in poor beam quality, whereas for root-mean-square deviations of $\Delta n_{eff} = 0.0001$, and all else being equal, filaments were nearly absent. Such random nonuniformities in the effective index can be caused by compositional and/or dimensional nonuniformities in the wafer epilayers [1.194] and nonuniform-heat dissipation [8.49]. As pointed out by [8.49], voids in the solder bond between the chip and the heat sink can create a random variation in the thermal impedance which gives rise to a nonuniform-heat dissipation. To illustrate the effects of thermal seeding on filamentation, a model calculation of a tapered amplifier with a localized hot-spot within the tapered amplifier was done by [8.49]. The hot spot (at a fixed longitudinal position) was incorporated into the model by multiplying the lateral temperature profile by a Gaussian with a $1/e^2$ width of $8\,\mu m$. The maximum value of the Gaussian was selected to correspond to a 10 % reduction in the unperturbed thermal conductivity. Figure 8.19 displays the result of the calculated intensity map of a tapered MOPA containing a localized nonuniformity in the thermal impedance that gives rise to a localized-hot spot. Calculations were done at a fixed-peak intensity for short-pulse, low-duty cycle pulsed operation in Figs. 8.19a and c, as well as, for cw operation Figs. 8.19b and d. Under short-pulse, low-duty cycle operation, the temporal duration of the optical pulses is much shorter than the thermal-time constant and the pulse period is long enough so that the heat from previous pulses has completely dissipated. By the time the heat has begun to flow to other parts of the amplifier, the optical pulse is gone; and by the time the amplifier is pulsed again with current, the amplifier is nearly cooled to the ambient temperature. Clearly then, for the short-pulse, low-duty cycle regime of operation, the thermal effects would be expected to have a negligible effect on the lateral-index profile. However, under cw operation, heat dissipation is a steady-state process, so that the epilayers of the amplifier and heatsink come into thermal equilibrium

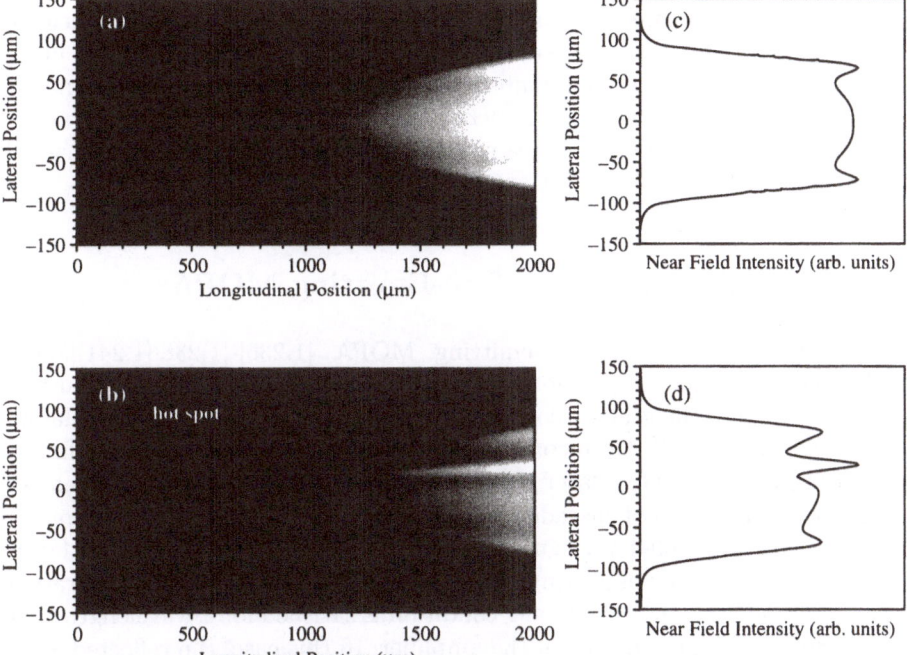

Fig. 8.19. Calculated near-field intensity maps of a tapered-semiconductor amplifier under short-pulse operation at a 10^{-5} duty cycle (**a**) and cw operating conditions (**b**). The corresponding lateral-intensity profiles at the output facets are displayed in (**c**) and (**d**), respectively [8.49].

and corresponding thermal gradients develop within the amplifier. As seen in Fig. 8.19b and d a significant distortion appears in the lateral-intensity profile at positions downstream of the hot spot itself, which is located at about the halfway point of the amplifier. At the exact location of the hot spot and immediate neighborhood, there does not appear to be any discernable change in the optical intensity. This is the case because initially the hot spot, which causes an increase in the local-refractive index, only effects the phase front of the traveling wave. Downstream the intensity increases because of the lateral thermal-lensing effect that is created by the hot spot.

Experimental observations of the near-field of tapered amplifiers as a function of pulse length and duty cycle are consistent with the assertion of [8.49] that thermal nonuniformities can act as seeds for beam filamentation. In experiments conducted by [8.50] on monolithic-tapered amplifiers [8.48], that contained nonresonant output-coupling gratings in the amplifier so that the near-field evolution within the amplifier could be viewed, it was found that increasing the pulse length and/or duty cycle resulted in strong filamentation of the intensity profile observed within the amplifiers. Although, monolithic-tapered amplifiers have demonstrated single-wavelength operation with a single-lobed, near-diffraction limited output beam at > 1 W cw power output, the sensitivity of the beam quality to the random-spatial nonuniformities within the amplifier tends to make the fabrication yield of beam-quality monolithic tapered MOPAs quite low. Moreover, it is likely that further improvements in material uniformity and in the maintainence of sufficiently uniform thermal conductivities will be needed so that the aforementioned desirable performance characteristics become routinely observed in the majority of devices that are fabricated from a wafer.

8.2 Grating-Coupled Surface-Emitting MOPAs

The grating-coupled, surface-emitting MOPA [1.236]-[1.238][1.241][8.51]-[8.53] was proposed as a solution to the mode discrimination problem that had been observed in the grating-coupled, surface-emitting, laser-array oscillators described in Chap. 7. By incorporating different-period gratings within the structure of a linear GSE laser array it was realized that a DBR laser master oscillator and a chain of cascaded-power amplifiers, with passive nonresonant grating output-coupled waveguides following each amplifier, could be fabricated. Each nonresonant grating output-coupled waveguide contains a grating with a period such that the Bragg condition is satisfied for a wavelength that is not within the gain profile of the amplifier. In this case, the reflected wave is almost negligible [1.194]. A modification of the cascaded GSE-MOPA, the active-grating surface-emitting amplifier, which can be thought of as the continuum limit of the cascaded GSE-MOPA, was considered as a scalable design that had a single large-area emitting section [1.236][1.242]-[1.245][8.54]. The projected benefit of having a single continuous-grating-output coupler over multiple grating output couplers is improved beam quality.

8.2.1 Cascaded Grating-Coupled Surface-Emitting MOPAs

A schematic diagram of the cascaded GSE-MOPA array is depicted in Fig. 8.20. A common single-mode waveguide spans the length of the device. At one end of the common single-mode waveguide, a pair of resonant first-or second-order Bragg reflectors are fabricated on each end of the gain section to form a DBR-laser oscillator [1.236]. The oscillator integrated into the single-mode waveguide injects light into a chain of cascaded-power amplifiers where passive grating-output-coupled waveguide sections are placed after each amplifier section. The output of the DBR laser is transmitted by the passive-DBR waveguide into the first adjacent gain section, where the light is amplified. The amplified forward-traveling wave enters a passive-grating output-coupler that deflects most of the amplified forward-traveling wave out of the waveguide into the air (and substrate), while reflecting very little light back towards the amplifier and oscillator. The light that remains at the end of the passive-waveguide section is injected into the second amplifier, where upon it is amplified, and the process of grating-output coupling followed by amplification is repeated down the amplifier chain. As already mentioned in Sect. 7.1.1, by selecting the period of the grating-output coupler so that the Bragg condition is not satisfied for any wavelength within the gain profile of the amplifier, a backward-traveling wave is not generated so that feedback is not provided to the amplifier sections. In nonresonant first- and second-order gratings, it is the first-diffracted order that is emitted into air. This corresponds to having $m = 1$ in the general expression (7.5) for the output-coupling angle; and in this instance, the light-output beam from the forward-traveling wave is emitted into air at an angle θ_a, indicated in Fig. 8.20, given by

$$\theta_a = \sin^{-1}\left(\frac{\lambda}{\Lambda} - n_e\right) \tag{8.3}$$

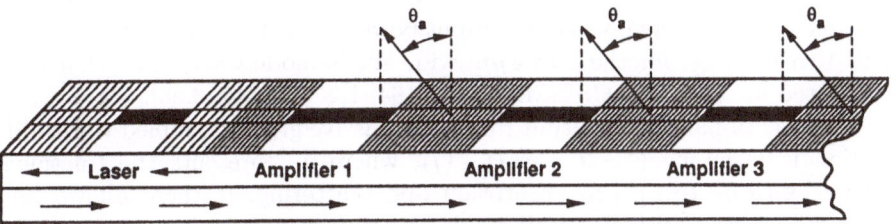

Fig. 8.20. Schematic diagram of a cascaded grating-coupled surface-emitting MOPA array. The emission angle of the output-coupled light is denoted by θ_a. The indicated emission direction is for a first-order output-coupling grating. A second-order output coupling grating would emit light on the other side of the dotted-surface normal. [1.236]

where λ is the wavelength of light, Λ is the grating period, and n_e is the effective index. A nonresonant first-order grating, that emits light into air, will have a period Λ that satisifies the condition [1.236],

$$\frac{\lambda}{n_e + 1} < \Lambda < \frac{\lambda}{n_e} \qquad (8.4)$$

and light will be emitted at an angle that is inclined towards the oscillator, as illustrated in Fig. 8.20. For a nonresonant second-order grating, surface emission occurs when the grating period satisfies $\lambda / (n_e - 1) > \Lambda > \lambda/n_e$. In this case, θ_c is inclined away from the master oscillator.

The conditions for optimizing a cascaded uniform GSE-MOPA array were elucidated by *Carlson* et al. [1.194, 1.236]. Maximum power output and efficiency from a chain of N identical amplifiers and output-coupling gratings requires that the strength of the grating-output coupler be selected such that the total losses (most of which are due to output coupling) of each grating section are balanced by the single-pass gain of each amplifier. Consider a spatially-uniform amplifier chain of sufficient length so that end effects can be neglected. Then the input power to each amplifier P_{in} is related to the output power of the adjacent amplifier by $P_{in} = T P_{out}$, where T is the transmission of a passive grating-output-coupled waveguide section. We also have the output power of an amplifier being related to the input power by $P_{out} = G P_{in}$, where G is the single-pass gain of an amplifier section. Combining these two equations, we find the mathematical expression $GT = 1$ that corresponds to the loss-gain balence condition in the amplifier grating-output-coupler chain. As a further illustration of the importance of the loss-gain balance, Fig. 8.21 displays the calculated power output as a function of the total number of amplifiers N in a GSE-MOPA chain for the indicated values of the grating transmission T. These curves were generated by employing the Rigrod model (Sect. 2.3.4) to calculate the power output $P_{out}(j)$ for $j = 1, 2 \ldots, N$ of each amplifier, assuming that only a single-traveling wave propagated in the amplifier-grating-output coupler chain. Here, each amplifier was assumed to have a small-signal gain of $g_0 = 100 \, cm^{-1}$, a saturation power of $P_S = 5 \, mW$, and the input power to the first amplifier in the chain was $P_{in}(1) = 5 \, mW$. These parameter values are representative of the GaAs based GSE-MOPAs that have been studied, where the input power needed for saturation is in the 1 to 10 mW range for the 2 to 4 μm wide single-mode waveguides that have been used for cascaded GSE-MOPAs [1.194, 1.236, 1.240, 8.55].

The fractional-power output from each passive grating-coupled waveguide section is given by $(1 - T - a) P_{out}(j)$, where a represents the fractional power loss due to optical absorption and scattering. It then follows that the total-power output from all passive grating- coupled waveguide sections P_{total} is expressed as,

$$P_{total} = \sum_{j=1}^{N} (1 - T - a) P_{out}(j) . \qquad (8.5)$$

It is seen from Fig. 8.21 that the value of the grating transmission T of the passive grating-coupled waveguide section can have a significant influence on the scaling properties of the cascaded MOPA. In those cases where the transmission $T \leq 0.01$ is too small in Fig. 8.21, we have $T \ll 1/G$ and the distributed losses exceed the distributed gain in the amplifier chain. Therefore, the input power to each successive amplifier in the chain is decreased, and the power output scales poorly, or not at all, with the number of amplifiers in the chain, as displayed in Fig. 8.21. This occurs because with $T \ll 1/G$ the input power to each successive amplifier is decreased, so we have $P_{in}(j) < P_{in}(j-1)$. Amplifiers further away from the oscillator are injected with less and less power, so that $P_{out}(j) < P_{out}(j-1)$. In this situation, amplified spontaneous emission that is traveling in both directions in the amplifier will build up and quench the gain so that amplification of the injected coherent light from the master oscillator is no longer possible. This type of gain-quenching effect is analyzed in more detail for the case of active-grating surface-emitting amplifiers in Sect. 8.2.2.

To optimize the power-output scaling, it is critical that the transmission of the grating-output couplers must be high enough to provide sufficient input power to each amplifier in the chain to saturate the gain at a level where spontaneous emission noise will be suppressed. A self-consistent analysis of the effects of spontaneous emission on the performance characteristics of cascaded GSE-MOPAs can be found in [1.240, 8.55]. The range of transmission values that allow for near-optimum power output can be quite broad. An analysis presented in [1.194], revealed that for grating-coupled waveguide

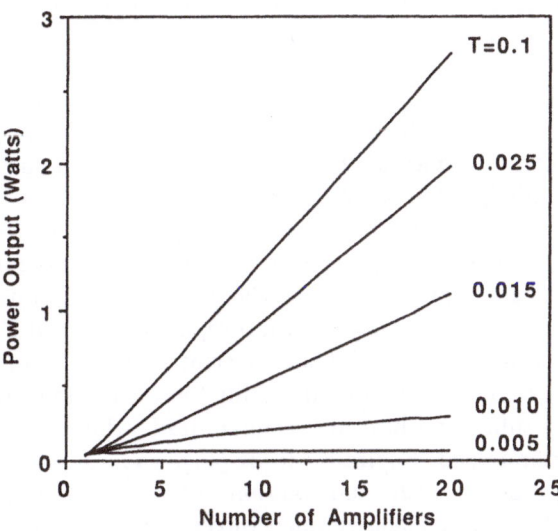

Fig. 8.21. Calculated power output as a function of the number of amplifiers in a cascaded GSE-MOPA array. The transmission of the grating-output coupler is indicated next to each curve [1.194].

transmissions in the range on 0.03 to 0.3, the total-power output P_{total} for a 20-amplifier chain drops to only 80 % of the maximum value at the extremes of the 0.03 to 0.3 range of transmissions. In general, it is not always possible to satisfy the loss-gain balence condition for the first few amplifiers in a cascaded GSE-MOPA array. However, in a uniform-amplifier chain as long as $GT \geq 1$, the power distribution will relax to where $GT = 1$ is satisfied for all but the first few amplifier stages [1.240].

Experimental studies of the performance characteristics of cascaded GSE-MOPAs have been reported by [1.236]-[1.240] which illustrate the aforementioned principles of operation. As an example, Fig. 8.22 presents the results of the experimental study of [1.236] on the spectral output of a GSE-MOPA (under short-pulse, low-duty cycle operation) as a function of the master oscillator drive current for a fixed current to the amplifier. When the laser oscillator was not operated, Fig. 8.22a, the spectral output of each amplifier was amplified spontaneous emission that extended over a broad 30 nm spectral bandwidth. However, as the laser oscillator current was increased in Fig. 8.22b-d, and coherent light was injected into the amplifier, the amplifier spectral output became increasingly contained within the much-narrower spectral bandwidth (< 1 Å) of the oscillator, and the spontaneous emission level was decreased to below the limit of sensitivity in the measurement. In monolithic MOPAs, it is not generally possible to directly measure the power that is injected into an amplifier. However, by operating an amplifier at transparency one can obtain an estimate of the injected power to the amplifiers. By employing this method, it was inferred by [1.236] that the amplifier gain saturated at an input power of ≈ 1 mW. Note that although the peak of the spontaneous emission in Fig. 8.22a occured at about 7990 Å, the laser oscillator operated at 8223 Å, as determined by the grating period selection. This was done because the GRINSCH-SQW structure (Sect. 2.1) used in the amplifiers of [1.236], typically experience maximum gain about $100 - 300$ Å towards the long-wavelength side of the spontaneous emission profile [1.451].

An important topic that has been addressed in the course of experimental work on cascaded GSE-MOPAs is the demonstration of a high-efficiency, passive-grating output coupler by *Mehuys* et al. [1.240]. To improve the output-coupling efficiency of passive grating-coupled waveguides, a multilayer reflector stack comprised of a superlattice was incorporated on the substrate side of the amplifier epilayer structure, as depicted in Fig. 8.23a. The idea here is to design the superlattice-substrate reflector so that the phase of the reflected grating orders are all modifed is such a away that the intensity of the order diffracted into air is maximized, while the intensity of all other orders is minimized. Figure 8.23b displays the calculated values of the transmission T, reflectivity R, and fraction of light emitted into air U all as a function length of the grating-coupled waveguide section. For this praticular example, the wavelength of light was 874 nm and the grating period was selected so that the second order Bragg condition was satisified at a wavelength of 904 nm. The reflectivity being $\leq 10^{-4}$ is much too low to register

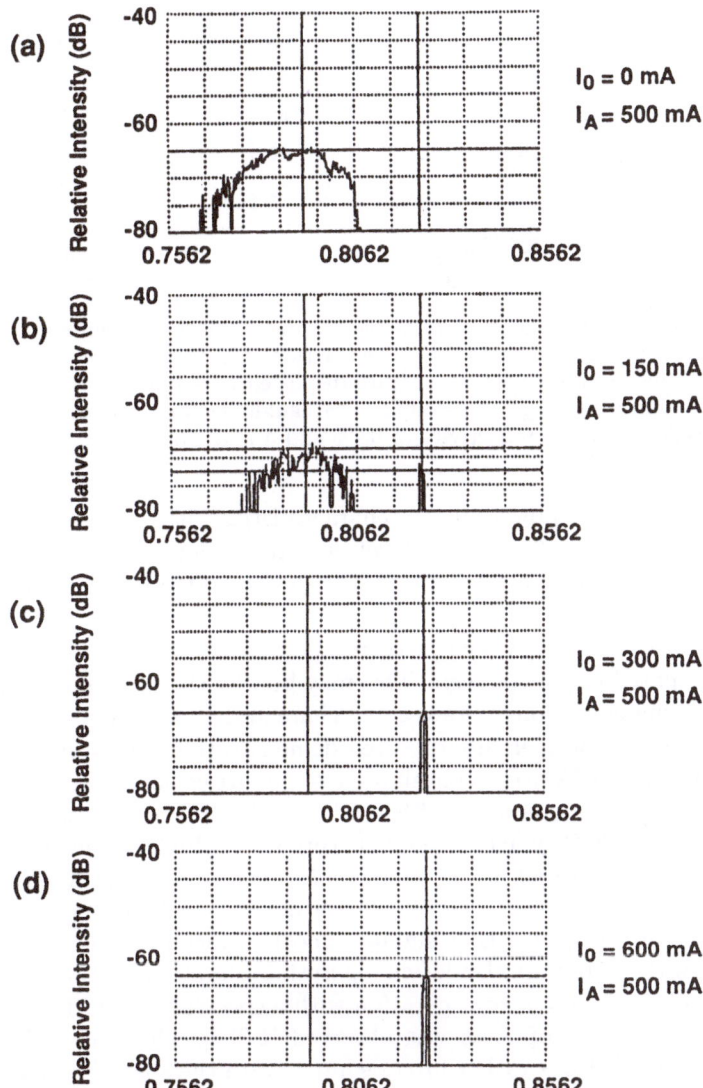

Fig. 8.22. The measured spectral output (on a log-intensity scale) of an amplifier in a monolithic GSE-MOPA chain is shown at a constant current of 500 mA for laser oscillator currents of 0 mA (**a**), 150 mA (**b**), 300 mA (**c**), 600 mA (**d**) [1.236].

on the scale in Fig. 8.23b. It is seen that the fraction of light diffracted into air by the grating can be almost as high as 80 %. Note that higher values of the fraction of light diffracted into air U are possible, but this lowers the transmission to levels where $GT < 1$, which as already mentioned, is undesirable. This was a consideration in the selection of parameters used in the model calculation of [8.55]. Experimental studies by *Mehuys* et al. [1.240] on cascaded GSE-MOPAs under short-pulse, low-duty cycle operating con-

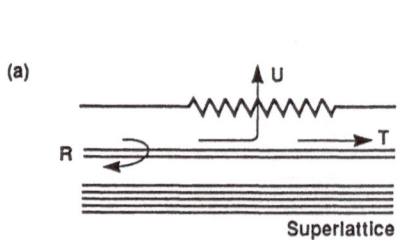

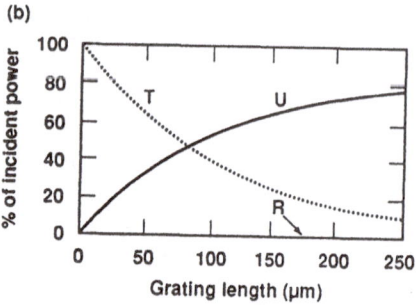

Fig. 8.23. Schematic diagram of a passive grating-coupled waveguide that contains a superlattice-substrate reflector (**a**). The fraction of incident power transmitted T, reflected R, and diffracted into air U are displayed as a function of the length of the passive nonresonant grating-coupled waveguide section (**b**). The superlattice substrate reflector had a reflectivity of 95 % and a phase of $\pi/2$. [8.55].

ditions have revealed power-current characteristics with differential-quantum efficiencies as high as 67 %. This level of output coupling efficiency is competitive with cleaved-facet edge-emitting devices. One disadvantage of the superlattice-substrate reflector is that it leads to an increase in the series resistance of the device, which in turn increases the level of thermal power that is dissipated in the epilayers. This may be why the short-pulse, low-duty cycle operating conditions were used to characterize the device in [1.240], as so far there have not been any reported demonstrations of cw operation of a cascaded GSE-MOPA array containing a superlattice substrate reflector.

Continuous-wave operation of cascaded GSE-MOPAs has been achieved by using a GRINSCH type epilayer structure with a InGaAs quantum well so that the GaAs substrate is transparent to the ≈ 950 nm operating wavelength of the laser and amplifiers [1.238]. These devices could be mounted junction side down, for better heat sinking, and the output-coupled light taken out through the substrate. Under cw-operating conditions, power-outputs of 300 mW and single-frequency operation with a spectral linewidth of 135 MHz from an array of 6 amplifiers was observed [1.238]. The observed spectral linewidth of 135 MHz is rather broad compared to the spectral linewidths of 5 to 10 MHz that have been measured for the design of GSE-DBR master oscillator that was used [1.187, 1.188]. The additional linewidth broadening from the amplifiers could be due to large levels of amplified spontaneous emission from the amplifier chain being injected into the oscillator. Broadening of the laser-oscillator linewidth in monolithic MOPAs, due to high levels of amplified spontaneous emission being injected into the laser oscillator, has been observed experimentally by [4.82] for active-grating surface-emitting amplifiers (Sect. 8.2.2).

Under short-pulse, low-duty-cycle operating conditions, a peak-power output of 1.2 W was observed from an array of 9 amplifiers [1.240], and more than 4.5 W was observed from a monolithic two-dimensional array of 9 parallel-amplifier chains, where each amplifier chain contained 7 cascaded ampli-

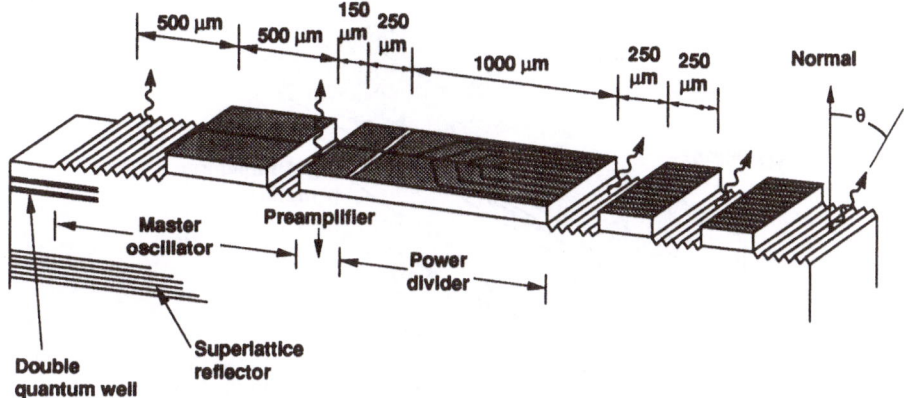

Fig. 8.24. Schematic diagram of the two-dimensional cascaded GSE-MOPA array of [1.239].

fiers, for a total of 63 amplifiers [1.239]. A schematic diagram of the two-dimensional MOPA array appears in Fig. 8.24.

A significant problem that was discovered with the cascaded GSE-MOPA arrays, for both high-power cw and pulsed operating conditions, was that it not possible to maintain good quality of the output beam. Near-diffraction limited beam quality was observed from cascaded GSE-MOPAs under pulsed operating conditions at peak powers up to ≈ 300 mW [1.240] by appropriately adjusting the current to the amplifier sections, as was done in the case of the GSE laser oscillators presented in Chap. 7. Much above this power level a rapid degradation in the beam quality was found to occur [1.240], as the gain in the amplifiers would saturate so that it was no longer possible to use the amplifier-current controls to effect changes in the relative phase of the light that was emitted from the different grating-coupled waveguides. The desire for a single-emitting section with a uniform phase of the emitted light led researchers to investigate the active-grating surface-emitting amplifier, which is the subject of the next subsection.

8.2.2 Active Grating-Coupled MOPAs: Operating Principles

In an effort to obtain improved beam quality (over that demonstrated by the cascaded GSE-MOPA), it was proposed that a nonresonant buried-output coupling grating be fabricated in the active-layer of a long single-mode index-guided semiconductor amplifier [1.241]. The nonresonant buried-output coupling grating acts as a distributed-output coupler by diffracting some of the amplified guided-traveling wave into a radiation mode. This class of device will be refered to as a *monolithic active grating-coupled* MOPA or *active-grating surface-emitting amplifier*. An illustration of this type of monolithic-MOPA structure is presented in Fig. 8.25. The structure comprises a buried-grating DFB laser oscillator [1.295, 7.49, 7.50] with a single-mode ridge-guide

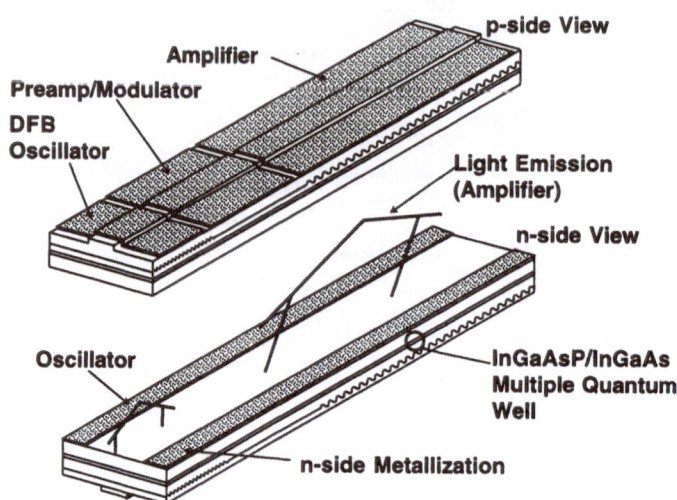

Fig. 8.25. Schematic diagram of an active-grating surface-emitting amplifier with a buried-grating-output coupler [6.25].

to provide single lateral-mode operation. The same single-mode ridge waveguide extends the entire length of the device. At the output end of the DFB oscillator, a length of gain section (without a grating of any kind) is fabricated to act as a preamplifier to boost the oscillator power up to the loss-limited saturation limit. Note some designs of monolithic active grating-coupled MOPAs have incorporated a single-mode DBR-laser oscillator section in place of the DFB laser [1.242, 1.490]. The amplified single-frequency signal from the oscillator is injected into the active-grating surface-emitting amplifier, which contains a nonresonant grating-output coupler that has been fabricated in close proximity to the active layer of the amplifier. Active-grating surface-emitting diode-laser amplifiers have been fabricated in the strained-layer InGaAsP/InP materials systems [1.245, 6.25, 8.56], the GaAs/AlGaAs materials system [1.242], and the strained-layer InGaAs/AlGaAs materials system [1.243, 1.247, 1.490, 8.57].

For reasons given below, it is important that the input light to the amplifier have a power density at or near the limit intensity (2.50) so that the available gain for the coherent light is maximized. As the injected traveling wave propagates along the power amplifier, it continually experiences both amplification and output coupling, so that the local gain is saturated down to the level of the losses, as occurs in a laser oscillator. In this situation, the carrier density is pinned along most of the length of the active-grating amplifier and the power density is a constant given by (2.52), where the losses in the denominator are due to both optical absorption and grating-output coupling. Ideally, the use of a continuous grating-output coupler in the uniformly-saturated amplifier section provides both a uniform intensity and phase tilt (due to the grating-output coupler) on the output beam. Of course, this structure requires that the operating wavelength be shifted to longer wavelengths so that output coupled light is not absorbed as it is transmitted through the substrate, as was the case for the GSE diode-laser arrays described in Sect.7.1.2. An alternative to the buried active-grating MOPA is to use a metalized-relief grating-output coupler [1.184, 1.484] in the active power-amplifier section [1.242].

In the active-grating surface-emitting MOPA, the gain and output-coupling functions are distributed continuously along the length of the amplifier, as opposed to the cascaded GSE-MOPA (presented in the previous section) where the segmented gain and passive grating-coupled waveguide sections are placed in alternating sequence along the length of the amplifier. The continuum limit of the cascaded GSE-MOPA corresponds to the active-grating surface-emitting MOPA. As was the case with the cascaded GSE-MOPA, the period of the buried-grating-output coupler is, for the same reasons, selected so that that the Bragg condition (7.3) is not satisified for any wavelength within the gain profile of the amplifier. Therefore, the output beam is emitted at an angle with respect to the surface normal that is also given by (8.3).

To model the characteristics and optimize the performance of single-spatial mode, active-grating surface-emitting amplifiers, we will follow the approach of *Carlson* [1.243, 1.244, 6.25, 8.54, 8.58] which employed a nonlinear, self-consistent analysis, as presented in Sect. 2.3.4, to account for the spatial depenencies of the carrier concentration $N(z)$, amplified coherent signal power in the forward-traveling amplifier-waveguide mode $P_C(z)$, and the *Amplified Spontaneous Emission* (ASE) power in the forward ($P_N^+(z)$) and backward ($P_N^-(z)$) traveling amplifier-waveguide modes. Under steady-state operating conditions, the coherent and amplified spontaneous noise powers in the waveguide mode of the active-grating amplifier can be modeled according to the set of differential equations expressed by

$$\frac{dP_C(z)}{dz} = [g(z) - \alpha - \alpha_0] P_C(z) \qquad (8.6)$$

and

$$\frac{dP_N^\pm(z)}{dz} = \pm \frac{\gamma E A_m N(z)}{\tau} \pm [g(z) - \alpha - \alpha_0] P_N^\pm(z) , \qquad (8.7)$$

where A_m is the cross-sectional area of the active-grating amplifier-waveguide mode, $g(z)$ is expressed by (2.33) or (2.36) (multiplied by an additional factor of n_w to account for the possibility of a multiple-quantum well active layer), E is the photon energy, τ is the electron-hole recombination time, γ is the spontaneous emission factor defined in Sect. 3.3.2, α_o is the grating-output coupling coefficient, and α is the optical loss coefficient due to absoprtion and scattering. The axial-dependent carrier density $N(z)$ is expressed as

$$N(z) = \tau \left[\frac{\eta J}{qd} - \frac{g(z) P(z)}{\Gamma_y E A_m} \right] , \qquad (8.8)$$

where J is the total injected current density, η is the efficiency for radiative recombination of injected carriers, q is the electronic charge, d is the active-layer thickness, Γ_y is the transverse-mode confinement factor (2.1), and

$$P(z) = P_C(z) + P_N^+(z) + P_N^-(z) \ . \tag{8.9}$$

In this analysis, the assumptions have been made that the gain profile is flat over the spectral bandwidth of the spontaneous emission and that the amplifier is terminated in such a manner so that reflections from the end of the amplifier can be neglected [1.244, 8.54]. The equations that describe the spatial dependence of the amplified-coherent light (8.6) and the forward and backward ASE traveling waves (8.7) are similar, with the distinction that the ASE traveling waves have a distributed-source term that is proportional to the local carrier concentration $N(z)$ or local gain $g(z)$.

The total-power output of an active-grating surface-emitting amplifier is expressed as

$$P_T(L) = \int_0^L \eta_{air}\alpha_o P(z)\, dz \ , \tag{8.10}$$

where η_{air} is the percentage of grating-output coupled light that is emitted into air. For most, grating-output couplers that have been fabricated in surface-emitting lasers and amplifiers, we have $\eta_{air} \approx 0.5$, because light is radiated equally into the substrate and superstrate. Unless appropriate measures are taken, about half the light will be lost when the device is mounted to a heat sink. An example of a grating-output coupler that emits most of the light into air (for the case of junction up mounting) with a substrate reflector of [1.240] is depicted in Fig. 8.23. For junction down mounting, which is required for cw and high-average power operation, radiation emitted into the superstrate can, in principle, be nearly eliminated by using a superstrate reflector or a blazed grating-output coupler [7.25, 7.27], as discussed in [6.25]. Also, at the air-substrate interface there is a 30 % Fresnel loss, because of the large difference in refractive indices, but this too can be eliminated by applying a suitable anti-reflection coating.

As seen from (8.10), light in the amplifier waveguide is output coupled by the grating, and emitted in a direction according to (8.3) that depends on the wavelength of light. The dispersive nature of the grating-coupled waveguide makes it possible to spatially separate, in the far-field zone, the narrow-spectral band ($\approx 10\,\mathrm{MHz}$) coherent amplified light from most of the broad-spectral band ($\approx 30\,\mathrm{nm}$) amplified spontaneous emission and spontaneous emission. Figure 8.26 illustrates how the dispersion of the grating-output coupler can spatially separate (in the longitudinal dimension) the coherent light from the directly-radiated spontaneous emission that does not interact with the grating (Fig. 8.26a), and the amplified spontaneous emission in the waveguide that is deflected into radiation modes by the grating-output coupler (Fig. 8.26b). Recall from Sect. 3.3.2, that the active layer radiates spontaneous emission isotropically into the surrounding volume of the epilayer

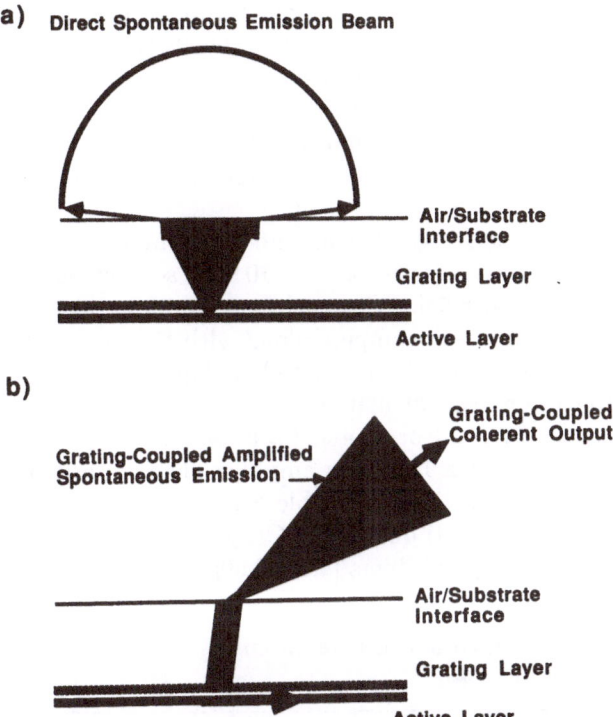

Fig. 8.26. A schematic drawing of the dispersive properties of an active-grating surface-emitting amplifier. The triangular region indicates the range of angles for direct spontaneous emission that does not suffer total internal reflection at the air-substrate interface (**a**). The relative emission directions of the grating coupled amplified spontaneous emission and amplified coherent output are indicated (**b**).

structure. Most of the directly-radiated spontaneous emission will not be deflected by the grating as the efficiency of semiconductor gratings is typically only a few percent [7.2]. Only directly-radiated spontaneous emission that is emitted at an angle within the $\sin^{-1}(1/n_{sub})$ cone indicated in Fig. 8.26a is emitted into air, while the rest undergoes total internal reflection. Spectrally-broad amplified spontaneous emission traveling in the waveguide is dispersed in the far field in accordance with (8.3).

The far-field beam divergence $\Delta\theta$ of a grating-output coupled beam of spectral bandwidth $\Delta\lambda_0$ centered on the wavelength λ_0 can be expressed as

$$\Delta\theta(\lambda_0) = \frac{2\lambda_0}{L} + D\Delta\lambda_0 \tag{8.11}$$

where L is the length of the grating-ouput coupled waveguide and D is the absolute value of the grating dispersion, which has been measured to be $\approx 5 \times 10^{-4}$ radians/Å for the second-order gratings that are commonly used in semiconductor lasers [8.59]. For a 1 cm long amplifier, we find that the coherent light beam divergence in the longitudinal dimension is $\approx 0.02°$ and the divergence of the amplified spontaneous emission is $\approx 9°$. Since the amplified spontaneous emission beam divergence subtends an angle more than two-orders of magnitude greater than the coherent light beam, spatial filtering can be employed to eliminate most of the noise in the far-field zone. This also applies to the directly radiated spontaneous emission, which is emitted into the air in a 2π steradian solid angle.

The length L of active-grating amplifier required for power outputs of ≈ 1 W is typically in the range of 5 to 10 mm, so that generally speaking $L > 1/(\alpha + \alpha_o)$ [1.244, 8.54]. With such long amplifier lengths, the ASE in the amplifier waveguide will compete along with the injected coherent light for the available gain. This is illustrated in Fig. 8.27 where the coherent power versus position was calculated from (8.6), and the noise powers versus position in the forward or $+$ and backward or $-$ traveling waves were calculated using (8.7) for a 1 cm long amplifier. The material parameters for this example calculation, listed in Table 8.1, are representative of some of the InGaAsP/InGaAs multi-quantum SCH structures that have been used for 1.5 μm edge-emitting amplifiers [8.18, 8.60].

Table 8.1. Parameters used to calculate the curves for $P_C(z)$, $P_N^+(z)$, and $P_N^-(z)$ that are depicted in Fig. 8.27.

Number of quantum wells	5
Transverse mode confinement factor (Γ_y)	0.016
Lateral mode dimension (w)	3 μm
Differential gain (σ)	1.66×10^{-16} cm^2
Carrier recombination time (τ)	0.25×10^{-9} s
Internal quantum efficiency (η)	0.7
Quantum-well thickness (d)	80 Å
Transparency carrier level per quantum well (N_{tr})	2×10^{17} cm^{-3}
Wavelength	1.55 μm
Optical loss coefficient (α)	15 cm^{-1}
Grating coupling coefficient (α_o)	22 cm^{-1}

When there is no input power to the amplifier (Fig. 8.27a) the ASE in the forward-traveling wave $P_N^+(z)$ and the ASE in the backward-traveling wave $P_N^-(z)$ build up over about a 1 mm distance on either side of the center of the amplifier and quickly level off as the gain is saturated down to the level of the losses, which is $\alpha + \alpha_o$. For the InGaAsP/InGaAs edge-emitting semiconductor amplifiers that have been studied experimentally, optimum performance has been observed for amplifier lengths ≤ 1 mm [1.333][8.18]-[8.20][8.61]. However, edge-emitting amplifiers that were longer than 1 mm have been observed to exhibit reduced gain and higher levels of ASE [8.18]. From the results of the model calculation displayed in Fig. 8.27a, it is seen that, when the amplifier length approaches one absorption length, the ASE power level rapidly increases to the point where it can completely saturate the gain, even in the absence of injected coherent light from the oscillator. For low-input power levels from the oscillator, presented in Fig. 8.27b, the ASE in the amplifier-waveguide mode still plays a dominant role. The injected coherent signal must travel a distance of $\approx 3\,\text{mm}$ before it reaches a level where it can compete for the available gain; and even then, the P_N^+ ASE wave is still ≈ 0.6 that of the amplified coherent wave P_C. Another matter of concern in Fig. 8.27b, is the $80\,\text{mW}$ of ASE power in the amplifier waveguide that is being injected back into the oscillator, which being at the limit intensity is sufficient to quench the oscillator. In Fig. 8.27c, the input power from the oscillator is $30\,\text{mW}$ and it is seen that the ASE power in the waveguide is mininized over most of the length of the amplifier. The ASE power in the waveguide that is directed into the oscillator is $\approx 6\,\text{mW}$. Power levels in this range can be sufficient to destabilize the operation of the single-frequency DFB laser oscillator in a monolithic MOPA [4.82]. There is not yet a waveguide-optical isolator that can be integrated into a monolithic MOPA. Until such time as a semiconductor waveguide isolator is realized, ASE will be a significant factor that limits the performance of monolithic active-grating surface-emitting amplifiers, and all other monolithic MOPAs.

To optimize the generation of coherent power in an active-grating surface-emitting amplifier, the only conclusion that can be reached from Fig. 8.27 is that the input power to the amplifier should be $P_C(0) = P_{\text{Lim}} = A_m I_{\text{Lim}}$, where I_{Lim} is given by (2.50). Also, for operation in the loss-limited saturate regime, the coherent and ASE powers will satisfy $P_C(z) + P_N^+(z) + P_N^-(z) = P_{\text{Lim}}$ everywhere along the length of the amplifier. Under these conditions (8.6-8.8) reduce to the more simplified forms

$$P_C(z) = P_{\text{Lim}} - \frac{\gamma E A_m N_S L}{2\tau} \quad , \tag{8.12}$$

$$P_N^+(z) = \frac{\gamma E A_m N_S z}{2\tau} \quad , \tag{8.13}$$

$$P_N^-(z) = \frac{\gamma E A_m N_S (L - z)}{2\tau} \quad , \tag{8.14}$$

$$N(z) = N_S = n_w N_{tr} + \frac{\alpha + \alpha_o}{\sigma \Gamma_y} \quad . \tag{8.15}$$

We see from (8.12) that, as the length of the amplifier L is increased, the amplified coherent-power output in the waveguide P_C decreases linearly. The reason for the decrease in P_C with increasing L is that the noise power coupled into the waveguide mode (in each direction) increases linearly with L. Recall from (8.7) that noise is coupled into the waveguide along the entire length of the amplifier. As the length of the amplifier is increased, the total noise power coupled into the waveguide is also increased, however, the coherent power that is being injected at the amplifier input $P_C = P_{\text{Lim}}$ remains fixed. When the length of the amplifier reaches the critical length L_{max}, the coherent

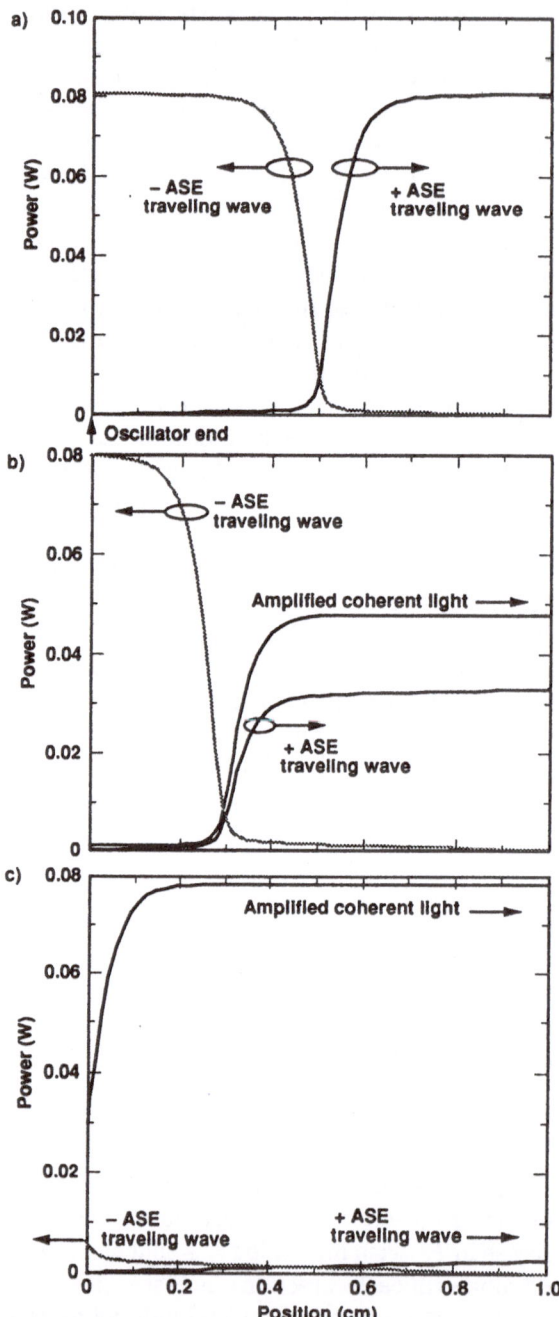

Fig. 8.27. The calculated powers (using parameters displayed in Table 8.1) $P_C(z)$, $P_N^+(z)$, and $P_N^-(z)$ for $J = 8\,\text{kA/cm}^2$ and **(a)** $P_C(0) = 0\,\text{mW}$, **(b)** $P_C(0) = 1\,\text{mW}$, and **(c)** $P_C(0) = 30\,\text{mW}$. The light from the oscillator is injected at $z = 0\,\text{cm}$ and the active-grating amplifier ends at $z = 1\,\text{cm}$.

power is zero everywhere in the amplifier ($P_C(z) = 0$), because the amplified sponataneous emisson has quenched the gain. From (8.12), we see that L_{max} can be expressed as

$$L_{max} = P_{Lim}\frac{2\tau}{\gamma E N_S A_m} = P_{Lim}\left(\frac{\partial P_N}{\partial L}\right)^{-1} \tag{8.16}$$

where the factor $\partial P_N/\partial L$ can be thought of as the noise power per unit length coupled into the amplfier waveguide. For the type of strained-layer InGaAsP/InP quantum-well structures [1.333, 8.62, 8.63] that have been investigated for active-grating surface-emitting amplifiers [1.245, 6.25, 8.56], the values for the differential gain are typically higher, $\sigma \approx 5 \times 10^{-16}\,\text{cm}^2$, and the losses are lower, $\alpha \leq 10\,\text{cm}^{-1}$ than in unstrained structures (see Table 8.2). Hence the noise per unit length coupled into a single spatial-mode, index-guide, strained-layer InGaAsP active-amplifier waveguide is $\partial P_N/\partial L \approx 30\,\text{mW/cm}$ at an operating level where $P_{Lim} \approx 100\,\text{mW}$, which corresponds to L_{max} being in the range of 3 to 4 cm [8.58]. Therefore, the effects of the amplified noise on the performance become more important as the amplifier length is scaled to get higher-power outputs.

To gain further insight into the impact of the amplifed spontaneous emission on the high-power performance, further consideration will be given to the grating-coupled output powers corresponding to the coherent light and the amplified spontaneous emission. Moreover, we need to optimize the design of the active-grating surface-emitting amplifier to maximize the power that is radiated into the amplified coherent output beam. To find the coherent-power output that is diffracted by the grating-output coupler when loss-limited saturation is realized everywhere within the amplifier, we substitute (8.12) into (8.10) and use (8.13) to obtain

$$P_C^{(out)}(L) = \eta_{air}\alpha_o L\left(P_{Lim} - P_N^+(L)\right) \ . \tag{8.17}$$

In an analogous manner, we find that the noise power, due to each traveling wave, that is diffracted into air by the grating-output coupler is expressed as

$$P_N^{(out)}(L) = \eta_{air}\alpha_o\frac{\gamma E A_m N_S L^2}{4\tau} = \frac{\eta_{air}\alpha_o}{2}P_N^{(+)}(L)\,L \ . \tag{8.18}$$

From (8.17 and 8.18), it is seen that the reduction in $P_C^{(out)}(L)$ relative to P_{Lim} corresponds to the amplified spontaneous emission noise power radiated away in the forward and backward traveling waves. Moreover, from (8.10) for loss-limited saturation, we have the condition that the sum of the power outputs due to coherent light and the noise in both traveling waves sum to $\eta_{air}\alpha_o L P_{Lim}$. Note that the amplified spontaneous emission in the backward- and forward-travelling waves are emitted at angles of equal magnitude but opposite sense with respect to the surface normal. The noise power output coupled from the forward-travelling wave will overlap the beam of amplified coherent light. However, the noise power output coupled from the backward-travelling wave is emitted off at an angle of $2\theta_c$ (where θ_a is given by (8.3)). Typically $\theta_a \approx 10°$, so in most cases it is safe to assume that none of the noise power output coupled from the backward-travelling wave is emitted into the same solid angle as the amplified coherent power.

The photometric brightness (Sect. 1.3.3) B_C corresponding to the coherent output beam of the active-grating surface-emitting amplifier is defined as the coherent-intensity output per steradian that is emitted into the direction θ_a and that is subtended by solid angle $\Delta\Omega_C = 4\lambda^2/(Lw)$. The solid angle $\Delta\Omega_C$ is equivalent to the beam divergence expected for uniform illumination of the active-grating amplifier aperture of length L and width w. From (8.17) and the definition of $\Delta\Omega_C$, it is revealed that the photometric brightness can be expressed as

$$B_C(L) = \frac{P_C^{(\text{out})}(L)}{4\lambda^2} , \qquad (8.19)$$

We from (8.17) and (8.19) that $P_C^{(\text{out})}(L)$ and $B_C(L)$ exhibit indentical scaling dependencies with amplifier length. Therefore, optimizing the amplifier structure for maximum power output will also optimize the power emitted into the dominant lobe of the far field.

Now let us proceed to maximize the power output $P_C^{(\text{out})}(L)$ with respect to the grating output coupling coefficient α_o. When the amplifier length is such that $P_{\text{Lim}} \gg P_N^+$ (which could occur for structures of short length or with very small γ), then maximizing only the $\alpha_o L P_{\text{Lim}}$ part of (8.17) with respect to α_o gives $\alpha_o = \sqrt{\alpha g_0} - \alpha$ [1.243, 1.244]. However, when $P_{\text{Lim}} \approx P_N^+$ the $P_N^+(L)$ term in (8.17) cannot be neglected. We see from (8.13) that $P_N^+(L)$ is proportional to the carrier concentration N_S, which has a linear dependence on α_o. Taking the dervative of (8.17) with respect to α_o and setting the result equal to zero yields a cubic equation for the optimum value of α_o that will maximize $P_C^{(\text{out})}(L)$. As an illustrative example of the effect of L on the optimum value of α_o and the corresponding theoretical-maximum power output, let us consider the results of a numerical solution to

$$\frac{\partial P_C^{(\text{out})}(L)}{\partial \alpha_o} = 0 . \qquad (8.20)$$

For this example, the parameters representative of the long-wavelength strained-layer InGaAsP/InGaAs materials system [1.333, 8.62, 8.63] that are listed in Table 8.2 were used. In addition, the spontaneous emission factor for the amplifier waveguide was calculated from the expression [2.29]

$$\gamma = \frac{\Gamma_y \lambda^2}{4\pi n_e^2 wd} . \qquad (8.21)$$

Note that (8.21) is an approximation that should be applicable to index-guided waveguide laser structures, such as those found in active-grating surface emitting lasers. In cases where one is dealing with gain-guided waveguide

Table 8.2. Parameters indicative of the strained-layer InGaAsP/InGaAs materials system for modeling active-grating surface-emitting amplifiers.

Number of quantum wells	5
Transverse mode confinement factor (Γ_y)	0.016
Lateral mode dimension (w)	$3\,\mu m$
Differential gain (σ)	$5 \times 10^{-16}\,cm^2$
Carrier recombination time (τ)	$0.25 \times 10^{-9}\,s$
Internal quantum efficiency (η)	0.7
Quantum-well thickness (d)	$80\,\text{Å}$
Transparency carrier level per quantum well (N_{tr})	$2 \times 10^{17}\,cm^{-3}$
Wavelength	$1.55\,\mu m$
Optical loss coefficient (α)	$10\,cm^{-1}$

laser amplifiers, then (8.21) is enhanced by the K factor which was described in Sect. 3.3.2. A more exact expression for γ than (8.21) would require a precise accounting of all radiation and guided modes of the waveguide structure [2.29] that are used in calculating γ (Sect. 3.3.2).

Figure 8.28 presents a comparison of the amplified-coherent and amplified spontaneous emission powers that are calculated by the numerical optimization of α_o using (8.20) (illustrated in Fig. 8.28a), and for a fixed value of $\alpha_o = 15\,cm^{-1}$ (displayed in Fig. 8.28b), which is perhaps an upper estimate of α_o for the buried-grating output couplers [1.243] that have been fabricated in InGaAsP/InGaAs active-grating surface-emitting amplifiers [6.25, 8.56]. Since we are interested in projecting the maximum-theoretical power output, the efficiency factor for grating emission into air is set to $\eta_{air} = 1$. The optimum values of α_o versus L that correspond to Fig. 8.28 appear in Fig. 8.29. A close inspection of Figs. 8.28a and 8.29 reveals that the optimized theoretical-maximum power output of 1.09 W is obtained at an amplifier length of 1.22 cm for $\alpha_o = 5.5\,cm^{-1}$. A similar inspection of Fig. 8.28b indicates for a fixed $\alpha_o = 15\,cm^{-1}$, that a maximum power of 0.733 W is provided from an amplifier of length 0.51 cm.

There is a significant difference in the noise powers $P_N^{out}(L)$ versus L for the case of α_o being optimized at each L and $\alpha_o = 15\,cm^{-1}$ for each L. We see from Fig. 8.28a and Fig. 8.29 that with increasing L, the optimum value of α_o continually decreases. When α_o is fixed, it is evident from (8.18) that the term $P_N^{out}(L)$ has an L^2 dependence, whereas when α_o is decreased as the length is increased, it reduces the level of spontaneous emission that is generated, which diminishes the rate of increase of $P_N^{out}(L)$ with L. In other words, the reduction in α_o with increasing L translates to an increase in P_{Lim} and a decrease in the noise power per unit length coupled into the amplifier waveguide $\partial P_N/\partial L$, defined by (8.16). The effect of decreasing $\partial P_N/\partial L$ with L is to, in effect, impart a sublinear growth rate to $P_N(L)$ with increasing L. The combination of increasing P_{Lim} and reducing the growth rate of $P_N(L)$ with increasing L raises the level of amplified coherent power in the wave-

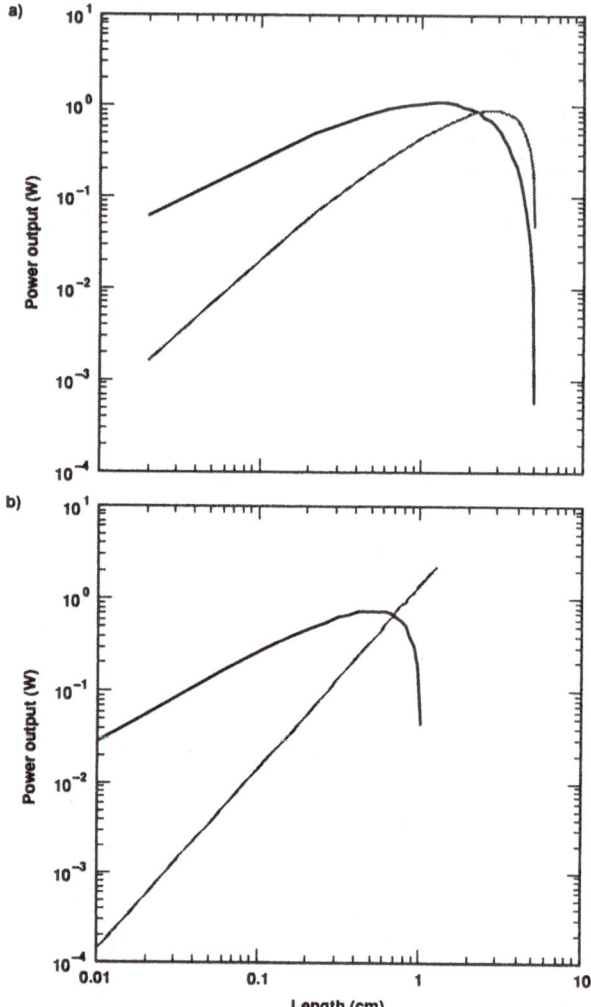

Fig. 8.28. Log-log plot of the calculated values of the output powers $P_C^{(\text{out})}(L)$ and $P_N^{(\text{out})}(L)$ (shaded line) as a function of length L for the case where α_o is optimized by the numerical solution of (8.20) for each value of L (**a**) and the case where $\alpha_o = 15\,\text{cm}^{-1}$ is fixed for all lengths (**b**).

guide (8.12) for grating-output coupling. The strength of the grating-output coupling is measured by the product $\alpha_o L$, so that lowering α_o with L does not necessarily mean that the total integrated strength of the grating output coupler is diminished. This is illustrated in Fig. 8.29 where both the optimized α_o and $\alpha_o L$ are plotted versus L. It is evident that the maximum integrated strength of the grating-output coupler corresponds closely to the theoretical maximum coherent power output of the active-grating amplifier exhibited in Fig. 8.28a. Note the difference between the scaling analysis exhibited in

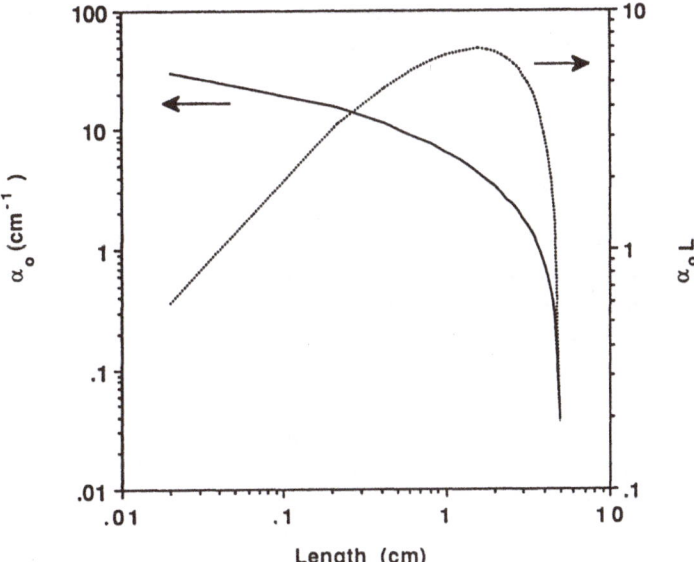

Fig. 8.29. Log-log plot of the optimum values of α_o and $\alpha_o L$ versus L as calculated by a numerical solution of (8.20), corresponding to the power output curves in Fig. 8.28.

Figs. 8.28 and 8.29 and that presented in [8.58] is that [8.58] assumed a fixed $\alpha_o = 20\,\mathrm{cm}^{-1}$ for spontaneous emission factors $\gamma = 0.001$ and 0.01. For the scaling analysis exhibited in Figs. 8.28 and 8.29, we find from (8.21) (using the parameters in Table 8.2) that $\gamma = 0.007$ for the optimized scaling analysis.

As seen from (8.12-8.14) and the numerical scaling analysis for optimizied grating-output coupling, presented in Fig. 8.28, amplified spontaneous emission noise plays a critical role in limiting the performance of active-grating surface-emitting MOPAs. Even when the coherent power output is optimized (Fig. 8.28a) versus length, the amplified spontaneous emission power output is only about 40 % less than the highest coherent power output, which occurs at $L = 1.22\,\mathrm{cm}$ in Fig. 8.28a. However, the dispersive properties of the grating-output coupler, illustrated in Fig. 8.26 and described on page 320, act as a filter for removing the amplified spontaneous emission from the solid-angle subtended by the coherent amplified light.

The figure-of-merit for quantifying the noise filtering properties of the grating-output coupler is the *ratio of amplified coherent light power to amplified spontaneous noise power within the solid angle* $\Delta\Omega_C$, where $\Delta\Omega_C$ is the solid-angle subtended by the coherent amplified light. As already mentioned on page 322, the amplified spontaneous emission noise is emitted into an angle with a longitudinal beam divergence that is more than two orders of magnitude larger than the beam divergence of the amplified coherent light. Therefore, even when the total amplified spontaneous emission power is comparable to the amplified coherent power, the amplified spontaneous emission power per unit bandwidth is much less so very little noise is emitted into the far field of the coherent light. Experimental examples of the noise filtering properties are presented in the next section.

8.2.3 Active Grating-Coupled MOPAs: Performance Characteristics

Experimental studies of active grating-coupled MOPAs [1.242, 1.243, 1.245, 1.490, 4.82, 6.25, 8.57, 8.56, 8.64] have been to a large extent consistent with the performance projections presented in Sect. 8.2.2 and [1.241, 1.244, 8.54]. The performance characteristics in terms of maximum power output and efficiency have been less than expected, and this is perhaps due in a large part to the difficulty in fabricating a high-efficiency grating-output coupler. Under pulsed operating conditions, an output power of 1.28 W was reported with a differential quantum efficiency of 35 % from a 5 mm long strained-layer InGaAs/GaAs index-guided active-grating amplifier that operated at a wavelength of 0.96 μm [1.490], however, the details on the design of this structure have not been reported in the technical literature. In strained-layer multi-quantum-well InGaAsP/InGaAs active-grating surface emitting lasers, a differential-quantum efficiency of 24 % has been reported [1.245], which compares favorably with the quantum efficiencies per facet reported for other edge-emitting strained-layer multi-quantum-well InGaAsP/InGaAs lasers with uncoated facets [1.333]. In the active-grating amplifier of [1.245], no attempt was made to recover the light emitted into the substrate, so there is clearly reason to think that differential-quantum efficiencies of $\approx$ 50 % may be possible.

The dispersive effects of the grating-output coupler have been studied experimentally by *Carlson* et al. [1.243, 4.82, 8.56, 8.57][8.64]-[8.67] and found to be consistent with theoretical predictions of [1.244, 8.54]. Figure 8.30 presents the measured far-field and spectral output characteristics of an InGaAs/GaAs active-grating surface-emitting MOPA. In Fig. 8.30a with the oscillator off, the free-running amplifier exhibits an approximately-symmetric double-lobed pattern, with each lobe being emitted at an $\approx$ 15 ° angle with respect to the surface normal. Each lobe has a width of $\approx$ 12°. Figure 8.30 also presents the far-field that was observed when the DFB-laser oscillator injected light into the amplifier. Note that only a single narrow lobe is observed at an angle of $\approx$ −14°, which was expected for the $\approx$ 3050 Å period that was used for the output-coupling grating. A high-resolution measurement of the narrow lobe revealed a longitudinal beam divergence of 0.01°. The far-field beam divergence was measured to be about 20°, which was characteristic of the 5 μm wide ridge waveguide structure [8.57].

The amplifier spectral output corresponding to the DFB-laser oscillator being on in Fig. 8.30b, contains a narrow double-peak structure corresponding to the spectrum of the DFB-laser oscillator. The reason for the double peak is that in this case the oscillator was operated in two modes [8.57].

Nevertheless, a faithful reproduction of the oscillator spectral characteristics was obtained along with a 40 dB ratio of coherent amplified light power to amplified spontaneous emission power. This large ratio of coherent power to noise power demonstrates that the injected coherent light from the DFB-laser oscillator has saturated the amplifier gain so that the amplified spontaneous emission is greatly suppressed relative to the free-running case.

Let us turn our attention back to the case when the amplifier is free-running. The amplified spontaneous distribution has built up symmetrically in each traveling-wave direction, resulting in the characteristic double-lobed far-field pattern observed in Fig. 8.30a. Note that the results of the model calculation of the power distributions P_C, P_N^+, and P_N^- for a free-running amplifier that were presented in Fig. 8.27a are consistent with the experimentally observed far fields in Fig. 8.30a. From the observed spectra displayed in Fig. 8.27b, we see that the amplified spontaneous emission spans a spectral bandwidth of about 500 Å. As already mentioned, the grating dispersion is expected to be $\approx 5 \times 10^{-4}$ radians/Å [8.59], which corresponds to an $\approx 15°$ beam divergence for the 500 Å spectral bandwidth of the amplified spontaneous emission. Here, we find a reasonable agreement between the model predictions of the amplified spontaneous emission beam divergence as calculated by (8.11) and the experimental observations in Fig. 8.30a.

Under cw operating conditions, the active-grating surface-emitting MOPAs reported in [8.57, 8.64] were found to operate single mode up to

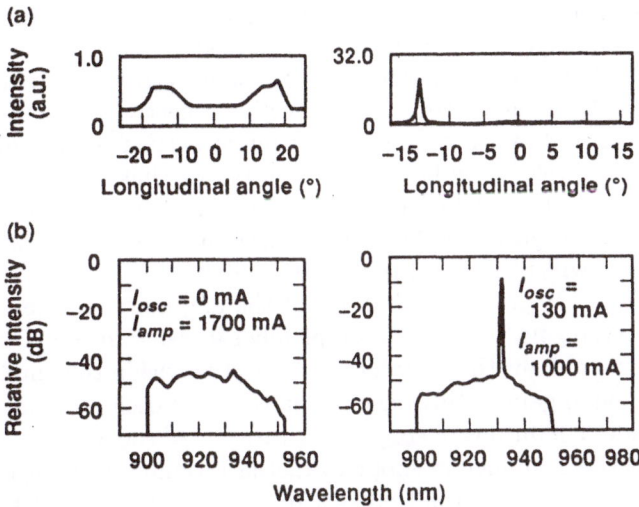

Fig. 8.30. The measured longitudinal far-field patterns are presented for a free-running active-grating amplifier (left-hand side) and the same active-grating amplifier when injected with light from the integrated DFB-laser oscillator (right-hand side) (**a**). Spectra measured for the respective far-field patterns are displayed in (**b**). Note that the abrupt drop in the spectral intensity at wavelengths < 900 nm is due to the absorption edge of the GaAs substrate [1.194, 8.57, 8.64].

power levels of 300 mW. A cylindrical lens to collimate the fast divergence of the output beam in the lateral direction [8.68] to obtain a nearly circularized spot with a divergence of 0.010°, with 100 mW being contained in the circularized spot. These power output levels were lower than expected. The problem with obtaining high coherent power outputs seemed to be strongly related to the poor degree of isolation provided between the DFB-laser oscillator and the active-grating power amplifier in a monolithic structure. Free-running active-grating surface-emitting MOPAs (1.3 cm in length) of the type studied in [1.243, 4.82, 8.56, 8.57][8.64]-[8.67] were observed to produce amplified spontaneous emission power outputs of ≈ 1 W, indicating the high-power capacity of these devices. However, as the current to the active amplifier was increased to give power levels much above several hundred mW, the DFB-laser oscillator was typically observed to become unstable and the amplifier would eventually self oscillate. The reasons for this behavior are described in the next paragraph.

As we already saw in Fig. 8.27, the level of amplified spontaneous emission that is injected into the laser oscillator by the amplifier can be high enough to quench oscillation. Another factor that can contribute to destabilizing the laser oscillator are back reflections in the amplifier that might occur because of a residual reflectivity at the amplifier termination or some inhomogeneity within the amplifier waveguide structure. The effect of the input and termination reflections on completely saturated active-grating amplifiers has been analyzed in [1.244]. When $P_{in} \geq P_{Lim}$, where P_{in} is the input power to the amplifier, the gain in the active-grating amplifier is clamped at $\alpha + \alpha_o$ along the entire length. The gain required for self-oscillation of the amplifier is $\alpha + \alpha_o - (2/L) \ln (R_{in} R_t)$, where R_{in} is the effective reflectivity the amplifier sees at the input and R_t is the effective reflectivity due the amplifier termination, as well as any feedback that might occur from the off-resonant grating-output coupled amplifier waveguide. We see that the losses that must be overcome for self-oscillation of the amplifier are larger than the losses for coherent amplification in the loss-limited saturated regime. Therefore, having $P_{in} \geq P_{Lim}$ will help to suppress self-oscillation of the amplifier. In the event that $P_{in} < P_{Lim}$, self-oscillations of the amplifier may be able to develop because the gain will not be completely saturated in the vicinity of the amplifier input, so light that is internally reflected within the amplifier could experience enough gain in this region to initiate self oscillation of the amplifier. Even under conditions of complete saturation of the amplifier, when self oscillation of the amplifier is suppressed, the termination reflectivity must satisfy $R_t \leq 10^{-3}$ so that the feedback to the DFB-laser oscillator will not destabilize the single-mode operation [1.244, 8.69, 8.70]. When the amplifier is completely saturated by the injected coherent signal, back reflected light will not experience any gain. This is yet another reason why it is important to have $P_{in} \geq P_{Lim}$. If $P_{in} < P_{Lim}$, back reflected light will experience gain in the region near the input, where the gain is not completely saturated. In this case, the termination reflectivity would have to be much much lower than

10^{-3} to keep the back reflected light below the level required to maintain stable single-mode operation of the DFB-laser oscillator [1.244, 8.69, 8.70].

The point of this digression is to emphasize the importance in maintaining complete saturation along the entire length of the active-grating amplifier for all levels of current injection. As the current level to the amplifier is increased, then the oscillator power needs to be increased to satisfy $P_{in} \geq P_{Lim}$ in order to avoid destabilization of the oscillator and self-oscillation of the amplifier. This was not possible in the device design used by [8.57], as these active-grating MOPAs did not contain the preamplifier section, depicted in Fig. 8.25, that is required to boost the input power to the amplifier up to the levels where the gain can be saturated at the input. Note the active-grating MOPAs studied in [1.490] contained a preamplifier section, and could be operated to higher-current levels (with higher-power output) than the active-grating MOPAs of [8.57], which did not contain a preamplifier section.

The active-grating MOPAs in [1.490] were found to exhibit longitudinal beam steering when operated at peak-power outputs $> 800\,\mathrm{mW}$ under pulsed conditions. Beam steering and beam stability of active-grating surface-emitting MOPAs has not yet been analyzed experimentally. However, the mechanisms that give rise to beam steering can be elucidated be considering the emission angle of the grating-coupled light, which is given by (8.3). Inspection of (8.3), reveals that the dominant sources for longitudinal beam steering are changes in λ the operating wavelength of the DFB-laser oscillator and temperature changes that cause changes in n_e. To avoid changes in the oscillator wavelength λ, stable single-mode operation must be maintained; and as already mentioned above, loss-limited saturation in the amplifier is critical factor in maintaining stability of the oscillator at high powers.

The effects of temperature are yet another issue. As discussed in Sect. 4.3.3, the temperature dependence of the refractive index of semiconductors is $dn_e/dT \approx 4 \times 10^{-4}\,\mathrm{K}^{-1}$. From (8.3) for $\theta_C \ll 1$ radian, we find that the beam steers with temperature changes ΔT according to $\Delta\theta_a = (dn/dT)\,\Delta T$. In other words, a $1\,\mathrm{K}$ rise in temperature in an active-grating amplifier waveguide will steer the output beam by $0.023°$. We see from the first term of (8.11) that the beam divergence per unit length of a narrow bandwidth active-grating amplifier is $\approx 0.01°\,\mathrm{cm}/L$, where L is the amplifier length in centimeters. In other words, a $1\,\mathrm{K}$ rise in temperature in a $1\,\mathrm{cm}$ long active-grating MOPA will steer the output beam through an angle that is more than twice as great as the output beam divergence. Clearly, if one is interested in obtaining $\geq 1\,\mathrm{W}$ power outputs from active-grating MOPAs, then the issue of thermal-induced beam steering will have to be addressed.

The primary sources of heat dissipation under loss-limited saturated operation are expected to be Joule heating and absorption of spontaneous emission (discussed in Sect. 2.8.1), with the added contribution of the absorption of grating-output coupled light in the structure epilayers. The reason for this is that under completely saturated operation the carrier density is pinned at $N(z) = N_S$ given by (8.15), which is independent of the injected current density. Therefore, the heat dissipated in the active layer is independent of the current. Note that N_S depends on σ and N_{tr}, which are both temperature dependent, so heating of the active layer will occur by conduction through the adjacent cladding layers, due to the Joule heating and optical absorption that occurs throughout the wafer structure. In amplifiers that are not completely saturated $N(z)$ is not constant along the amplifier length, and one must contend with longitudinal temperature gradients. The degree to which thermal beam steering can be controlled (if at all) in active-grating surface-emitting MOPAs has not yet been established.

Let us turn our attention now to the question of the spectral linewidth of the amplified coherent output light from an active-grating surface-emitting amplifier. This subject has been studied experimentally by *Liew* et al. [4.82]. Interestingly, it was discovered by *Liew* et al. [4.82] that, depending on operating conditions, different values of the spectral linewidth were measured in the near- and far-field zones of active-grating surface-emitting MOPAs (of the type in [1.243, 8.57, 8.64]), with a narrower linewidth being observed in the far field. The cause of this linewidth narrowing was attributed to the dispersive properties of the grating-output coupler that effect spatial filtering in the far-field zone, which has been described on page 320. To explain the linewidth narrowing, we again refer to Fig. 8.26. The angular spreads of the directly-radiated spontaneous emission (Fig. 8.26a) and grating-coupled amplified spontaneous emission (Fig. 8.26b) are orders of magnitude larger than the narrow-beam divergence of the grating-coupled coherent output. As a result, the directly-radiated spontaneous emission power and amplified spontaneous emission power contained within the coherent output beam profile (or footprint) will decrease on propagation from the near-field zone to the far-field zone, leading to a spectral linewidth narrowing in the far-field zone.

The difference in the emission characteristics of the directly-radiated spontaneous emission and the grating-coupled amplified coherent light influences the spectral-linewidth measurements of active-grating surface-emitting MOPAs. This is a general property of these devices. As illustrated in Fig. 8.26a, only directly-radiated spontaneous emission within an $\approx 18°$ cone with respect to the surface normal is emitted into air. Note that the air/substrate interface was $\approx 100\,\mu m$ above the active-layer, so the directly-radiated spontaneous emission was spread over an $\approx 50\,\mu m$ lateral dimension in the near field at the wafer surface [4.82]. In contrast, the grating-coupled coherent light spanned an $\approx 4\,\mu m$ lateral dimension in the near field at the wafer surface. Now consider a detector of fixed size, such as the $\approx 5\,\mu m$ core diameter single-mode fiber used in the experiments of *Liew* et al. [4.82]. As this detector is moved away from the surface of the wafer (along the emission direction of the coherent light), and into the far-field zone more amplified coherent power, less spontaneous emission will be incident on the surface of the detector. In this sense, propagation or diffraction effects the spectral linewidth measurement. However, as seen from (4.80) and the discussion of Parseval's Theorem on page 168, if a sufficiently large aperture is used in the far-field measurement then one should observe the same power spectra in the near- and far-field zones of the active-grating amplifier. For most applications purposes it is only the grating-coupled amplified coherent light that is useful, so that there would no need (and certainly no benefit in terms of the spectral linewidth) to consider apertures larger than the coherent beam divergence.

Figure 8.31 displays the spectral linewidth measurement data of *Liew* et al. [4.82]. Higher current to the oscillator corresponded to a higher input power being injected into the amplifier. The measured amplifier linewidths in the near field (Fig. 8.31a) matched that of the DFB oscillator only at the highest oscillator powers, where the spontaneous emission in the amplifier is most strongly suppressed. At higher amplifier bias currents, more input power is required to suppress the spontaneous emission. On the other hand, as the

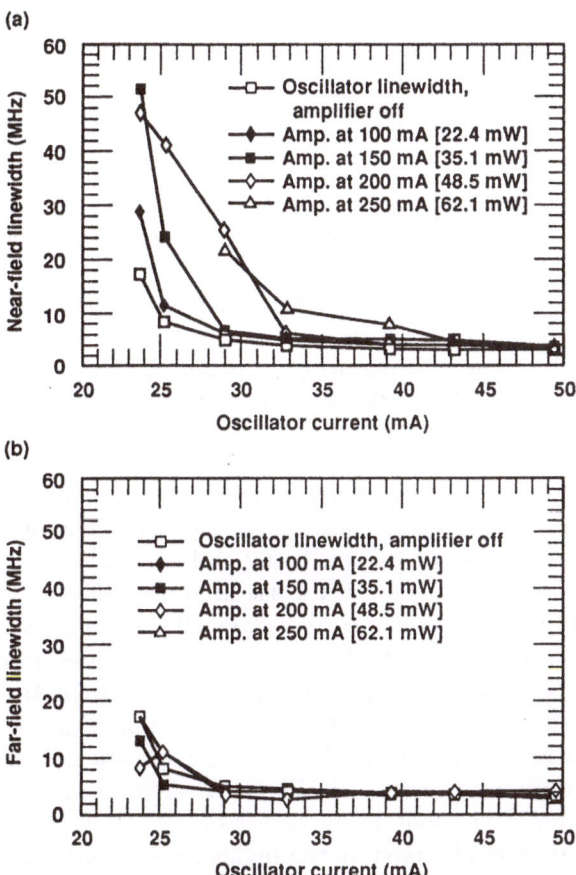

Fig. 8.31. Near-field measurement of active-grating amplifier spectral linewidth versus oscillator current (**a**). Far-field measurement of active-grating amplifier spectral linewidth versus oscillator current (**b**). The spectral linewidth of the DFB-laser oscillator alone (with the amplifier off) is designated in each plot [4.82].

input powers to the amplifier were decreased, the spectral linewidth of the amplifier was observed to be significantly larger as compared to the spectral linewidth of the DFB-laser oscillator when the amplifier was off. This occured because at lower oscillator powers the amplifier gain was not as well saturated, and therefore, higher levels of amplified spontaneous emission power were injected into the oscillator, as was illustrated in Fig. 8.27. For low input power from the oscillator, *Liew* et al. [4.82] observed the following: 1) a broadening of the oscillator spectral linewidth (relative to that when the amplifier was off), 2) a lower ratio of amplified coherent power output to amplified spontaneous emission power output, and 3) in some cases multimode operation of the DFB-laser oscillator.

The far-field linewidth measurements, which are displayed in Fig. 8.31b, exhibit qualitatively the same trends as the near-field measurements in Fig. 8.31b. The noteable exception being that largest linewidths observed in the far field were about a factor of ten narrower than the linewidth observed in the near field for the same operating conditions. The observed narrowing between of the far-field linewidth and the near-field linewidth was most prevalent at the lowest oscillator powers, where the spontaneous emission power was highest. As the oscillator current was increased in Fig. 8.31b, the observed far-field linewidth rapidly approached that of the DFB-laser oscillator with the amplifier off. Spatially resolved measurements of the near field linewidths revealed that the linewidth was constant along the 7 mm length active-grating amplifier for all single-frequency operating conditions. Futhermore, at the highest oscillator power, a maximum power output of 200 mW was measured from the active-grating amplifier in [4.82] with the near-field linewidth measured to be 3 MHz and the corresponding far-field linewidth was observed to be 2.1 MHz ± 0.3MHz. These measurements indicated that spontaneous emission was greatly suppressed.

The active-grating surface-emitting MOPA has also been demonstrated by *Carlson* et al. in the strained-layer InGaAsP/InGaAs materials system at operating wavelengths of 1.6 μm [1.245] and 1.7 μm [8.56]. A principle interest in wavelenghts > 1.54 μm is for use in applications and environments where eye safety is an important consideration. The designs that were used in [1.245, 8.56] contained a preamplifier section, as illustrated in Fig. 8.25. The preamplifier is of particular importance in InGaAsP/InGaAs active-grating amplifiers, because as we saw in the previous section $P_{Lim} \approx 100$ mW. The measured performance characteristics under cw operation of the devices that operated at 1.7 μm are presented in Fig. 8.32. At maximum ratio of coherent power to amplified spontaneous power or *Side-Mode-Suppression-Ratio* (SMSR) of 38 dB was observed at the highest power outputs, and at lower output powers, a ratio of 30 dB was observed. This observed behavior of increasing SMSR with increasing current to the amplifier is expected when the amplifier is completely saturated and the carrier density is pinned. The same behavior is observed in all lasers operated above threshold. We see then that one can think of an active-grating amplifier as a single-pass laser [1.244].

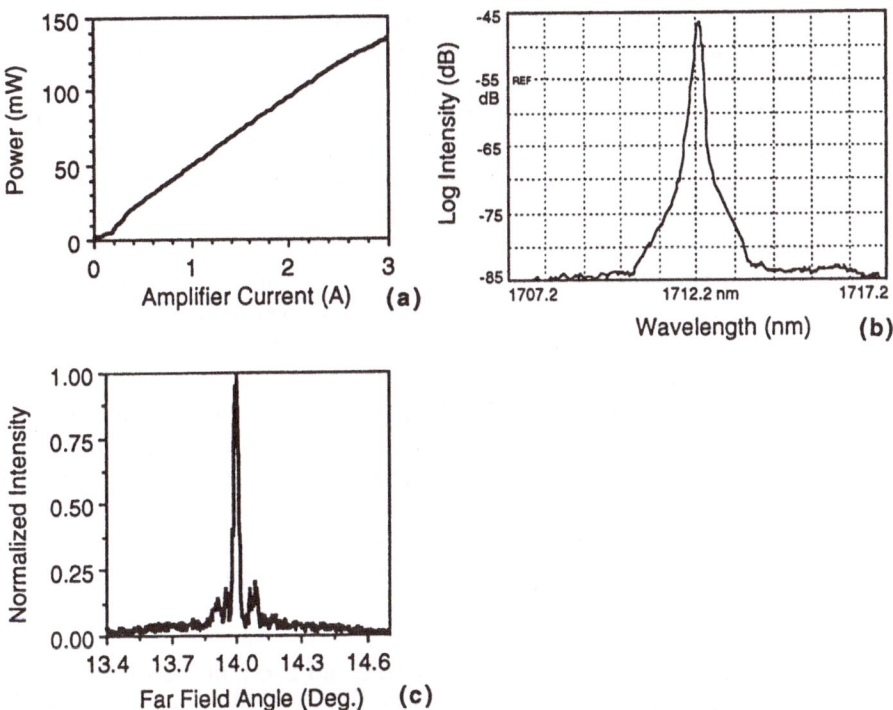

Fig. 8.32. The power-current characteristic (a), spectral output (b), and far-field pattern (c) for the strained-layer multi-quantum-well InGaAsP/InGaAs active grating surface-emitting amplifiers of [8.56].

In all respects, except for perhaps the projected maximum power output, the active-grating surface-emtting amplifiers have exhibited behavior consistent with the operating principles detailed in Sect. 8.2.2. The observed limitations in power output were due likely due to nonoptimized grating-output couplers, as well as failure to maintain complete saturation in the power amplifiers. Another area of concern is material damage due to fabrication of the grating output coupler and subsequent regrowth over the corrugrated wafer surface. For example, defect generation associated with the interface of an InGaAsP relief grating and an MOVPE grown InP layer has been observed for grating depths ≥ 40 nm [8.71]. As discussed by *Sato* et al. [8.71] these defects, which appeared as bundles of straight line defects, were observed to degrade the performance of DFB lasers. The issue of material damage due to the fabrication of the buried grating in active-grating MOPAs, and how such damage impacts performance, has not been investigated in active-grating MOPAs.

In order for active-grating surface-emitting amplifiers to meet theoretical predictions, a high-efficiency grating-output coupler is required. Of the devices demonstrated to date (with the possible exception of [1.242] which used a metalized relief grating as opposed to a buried-semiconductor grating), a sizeable fraction of the grating-output coupled light was radiated into the superstrate and not recovered. As was the case with the GSE-laser arrays of Chap. 7 and the cascaded GSE-MOPAs of Sect. 8.2.1, a high-efficiency grating output coupler is also required for the active-grating MOPAs. The issue of maintaining loss-limited saturated operation, over all ranges of currents to the amplifier, and minimizing the level of amplified spontaneous emission injected into the oscillator is indeed a challenging problem. The ideal solution to this would be an integrable waveguide optical isolator, since it could presumably handle reflections due to the structural and compositional inhomogeneities which would likely be present in the long-cavity of the active-grating amplifier. However, at present there do not appear to be any near-term solutions for an integrable waveguide optical isolator for semiconductor material. This may well prove to be an interesting area of study for the future. Having an integrated waveguide optical isolator would not only be very desirable for all of the monolithic MOPAs presented in this chapter, but it would also be of great utility for the whole field of integrated optics.

Given that all of the above issues can be resolved, one is still faced with the prospect of thermal-induced beam steering, as discussed on page 333. It is possible that the length of active-grating surface-emitting amplifiers may need to be limited to a few mm in order to avoid thermal-induced instabilities in the output beam. As seen from Fig. 8.28, at this range of amplifier lengths, the maximum power output could be limited to be at most $\approx 1\,\mathrm{W}$. A solution to this problem, proposed and analyzed for InGaAsP/InGaAs structures by [6.25], is to replace the single-mode lateral waveguide structure with an array of near-resonant leaky-wave-coupled array of antiguided lasers (Sect. 6.3). Subsequent analysis [8.72] of single-pass MOPAs based on resonant antiguided array structures has predicted output beams (from edge-emitting structures) with plane wavefronts. Note a MOPA antiguided array, as with the antiguide laser-array oscillators of Sect. 6.3, would also have a large lateral index differential ($\Delta n \geq 0.02$). Therefore, the MOPA antiguided laser array, being a strong index-guided structure, might be expected to be more robust than the gain-guided MOPA approaches (Sects. 8.1.2-8.1.4) to destabilization of the output beam (Sect. 4.3.5) due to carrier- and thermal-induced changes in the lateral-index profile.

References

Chapter 1

1.1 R. N. Hall, G. E. Fenner, J. D. Kinglsey, T. J. Soltys, R. O. Carlson: Coherent light emission from GaAs junctions. Phys. Rev. Lett. **9**, 366–378 (1962)

1.2 M. I. Nathan, W. P. Dumke, G. Burns, Jr., G. Lasher F. H. Dill: Stimulated emission of radiation from GaAs p-n junctions. Appl. Phys. Lett. **1**, 62–64 (1962)

1.3 N. Holonyak, S. F. Bevacqua: Coherent (visible) light emission from $Ga(As_{1-x}P_x)$ junctions. Appl. Phys. Lett. **1**, 82–83 (1962)

1.4 T. M. Quist, R. H. Rediker, R. J. Keyes, W. E. Krag, B. Lax, A. L. McWorter, H. J. Ziegler: Semiconductor maser of GaAs. Appl. Phys. Lett. **1**, 91–92 (1962)

1.5 N. G. Basov, E. M. Belenov, V. S. Letokhov: Diffraction synchronization of lasers. Sov. Phys.–Tech. Phys. **10**, 845–805 (1965)

1.6 N. G. Basov, E. M. Belenov, V. S. Letokov: Synchronization of oscillations in a semiconducting laser having several $p - n$ junctions. Sov. Phys.–Solid State **7**, 275–276 (1965)

1.7 V. S. Letokhov: Diffraction losses of an open resonator formed by mirrors with absorbing strips. Sov. Phys.–Tech. Phys. **10**, 629–632 (1965)

1.8 E. M. Belenov, V. S. Letokov: Theory of coupled optical masers. Sov. Phys.–Tech. Phys.. **10**, 1628–1629 (1966)

1.9 R. Adler: A study of locking phenomena in oscillators. Proc. IRE **34**, 351–357 (1946)

1.10 N. Minorsky: *Introduction to Nonlinear Mechanics* (Edwards, Ann Arbor, MI 1947)

1.11 A. B. Fowler: Quenching of gallium-arsenide injection lasers. Appl. Phys. Lett. **3**, 1–3 (1963)

1.12 A. B. Fowler: Cooperative effect in GaAs lasers. J. Appl. Phys. **35**, 2275–2276 (1964)

1.13 C. E. Kelly: Interaction between closely coupled GaAs injection lasers. IEEE Trans. ED-**12**, 1–4 (1965)

1.14 R. H. Pantell: The laser oscillator with an external signal. Proc. IEEE **53**, 474–477 (1965)

1.15 R. Vuilleumier, N. E. Collins, J. M. Smith, J. C. S. Kim, H. Raillard: Coherent amplification characteristics of a GaAs phased array. Proc. IEEE **55**, 1420–1425 (1967)

1.16 J. W. Crowe, W. E. Ahearn: External cavity coupling and phase locking of gallium arsenide injection lasers. IEEE J. QE-**4**, 169–172 (1968)

1.17 E. M. Philipp-Rutz, H. D. Edmonds: Diffraction-limited GaAs laser with external resonator. Appl. Opt. **8**, 1859–1865 (1969)

1.18 R. M. Craig, J. W. Crowe: High average-power gallium arsenide illuminators. IEEE J. QE-**6**, 373–376 (1970)

1.19 W. E. Ahearn, R. M. Craig: High peak-power room-temperature GaAs laser arrays. IEEE J. QE-**6**, 377–382 (1970)

1.20 D. G. Herzog, H. Kressel: Thermoelectrically cooled GaAlAs laser illuminator. Appl. Opt. **9**, 2249–2255 (1970)

1.21 C. J. Nuese, M. Ettenberg, R. E. Enstrom, H. Kressel: Efficient laser diodes of $In_xGa_{1-x}As$ for 1.06 μm emission. In International Electron Devices Meeting 1973. Technical Digest, pp. 320–323

1.22 E. M. Philipp-Rutz: Spatially coherent radiation from an array of GaAs lasers. Appl. Phys. Lett. **26**, 475–477 (1975)

1.23 E. M. Philipp-Rutz: Single laser beam of spatial coherence from an array of GaAs lasers: free-running mode. J. Appl. Phys. **46**, 4552–4556 (1975)

1.24 W. F. Kosonocky, R. H. Cornely, I. J. Hegyi: Multilayer GaAs injection laser. IEEE J. QE-**4**, 176–179 (1968)

1.25 H. C. Casey, M. B. Panish: *Heterojunction Lasers Part A Fundamental Principles*: (Academic, New York 1978)

1.26 G. H. B. Thompson: *Physics of Semiconductor Laser Devices*: (Wiley, Chichester 1980)

1.27 J. I. Pankove: Integrated phase coherent laser array. RCA Technical Notes **21**, TN760-2 (1968)

1.28 D. Botez, L. Figueroa, S. Wang: Optically pumped GaAs − $Ga_{1-x}Al_xAs$ half-ring laser fabricated by liquid-phase epitaxy over chemically etched channels. Appl. Phys. Lett. **29**, 502–504 (1976)

1.29 T. Krauss, P. J. R. Laybourn: Very low theshold current operation of semiconductor ring lasers. IEE Proc. J **139**, 383–388 (1992)

1.30 N. Bar-Chaim, K. Y. Lau, M. A. Mazed, M. Mittelstein, S. Oh, J. E. Ungar, I. Ury: Half-ring geometry quantum well GaAlAs lasers, Appl. Phys. Lett. **57**, 966–967 (1992)

1.31 G. Burns, M. I. Nathan: P-N junction lasers. Proc. IEEE **52**, 770–794 (1964)

1.32 G.E. Fenner, J. D. Kingsley: Spatial distribution of radiation from GaAs lasers. J. Appl. Phys. **34**, 3204–3208 (1963)

1.33 J. C. Dyment: Hermite-gaussian mode patterns in GaAs junction lasers. Appl. Phys. Lett. **10**, 84–86 (1967)

1.34 J. E. Ripper, T. L. Paoli: Optical coupling of adjacent stripe-geometry junction lasers. Appl. Phys. Lett. **17**, 371–373 (1970)

1.35 V. N. Luk'yanov, N. V. Shelkov, S. D. Yakubovich: Semiconductor laser with a distributed feedback. Sov. J. Quant. Electron. **5**, 99–100 (1975)

1.36 S. Wang, S. Sheem: Two-dimensional distributed-feedback lasers and their applications. Appl. Phys. Lett. **22**, 460–462 (1973)

1.37 L. I. Luk'yanova, V. N. Luk'yanov, N. V. Shelkov, S. D. Yakubovich: Thin-film laser with a two-dimensional diffraction grating. Sov. J. Quant. Electron. **9**, 496–498 (1979)

1.38 V. N. Luk'yanov, N. V. Shelkov, S. D. Yakubovich: Investigation of multi-beam semiconductor laser with an emitting array. Sov. J. Quant. Electron. **4**, 98–99 (1974)

1.39 K. Otsuka: A proposal on coupled waveguide lasers. IEEE J. QE-**17**, 895–898 (1977)

1.40 D. R. Scifres, R. D. Burnham, W. Streifer: Phase-locked semiconductor laser array. Appl. Phys. Lett. **33**, 1015–1017 (1978)

1.41 W. T. Tsang, R. A. Logan, R. P. Salathe: A densely packed monolithic linear array of GaAs-AlGaAs strip buried heterostructure laser. Appl. Phys. Lett. **34**, 162–165 (1979)

1.42 D. R. Scifres, W. Streifer, R. D. Burnham: Beam scanning and wavelength and modulation with branching waveguide stripe injection lasers. Appl. Phys. Lett. **33**, 616–618 (1978)

1.43 D. R. Scifres, W. Streifer, R. D. Burnham: High-power coupled-multiple-stripe phase-locked injection laser. Appl. Phys. Lett. **34**, 259–261 (1979)

1.44 D. R. Scifres, W. Streifer, R. D. Burnham: Beam scanning with twin-stripe injection lasers. Appl. Phys. Lett. **33**, 702–704 (1978)

1.45 D.R. Scifres, R.D. Burnham, W. Streifer: High power coupled multiple stripe quantum well injection lasers. Appl. Phys. Lett. **41**, 118–120 (1982)

1.46 D.R. Scifres, R.D. Burnham, W. Streifer, M. Bernstein: Lateral beam collimation of a phased array semiconductor laser. Appl. Phys. Lett. **41**, 614–616 (1982)

1.47 D.R. Scifres, R.D. Burnham, W. Streifer: Continuous wave high-power, high-temperature semiconductor laser phase-locked arrays. Appl. Phys. Lett. **41**, 1030–1032 (1982)

1.48 D.R. Scifres, R. A. Sprague, W. Streifer, R.D. Burnham: Focusing of a 7700 Å high power phased array semiconductor laser. Appl. Phys. Lett. **41**, 1121–1123 (1982)

1.49 D.R. Scifres, C. Lindstrom, R.D. Burnham, W. Streifer, T. L. Paoli: Phase-locked (GaAl) As laser diode emitting 2.6 W CW from a single mirror. Electron. Lett. **19**, 169–171 (1983)

1.50 H. Temkin, R. D. Dupuis, R. A. Logan, J. P. van der Ziel: Schottky barrier restricted arrays of phase-coupled AlGaAs quantum well lasers. Appl. Phys. Lett. **44**, 473–475 (1984)

1.51 T. L. Paoli, W. Streifer, R. D. Burnham: Observations of supermodes in a phase-locked diode laser array. Appl. Phys. Lett. **45**, 217–219 (1984)

1.52 J. P. van der Ziel, R. M. Mikulyak, H. Temkin, R. A. Logan, R. D. Dupuis: Optical beam characteristics of Schottky barrier confined arrays of phase-coupled multiquantum well GaAs lasers. IEEE J. QE-**20** , 1259–1266 (1984)

1.53 G. L. Harnagel, P. S. Cross, D. R. Scifres, D. P. Worland: 11 W quasi-CW monolithic laser diode arrays. Electron. Lett. **22**, 231–233 (1986)

1.54 C. A. Zmudzinski, L. J. Mawst, M. E. Givens, M.A. Emanuel, J. J. Coleman: Phase locked narrow zinc diffused stripe laser arrays. Appl. Phys. Lett. **48**, 1424–1426 (1986)

1.55 G. L. Harnagel, P. S. Cross, D. R. Scifres, D.F. Welch, C. R. Lennon: High-power quasi-CW monolithic laser diode linear arrays. Appl. Phys. Lett. **49**, 1418–1419 (1986)

1.56 Y. Seiwa, T. Aoyagi, S. Hinata, T. Kadowaki, N. Kaneno, K. Ikeda, W. Susaki: High power CW operation over 400 mW on five-stripe phase-locked laser arrays assembled by new junction down mounting. J. Appl. Phys. **61**, 440–442 (1987)

1.57 S. C. Wang, K. M. Leung, L. Z. Gacusan: Observation of single-mode oscillation in gain-guided GaAlAs laser array. J. Appl. Phys. **61**, 2686–2688 (1987)

1.58 G. L. Harnagel, P. S. Cross, C. R. Lennon, D. R. Scifres: Ultra-high-power quasi-CW monolithic laser diode arrays with high power conversion efficiency. Electron. Lett. **23**, 743–744 (1986)

1.59 D. F. Welch, M. Devito, M. Cardinal, M. Abraham, H. Kung, G. Harnagel, P. Cross, D. Scifres, W. Streifer: Performance characteristics of high-brightness, CW, diode laser arrays. Electron. Lett. **23**, 892–894 (1987)

1.60 D. F. Welch, M. Cardinal, W. Streifer, D. R. Scifres, P. S. Cross: High-brightness, high-efficiency, single-quantum-well laser diode array. Electron. Lett. **23**, 1240–1241 (1987)

1.61 T. Aoyagi, S. Hinata, K. Shigihara, Y. Seiwa, K. Ikeda, W. Susaki: High-power operation of long-cavity phase-locked laser arrays. Electron. Lett. **23**, 1396–1397 (1987)

1.62 D. F. Welch, B. Chan, W. Streifer, D. R. Scifres: High-power, 8 W CW, single-quantum-well laser diode array. Electron. Lett. **24**, 113–115 (1988)

1.63 G. S. Jackson, D. C. Hall, L.J. Guido, W. E. Plano, N. Pan, N. Holonyak: High-power gain-guided coupled-stripe quantum well laser array by hydrogenation. Appl. Phys. Lett. **52**, 691–693 (1988)

1.64 M. Sakamoto, G. L. Harnagel, D. F. Welch, C. R. Lennon, W. Streifer, H. Kung, D. R. Scifres: 12.5 W continuous-wave monolithic laser-diode arrays. Optics Lett. **13**, 378–379 (1988)

1.65 M. Sakamoto, D. F. Welch, G. L. Harnagel, W. Streifer, H. Kung, D. R. Scifres: Ultrahigh power 38 W continuous-wave monolithic laser diode arrays. Appl. Phys. Lett. **52**, 2220–2221 (1988)

1.66 D. R. Scifres, D. F. Welch, G. Harnagel, M. Sakamoto, H. Kung, W. Streifer, J. Berger: Power limits, efficiency, and reliability of 1-D and 2-D laser diodes and diode arrays. Proc. SPIE **893**, 2–9 (1988)

1.67 M. Sakamoto, D. F. Welch, J. G. Endriz, D. R. Scifres, W. Streifer: 76 W CW monolithic laser diode arrays. Appl. Phys. Lett. **54**, 2299–2300 (1989)

1.68 M. Sakamoto, D. F. Welch, J. G. Endriz, E. P. Zucker, D. R. Scifres: 10 watt CW, 5000 h lifetime monolithic AlGaAs laser diode arrays. Electron. Lett. **25**, 972–973 (1989)

1.69 M. Sakamoto, J. G. Endriz, D. R. Scifres: 120 W CW output power from monolithic AlGaAs(800 nm) laser diode array on diamond heatsink. Electron. Lett. **28**, 197, January (1992)

1.70 D. E. Ackley, D. Botez, B. Bogner: Phase-locked injection laser arrays with integrated phase shifters. RCA Rev. **44**, 625–633 (1983)

1.71 D. Botez, J. C. Connolly: High power phase locked arrays of index guided diode lasers. Appl. Phys. Lett. **43**, 1096–11098 (1983)

1.72 M. Matsumoto, M. Taneya, S. Matsui, S. Yano, T. Hijikata: Single-lobed far-field pattern operation in a phased array with an integrated phase shifter. Appl. Phys. Lett. **50**, 1541–1543 (1987)

1.73 M. Taneya, M. Matsumoto, S. Matsui, S. Yano, T. Hijikata: Stable quasi 0-degree phase mode operation in a laser array diode nearly aligned with a phase shifter. Appl. Phys. Lett. **50**, 783–785 (1987)

1.74 D. E. Ackley, R. W. H. Engelmann: High-power leaky-mode multiple-stripe laser. Appl. Phys. Lett. **39**, 27–29 (1981)

1.75 J. Ohsawa, K. Ikeda, K. Takahashi, W. Susaki: A dual-stripe phase-locked diode laser. Jpn. J. Appl. Phys. **22**, L230–L232 (1983)

1.76 S. Mukai, C. Lindsey, J.Katz, E. Kapon, Z. Rav-Noy, S. Margalit, A. Yariv: Fundamental mode oscillation of a buried ridge waveguide laser array. Appl. Phys. Lett. **45**, 834–835 (1984)

1.77 Y. Twu, A. Dienes, S. Wang, J. R. Whinnery: High power coupled ridge waveguide semiconductor laser arrays. Appl. Phys. Lett. **45**, 709–711 (1984)

1.78 Y. Twu, K. L. Chen, A. Dienes, S. Wang, J. R. Whinnery: High-performance index-guided phase-locked semiconductor laser arrays. Electron. Lett. **21**, 324–325 (1985)

1.79 F. Kappeler: Monolithic phase-locked GaAlAs laser arrays. Siemens Forschungs-und Entwicklungsberichte. **14**, 289–294 (1985)

1.80 D. F. Welch, D. Scifres, P. Cross, H. Kung, W. Streifer, R. D. Burnham, J. Yaeli: High-power (575 mw) single-lobed emission from a phased-array laser. Electron. Lett. **21**, 603–605 (1985)

1.81 D. F. Welch, P. S. Cross, D. R. Scifres, W. Streifer, R. D. Burnham: Properties of AlGaAs buried heterostructure lasers and laser arrays grown by a two-step metalorganic chemical vapor deposition. Appl. Phys. Lett. **48**, 1716–1718 (1986)

1.82 L. J. Mawst, D. Botez, T. J. Roth: High-power, diffraction-limited-beam operation from diode-laser phase-locked arrays operating in coupled first-order modes. Appl. Phys. Lett. **53**, 1236–1238 (1988)

1.83 R. Y. Hwang, C. P. Lee, T. F. Lei: GaAs/AlGaAs laser arrays with and without proton isolation. J. Chinese Inst. Eng. **12**, 255–261 (1989)

1.84 K. Shinozaki, R. Furukawa, T. Fukunaga, N. Watanabe: Phase front measurements of AlGaAs 830 nm phase-locked lasers with a real-refractive-index waveguide. Appl. Phys. Lett. **54**, 2654–2655 (1989)

1.85 K. Shinozaki, R. Furukawa, T. Fukunaga, N. Watanabe: Supermode control and phase front measurements of phase-locked offset-coupled laser arrays with a large optical waveguide structure. J. Appl. Phys. **66**, 1057–1064 (1989)

1.86 M. Sagawa, T. Kajimura: Complete single lateral 180 degrees phase mode operation for AlGaAs phased array lasers. Appl. Phys. Lett. **55**, 1376–1377 (1989)

1.87 K. Shinozaki, R. Furukawa, T. Fukunaga, N. Watanabe: Low-phase-aberration output of 830 nm AlGaAs offset-coupled laser arrays. Jpn. J. Appl. Phys. **28**, L1426–L1428, (1989)

1.88 D. E. Ackley: Phase-locked injection laser arrays with nonuniform stripe spacing. Electron. Lett. **20**, 695–697 (1984)

1.89 D. E. Ackley, J. K. Butler, M. Ettenberg: Phase-locked injection laser arrays with variable stripe spacing. IEEE J. QE-**22**, 2204–2212 (1986)

1.90 J. Katz, E. Kapon, C. Lindsey, S. Margalit, U. Shreter, A. Yariv: Phase-locked semiconductor laser array with separate contacts. Appl. Phys. Lett. **43**, 521–523 (1983)

1.91 E. Kapon, J. Katz, S. Margalit, A. Yariv: Controlled fundamental supermode operation of phase-locked arrays of gain-guided diode lasers. Appl. Phys. Lett. **45**, 600–602 (1984)

1.92 E. Kapon, J. Katz, S. Margalit, A. Yariv: Longitudinal-mode control in integrated semiconductor laser phased arrays by phase velocity matching. Appl. Phys. Lett. **44**, 157–159 (1984)

1.93 Y. Twu, K. L. Chen, A. Dienes, S. Wang, J. R. Whinnery: High-performance index-guided phase-locked semiconductor laser arrays. Electron. Lett. **21**, 324–325 (1985)

1.94 J. Ohsawa, S. Hinata, T. Aoyagi, T. Kadowaki, N. Kaneno, K. Ikeda, W. Sasaki: Triple-stripe phase-locked diode lasers emitting 100 mW CW with single-lobed far-field patterns. Electron. Lett. **21**, 779–780 (1985)

1.95 L. D. Zhu, G. B. Feak, J. M. Ballantyne, D. K. Wagner, P. Tihanyi: In-phase coupling between ridge guide lasers by introducing distributed saturable absorption regions in subordinate laser cavities. Appl. Phys. Lett. **50**, 1550–1552 (1987)

1.96 T. R. Chen, K. L. Yu, Chang Y. B, A. Hasson, S. Margalit, A. Yariv: Phase-locked InGaAsP laser array with diffraction coupling. Appl. Phys. Lett. **43**, 136–137 (1983)

1.97 S. Wang, J. Z. Wilcox, M. Jansen, J. J. Yang: In-phase locking in diffraction-coupled phased-array diode lasers. Appl. Phys. Lett. **48**, 1770–1772 (1986)

1.98 M. Jansen, S. S. Ou, J. J. Yang, J. Wilcox, M. Sergant, L. Eaton, W. Simmons: Large optical cavity (LOC) semiconductor laser arrays. Electron. Lett. **22**, 1083–1084 (1986)

1.99   E. Towe, C. G. Fonstad: Mixed-mode phase-locked quantum well laser arrays: In International Electron Devices Meeting (1986). Technical Digest, pp. 626–629

1.100 J. Z. Wilcox, M. Jansen, J. J. Yang, G. Peterson, A. Silver, W. Simmons, S. S. Ou, M. Sergant: Supermode discrimination in diffraction-coupled laser arrays with separate contacts. Appl. Phys. Lett. **51**, 631–633 (1987)

1.101 J. Z. Wilcox, M. Jansen, J. J. Yang, S. S. Ou, M. Sergant, W. Simmons: Supermode selection in diffraction-coupled semiconductor laser arrays. Appl. Phys. Lett. **50**, 1319–1321 (1987)

1.102 L. J. Mawst, D. Botez, T. J. Roth, J. J. Yang: Diffraction-limited beam operation from quantum-well laser phase-locked array grown by metalorganic chemical vapour deposition. Electron. Lett. **24**, 570–571 (1988)

1.103 D. Mehuys, K. Mitsunaga, L Eng, W. K. Marshall, A. Yariv: Supermode control in diffraction-coupled semiconductor laser arrays. Appl. Phys. Lett. **53**, 1165–1167 (1988)

1.104 D. Yap, J. N. Walpole, Z. L. Liau: Novel scalloped-mirror diffraction-coupled InGaAsP/InP buried heterostructure laser arrays. Appl. Phys. Lett. **54**, 687–689 (1989)

1.105 I. I. Vinogradov, A. E. Kosykh, A. S. Logginov: Analysis of an array of diffraction-coupled injection lasers arrays. Sov. J. Quantum Electron **20**, 515–517 (1990)

1.106 D. Botez, T. Pham, D. Tran: 0 degree-phase-shift, single-lobe operation from wide-waveguide interferometric (WWI) phase-locked arrays of InGaAsP/InP ($\lambda = 1.3\,\mu$m) diode lasers. Electron. Lett. **23**, 416–417 (1987)

1.107 J. Z. Wilcox, W. W. Simmons, D. Botez, M. Jansen, L. J. Mawst, G. Peterson, T. J. Wilcox, J. J. Yang: Design considerations for diffraction coupled arrays with monolithically integrated self-imaging cavities. Appl. Phys. Lett. **54**, 1848–1850 (1989)

1.108 M. Jansen, J. J. Yang, S. S. Ou, D. Botez, J. Z. Wilcox, L. J. Mawst: Diffraction-limited operation from monolithically integrated diode laser array and self-imaging (Talbot) cavity. Appl. Phys. Lett. **55**, 1949–1951 (1989)

1.109 M. Taneya, S. Matsumoto, S. Matsui, S. Yano, T. Hijikata: 0 degrees phase mode operation in phased-array laser diode with symmetrically branching waveguide. Appl. Phys. Lett. **47**, 341–343 (1985)

1.110 M. Taneya, S. Matsumoto, H. Kawanishi, S. Matsui, S. Yano, T. Hijikata: Phased-array with the 'YY' shaped symmetrically branching waveguide (SBW). Jpn. J. Appl. Phys. **47**, L432–L443 (1986)

1.111 D. F. Welch, P. S. Cross, D. R. Scifres, W. Streifer, R. D. Burnham: High-power (CW) in-phase locked 'Y' coupled laser arrays. Appl. Phys. Lett. **49**, 1632–1634 (1986)

1.112 D. F. Welch, W. Streifer, P. S. Cross, D. R. Scifres: Y-junction semiconductor laser arrays: Part *II* – experiments. IEEE J. QE-**23**, 752–756 (1987)

1.113 A. E. Bazarov, I. S. Goldobin, P. G. Eliseev, O. A. Kobilzhanov, G. T. Pak, T. V. Petakova, A. T. Semenov: Sov. J. Quantum Electron **17**, 551 (1987)

1.114 I. S. Goldobin, G. T. Pak T. V. Petakova, T. N. Pushkina, A. T. Semenov, S. I. Filimonov, S. D. Yakubovich: Phase-locked generation of radiation with the aid of a regular array of active symmetric meastripe couplers based on AlGaAs/GaAs double heterostructures. Sov. J. Quantum Electron **17**, 1108–1109 (1987)

1.115 D. Botez, P. Hayashida, L. Mawst, T. J. Roth: Diffraction-limited-beam, high-power operation from X-junction coupled phase-locked arrays of AlGaAs/GaAs diode lasers. Appl. Phys. Lett. **53**, 1366–1368 (1988)

1.116 J. E. A. Whiteaway, D. J. Moule, S. J. Clements: Tree array lasers. Electron. Lett. **25**, 779–781 (1989)

1.117 M. Matsumoto, M. Taneya, S. Matsui, S. Yano, T. Hijikata: Stable supermode operation in phase-locked laser diode arrays with three index waveguides. J. Appl. Phys. **58**, 2783–2785 (1985)

1.118 L. J. Mawst, M. E. Givens, M. A. Emanuel, C. A. Zmudzinski, J. J. Coleman: Complementary self-aligned laser arrays by metalorganic chemical vapor deposition. J. Appl. Phys. **60**, 2633–2635 (1986)

1.119 N. W. Carlson, V. J. Masin, M. Lurie, B. Goldstein, G. A. Evans: Measurement of the coherence of a single-mode phase-locked diode laser array. Appl. Phys. Lett. **51**, 643–645 (1987)

1.120 B. Goldstein, N. Dinkel, N. W. Carlson, G. A. Evans, V. J. Masin: Performance of a channeled-substrate-planar high power phase-locked array operating in the diffraction limit. Electron. Lett. **23**, 1136 (1987)

1.121 M. E. Givens, C. A. Zmudzinski, R. P. Bryan, J. J. Coleman: High-power nonplanar quantum well heterostructure periodic laser arrays. Appl. Phys. Lett. **53**, 1159–1161 (1988)

1.122 C. A. Zmudzinski, M. E. Givens, R. P. Bryan, J. J. Coleman: Nonplanar index-guided quantum well heterojunction periodic laser array. Appl. Phys. Lett. **53**, 350–352 (1988)

1.123 H. Hosoba, M. Matsumoto, S. Matsui, S. Yano, T. Hijikata: Phased array laser diode-SAWTOOTH channeled array. Rev. of Laser Eng. **17**, 32–42 (1989)

1.124 C. A. Zmudzinski, M. E. Givens, R. P. Bryan, J. J. Coleman: Optical characteristics of high-power nonplanar periodic laser arrays. IEEE J. QE-**25**, 1539–1546 (1989)

1.125 F. Zhao, G. Du, X. Zhang, D. Gao: Trapezoidal channeled substrate inner stripe phase-locked semiconductor laser arrays. J. Appl. Phys. **66**, 5637–5639 (1989)

1.126 R. L. Thornton, R. D. Burnham, T. L. Paoli, N. Holonyak, D. G. Deppe: Highly efficient multiple emitter index guided array lasers fabricated by silicon impurity induced disordering. Appl. Phys. Lett. **48**, 7–9 (1986)

1.127 R. L. Thornton, D. F. Welch, R. D. Burnham, T. L. Paoli, P. S. Cross: High power (2.1W) 10-stripe AlGaAs laser arrays with Si disordered facet windows. Appl. Phys. Lett, **49**, 1572–1574 (1986)

1.128 D. G. Deppe, G. S. Jackson, N. Holonyak: Impurity-induced layer-disordered buried heterostructure $Al_xGa_{1-x}As$ – GaAs quantum well edge-injection laser array. Appl. Phys. Lett. **50**, 392–394 (1987)

1.129 D. G. Deppe, G. S. Jackson, N. Holonyak, R. D. Burnham, R. L. Thornton: Coupled stripe $Al_xGa_{1-x}As$ – GaAs quantum well lasers defined by impurity-induced (Si) layer disordering. Appl. Phys. Lett. **50**, 632–634 (1987)

1.130 L. J. Guido, W. E. Plano, G. S. Jackson, N. Holonyak, R. D. Burnham, J. E. Epler: Coupled stripe $Al_xGa_{1-x}As$ – GaAs quantum well lasers defined by vacancy-enhanced impurity-induced layer disordering from $(Si_2)_y GaAs_{1-y}$ barriers. Appl. Phys. Lett. **50**, 757–759 (1987)

1.131 J. E. Epler, R. L. Thornton, T. L. Paoli: Laser-assisted processing of GaAs – AlGaAs optoelectronic devices. Proc. SPIE **893**, 172–180 (1988)

1.132 D. F. Welch, W. Streifer, R. L. Thornton, T. Paoli: 2.4 W CW, 770 nm laser arrays with nonabsorbing mirrors. Electron. Lett. **23**, 525–527 (1987)

1.133 M. Kume, H. Naito, I. Ohta, H. Shimuzu: High-power laser diodes with nonabsorbing mirrors. Review of Laser Engineering **18**, 525–527 (1990)

1.134 F. Koyama, K. Tomomatsu, K. Iga: GaAs surface emitting lasers with circular buried heterostructure grown by metalorganic chemical vapor deposition and two-dimensional laser array Appl. Phys. Lett. **52**, 528–529 (1988)

1.135 K. Iga, F. Koyama, S. Kinoshita: Surface emitting semiconductor laser array: its advantages and future. J. Vacuum Sci. & Techn. A (Vacuum, Surfaces, and Films) **7**, 842–846 (1989)

1.136 H. J. Yoo, J. R. Hayes, E. G. Paek, A. Scherer, Y. S. Kwon: Array mode analysis of two-dimensional phased arrays of vertical cavity surface emitting lasers. IEEE J. QE-**26**, 1039–1051 (1990)

1.137 H. J. Yoo, J. R. Hayes, E. G. Paek, J. P. Harbison, L. T. Florez, Y. S. Kwon: Phase-locked two-dimensional arrays of implant isolated vertical cavity surface emitting lasers. Electron. Lett. **26**, 1944–1946 (1990)

1.138 H. J. Yoo, J. R. Hayes, Y. S. Kwon: Analysis of coupling coefficient between two vertical cavity surface emitting lasers for two-dimensional phase-locked array. Electron. Lett. **26**, 896–897 (1990)

1.139 J. P. van Der Ziel, D. G. Deppe, N. Chand, G. J. Zydzik, S. N. G. Chu: Characteristics of single- and two-dimensional phase coupled arrays of vertical cavity surface emitting GaAs – AlGaAs lasers. IEEE J. QE-**26**, 1873–1882 (1990)

1.140 J. Y. Hoi, J. R. Hayes, E. G. Paek, J. P. Harbison, L. T. Florez, S. K. Young: Phase-locked two-dimensional arrays of vertical cavity surface emitting lasers. Jpn. J. Appl. Phys. **29**, L2423–L2426 (1990)

1.141 D. L. McDaniel Jr., J. G. McInerney, M. Y. A. Raja, C. F. Schaus, S. R. J. Brueck: Vertical cavity surface-emitting semiconductor laser with CW injection laser pumping. IEEE Photon. Technol. Lett. **2**, 156–158 (1990)

1.142 K. Iga, S. Kinoshita, J. Koyama: Two-dimensionally arrayed laser and its application: *Optical computing in Japan*, (Nova, 1990)

1.143 K. Iga: Surface emitting semiconductor laser: J. Institute of Electronics. Information and Communication Engineers **73**, 882–885 (1990)

1.144 C. J. Chang-Hasnain, M. W. Maeda, N. G. Stoffel, J. P. Harbison, L. T. Florez: Surface emitting laser arrays with uniformly separated wavelengths. Electron. Lett. **26**, 940–942 (1990)

1.145 A. C. von Lehmen, C. J. Chang-Hasnain, M. Orenstein, N. G. Stoffel, L. T. Florez, J. P. Harbison: Large integrated electronically addressed surface emitting laser arrays: In LEOS Summer Topical on Optical Multiple Access Networks (1990) Digest, pp. 71–72

1.146 Y. H. Lee, J. L. Jewell, J. Jahns, A. Scherer andJ. P. Harbison, L.T. Florez: Microlasers for photonic switching and interconnection. Proc. SPIE **1319**, 683–684 (1990)

1.147 Y. H. Lee, B. Tell, K. Brown-Goebeler, J. L. Jewell, C. A. Burrus, J. M. V. Hove: Characteristics of top-surface-emitting GaAs quantum-well lasers. IEEE Photon. Technol. Lett. **2**, 686–688 (1990)

1.148 R. A. Morgan, M. C. Robinson, L. M. F. Chirovsky, M. W. Focht, G. D. Guth, R. E. Leibenguth, K. G. Glogovsky, G. J. Przybylek, L. E. Smith: Uniform 64×1 arrays of individually-addressed vertical cavity top surface emitting lasers. Electron. Lett. **27**, 1400–1402 (1991)

1.149 J. L. Jewell, Y. H. Lee, S. L. McCall, A. Scherer, J. P. Harbison, S. T. Florez, N. A. Olsson, R. S. Tucker, C. A. Burrus, C. J. Sandrof: Vertical cavity lasers for optical interconnects. Proc. SPIE **1389**, 401–407 (1991)

1.150 A. C von Lehmen, I. C. Banwell, R. Cordell, C. Chang-Hasnain, J. W. Mann, J. Harbison: High speed operation of hybrid CMOS vertical cavity surface emitting laser array. Electron. Lett. **27**, 1189–1191 (1991)

1.151 C. J. Chang-Hasnain, J. P. Harbison, C. E. Zah, M. W. Maeda, L. T. Florez, N. G. Stoffel, T. P. Lee: Multiple wavelength tunable surface-emitting laser arrays. IEEE J. QE-**27**, 1368–1376 (1991)

1.152 K. Rastani, M. Orenstein, E. Kapon, A. C. von Lehmen: Integration of planar Fresnel microlenses with vertical-cavity surface-emitting laser arrays. Opt. Lett. **16**, 919–921 (1991)

1.153 L. A. Coldren, S. W. Corzine, R. S. Geels, A. C. Gossard, K. K. Law, J. L. Merz, J. W. Scott, R. J. Simes, R. H. Yan: High-efficiency vertical cavity lasers and modulators. Proc. SPIE, **1362**-part 1, 24–37 (1991)

1.154 M. Orenstein, A. C. von Lehmen, C. Chang-Hasnain, N. G. Stoffel, J. P. Harbison, L. T. Florez: Matrix addressable vertical cavity surface emitting laser array. Electron. Lett. **27**, 437–438 (1991)

1.155 R. R. A. Syms, E. M. Yeatman: Optical layout for single-transverse-mode operation of 2D arrays of vertical cavity surface-emitting lasers. Electron. Lett. **27**, 349–350 (1991)

1.156 A. Ibaraki, K. Furusawa, T. Ishikawa, K. Yodoshi, T. Yamaguchi, T. Niina: GaAs buried heterostructure vertical cavity top-surface emitting lasers. IEEE J. QE-**27**, 1386–1390 (1991)

1.157 R. D. Dupuis, D. G. Deppe, C. J. Pinzone, N. D. Gerrard, S. Singh, G. J. Zydzik, J. P. van der Ziel, C. A. Green: $In_{0.47}Ga_{0.53}As$ − InP heterostructures for vertical cavity surface emitting lasers at $1.65\,\mu m$ wavelength. J. Crystal Growth **107**, 790–795 (1991)

1.158 K. Iga, F. Koyama: Vertical cavity surface emitting lasers and arrays, in *Surface Emitting Semiconductor Laser and Arrays*, ed. by G. A. Evans, J. M. Hammer (Academic, San Diego 1993)

1.159 J. N. Walpole, Z. L. Liau: Monolithic two-dimensional arrays of high-power GaInAsP/InP surface-emitting diode lasers. Appl. Phys. Lett. **48**, 1636–1638 (1986)

1.160 T. H. Windhorn, W. D. Goodhue: Monolithic GaAs/AlGaAs diode laser/ deflector devices for light emission normal to the surface. Appl. Phys. Lett. **48**, 1675–1677 (1986)

1.161 J. J. Yang, M. Sergant, M. Jansen, S. S. Ou, L. Eaton, W. Simmons: Surface-emitting GaAlAs/GaAs linear laser arrays with etched mirrors. Appl. Phys. Lett. **49**, 1138–1139 (1986)

1.162 Z. L. Liau, J. N. Walpole: Large monolithic two-dimensional arrays of GaInAsP/InP surface-emitting diode lasers. Appl. Phys. Lett. **50**, 528 (1987)

1.163 J. N. Walpole, Z. L. Liau, V. Diadiuk: Monolithic two-dimensional diode laser arrays. Proc. SPIE **783**, 42–48 (1987)

1.164 J. P. Donnelly, W. D. Goodhue, T. H. Windhorn, R. J. Bailey, S. A. Lambert: Monolithic two-dimensional surface-emitting arrays of GaAs/AlGaAs diode laser. Appl. Phys. Lett. **51**, 1138 (1987)

1.165 J. P. Donnelly, R. J. Bailey, C. A. Wang, G. A. Simpson, K. Rauschenbach: Hybrid approach to two-dimensional surface-emitting diode laser arrays, Appl. Phys. Lett. **53**, 938 (1988)

1.166 J. P. Donnelly, K. Rauschenbach, C. A. Wang, W. D. Goodhue, R. J. Bailey: Two-dimensional surface-emitting arrays of GaAs/AlGaAs diode lasers. Proc. SPIE **1043**, 92 (1989)

1.167 J. P. Donnelly: Two-dimensional surface-emitting arrays of GaAs/AlGaAs diode lasers. Lincoln Laboratory J. **3**, 361 (1990)

1.168 W. D. Goodhue, K. Rauschenbach, C. A. Wang, J. P. Donnelly, R. J. Bailey, G. D. Johnson: Monolithic two-dimensional GaAs/AlGaAs laser arrays fabricated by chlorine ion-beam-assisted micromachining. J. Electronic Materials **19**, 463–469 (1990)

1.169 W. D. Goodhue, J. P. Donnelly, C. A. Wang, G. A. Lincoln, K. Rauschen-bach, R. J. Bailey, G. D. Johnson: Monolithic two-dimensional surface-emitting strained-layer InGaAs/AlGaAs and AlInGaAs/AlGaAs diode laser arrays. Appl. Phys. Lett. **59**, 632 (1991)

1.170 R. C. Williamson, J. P. Donnelly, Z. L. Liau, W. D. Goodhue, J. N. Walpole: Horizontal-cavity surface emitting lasers with integrated beam deflectors, in *Surface Emitting Semiconductor Laser and Arrays*, ed. by G. A. Evans, J. M. Hammer (Academic, San Diego 1993)

1.171 V. N. Luk'yanov, A. T. Semenov, N. V. Shelkov, S. D. Yakubovich: Laser with distributed feedback (review). Sov. J. Quant. Electron. **5** , 1293–1307 (1976)

1.172 K. Mitsunaga, K. Kojima, S. Noda, M. Kameya, K. Kyuma, K. Hamanaka, T. Nakayama: Surface-emitting grated-coupled lasers. Optoelectronic- Devices and Technologies **2**, 247–263 (1987)

1.173 N. W. Carlson, G. A. Evans, J. M. Hammer, M. Lurie, S. L. Palfrey, A. Dholakia: Phase-locked operation of a grating-surface-emitting diode laser array. Appl. Phys. Lett. **13**, 1301–1303 (1987)

1.174 S. H. Macomber, J. S. Mott, R. J. Noll, G. M. Gallatin, E. J. Gratix, S. L. O'Dwyer, S. A. Lambert: Surface-emitting distributed feedback semiconductor laser. Appl. Phys. Lett. **51**, 472–474 (1987)

1.175 N. W. Carlson, G. A. Evans, J. M. Hammer, M. Lurie, J. K. Butler, S. L. Palfrey, M. Ettenberg, L. A. Carr, F. Z. Hawrylo, E. A. James, C. J. Kaiser, J. B. Kirk, W. F. Reichert, J. R. Shealy, J. W. Sprague, S. R. Chinn, P. S. Zory: Dynamically stable 0° degrees phase mode operation of a grating-surface-emitting diode-laser array. Opt. Lett. **13**, 312–314 (1988)

1.176 N. W. Carlson, G. A. Evans, J. M. Hammer, M. Lurie, L. A. Carr, F. Z. Hawrylo, E. A. James, C. J. Kaiser, J. B. Kirk, W. F. Reichert, D. A. Truxal: High-power seven-element grating surface emitting diode laser array with 0.012 degrees far-field angle. Appl. Phys. Lett. **52**, 939–941 (1988)

1.177 G. A. Evans, N. W. Carlson, J. M. Hammer, M. Lurie, J. K. Butler, R. Amantea, L. A. Carr, F. Z. Hawrylo, E. A. James, C. J. Kaiser, J. B. Kirk, W. F. Reichert, S. R. Chinn, J. R. Shealy, P. S. Zory: Coherent, monolithic two-dimensional (10 × 10) laser arrays using grating surface emission. Appl. Phys. Lett. **53**, 2123–2125 (1988)

1.178 G. A. Evans, N. W. Carlson, D. P. Bour, J. M. Hammer, M. Lurie, J. K. Butler, S. L. Palfrey, Amantea R, F. A. Bibby T, R. E. Farkas, D. B. Gilbert, D. A. Truxal: Grating-surface-emitting laser arrays with > 1 cm output apertures. Microwave and Optical Technology Lett. **2**, 334–336 (1989)

1.179 J. M. Hammer, G. A. Evans, N. W. Carlson, D. P. Bour, M. Lurie, S. L. Palfrey, Amantea R, S. K. Liew, L. A. Carr, E. A. James, J. B. Kirk, W. F. Reichert: Lateral beam steering in mutual injection coupled Y-branch grating-surface-emitting diode laser arrays. Appl. Phys. Lett. **56**, 224–226 (1990)

1.180 K. Kojima, M. Kameya, S. Noda, K. Kyuma: High efficiency surface-emitting distributed bragg reflector laser array. Electron. Lett. **24**, 283–284 (1988)

1.181 G. A. Evans, D. P. Bour, N. W. Carlson, J. M. Hammer, M. Lurie, J. K. Butler, S. L. Palfrey, R. Amantea, L. A. Carr, F. Z. Hawrylo, E. A. James, J. B. Kirk, S. K. Liew, W. F. Reichert: Coherent, monolithic two-dimensional strained InGaAs/AlGaAs quantum well laser arrays using grating surface emission. Appl. Phys. Lett. **55**, 2721–2723 (1989)

1.182 G. A. Evans, N. W. Carlson, J. M. Hammer, M. Lurie, J. K. Butler, J. Connolly, L. A. Carr, F. Z. Hawrylo, E. A. James, C. J. Kaiser, J. B. Kirk, W. F. Reichert: Two-dimensional, coherent Y-coupled grating surface-emitting laser arrays. Electron. Lett. **25**, 597–599 (1989)

1.183 J. Buus, P. J. Williams, I. Goodridge, D. J. Robbins, J. Urquhart, A. P. Webb, T. Reid, R. Nicklin, P. Charles, D. C. J. Reid, A. C. Carter: Surface-emitting two-dimensional coherent semiconductor laser array. Appl. Phys. Lett. **52**, 331–333 (1989)

1.184 J. S. Mott, S. H. Macomber: Two-dimensional surface emitting distributed feedback laser arrays. IEEE Photon. Technol. Lett. **1**, 202–204 (1989)

1.185 J. Buus: A large area surface emitting semiconductor laser. GEC Rev. **6**, 107–111 (1990)

1.186 G. A. Evans, N. W. Carlson, D. P. Bour, J. M. Hammer, M. Lurie, J. K. Butler, R. Amantea, S. K. Liew, J. B. Kirk, W. F. Reichert, R. K. DeFreez, D. J. Bossert: Two dimensional grating surface emitting laser arrays with wide lateral extent. Electron. Lett. **26**, 907–908 (1990)

1.187 N. W. Carlson, D. P. Bour, G. A. Evans, R. Amantea, S. K. Liew: Stable single mode operation of grating-surface-emitting laser arrays under frequency-modulated operation. Appl. Phys. Lett. **57**, 756–758 (1990)

1.188 N. W. Carlson, D. P. Bour, G. A. Evans, S. K. Liew: Spectral linewidth narrowing in monolithic grating-surface-emitting laser arrays. IEEE Photon. Technol. Lett. **2**, 242–243 (1990)

1.189 N. W. Carlson, G. A. Evans, M. Lurie, J. M. Hammer, C. J. Kaiser, S. K. Liew: Coherent coupling of independent grating-surface-emitting diode laser arrays using an external prism. Appl. Phys. Lett. **56**, 114–116 (1990)

1.190 G. A. Evans, N. W. Carlson, J. M. Hammer, M. Lurie, J. K. Butler, J. Connolly, L. A. Carr, F. Z. Hawrylo, E. A. James, C. J. Kaiser, J. B. Kirk, W. F. Reichert: Two-dimensional, coherent Y-coupled grating surface-emitting laser arrays. Electron. Lett. **25**, 597–599 (1989)

1.191 R. G. Waarts, D. F. Welch, R. Parke, A. Hardy, W. S. Streifer: Coherent linear arrays of grating coupled surface-emitting lasers. Electron. Lett. **26**, 129–130 (1990)

1.192 R. Parke, R. G. Waarts, D. F. Welch, A. Hardy, W. S. Streifer: Optical characteristics of multiple grating surface emitting semiconductor lasers. Proc. SPIE **1219**, 242–245 (1990)

1.193 D. J. Bossert, R. K. DeFreez, H. Ximen, J. M. Hunt, G. A. Wilson, J. Orloff, G. A. Evans, N. W. Carlson, M. Lurie, J. M. Hammer, D. P. Bour, S. L. Palfrey, R. Amantea: Spectral locking in an extended area two-dimensional coherent grating surface emitting laser array. Appl. Phys. Lett. **56**, 2068–2070 (1990)

1.194 G. A. Evans, N. W. Carlson, J. M. Hammer, J. K. Butler: Grating-coupled surface emitting semiconductor lasers, in *Surface Emitting Semiconductor Laser and Arrays*, ed. by G. A. Evans, J. M. Hammer (Academic, San Diego 1993)

1.195 A. A. Hardy, D. F. Welch, W. Streifer: Second order grating surface emitting theory, in *Surface Emitting Semiconductor Laser and Arrays*, ed. by G. A. Evans, J. M. Hammer (Academic, San Diego 1993)

1.196 A. P. Bogatov, P. G. Eliseev, M. A. Man'ko, G. T. Mikaelyan, Yu. M. Popov: Injection laser with an unstable resonator. Soviet J. Quantum Electron. **10**, 620–622 (1980)

1.197 R. R. Craig, L. W. Casperson, O. M. Stafsudd, J. J. Yang, G. A. Evans, R. A. Davidheiser: Etched-mirror unstable-resonator semiconductor lasers. Electron. Lett. **53**, 62–63 (1985)

1.198 J. Salzman, T. Venkatesan, R. Lang, M. Mittelstein, A. Yariv: Unstable resonator cavity semiconductor lasers. Appl. Phys. Lett. **46** , 218–220 (1985)

1.199 M. Mittelstein, J. Salzman, T. Venkatesan, R. Lang, A. Yariv: Coherence and focusing properties of unstable resonator semiconductor lasers. Appl. Phys. Lett. **46**, 923–925 (1985)

1.200 O. V. Bogdankevich, N. D. Vorob'ev, M. M. Zverev, S. P. Kopyt, E. M. Krasavina, I. V. Kryukova, V. F. Pevtsov, V. A. Ushakhin, V. K. Yakushin: Pulsed multielement semiconductor laser with an unstable resonator. Sov. J. Quantum Electron.15, 1002–1003 (1985)

1.201 J. Salzman, R. Lang, A. Yariv: Efficiency of unstable resonator semiconductor lasers. Electron. Lett. **21**, 821–823 (1985)

1.202 J. Salzman, R. Lang, T. Venkatesan, M. Mittelstein, A. Yariv: Modal properties of unstable resonator semiconductor lasers. Appl. Phys. Lett. **47**, 445–447 (1985)

1.203 L. W. Casperson: Power characteristics of high magnification semiconductor lasers. Opt. and QE-**18**, 155–167 (1986)

1.204 R. Lang, J. Salzman, A. Yariv: Modal analysis of semiconductor lasers with nonplanar mirrors. IEEE J. QE-**22**, 463–470 (1986)

1.205 J. Salzman, A. Yariv: Phase-locked arrays of unstable resonator semiconductor lasers. Appl. Phys. Lett. **49**, 440–442 (1986)

1.206 J. Salzman, R. Lang, A. Larson, A. Yariv: Confocal unstable-resonator semiconductor lasers. Opt. Lett. **11**, 587–589 (1986)

1.207 R. Lang, M. Mittelstein,, A. Yariv, J. Salzman: Unstable resonator semiconductor lasers Part 1: Theory. IEE Proc. J **134**, 69–75 (1987)

1.208 J. Salzman, T. Venkatesan, R. Lang, M. Mittelstein, A. Yariv: Unstable resonator semiconductor lasers Part 2: Experiment. IEE Proc. J **134**, 76–86 (1987)

1.209 J. Salzman, R. Lang, A. Yariv: Eigenvalues of unstable resonator semiconductor lasers. Opt. Commun. **61**, 332–336 (1987)

1.210 H. Wang, Y. Y. Liu, M. Mittelstein, T. R. Chen, A. Yariv: Confocal unstable-resonator semiconductor lasers. Electron. Lett. **23** , 949–951 (1987)

1.211 M. L. Tilton, G. C. Dente, A. H. Paxton: Mode control of broad area semiconductor lasers using unstable resonator. Proc. SPIE **1219**, 423–434 (1990)

1.212 C. Largent, D. Gallant, J. Yang, M. Allen, M. Jansen: Fabrication of unstable resonator diode lasers. Proc. SPIE **1418**, 40–45 (1990)

1.213 M. L. Tilton, G. C. Dente, A. H. Paxton, J. Cser, R. K. DeFreez, C. Moeller, D. Depatie: High power, nearly diffraction-limited output from a semiconductor lasers using unstable resonator. IEEE J. QE-**27**, 2098–2108 (1991)

1.214 R. J. Lang: Geometric formulation of unstable-resonator design and application to self-collimating unstable-resonator diode lasers. Optics Lett. **16**, 1319–1321 (1991)

1.215 S. T. Srinivasan, C. F. Schaus, S. Z. Sun, E. A. Armour, S. D. Hersee, J. G. McInerney, A. H. Paxton, D. J. Gallant: High-power spatially coherent operation of unstable resonator semiconductor lasers with regrown lens trains. Appl. Phys. Lett. **61**, 1272–1274 (1992)

1.216 J. P. Hohimer, D. R. Meyers, T. M. Brennan, B. E. Hammons: Integrated injection-locked high-power cw diode laser arrays. Appl. Phys. Lett. **55**, 531–533 (1989)

1.217 J. P. Hohimer, D. R. Meyers, T. M. Brennan, B. E. Hammons: Injection-locking characteristics of gain-guided diode laser arrays with an 'on-chip' master laser. Appl. Phys. Lett. **56**, 1521–1523 (1990)

1.218 D. Botez, L. J. Mawst, P. Hayashida, G. Peterson, T. J. Roth: High-power, diffraction-limited-beam operation from phase-locked diode-laser arrays of closely spaced 'leaky' waveguides (antiguides). Appl. Phys. Lett. **53**, 464–466 (1988)

1.219 D. Botez, G. Peterson: Modes of phase-locked diode-laser arrays of closely spaced antiguides. Electron. Lett. **24**, 1042–1044 (1988)

1.220 D. Botez, L. J. Mawst, G. Peterson: Resonant leaky-wave coupling in linear arrays of antiguides. Electron. Lett. **24**, 1328–1330 (1988)

1.221 L. J. Mawst, D. Botez, T. J. Roth, Peterson G, J. J. Yang: Diffraction-coupled, phase-locked arrays of antiguided, quantum-well lasers grown by metalorganic chemical vapour deposition. Electron. Lett. **24**, 958–959 (1988)

1.222 L. J. Mawst, D. Botez, T. J. Roth, W. W. Simmons, G. Peterson, M. Jansen, T. J. Wilcox, J. J. Yang: Phase locked array of anti-guided lasers with monolithic spatial filter. Electron. Lett. **25**, 365–366 (1989)

1.223 D. Botez, L. J. Mawst, G. Peterson, T. J. Roth: Resonant optical transmission and coupling in phase-locked diode laser arrays of antiguides: the resonant optical waveguide array. Appl. Phys. Lett. **54**, 2183–2185 (1989)

1.224 L. J. Mawst, D. Botez, P. Hayashida, M. Jansen, G. Peterson, T. J. Roth, J. Z. Wilcox, J. J. Yang: Stabilized in-phase-mode operation from monolithic antiguided diode laser arrays. Proc. SPIE **1219**, 156–171 (1990)

1.225 T. H. Shiau, S. Sun, C. F. Schaus, K. Zheng, G. R. Hadley: Highly stable strained layer leaky-mode diode laser arrays. IEEE Photon. Technol. Lett. **2**, 534–536 (1990)

1.226 L. J. Mawst, D. Botez, M. Jansen, T. J. Roth, J. Rozenbergs: 1.5 W diffraction-limited-beam operation from resonant-optical-waveguide (ROW) array. Electron. Lett. **27**, 369–371 (1991)

1.227 L. J. Mawst, D. Botez, T. J. Roth, G. Peterson, J. Rozenbergs: CW high-power diffraction-limited-beam operation from resonant optical waveguide arrays of diode lasers. Appl. Phys. Lett. **58**, 22–24 (1991)

1.228 L. J. Mawst, D. Botez, M. Jansen, T. J. Roth, C. Tu, C. Zmudzinski: 0.5 W CW diffraction-limited-beam operation from high-efficiency resonant-optical-waveguide diode-laser arrays. Electron. Lett. **27**, 1586–1588 (1991)

1.229 G. R. Hadley, D. Botez, L. I. Mawst: Modal discrimination in leaky-mode (antiguided) arrays. IEEE J. QE-**27**, 921–930 (1991)

1.230 D. Botez, M. Jansen, L. J. Mawst, G. Peterson, T. J. Roth: Watt-range, coherent, uni-phase powers from phase-locked arrays of anti-guided diode lasers. Appl. Phys. Lett. **58**, 2070–2072 (1991)

1.231 J. S. Major, D. Mehuys, D. F. Welch, D. R. Scifres: High power, high efficiency antiguide laser arrays. Appl. Phys. Lett. **59**, 2210–2212 (1991)

1.232 J. P. Hohimer, G. R. Hadley, D. C. Craft, T. H. Shiau, S. Sun, C. F. Schaus: Stable-mode operation of leaky-mode diode laser arrays at high pulsed and CW currents. Appl. Phys. Lett. **58**, 452–454 (1991)

1.233 C. Zmudzinski, L. J. Mawst, D. Botez, C. Tu, C. A. Wang: 1 W diffraction-limited-beam operation of resonant-optical-waveguide diode laser arrays at 0.98 μm. Electron. Lett. **28**, 1543–1544 (1992)

1.234 D. Botez, L. J. Mawst: Γ effect: Key intermodal-discrimination mechanism in arrays of antiguided diode lasers. Appl. Phys. Lett. **60**, 3096–3098 (1992)

1.235 D. Botez: High-power monolithic phase-locked arrays of antiguided semiconductor diode lasers. IEE Proc. J **139**, 14–23 (1992)

1.236 N. W. Carlson, J. H. Abeles, D. P. Bour, S. K. Liew, P. S. D. Lin, A. S. Gozdz: Demonstration of a monolithic, grating-surface-emitting master oscillator-cascaded power amplifier laser array. IEEE Photon. Technol. Lett. **2**, 708–710 (1990)

1.237 D. F. Welch, D. Mehuys, R. Parke, R. G. Waarts, D. Scifres, W. Streifer: Coherent operation of monolithically integrated master oscillator amplifiers. Electron. Lett. **26**, 1327–1329 (1990)

1.238 N. W. Carlson, S. K. Liew, G. A. Evans, D. P. Bour, J. H. Abeles, R. Amantea: CW operating characteristics of grating-surface-emitting master oscillator power amplifier laser arrays. CLEO 1991 Technical Digest **11**, 250–251 (1991)

1.239 R. Parke, D. F. Welch, D. Mehuys: Coherent operation of 2-D monolithically integrated master oscillator power amplifier. Electron. Lett. **27**, 2097–2098 (1991)

1.240 D. Mehuys, R. Parke, R. G. Waarts, D. F. Welch, A. Hardy, W. Streifer, D. Scifres: Characteristics of multistage monolithically integrated master oscillator power amplifiers. IEEE J. QE-**27**, 1574–1581 (1991)

1.241 N. W. Carlson, R. Amantea, G. A. Evans, D. P. Bour, S. K. Liew: Applications of surface emitting lasers to coherent communications systems. LEOS '90 Conference Proc., pp. 406–409

1.242 D. Mehuys, D. F. Welch, R. Parke, R. G. Waarts, A. Hardy, D. Scifres: High power, diffraction-limited emission from monolithically integrated active grating master oscillator power amplifier. Electron. Lett. **27**, 492–494 (1991)

1.243 N. W. Carlson, J. H. Abeles, R. Amantea, J. K. Butler, G. A. Evans, S. K. Liew: Characteristics of active grating-surface-emitting amplified lasers. Proc. SPIE **1634**, 39–48 (1992)

1.244 N. W. Carlson: Design considerations and operating characteristics of high-power active-grating-surface-emitting amplifiers. IEEE J. QE-**28**, 1884–1893 (1992)

1.245 N. W. Carlson, P. Gardner, R. Menna, J. Andrews, R. Stolzenberger, A. Triano, E. Vangieson, D. Bour, G. A. Evans, S. K. Liew, J. Kirk, W. Reichert: Demonstration of an InGaAsP/InGaAs multi-quantum well active-grating surface emitting amplifier. IEEE Photon. Technol. Lett. **4**, 988–990 (1992)

1.246 D. F. Welch, R. Parke, D. Mehuys, A. Hardy, R.Lang, S. O'Brien, D. S. Scifres: 1.1 W CW diffraction-limited operation of a monolithically flared-amplifier master oscillator power-amplifier. Electron. Lett. **28**, 2011–2013 (1992)

1.247 J. H. Abeles, P. K. York, N. W. Carlson, J. T. Andrews, W. F. Reichert, J. B. Kirk, N. A. Dinkel, C. G. Dupuy, J. T. McGinn, J. H. Thomas, T. J. Zamerowski, S. K. Liew, J. C. Connolly, G. A. Evans, J. K. Butler: High-power cw distributed out-coupled surface emitting laser-amplifiers. 13th IEEE Int'l Semiconductor Laser Conference Digest, (1992)

1.248 K. Kubodera, K. Otsuka: Diode-pumped miniature solid-state laser: design considerations. Appl. Opt. **16**, 2747–2752 (1977)

1.249 R. J. Smith, R. R. Rice, L. B. Allen Jr.: 100 mW laser diode pumped Nd:YAG laser. Proc. SPIE **247**, 144–148 (1980)

1.250 M. Ettenberg: Laser diode arrays for pumping Nd:YAG. In *Tunable Solid State Lasers for Remote Sensing*, ed. by R. L. Byer, E. K. Gustafson, R. Trebino, Springer Ser. Opt. Sci. Vol. 51 (Springer, Berlin, Heidelberg 1984)

1.251 D. L. Sipes: Highly efficient neodymium: yttrium aluminium garnet laser end pumped by a semiconductor laser array. Appl. Phys. Lett. **47**, 74–76 (1985)

1.252 R. Allen, L. Esterowitz, L. Goldberg, J. F. Weller, M. Storm: Diode-pumped $2 \mu m$ holmium laser. Electron. Lett. **22**, 947 (1986)

1.253 T. M. Baer: Diode laser pumping of solid-state lasers. Laser Focus/Electro-Optics **22**, 82–84 (June 1986)

1.254 L. B. Allen: Scaling and configuring diode pumped slab lasers. Proc. SPIE **736**, 45–52 (1987)

1.255 M. S. Zediker, D. J. Krebs, J. L. Levy, R. R. Rice, G. M. Bender, D. Begley: Two-dimensional laser array technology comparison: 'stack-and-rack' vs. monolithic. Proc. SPIE **893**, 21–24 (1988)

1.256 C. A. Krebs, B. D. Vivian: High power quasi-CW and CW laser diode bar arrays. Proc. SPIE **893**, 35–37 (1988)

1.257 F. Baberg, J. Luft: High-power GaAlAs semiconductor lasers. Siemens Components (English Edition) **23**, 154–1594 (1988)

1.258 R. Scheps, J. Meyers, E. J. Schimitschek, D. F. Heller: Nd:BEL laser pumped by laser diodes. Proc. SPIE **898**, 91–98 (1988)

1.259 R. L. Byer: Solid state lasers–the next 10 years. AIP Conf. Proc. **172**, 6 (1988)

1.260 R. L. Byer: Diode laser-pumped solid-state lasers. Science **239**, 742–747 (1988)

1.261 J. Berger, D. F. Welch, W. Streifer, D. R. Scifres, N. J. Hoffman, J. J. Smith, D. Radecki: Fiber-bundle coupled, diode end-pumped Nd:YAG laser. Optics Lett. **13**, 306–308 (1988)

1.262 G. J. Dixon, L. S. Lingvay, R. H. Jarman: Lithium neodymium tetraphosphate lasers pumped via close-coupling to high-power laser diode arrays. IEEE Photon. Technol. Lett. **1**, 97–99 (1989)

1.263 R. A. Fields, C. L. Fincher: Performance comparisons of diode-pumped neodymium laser materials. Proc. SPIE **1104**, 277–282 (1989)

1.264 T. Y. Fan, C. L. Fincher: Performance comparisons of diode-pumped neodymium laser materials. Proc. SPIE **1104**, 277–282 (1989)

1.265 F. Hanson: Laser-diode side-pumped Nd:YAlO$_3$ laser at 1.08 and 1.34 μm. Optics Lett. **14**, 674–676 (1989)

1.266 H. Hemmati: 2.07 μm CW diode-laser-pumped Tm,Ho:YLiF$_4$ room-temperature laser. Optics Lett. **14**, 435–437 (1989)

1.267 R. Scheps, J. Meyers: Performance and aging of a high power 2-D laser diode array. Appl. Opt. **29**, 341–347 (1990)

1.268 T. Y. Fan: Diode-pumped solid state lasers. The Liconln Laboratory J. **3**, 4136–425 (1990)

1.269 T. Gray, C. Frederickson: Pumping Nd:YAG lasers: lamp or diode array ? Lasers & Optronics **9**, 40–42 (1990)

1.270 N. N. Groshkova, M. N. Gruden, V. D. Vetrov, V. R. Kushnir, E. I. Lebedeva, A. A. Pleshkov, M. N. Shkunov: Pulsed solid-state laser pumped optically by laser diode arrays. Sov. J. QE-**21**, 263–264 (1991)

1.271 P. F. Moulton: Pumping with diodes. IEEE Circuits and Devices Magazine **7**, 36–40 (1991)

1.272 W. Koechner: *Solid-State Laser Engineering, 3rd edn*, Springer Ser. Opt. Sci. Vol. 1 (Springer, Berlin, Heidelberg 1992)

1.273 D. B. Tuckerman, R. F. W. Pease: High-performance heat sinking for VLSI. IEEE Electron Device Lett. **EDL-2**, 126 (1981)

1.274 J. N. Walpole, L. J. Missaggia: Microchannel heat sinks for two-dimensional laser arrays, in *Surface Emitting Semiconductor Laser and Arrays*, ed. by G. A. Evans, J. M. Hammer (Academic, San Diego 1993)

1.275 R. Beach, W. J. Benett, B. L. Freitas, D. Mundinger, B. J. Comaskey, R. W. Solarz, M. A. Emanuel: Modular microchannel cooled heatsinks for high average power laser diode arrays. IEEE J. QE-**28**, 966–976 (1992)

1.276 R. Solarz, G. Albrecht, S. Sutton, R. Beach: Packaging of high-power semiconductor laser arrays for pumping solid-state lasers, in *Diode-Laser Arrays*,ed. by D. Botez, D. Scifres (Cambridge Univ. Press, Cambridge 1994)

1.277 N. K. Dutta, S. G. Napholtz, R. B. Wilson, R. L. Brown, T. Cella, D. C. Craft: 1.3 μm InGaAsP index-guided multirib waveguide laser array. Appl. Phys. Lett. **45**, 941–943 (1984)

1.278 N. K. Dutta, T. Cella, S. G. Napholtz, D. C. Craft: 1.3 μm InGaAsP index-guided multirib waveguide laser array. Electron. Lett. **21**, 326–327 (1985)

1.279 N. K. Dutta, L. A. Koszi, B. P. Senger, D. C. Craft, S. G. Napholtz: High-power index-guided multiridge waveguide laser array. Appl. Phys. Lett. **46**, 803–804 (1985)

1.280 S. Uchiyama, K. Iga: Two-dimensional array of GaInAsP/InP surface-emitting lasers. Electron. Lett. **21**, 162–164 (1985)

1.281 N. K. Dutta, L. A. Koszi, B. P. Senger, S. G. Napholtz: InGaAsP ridge waveguide laser array with nonuniform spacing. Appl. Phys. Lett. **48**, 312–314 (1986)

1.282 M. Razeghi, R. Blondeau: First fabrication of CW high power phase-locked laser arrays emitting at 1.3 μm grown by LP-MOCVD. Gallium Arsenide and Related Compounds 1986, Proceedings of the 13th Int'l Symp., pp. 391–394

1.283 Z. L. Liau, J. N. Walpole: Monolithic two-dimensional GaInAsP/InP laser arrays: International Electron Devices Meeting 1986. Technical Digest, pp. 622–625

1.284 N. K. Dutta, T. M. Shen, S. G. Napholtz, T. Cella: InGaAsP high power laser array. Proc. SPIE **723**, 25–28 (1987)

1.285 D. Yap, Z. L. Liau, D. Z. Tsang, J. N. Walpole: High-performance InGaAsP/InP buried-heterostructure lasers and arrays defined by ion-beam-assisted etching. Appl. Phys. Lett. **52**, 1464–1466 (1988)

1.286 Y. Twu, N. K. Dutta, C. A. Green, J. D. Wynn: GaInAsP distributed feedback laser array. Electron. Lett. **24**, 743–744 (1988)

1.287 D. Yap, J. N. Walpole: Mass-transported InGaAsP/InP lasers. The Lincoln Laboratory J.**2**, 77–93 (1989)

1.288 H. Saito, Y. Noguchi: A reflection-type surface-emitting 1.3μm InGaAsP/InP laser array with microcoated reflector. Jpn. J. Appl. Phys. **28**, L1239–L1241 (1989)

1.289 S. L. Palfrey, R. Enstrom, E. Vangieson, J. M. Hammer, R. U. Martinelli, J. T. Andrews, J. Appert, R. Stolzenberger, A. Triano, N. W. Carlson, G. A. Evans: Coherent mutually-injection-coupled linear and two-dimensional arrays of InGaAs/InGaAsP/InP multi-quantum-well grating-surface-emitting diode lasers operating at 1.5 μm. Appl. Phys. Lett. **57**, 2753–2755 (1990)

1.290 E. A. Vangieson, S. L. Palfrey, R. Enstrom, J. M. Hammer, R. U. Martinelli, N. W. Carlson, G. A. Evans, J. T. Andrews, R. Stolzenberger, A. Triano: Coherent high power arrays of InGaAs/InGaAsP multi-quantum-well grating-surface-emitting diode lasers operating at $\lambda = 1.5$ μm. Appl. Phys. Lett. **59**, 2790–2792 (1991)

1.291 A. Valster, J. P. Andre, E. Dupont-Nivet, G. M. Martin: High-power AlGaInP three-ridge type laser diode array. Electron. Lett. **24**, 326–327 (1988)

1.292 D. P. Bour: AlGaInP quantum well lasers, in *Quantum Well Lasers*, ed. by P. Zory (Academic, Orlando 1993)

1.293 K. J. Linden, R. E. Reeder: Diode laser arrays with high power in the 4 to 5 μm infrared region. Optical Engineering **23**, 685–686 (1984)

1.294 J. N. Baillargeon, P. K. York, C. A. Zmudzinski, G. E. Fernandez, K. J. Beernink, J. J. Coleman: High-power phase-locked InGaAs strained-layer quantum well heterostructure periodic laser array. Appl. Phys. Lett. **53**, 457–459 (1988)

1.295 D. P. Bour, N. W. Carlson, G. A. Evans, S. K. Liew, J. B. Kirk, W. F. Reichert: Surface-emitting, distributed feedback InGaAs/AlGaAs lasers by organometallic vapor phase epitaxy. J. Appl. Phys. **70**, 4687–4693 (1991)

1.296 J. J. Coleman: Strained layer quantum well heterostructure lasers, in *Quantum Well Lasers*, ed. by P. Zory (Academic, Orlando 1993)

1.297 V. J. Corcoran, I. A. Crabbe: Electronically scanned waveguide laser arrays. Appl. Opt. **13**, 1755–1757 (1974)

1.298 V. J. Corcoran: Far-infrared-submillimeter phased arrays and applications. IEEE Trans.MTT-**22**, 1103–1107 (1974)

1.299 E. M. Philipp-Rutz: Spatially coherent beam formation and mode locking of an array of solid-state lasers. IEEE J. QE-**14**, 112–118 (1978)

1.300 D. G. Youmans: Phase locking of adjacent channel leaky waveguide CO_2 lasers. Appl. Phys. Lett. **44**, 365–367 (1984)

1.301 A. F. Glova, Yu. A. Dreizin, O. R. Kachurin, F. V. Lebedev, V. D. Pis'mennyi: Phase locking of a two-dimensional array of CO_2 waveguide lasers. Sov. Tech. Phys. Lett. **11**, 102–103 (1985)

1.302 V. V. Antyukov, A. F. Glova, O. R. Kachurin, F. V. Lebedev, V. V. Likhanskii, A. P. Napartovich, V. D. Pis'mennyi: Effective phase locking of an array of lasers. JETP Lett. **44**, 79–81 (1986)

1.303 L. A. Newman, R. A. Hart, J. T. Kennedy, A. J. Cantor, A. J. DeMaria, W. B. Bridges: High power coupled CO_2 waveguide laser arrays. Appl. Phys. Lett. **48**, 1701–1703 (1986)

1.304 G. L. Bourdet, G. M. Mullot, J. Y. Vinet: Linear array of self-focusing CO_2 waveguide lasers. IEEE J. QE-**26**, 701–710 (1990)

1.305 K. M. Abramski, A. D. Colley, H. J. Baker, D. R. Hall: Offset frequency stabilization of RF excited waveguide CO_2 laser arrays. IEEE J. QE-**26**, 711–717 (1990)

1.306 K. M. Abramski, A. D. Colley, H. J. Baker, D. R. Hall: Phase-locked CO_2 laser array using diagonal coupling of waveguide channels. Appl. Phys. Lett. **60**, 530–532 (1992)

1.307 A. F. Vasil'ev, A. A. Mak andV. M. Mit'kin, V. A. Serebryakov, V. E. Yashin: Correction of thermally induced optical aberrations and coherent phasing of beams during stimulated brillouin scattering. Sov. Phys.–Tech. Phys. **31**, 191–193 (1986)

1.308 L. E. Zapata: Medium power Nd^{3+} : glass array laser. Proc. SPIE **783**, 53–59 (1987)

1.309 H. Tajima, T. Yamashita, T. Mochizuki: High-average-power fiber bundle laser. CLEO '88 Technical Digest **7**, paper THH5

1.310 K. P. Driedger, R. M. Ifflander, H. Weber: Multirod resonators for high-power solid-state lasers with improved beam quality. IEEE J. QE-**24**, 665–673 (1988)

1.311 J. M. Eggleston: Periodic resonators for average-power scaling of stable-resonator solid-state lasers. IEEE J. QE-**24**, 1821–1824 (1988)

1.312 E. H. C. Parker (ed): *The Technology and Physics of Molecular Beam Epitaxy:* (Plenum, New York 1985)

1.313 L. M. Miller, J. J. Coleman: *CRC Critical Reviews in Solid State and Materials Sciences*, in Metalorganic Chemical Vapor Deposition: (CRC, Boca Raton 1988)

1.314 G. B. Stringfellow: *Organometallic Vapor-Phase Epitaxy: Theory and Practice:* (Academic, San Diego 1989)

1.315 M. A. Herman, H. Sitter: *Molecular Beam Epitaxy*, Springer Ser. Mater. Sci., Vol. 7 (Springer, Berlin, Heidleberg 1989)

1.316 C. A. Wang: A new organometallic vapor phase epitaxy reactor for highly uniform epitaxy. The Linconl Laboratory Journal **3**, 3–21 1990

1.317 W. Richter, J. B. Mullin (eds.): Metalorganic Vapor Phase Epitaxy: Proc. 5th Int'l Conf. on Metalorganic Vapor Phase Epitaxy and Workshop on MOMBE, CBE, GSMBE, and Related Techniques, Aachen, Germany (1990) (Elsevier, Amsterdam 1991)

1.318 A. Y. Cho: Advances in molecular beam epitaxy (MBE). J. of Crystal Growth **111**, 1 1991

1.319 M. B. Panish H. Temkin: *Gas Source Molecular Beam Epitaxy*, Springer Ser. Mater. Sci., Vol. 26 (Springer, Berlin, Heidleberg 1993)

1.320 S. L. Bernasek, T. Venkatesan, H. Temkin (eds.): MRS Proc. **126** (1988)

1.321 LEOS 1991 Topical Meeting on Microfabrication for Photonics and Opto-electronics, (1991)

1.322 K. Iga, S. Kinoshita: *Liquid Phase Epitaxy for Semiconductor Lasers*, Springer Ser. in Mater. Sci., Vol. 30 (Springer, Berlin, Heidelberg 1994)

1.323 Kressel J.K. Butler: *Semiconductor Lasers and Heterojunction LEDs* (Academic, New York 1977)

1.324 G. P. Agrawal, N. K. Dutta: *Long-Wavelength Semiconductor Lasers* (Van Nostrand Reinhold, New York 1986)

1.325 H. C. Casey, M. B. Panish: *Heterojunction Lasers Part B Materials and Operating Characteristics* (Academic, New York 1978)

1.326 H. K. Choi, C. A. Wang, S. J. Eglash: III − V diode lasers for new emission wavelengths. The Lincoln Laboratory J. **3**, 395–411 (1990)

1.327 Y. Arakawa, A. Yariv: Quantum well lasers- gain, spectra, dynamics. IEEE J. QE-**22**, 1887 (1986): and references therein.

1.328 H. Okamoto: Jpn. J. Appl. Phys. **26**, 315 (1987): and references therein.

1.329 J. R. Shealy: Optimizing the performance of AlGaAs graded index separate confining heterostructure quantum well lasers. Appl. Phys. Lett. **50**, 1634–1636 (1988)

1.330 R.G. Waters, D. K. Wagner, D. S. Hill, P. L. Tihanyi, B. J. Vollmer: High-power conversion efficiency quantum well diode lasers. Appl. Phys. Lett. **51**, 1318–1319 (1987)

1.331 M. Kondo, T. Suyama, M. Hosoda, T. Hayakawa, T. Hijikata: 3.7 W CW operation of (111)- oriented GaAs/AlGaAs quantum well lasers prepared by molecular beam epitaxy: Int'l Electron Devices Meeting 1988. Technical Digest, pp. 311–314

1.332 C. A. Wang, H. K. Choi: Organometallic vapor phase epitaxy of high-performance strained-layer InGaAs − AlGaAs diode lasers. IEEE J. QE-**27**, 681 (1991)

1.333 P. J. A. Thijs, L. F. Tiemeijer, P. I. Kuindersma, J. J. M. Binsma, T. Van Dongen: High-performance 1.5 μm wavelength InGaAs − InGaAsP strained quantum well lasers and amplifiers. IEEE J. QE-**27**, 1426–1439 (1991)

1.334 A. R. Adams: Band structure engineering for low threshold high efficiency semiconductor lasers. Electron. Lett. **22**, 249–250 (1986)

1.335 E. Yablonovitch, E. O. Kane: Reduction of lasing threshold current density by the lowering of the valence band effective mass. IEEE J. LT-**4**, 504–506 (1986): see correction in IEEE J. LT-**4**, 961 (1986)

1.336 E. Yablonovitch, E. O. Kane: Band structure engineering of semiconductor lasers for optical communications. IEEE J. LT-**6**, 1292–1299 (1988)

1.337 R. J. Fu, C. S. Hong, E. Y. Chan, D. J. Booher, L. Figueroa: High-temperature operation of InGaAs strained quantum-well lasers. IEEE Photon. Technol. Lett. **3**, 308–310 (1991)

1.338 C. S. Hong, R. J. Fu, L. Figueroa: Characteristics and reliablity of high temperature strained quantum-well lasers. Proc. SPIE **1634**, 350–360 (1992)

1.339 P. L. Derry, R. J. Fu, C. S. Hong, E. Y. Chan, K. Chiu, H. E. Hager, L. Figueroa: Design of InGaAs strained quantum well lasers for high temperature operation. Proc. SPIE **1634**, 374–385 (1992)

1.340 K. Itaya, G. Hatakoshi, M. Ishikawa, Y. Nishikawa, S. Saito, M. Okajima: IEEE J. QE-**29**, 2068–2073 (1993)

1.341 H. Naito, M. Kume, K. Hamada, H. Shimizu, G. Kano: Highly-reliable CW operation of 100 mW GaAlAs buried twin ridge substrate lasers with nonabsorbing mirrors. IEEE J. QE-**25**, 1495–1499 (1989)

1.342 D. Welch, R. Craig, W. Streifer, D. Scifres: High reliability, high power, single mode laser diodes. Electron. Lett. **26**, 1481–1482 (1990)

1.343 S. L. Yellen, R. G. Waters, Y. C. Chen, B. A. Stoltz, S. E. Fischer, D. Fekete, J. M. Ballantyne: 20,000 h InGaAs quantum well lasers. Electron. Lett. **26**, 2083–2084 (1990)

1.344 S. L. Yellen, R. G. Waters, H. B. Serreze, J. A. Baumann, R. J. Dalby: Reliability of wide bandgap semiconductor diode lasers. Proc. SPIE **1634**, 229–240 (1992)

1.345 A. Larsson, M. Mittelstein, Y. Arakawa, A. Yariv: High-efficiency broad-area single-quantum-well lasers with narrow single-lobed far-field patterns prepared by molecular beam epitaxy. Electron. Lett. **22**, 79–81 (1986)

1.346 W. Streifer, D. R. Scifres, G. L. Harnagel, D. F. Welch, J. Berger, M. Sakamoto: Advances in diode laser pumps. IEEE J. QE-**24**, 883–894 (1988)

1.347 D. K. Wagner, R. G. Waters, P. L. Tihanyi, D. S. Hill, A. J. Roza, H. J. Vollmer, M. M. Leopold: Operating characteristics of single-quantum well AlGaAs/GaAs high-power diode lasers. IEEE J. QE-**24**, 1258–1264 (1988)

1.348 R. H. Yan, S. W. Corzine, L. A. Coldren: Gain expressions for III-V bulk and quantum well semiconductors, in *Quantum Well Lasers*, ed. by P. Zory (Academic, Orlando 1993)

1.349 R. Engelmann: Multiquantum well lasers, in *Quantum Well Lasers*, ed. by P. Zory (Academic, Orlando 1993)

1.350 H. Naito, M. Kume, K. Hamada, H. Shimizu, M. Kazumura, G. Kano, I. Teramoto: Power-independent degradation of high-power GaAlAs lasers with nonabsorbing mirrors. IEEE J. QE-**27**, 1550–1554 (1991)

1.351 H. Naito, O. Imafuji, M. Kume, H. Shimizu, M. Kazumura: High-power single mode operation of long cavity GaAlAs lasers with nonabsorbing mirror buried twin ridge substrate structure. Appl. Phys. Lett. **61**, 515–516 (1992)

1.352 J. Ungar, N. Bar-Chaim, M. Mazed, M. Mittelstein, S. Oh, I. Ury: GaAlAs window laser emitting 500 mW CW in fundamental mode. Electron. Lett. **26**, 1441 (1990)

1.353 E.-E. Latta, A. Moser, A. Oosenbrug, C. Harder, M. Gasser, T. Forster: Operational limits of GaAs-based single quantum well laser diodes: Research report RZ 2227, IBM Research Division, October 1991.

1.354 H. Jaeckel, G.-L Bona, P. Buchmann, H. P. Meier, P. Vettiger, W. J. Kozlovsky, W. Lenth: Very high-power (425 mW) AlGaAs SQW-GRINSCH ridge laser with frequency-doubled output (41 mW at 428 nm). IEEE J. QE-**27**, 1560–1567 (1991)

1.355 O. Imafuji, T. Takayama, H. Sugiura, M. Yuri, H. Naito, M. Kume, K. Itoh: 600 mW CW single mode GaAlAs triple-quantum-well laser with a new index guided structure. 13th IEEE Int'l Semiconductor Laser Conf. Digest: Post-Deadline Papers, pp. 25–26 (September 1992)

1.356 M. Born, E. Wolf: *Principles of Optics*: (Pergamon, Oxford 1975)

1.357 R. W. Boyd: *Radiometry and the Detection of Optical Radiation*: (Wiley, New York 1983)

1.358 J. W. Goodman: *Introduction to Fourier Optics* (McGraw-Hill, New York 1968)

1.359 H. Wieder, H. Werlich: Characteristics of GaAs laser arrays designed for beam addressable memories. IBM J. of Res. and Develop. **15**, 272–277 (1971)

1.360 A. Limm, P. Nyul, R. Gill, T. Gonda: Moderate-power GaAlAs laser diode array light sources. Electrical Optical Systems Design Conf., pp. 13–19 (1971)

1.361 D. G. Herzog: Application of semiconductor laser diode arrays. Proc. 1973 IEEE/OSA Conf. on Laser Engineering and Applications, Digest of Technical Papers, p. 55

1.362 T. Gonda, R. Gill, H. Kressel, F. Z. Hawrylo: High power LOC heterojunction laser diodes and arrays. 17th Int'l Electron Devices Meeting (abstracts), p. 144

1.363 J. D. Crow, L. D. Comerford,, J. S. Harper, M. J. Brady, R. A. Laff: Gallium arsenide laser-array-on-silicon package. Appl. Opt. **17**, 479–485 (1978)

1.364 E. G. Lean: GaAs laser array and fiber-optic dectector array for disc applications. IBM Technical Disclosure Bulletin **23**, 2992–2993 (1980)

1.365 D. Botez, J. C. Connolly, D. B. Gilbert, M. G. Harvey, M. Ettenberg: High-power individually addressable monolithic array of constricted double heterojunction large-optical-cavity lasers. Appl. Phys. Lett. **41**, 1040–1042 (1982)

1.366 C. W. Reno: Optical disk recording techniques for data rates beyond 100 Mbps. Proc. SPIE **421**, 156–162 1983

1.367 D. B. Carlin, J. P. Bednarz, C. J. Kaiser, J. C. Connolly, M. G. Harvey: Multichannel optical recording using monolithic arrays of diode lasers. Appl. Opt. **23**, 3994 (1984)

1.368 Yu. V. Vovk: Methods of rapid optical recording for binary data. Avtometriya, no. 3, 3–12 (1984)

1.369 M. Kume, M. Horose, N. Yoshikawa, H. Shimizu, M. Wada, K. Itoh, G. Kano, I. Teramoto: A new monolithic dual GaAlAs laser array for read/write optical disk applications. IEEE J. QE-**23**, 898–902 (1987)

1.370 J. J. Lee, N. R. Strader: CMOS ROM arrays programmable by laser beam scanning. IEEE J. SC-**22**, 622–624 (1987)

1.371 S. Nakamura, M. Ojima, T. Nakao, T. Kato, K. Mizuishi: Compact two-beam head with a hybrid two-wavelength laser array for magneto-optic recording. Jpn. J. Appl. Phys. **26**, suppl.26-4, 117–120 (1987)

1.372 K. Torazawa, S. Sumi, S. Murata, S. Minechika, Y. Ishii: Real-time recording and erasure of information by optical head with laser diode array in magneto-optical disk. IEEE Translation J. on Magnetics Jpn. **3**, 132–138 (1988)

1.373 C. J. Hwang, J. S. Chen,, R. J. Fu, D. H. Wu, C. S. Wang: Incoherent GaAlAs/GaAs semiconductor laser arrays. Proc. SPIE **893**, 30–34 (1988)

1.374 M. Tsunakane, K. Endo, M. Nido, I. Komazaki, R. Katayama, K. Yoshihara, Y. Yamanaka, T. Yuasa: High-power individually addressable monolithic laser diode array. Electron. Lett. **25**, 1091–1092 (1989)

1.375 S. Murata, K. Nishimura: Improvement in thermal properties of a multibeam laser diode array. Jpn. J. Appl. Phys. **28**, suppl.28-3, 165–170 (1989)

1.376 M. Tsunakane, K. Endo, S. Ishikawa, R. Katayama, K. Yoshihara, K. Kubota, T. Yuasa: Monolithic eight-channel high-power low-astigmatism AlGaAs laser diode array. Jpn. J. Appl. Phys. **28**, L468–L469 (1989)

1.377 D. W. Nam, R. R. Craig, D. G. Mehuys, D. F. Welch: Uniform high power nine and 18 element individually addressable laser diode array. Electron. Lett. **27**, 464–465 (1991)

1.378 D. B. Carlin: Individually addressed arrays of diode lasers, in *Diode-Laser Arrays*, ed. by D. Botez, D. Scifres (Cambridge Univ. Press, Cambridge 1994)

1.379 A. G. Dewey, R. C. Durbeck: Shared technology for printing and display. IBM Technical Disclosure Bulletin **18**, 245 (1975)

1.380 J. D. Crow, A. G. Dewey, R. C. Durbeck, B. G. Huth, E. G. Lean: Laser-liquid crystal display system. IBM Technical Disclosure Bulletin bf 23, 1633–1634 (1980)

1.381 A. G. Dewey: Design of a fiber-optic laser scanning system for a smectic liquid crystal display. Proc. SPIE **396**, 156–161 (1983)

1.382 J. Shmulovich: Electron-beam-pumped one-dimensional array of light emitters. IEEE Trans. ED-**33**, 1133–1149 (1986)

1.383 P. N. Nasibov: Laser cathode ray tubes and their applications. Proc. SPIE **893**, 200–202 (1988)

1.384 B. Mitchell: Optoelectronics: progress in systems. Electron **77**, 29 (1975)

1.385 M. H. Coden, F. W. Scholl: Lasers move in on information processing. Optical Spectra **15**, 50–52 (1981)

1.386 I. Teramoto, T. Sugino: Laser diode and light emitting diode for optical writing. Electrophotography **23**, 131–139 (1984)

1.387 T. Yagi, K. Yamashita, R. Hattori, M. Kubota, M. Ishii, S. Takamiya: Laser diode array for a laser printer. Proc. SPIE **610**, 143–147 (1986)

1.388 B. Fischer, G. Mader, H. Meixner, P. Kleinschmidt: Laser transfer printing using microencapsulated dyes. Siemens Forschungs- und Entwicklungsberichte **17**, 291–297 (1988)

1.389 R. J. S. Bates: Using monolithic laser arrays for improved transmitter availability in computer data links. Proc. SPIE **842**, 80–85 (1988)

1.390 E. Bradley, P. K. L. Yu: Laser diode requirements and limitations for VLSI holographic optical interconnects. Proc. SPIE **835**, 298–306 (1988)

1.391 M. H. Brodsky: Gallium arsenide optoelectronic IC's for computer networks. Superlattices and Microstructures **8**, 293–296 (1990)

1.392 J. L. Jewell, Y. H. Lee, S. L. McCall, A. Scherer, J. P. Harbison, L. T. Florez, N. A. Olsson, R. S. Tucker, C. A. Burrus, C. J. Sandroff, A. C. Gossard, J. H. English: Two-dimensional array microlasers for photonic switching. In Photonic Switching II. Proc., pp. 144–154 (April 1990)

1.393 M. Osinski: Vertical-cavity surface-emitting semiconductor lasers for optical interconnections. Proc. 1st Int'l Workshop on Photonic Networks, Components and Applications, pp. 70–80 (October 1990)

1.394 D. H. Hartman, L. A. Reith, S. F. Habiby, G. R. Lalk, B. L. Booth, J. E. Marchegiano, J. L. Hohman: Power economy using point-to-point optical interconnect links. Proc. SPIE **1390**, 368–376 (1991)

1.395 H. Yamanaka, M. Sasaki, S. Kikuchi, T. Takada, M. Idda: A gigabit-rate five-highway GaAs OE-LSI chipset for high-speed optical interconnections between modules or VLSIs. IEEE J. on Selected Areas in Communications **9**, 689–697 (1991)

1.396 M. R. Feldman: Holographic optical interconnects for multichip modules. Proc. SPIE **1390**, 427–433 (1991)

1.397 A. C. vonLehman, I. C. Banwell, R. Cordell, C. Chang-Hasnain, J. W. Mann, J. Harbison: High speed operation of hybrid CMOS vertical cavity surface emitting laser array. Electron. Lett. **27**, 1189–1191 (1991)

1.398 Y. Mori: Characteristics of 4 ∗ 4 photonic switch array with gain and high contrast. Appl. Phys. Lett. **58**, 438–440 (1991)

1.399 A. P. Goutzoulis, D. K. Davies, E. C. Malarkey: Prototype position-coded residue look-up table using laser diodes. Opt. Commun. **6**, 302–308 (1987)

1.400 P. A. Molley, K. T. Stalker, W. C. Sweatt: A compact real-time acousto-optic image correlator. Proc. SPIE **16**, 123–130 (1990)

1.401 N. N. Evtihiev, V. V. Perepelitsa, N. A. Esepkina, S. V. Pruss-Zhukovsky, O. N. Vlasov, S. K. Kruglov: A hybrid acousto-optic spectrum analyser for radio-astronomy with semiconductor lasers. J. of Mod. Opt. **36**, 1551–1557 (1989)

1.402 E. G. Paek, J. R. Wullert, A. Von Lehman, J. S. Patel, A. Scherer, J. Harbison, H. J. Yu, R. Martin: Vanderlugt correlator and neural networks. 1989 IEEE Int'l Conf. on Systems, Man and Cybernetics. Conference Proceeding, pp. 408–414

1.403 J. Hong, P. Yeh: Photorefractive parallel matrix-matrix multiplier. Optics Lett. **16**, 1343–1345 (1991)

1.404 R. Newman: Excitation of the Nd^{3+} fluorescence in $CaWO_4$ by recombination radiation in GaAs. J. Appl. Phys. **34**, 437 (1963)

1.405 R. J. Keyes, T. M. Quist: Injection luminescent pumping of $CaF_2:U^{3+}$ with GaAs diode lasers. Appl. Phys. Lett. **4**, 50 (1964)

1.406 C. Larat, M. Schwarz, J. P. Pocholle: High power surface emitting laser diode pumping of Nd:YAG slab. Electron. Lett. **28**, 1630–1631 (1992)

1.407 A. Rosen, P. J. Stabile, F. J. Zutavern, G. M. Loubriel, W. D. Helgeson, M. W. O'Malley, D. L. McLaughlin: 8.5 MW GaAs pulse biased switch optically controlled by 2-D laser diode arrays. IEEE Photon. Technol. Lett. **2**, 525–526 (1990)

1.408 A. Rosen, P. J. Stabile, W. M. Janton, A. M. Gombar, J. Delmaster, R. Hurwitz, P. Herczfeld, A. Bahasadri: Switching technology from DC to GHz using 2-D semiconductor laser arrays. Proc. SPIE **1219**, 517–524 (1990)

1.409 G. A. Evans, A. Rosen, P. J. Stabile, D. P. Bour, N. W. Carlson, J. C. Connolly: Two dimensional edge- and surface-emitting semiconductor laser arrays for optically activated switching. Proc. SPIE **1378**, 146–161 (1991)

1.410 A. Kim, R. Zeto, R. Youmans, C. Kondek, M. Weiner, B. Lalevic: Triggering GaAs lock-on switches with laser diode arrays. Proc. SPIE **1378**, 173–178 (1991)

1.411 G. M. Loubriel, M. T. Buttram, W. D. Helgeson, D. L. McLaughlin, M. W. O'Malley, F. J. Zutsvern, A. Rosen, P. J. Stabile: Triggering GaAs lock-on switches with laser diode arrays. Proc. SPIE **1378**, 179–186 (1991)

1.412 A. Rosen, P. J. Stabile, F. J. Zutavern, G. M. Loubriel: Generic applications for Si and GaAs optical switching devices utilizing semiconductor lasers as an optical source. Proc. SPIE **1378**, 187–194 (1991)

1.413 R. J. De Young, J. H. Lee, M. D. Williams, G. Shuster, E. J. Conway: Comparison of electrically driven lasers for space power transmission: NASA technical memorandum 4045, NASA (June 1988)

1.414 G. A. Landis: Satellite eclipse power by laser illumination. Acta Astronautica **25**, 229–233 (1991)

1.415 R. J. De Young, M. D. Williams, G. H. Walker, G. Shuster, J. H. Lee: A lunar rover powered by an orbiting laser diode array. Space Power **10**, 103–127 (1991)

1.416 V. J. Corcoran: Far-infrared-submillimeter phased arrays and applications. IEEE Trans. MTT-**22**, 1103–1107 (1974)

1.417 V. J. Corcoran: Long range pointer and tracker using an electronically scanned laser phased array. In Laser 77 Opto-Electronics, Digest pp. 700–706

1.418 R. Salathe, W. Bolleter, H. Gilgen: Long range injection laser radar. Appl. Opt. **16**, 2621–2623 (1977)

1.419 P. Akkapeedi, S. H. Macomber: Surface emitting distributed feedback laser as a source for laser radar. Proc. SPIE **1416**, 44–49 (1991)

1.420 M. F. Cullen, P. J. deGroot, G. M. Gallatin: Laser radar array used for improving image analysis algorithms. Proc. SPIE **1002**, 338–343 (1989)

1.421 P. J. deGroot, G. M. Gallatin: Three-dimensional imaging coherent laser radar array. Opt. Eng. **28**, 456–460 (1989)

1.422 W. J. Hannan: Laser mobility aid for the blind. RCA Technical Notes **21**, TN752 (1968)

1.423 W. T. Cathey, W. C. Davis: Vision system with ranging for maneuvering in space. Opt. Eng. **25**, 821–824 (1986)

1.424 D. Dopheide, M. Faber, G. Reim, G. Taux: New optoelectronic velocity and flowrate measuring methods using semiconductor lasers and photodiode. VDI Berichte, Nr. 368, 341–351 (1989)

1.425 D. Dopheide, V. Strunck, M. Faber: Phased diode arrays for velocity measurements and signal processing. Int'l Cong. on Instrumentation in Aerospace Simulation Facilities – ICIASF '89, Proc. pp. 12–18

1.426 J. Katz: Semiconductor optoelectronic devices for free-space optical communications. IEEE Commun. Magazine **21**, 20–27 (1983)

1.427 P. L. Fuhr, F. M. Davidson: Determination of a variable-spacing array's applicability as the NASA-ACTS intersatellite communication system transmitter. GLOBECOM '86: IEEE Global Telecommunications Conference. Communications Broadening Technology Horizons. Conference Record **3**, 1384–1388

1.428 K. Pobl: Semiconductor-laser: towards new applications. Technica **37**, 30–33 (1988)

1.429 M. Ross: Past progress and future advances in space laser communications, MILCOM 88. 21st Century Military Communications – What's Possible? 1988 IEEE Military Commun. Conf. **2**, Proc. 527–532

1.430 H. Kung, D. P. Worland, H. Nguyen, W. Streifer, D. R. Scifres, D. F. Welch, L. Wood, S. Daudt: Semiconductor lasers for space beacons and communications. Proc. SPIE **1044**, 2–10 (1989)

1.431 M. Lucente, E. S. Kintzer, S. B. Alexander, J. G. Fujimoto, V. W. S. Chan: Coherent optical communication with injection-locked high-power semiconductor laser array. Electron. Lett. **25**, 1112–1114 (1989)

1.432 D. K. Probst, R. R. Rice: Potential phased-array semiconductor laser source for coherent laser communications. Proc. SPIE **1218**, 346–357 (1990)

1.433 S. C. Wang, R. E. Stone: Single mode high-power diode laser array for optical communication. Proc. SPIE **1218**, 278–284 (1990)

1.434 D. K. Probst, R. R. Rice: Performance of a phased array semiconductor laser source for coherent laser communications. Proc. SPIE **1417**, 346–357 (1981)

1.435 P. Greulich, B. Hespeler, Th. Spatscheck: Lifetest on a high-power laser diode array transmitter. Proc. SPIE **1522**, 144–153 (1991)

1.436 M. Nakao, K. Sato, T. Nishida, T. Tamamura: Experimental analysis on lasing wavelength controllability of DFB laser array for WDM. Electronics and Communications in Japan, Part 2 (Electronics) **74**, 31–37 (1991)

1.437 S. S. Alexander, E. S. Kintzer, J. C. Livas, J. N. Walpole, C. A. Wang, L. J. Missaggia, S. R. Chinn: 1 Gbit/s coherent optical communication system using a 1 w optical power amplifier. Electron. Lett. **29**, 114–115 (1993)

1.438 S. Nakasuta, K. Tatsuno: Fundamental lateral-mode operation in broad-area lasers having built-in lenslike refractive index distributors. Jpn. J. Appl. Phys. **28**, L1003–L1005 (1992)

1.439 J. N. Walpole, E. S. Kintzer, S. R. Chinn, C. A. Wang, L. J. Missaggia: High-power strained-layer InGaAs/AlGaAs tapered traveling wave amplifier. Appl. Phys. Lett. **61**, 740–742 (1992)

1.440 J. N. Walpole, E. S. Kintzer, S. R. Chinn, C. A. Wang, L. J. Missaggia: High power tapered semiconductor amplifiers and oscillators at 980 nm: IEEE LEOS '92 Annual Meeting: Post Deadline Papers, pp PD–2

1.441 E. S. Kintzer, J. N. Walpole, S. R. Chinn, C. A. Wang, L. J. Missaggia: High-power, strained-layer amplifiers and lasers with tapered gain regions. IEEE Photon. Technol. Lett. **5**, 605–608 (1993)

1.442 S. Chinn: Review of edge-emitting coherent laser arrays, in *Surface Emitting Semiconductor Laser and Arrays*, ed. by G. A. Evans, J. M. Hammer (Academic, San Diego 1993)

1.443 D. Botez: Monolithic phase-locked semiconductor laser arrays, in *Diode-Laser Arrays*, ed. by D. Botez, D. Scifres (Cambridge Univ. Press, Cambridge 1994)

1.444 K. L. Chen, S. Wang: Single-lobe symmetric coupled laser arrays. Electron. Lett. **21**, 347–349 (1985)

1.445 K. L. Chen, S. Wang: Analysis of symmetric Y-junction laser arrays with uniform near-field distribution. Electron. Lett. **22**, 644–645 (1986)

1.446 D. F. Welch, P. S. Cross, D. R. Scifres, W. Streifer, R. D. Burnham: In-phase emission from index guided laser array up to 400 mW. Electron. Lett. **22**, 293–295 (1986)

1.447 B. Hermansson D. Yevick: Analysis of Y-junction and coupled laser arrays. Appl. Opt. **28**, 66–73 (1989)

1.448 W. Streifer, A. Hardy, D. F. Welch, D. R. Scifres, P.S. Cross: Improved Y-X junction laser array. Electron. Lett. **26**, 1730–1731 (1990)

1.449 W. Streifer, P. S. Cross, D. F. Welch, D. R. Scifres: Analysis of a Y-junction semiconductor laser array. Appl. Phys. Lett. **49**, 58–60 (1986)

1.450 W. Streifer, D. F. Welch, P. S. Cross, D. R. Scifres: Y-Junction semiconductor laser arrays: Part *I* – theory. IEEE J. QE-**23**, 744–751 (1987)

1.451 S. R. Chinn, P. S. Zory, A. R. Reisinger: A model for GRIN-SCH-SQW diode lasers. IEEE J. QE-**24**, 2191 (1988)

1.452 N. N. Evtikhiev, M. Sh. Kobyakova, V. N. Lazarev, G. T. Pak: Analysis of the influence of the coupling between channels on optical characteristics of laser Y-junction arrays. Sov. J. Quantum Electron. **19**, 774–776 (1989)

1.453 R. R. Drenten, J. Opschoor, C. J. van der Poel, C. J. Reinhoudt, A. Valster, G. A. Acket: Phase-locked index-guided semiconductor laser arrays. Proc. SPIE **1025**, 57–59 (1989)

1.454 C. J. van der Poel, J. Opschoor, A. Valster, R. R. Drenten, J. P. Andre: Visible Y-junction diode laser with mixed coupling. J. Appl. Phys. **68**, 868–870 (1990)

1.455 A. Yu. Mikhal'kov, A. G. Plyavenek: Hybrid coupling in integrated injection laser arrays. Sov. J. Quantum Electron.**21**, 213–217 (1991)

1.456 T. S. Lay, S. C. Lee, H. H. Lin: Y-junction and misaligned-stripe diode laser arrays with nonuniform reflective diffraction coupler. IEEE J. QE-**25**, 689–695 (1989)

1.457 J. Berger, D. F. Welch, W. Streifer, D. R. Scifres: Narrowing the far field of a Y-junction laser array using a customized spatial filter in an external cavity. Appl. Phys. Lett. **52**, 1560–1562 (1988)

1.458 W. Streifer, D. F. Welch, J. Berger, P. S. Cross, D. R. Scifres: Losses in Y-junction semiconductor laser arrays. Appl. Phys. Lett. **52**, 1297–1299 (1988)

1.459 W. Streifer, D. F. Welch, J. Berger, D. R. Scifres: Nonlinear analysis of Y-junction laser arrays. IEEE J. QE-**25**, 1617–1624 (1989)

1.460 D. Krebs, R. Herrick, K. No, W. Harting, F. Struemph, D. Driemeyer, J. Levy: 22 W coherent GaAlAs amplifier array with 400 emitters. IEEE Photon. Technol. Lett. **3**, 292–295 (1991)

1.461 S. A. Darznek, M. M. Zverev, V. A. Ushakhin: Investigation of a multielement electron-beam-pumped semiconductor laser with an external mirror. Sov. J. Quat. Electron. **4**, 1272–1274 (1975)

1.462 J. Katz, S. Margalit, A. Yariv: Diffraction coupled phase-locked semiconductor laser array. Appl. Phys. Lett. **42**, 554–556 (1983)

1.463 C. G. Dupuy, M. Lurie, J. M. Hammer, G. A. Evans, R. K. DeFreez: Lateral mode control of grating surface emitting laser diode arrays by monolithic Talbot filtering. IEEE J. QE-**28**, 1305–1308 (1992)

1.464 G. A. Evans, N. W. Carlson, J. M. Hammer, M. Lurie, J. K. Butler, S. L. Palfrey, R. Amantea, L. A. Carr, F. Z. Hawrylo, E. A. James, C. J. Kaiser, J. B. Kirk, W. F. Reichert: Two-dimensional coherent laser arrays using grating surface emission. IEEE J. QE-**25**, 1525–1538 (1989)

1.465 H. Talbot: Phil. Mag. **9**, 401 (1836)

1.466 J. T. Winthrop, C. R. Worthington: Theory of Fresnel images. I. plane periodic objects in monochromatic light. J. Opt. Soc. Am. **55**, 373–381 (1965)

1.467 A. A. Golubentsev, V. V. Likhanskii, A. P. Napartovich: Theory of phase locking of an array of lasers. Sov. Phys. JETP **66**, 676–682 (1987)

1.468 J. A. Leger, M. L.Scott, W. B. Veldkamp: Coherent addition of AlGaAs laser using microlenses and diffractive coupling. Appl. Phys. Lett. **52**, 1771–1773 (1988)

1.469 F. X .D'Amato, E. T. Siebert, C. Roychoudhuri: Coherent operation of an array of diode lasers using a spatial filter in a talbot cavity. Appl. Phys. Lett. **55**, 816–818 (1989)

1.470 J. Leger: External methods of phase locking and coherent beam addition of diode lasers, in *Surface Emitting Semiconductor Laser and Arrays*, ed. by G. A. Evans, J. M. Hammer. (Academic, San Diego 1993)

1.471 L. R. Harriott, R. E. Scotti, K. D. Cummings, A. F. Ambrose: Micromachining on integrated optical structures. Appl. Phys. Lett. **48**, 1704-1706 (1986)

1.472 R. K. DeFreez, J. Puretz, R. A. Elliot, G. A. Crow, H. Ximen, D. J. Bossert, G. A. Wilson, J. Orloff: Focused ion-beam micromachined diode laser mirrors. Proc. SPIE **1043**, 25–35 (1989)

1.473 J. P. Donnelly, K. K. Anderson, J. D. Woodhouse, W. D. Goodhue, D. Yap, M. C. Gaidis, C. A. Wang: Some applications of ion beams in III-V compound semiconductor device fabrication. Mat. Res. Soc. Symp. Proc. **144**, 421–432 (1989)

1.474 G. A. Evans, D. P. Bour, N. W. Carlson, R. Amantea, J. M. Hammer, H. Lee, M. Lurie, R. C. Lai, P. F. Pelka, R. E. Farkas, J. B. Jirk, S. K. Liew, W. F. Reichert, C. A. Wang, H. K. Choi, J. N. Walpole, J. K. Butler, W. F. Ferguson, R. K. DeFreez, M. Felisky: Characteristics of coherent two-dimensional grating surface emitting diode laser arrays during CW operation. IEEE J. QE-**27**, 1594–1608 (1991)

1.475 M. Jansen, J. J. Yang, L. Heflinger, S. S. Ou, M. Sergant, J. Huang, J. Wilcox, L. Eaton, W. Simmons: Coherent operation of injection-locked monolithic surface-emitting diode laser arrays. Appl. Phys. Lett. **54**, 2634–2636 (1989)

1.476 J. Z. Wilcox, W. W. Simmons, G. P. Peterson, J. J. Yang, M. Jansen, S. S. Ou: Design of multi quantum well lasers for surface emitting arrays. Proc. SPIE **1043**, 192–196 (1989)

1.477 H. Saito, Y. Kondo: 4×4 surface-emitting $1.5\,\mu$m InGaAsP/InP laser array with microcoated reflectors fabricated by reactive ion etching. Jpn. J. Appl. Phys. **30**, L599 (1991)

1.478 D. W. Nam, R. G. Waarts, D. F. Welch, D. R. Scifres: High power monolithic two-dimensional arrays of single mode surface emitting lasers. IEEE Laser and Electro-Optics Society 1991 Annual Meeting Conf. Digest, p. 15

1.479 M. Wu, Y. J. Chen, J. Hryniewicz, Y. P. Ho, D.Mergerian: A monolithic individually addressable two-dimensional surface-emitting diode laser array. Digest of the IEEE Laser and Electro-Optics Society Summer Topical Meetings, p. 15 (1991)

1.480 D. W. Nam, R. G. Waarts, D. F. Welch, D. R. Scifres: High power cw monolithic 2-d arrays of surface-emitting lasers. Conf. on Laser and Electro-Optics 1992 Technical Digest **12**, pp 346–347

1.481 D. R. Scifres, R. D. Burnham, W. Streifer: Grating-coupled GaAs single heterostructure ring laser. Appl. Phys. Lett. 28, 681–683 (1976)

1.482 T. Kajimura, K. Saito, N. Shige, R. Ito: Leaky-mode buried-heterostructure AlGaAs injection lasers, Appl. Phys. Lett. **30**, 590–591 (1977)

1.483 ZH. I. Alferov, V. M. Andreyev, S. A. Gurevich, R. F. Kazarinov, V. R. Larionov, M. N. Mizerov, E. L. Portnoy: Semiconductor lasers with the light output through the diffraction grating on the surface of the waveguide layer. IEEE J. QE-**11**, 449–451 (1975)

1.484 P. Zory, L. D. Comerford: Grating-coupled double-heterostructure AlGaAs diode lasers. IEEE J. QE-**11**, 451–457 (1975)

1.485 R. D. Burnham, D. R. Scifres, W. Streifer: Low-divergence beams from grating-coupled composite guide heterostructure GaAlAs diode lasers. Appl. Phys. Lett. **26**, 644–647 (1975)

1.486 D. R. Scifres, R. D. Burnham, W. Streifer: Highly collimated laser beams from electrically pumped SH GaAs/GaAlAs distributed-feedback lasers. Appl. Phys. Lett. **26**, 48–50 (1975)

1.487 D. R. Scifres, W. Streifer, R. D. Burnham: Leaky wave room-temperature double heterostructure GaAs:GaAlAs diode laser. Appl. Phys. Lett. **29**, 23–25 (1976)

1.488 F. K. Reinhart, R. A. Logan, C. V. Shank: GaAs – $Al_xAl_{1-x}As$ injection lasers with distributed Bragg reflectors. Appl. Phys. Lett. **27**, 45–48 (1975)

1.489 W. Ng, A. Yariv: Highly collimated broad side emission from room-temperature GaAs Bragg reflector lasers. Appl. Phys. Lett. **31**, 613–615 (1977)

1.490 D. F. Welch, R. Parke, D. Mehuys, D. Scifres: High-power diffraction-limited operation of monolithically integrated active grating master oscillator power amplifiers. CLEO 1992, Technical Digest **12**, pp. 342–343

Chapter 2

2.1 W. W. Chow, W. Koch, M. Sargent III: *Semiconductor-Laser Physics* (Springer, Berlin, Heidelberg 1994)

2.2 T. C. Wu, S. C. Kan, D. Vassilovski, K. Y. Lau, C. E. Zah, B. Pathak, T. P. Lee: Gain compression in tensile-strained 1.55 μm quantum well lasers operating at first and second quantized states. Appl. Phys. Lett. **60**, 1794–1796 (1992)

2.3 M. C. Tatham, C. P. Seltzer, S. D. Perrin, D. M. Cooper: Frequency response and differential gain in strained and unstrained InGaAs/InGaAsP quantum well lasers. Electron. Lett. **27**, 1278–1280 (1991)

2.4 A. Grabmaier, G. Fuchs, A. Hangleiter, R. W. Glew, P. D. Greene, J. E. A. Whiteaway: Linewidth enhancement factor and carrier-induced differential index in InGaAs separate confinement multi-quantum-well lasers. J. Appl. Phys. **70**, 2467–2469 (1991)

2.5 W. Rideout, B. Yu, J. LaCourse, P. K. York, K. J. Beerink, J. J. Coleman: Measurement of the carrier dependence of differential gain, refractive index, and linewidth enhancement factor in strained-layer quantum well lasers. Appl. Phys. Lett. **56**, 706–708 (1990)

2.6 T. Takahashi, M. Nishioka, Y. Arakawa: Differential gain of GaAs/AlGaAs quantum well and modulation-doped quantum well lasers. Appl. Phys. Lett. **58**, 4–6 (1991)

2.7 C. A. Zmudzinski, P. S. Zory, G. G. Lim, L. M. Miller, K. J. Beerink, T. L. Cockerill, J. J. Coleman, C. S. Hong, L. Figueroa: Differential gain in bulk and quantum well diode lasers. IEEE Photon. Technol. Lett. **3**, 1057–1060 (1991)

2.8 P. W. A. McIlroy, A. Kurobe, Y. Uematsu: Analysis and application of theoretical gain curves to the design of multiquantum-well lasers., IEEE J. QE-**21**, 1958–1963 (1985)

2.9 A. Kurobe, H. Furuyama, S. Naritsuka, N. Sugiyama, Y. Kokubun, M. Nakamura: Effects of well number, cavity length, and facet reflectivity on the reduction of threshold current of GaAs/AlGaAs multiquantum well lasers. IEEE J. QE-**24**, 635–640 (1988)

2.10 T. A. DeTemple, C. M. Herzinger: On the semiconductor laser logarithmic gain-current density relationship. IEEE J. QE-**29**, 1246–1252 (1993)

2.11 M. Mittelstein, Y. Arakawa, A. Larsson, A. Yariv: Second quantized state lasing of a current pumped single quantum well lasers. Appl. Phys. Lett. **49**, 1689–1691 (1986)

2.12 P. S. Zory, A. R. Reisinger, L. J. Mawst, G. Costrini, C. A. Zmudzinski, M. A. Emanuel, M. E. Givens, J. J. Coleman: Anomalous length dependence of threshold for thin quantum well AlGaAs lasers. Electron. Lett. **22**, 475–477 (1986)

2.13 R. Nagarajan, T. Kamiya, A. Kurobe: Bandfilling in GaAs/AlGaAs multiquantum well lasers and its effect on the threshold current. IEEE J. QE-**25**, 1161–1170 (1989)

2.14 J. Nagle, S. Hersee, M. Krakowski, T. Weil, C. Weisbuch: Threshold current of single quantum well lasers: The role of the confining layers. Appl. Phys. Lett. **49**, 1325–1327 (1986)

2.15 J. E. A. Whiteaway, G. H. B Thompson, P. D. Greene, R. W. Glew: Logarithmic gain/current-density characteristics of InGaAs/InGaAsP/InP multiquantum-well separate-confinement-heterostructure lasers. Electron. Lett. **27**, 340–342 (1991)

2.16 J. Katz: Power conversion efficiency of semiconductor injection lasers and laser arrays in cw operation. IEEE J. QE-**21**, 1854–1857 (1985)

2.17 D. Botez, P. Zory: Constricted double-heterostructure (AlGa)As diode lasers. Appl. Phys. Lett. **32**, 261–263 (1978)

2.18 J. R. Biard, W. N. Carr, B. S. Reed: Trans. AIME **230**, 286 (1964)

2.19 A. R. Reisinger, P. S. Zory, R. G. Waters: IEEE J. QE-**23**, 993 (1987)

2.20 U. Koren, B. I. Miller, Y. K. Su, T. L. Koch, J. E. Bowers: Low internal loss separate confinement heterostructure InGaAs/InGaAsP quantum well laser. Appl. Phys. Lett. **51**, 1744–1746 (1987)

2.21 R. A. Smith: *Semiconductors*: (Cambridge Univ. Press, Cambridge 1959)

2.22 A. E. Mozer, S. Hausser, M. H. Pilkuhn: Quantitative evaluation of gain and losses in quaternary lasers. IEEE J. QE-**21**, 719–724 (1985)

2.23 P. A. Andrekson, N. A. Olsson, T. Tanbun-ek, R. A. Logan, D. Coblentz, H. Temkin: Novel technique for determining internal loss of individual semiconductor lasers. Electron. Lett. **28**, 171–172 (1992)

2.24 D. P. Bour, A. Rosen: Optimum cavity length for high conversion efficiency quantum well diode lasers. J. Appl. Phys. **66**, 2813–2818 (1989)

2.25 G. P. Agrawal, W. B. Joyce, R. W. Dixon, M. Lax: Beam–propagation analysis of stripe–geometry semiconductor lasers: Threshold behavior. Appl. Phys. Lett. **43**, 11–13 (1983)

2.26 P. Meissner, E. Patzak, D. Yevick: A self–consistent model of stripe geometry lasers based on the beam propagation method. IEEE J. QE-**20**, 899–905 (1984)

2.27 R. Baets, J. P. van de Capelle, P. E. Lagasse: Longitudinal analysis of semiconductor lasers with low reflectivity facets. IEEE J. QE-**21**, 693–669 (1985)

2.28 J. K. Butler, G. A. Evans: Self-consistent analysis of gain saturation in channeled-substrate-planar lasers. Appl. Phys. Lett. **51**, 1792–1794 (1987)

2.29 D. Marcuse: Computer model of an injection laser amplifier. IEEE J. QE-**19**, 63–73 (1983)

2.30 W. W. Rigrod: Homogeneously broadened cw lasers with uniform distributed loss. IEEE J. QE-**14**, 377–381 (1978)

2.31 A. E. Siegman: *Lasers* (University Sicence Books, Mill Valley, CA 1987)

2.32 W. J. Witteman: *The CO_2 Laser*, Springer Ser. Opt. Sci., Vol. 53 (Springer, Berlin, Heidleberg 1986)

2.33 J.K. Butler, G. A. Evans, N. W. Carlson: Nonlinear characterization of modal gain and effective index saturation in channeled-substrate-planar heterojunction lasers. IEEE J. QE-**25**, 1646–1651 (1989)

2.34 L. W. Casperson: Laser power calculations: sources of error. Appl. Opt. **19**, 422–433 (1980)

2.35 G. M. Schindler: Optimum output efficiency of homogeneously broadened lasers with constant loss. IEEE J. QE-**16**, 546 (1980)

2.36 4th Edition: *CRC Handbook of Laser Science and Technology 1: Lasers and Masers* (CRC Press, Boca Raton 1987)

2.37 W. W. Duley: CO_2 *Lasers Effects and Applications*: (Academic, New York 1976)

2.38 F. J. Duarte (ed.): *High-Power Dye Lasers* (Springer, Berlin, Heidleberg 1991)

2.39 A. Yariv: *Quantum Electronics*, 2nd edn. (Wiley, New York 1975)

2.40 D. Mehuys, D. F. Welch, L. Goldberg, J. Weller: 4.5 W CW near-diffraction-limited tapered-stripe semiconductor optical amplifier. Electron. Lett. **29**, 219–221 (1993)

2.41 C. Zmudzinski, D. Botez, L. J. Mawst, C. Tu, L. Frantz: Coherent 1 W continuous wave operation of large-aperture resonant arrays of antiguided diode lasers. Appl. Phys. Lett. **62**, 2914–2916 (1993)

2.42 L. Goldberg, D. Mehuys, D. C. Hall: 3.3 W CW diffraction-limited broad area semiconductor amplifier. Electron. Lett. **28**, 1082–1084 (1992)

2.43 G. H. B. Thompson: A theory for filamentation in semiconductor lasers including the dependence of dielectric constant on injected carrier density. Opto-electronics **4**, 257–311 (1972)

2.44 B. R. Bennett, R. A. Soref, J. A. Del Alamo: Carrier-induced change in refractive index of InP, GaAs, and InGaAsP. IEEE J. QE-**26**, 113 (1990)

2.45 L. D. Landau, E. M. Lifshitz: *Electrodynamics of Continuoud Media*: (Pergamon Press, Oxford 1975)

2.46 F. Stern: Dispersion of the index of refraction near the absorption edge of semiconductors. Phys. Rev. A **133**, 1653–1664 (1964)

2.47 S. Tarucha, H. Kobayashi, Y. Horikoshi, H. Okamoto: Carrier-induced energy-gap shrinkage in current-injection GaAs/AlGaAs MQW heterostuctures. Jpn. J. Appl. Phys. **23**, 874–878 (1984)

2.48 E. Zielinski, H. Schweizer, S. Hausser, R. Stuber, M. K. Pilkuhn, G. Weimann: Systematics of laser operation in GaAs/AlGaAs multiquantum well lasers. IEEE J. QE-**23**, 969–975 (1987)

2.49 P. Blood: Stimulated emission in quantum well laser diodes. Appl. Phys. Lett. **55**, 1–3 (1989)

2.50 S. H. Park, J. I. Shim, K. Kudo, M. Asada, S. Arai: Band gap shrinkage in GaInAs/GaInAsP/InP multi-quantum well lasers. J. Appl. Phys. **72**, 279–281 (1992)

2.51 M. Asada: Intraband relaxation effect on optical spectra, in *Quantum Well Lasers*, ed. by P. Zory (Academic, Orlando 1993)

2.52 J. M. Ziman: *Principles of the Theory of Solids.* (Cambridge Univ. Press, Cambridge 1972)

2.53 C. H. Henry, R. A. Logan, F. R. Merritt, J. P. Luongo: The effect of intervalence band absorption on the thermal behavior of InGaAsP lasers. IEEE J. QE-**19**, 947–952 (1983)

2.54 M. Takeshima: Intervalence-band absorption in relation to Auger recombination in laser materials. Jpn. J. Appl. Phys. **23**, 428–435 (1984)

2.55 M. Asada, A. Kameyama, Y. Suematsu: Gain and intervalence band absorption in quantum–well lasers. IEEE J. QE-**20**, 745–753 (1984)

2.56 C. H. Henry, R. A. Logan, K. A. Bertness: Spectral dependence of the change in refractive index due to carrier injection in GaAs lasers. J. Appl. Phys. **52**, 4457–4461 (1981)

2.57 B. W. Hakki, T. L. Paoli: Gain spectra in GaAs double-heterostructure injection lasers. J. Appl. Phys. **46**, 1299–1306, (1975)

2.58 S. E. H. Turley, G. H. B. Thompson, D. F. Lovelace: Effect of injection current on dielectric constant of an inbuilt waveguide in twin-transverse-junction stripe lasers. Electron. Lett. **15**, 256–257 (1979)

2.59 I. D. Henning, J. V. Collins: Measurements of the semiconductor laser linewidth broadening factor. Electron. Lett. **19**, 927–929 (1983)

2.60 J. Manning, R. Olshansky, C. B. Su: The carrier-induced index change in AlGaAs and 1.3 μm InGaAsP diode lasers. IEEE J. QE-**19**, 1525–1520 (1983)

2.61 N. K. Dutta, N. A. Olsson, W. T. Tsang: Carrier induced refractive index change in AlGaAs quantum well lasers. Appl. Phys. Lett. **45**, 836–837 (1984)

2.62 N. Ogasawara, R. Ito, R. Morita: Linewidth enhancement factor in GaAs/AlGaAs multi-quantum-well lasers. Jpn. J. Appl. Phys. **24**, L519–L521 (1985)

2.63 S. Hausser, W. Idler, E. Zielinski, M. K. Pilkuhn, G. Weimann, W. Schlapp: Spontaneous emission factor and waveguiding in GaAs/AlGaAs MQW lasers. IEEE J. QE-**25**, 1469–1476 (1989)

2.64 P. Brosson, J. Jacquet, A. Perales, B. Mersali, D. Leclerc: Carrier induced differential refractive index detuning effect in GaInAsP SCMQW lasers with 3, 5, and 9 wells. 12th IEEE Int'l Semiconductor Laser Conf. Digest, pp. 88–89 (1990)

2.65 M. Osinski, J. Buus: Linewidth broadening factor in semiconductor lasers - an overview. IEEE J. QE-**19**, 9–29 (1987)

2.66 S. Banerjee, A. K. Srivastava, N. Chand: Reduction of the linewidth enhancement factor for $In_{0.2}Ga_{0.8}As/GaAs/Al_{0.5}Ga_{0.5}As$ strained quantum well lasers. Appl. Phys. Lett. **58**, 2198–2199 (1991)

2.67 N. Storkfelt, M. Yamaguchi, B. Mikkelsen, K. E. Stubkjaer: Recombination constants and α factor in $1.5\,\mu m$ MQW optical amplifiers taking carrier overflow into account. Electron. Lett. **28**, 1774–1776 (1992)

2.68 M. Asada, A. R Adams, K. E. Stubkjaer, Y. Suematsu, Y. Itaya, S. Arai: The temperature dependence of the threshold current of GaInAsP/InP DH lasers. IEEE J. QE-**17**, 611–618 (1981)

2.69 Y. C. Chen, A. R. Reisinger, S. R. Chinn: Thermal waveguiding in oxide-defined, narrow-stripe, large-optical-cavity lasers. Appl. Phys. Lett. **15**, 129–131 (1982)

2.70 P. Blood, S. Colak, A. I. Kucharska: Temperature dependence of threshold current in GaAs/AlGaAs quantum well lasers. Appl. Phys. Lett. **52**, 599–601 (1988)

2.71 N. K. Dutta: Calculated temperature dependence of threshold current of $GaAsAl_xGa_{1-x}As$ double heterostructure lasers. J. Appl. Phys. **52**, 70–73 (1981)

2.72 P.S. Zory, A.R. Reisinger, R. G. Waters, L.J. Mawst, C.A. Zmudzinski, M.A. Emanuel, M.E. Givens, J.J. Coleman: Anomalous temperature dependence of threshold for thin quantum well AlGaAs diode lasers. Appl. Phys. Lett. **49**, 16–18 (1986)

2.73 M. M. Leopold, A. P. Specht, C.A. Zmudzinski, M.E. Givens, J.J. Coleman: Temperature-dependent factors contributing to T_0 in graded-index separate-confinement-heterostructure single quantum well lasers. Appl. Phys. Lett. **50**, 1403–1405 (1987)

2.74 W. B. Joyce, R. W. Dixon: Thermal resistance of heterostructure lasers. J. Appl. Phys. **46**, 855–862 (1975)

2.75 T. Kobayashi, Y. Furukawa: Temperature distributions in the GaAs–GaAlAs double-heterostructure laser below and above the threshold current. Jpn. J. Appl. Phys. **14**, 1981 (1975)

2.76 H. Yonezu, T. Yuasa, T. Shinohara, T. Kamejima, I. Sakuma: CW optical power from (AlGa) As double heterostructure lasers. Jpn. J. Appl. Phys. **15**, 2393 (1976)

2.77 A. G. Stevenson, P. J. Fiddyment, D. H. Newman: Low threshold current proton-isolated (GaAlAs) double heterostructure lasers. Opt. Quantum Electron. **9**, 519–525 (1977)

2.78 D. H. Newman, D. J. Bond, J. Stefani: Thermal-resistance models for proton-isolated double heterostructure lasers. Solid State Electron Devices **2**, 41–46 (1978)

2.79 E. Duda, J. C. Carballes, J. Apruzzese: Thermal resistance and temperature distribution in double heterostructure lasers: Calculation and experimental results. IEEE J. QE-**15**, 812–817 (1979)

2.80 M. Ito, T. Kimura: Stationary and transient thermal properties of semiconductor laser diodes. IEEE J. QE-**17**, 787–795 (1981)

2.81 J. S. Manning: J. Appl. Phys. **52**, 3179 (1981)

2.82 W. B. Joyce: Current-crowded carrier confinement in double heterostructure lasers. J. Appl. Phys. **51**, 2394–2401 (1980)

2.83 W. B. Joyce: Carrier transport in double-heterostructure active layers. J. Appl. Phys. **53**, 7235–7239 (1982)

2.84 G. Lengyel, P. Meissner, E. Patzak, K. H. Zschauer: IEEE J. QE-**18**, 618 (1982)

2.85 R. Papannareddy, W. Ferguson, J. K. Butler: Current spreading and carrier diffusion in zinc-diffused multiple-stripe-geometry lasers. Appl. Phys. Lett. **50**, 1316–1318 (1987)

2.86 R. Papannareddy, W. Ferguson, J. K. Butler: A generalized thermal model for stripe-geometry injection lasers. J. Appl. Phys. **62**, 3565–3569 (1987)

2.87 W. Nakwaski: Thermal properties of buried-heterostructure laser diodes. IEE Proc. Pt. J **134**, 87–94 (1987)

2.88 E. M. Garmire, M. T. Travis: Heatsink requirements for coherent operation of high-power laser arrays. IEEE J. QE-**20**, 1277–1281 (1984)

2.89 Z. L. Liau, J. N. Walpole, D. Z. Tsang, V. Diadiuk: Characterization of mass-transported p-substrate GaInAsP/InP buried-heterostructure lasers with analytical solutions for electrical and thermal resistances. IEEE J. QE-**24**, 36–42 (1988)

2.90 J. G. Endriz, M. Vakili, G. S. Browder, M. DeVito, J. M. Haden, G. L. Harnagel, W. E. Plano, M. Sakamoto, D. F. Welch, S. Willing, D. P. Worland, H. C. Yao: High power diode laser arrays. IEEE J. QE-**28**, 952–965 (1992)

2.91 G. L. Harnagel, M. Vakili, K. R. Anderson, D. P. Worland, J. G. Endriz, D. R. Scifres: High-duty cycle, high-power two-dimensional laser diode arrays. Electron. Lett. **29**, 1008–1010 (1993)

2.92 V. Krause, F. Robert, B. Ollier, J. Buschke, T. Kimpel, H. -G. Treusch, P. Lossen, E. Beyer: Copper coolers for high-power laser diode in copper technology. Proc. SPIE **2148**, (1994)

2.93 R. J. Phillips: Microchannel heat sinks. Lincoln Laboratory J. **1**, 31 (1988)

2.94 D. Mundinger, R. Beach, W. Benett, R. Solarz, W. Krupke, R. Staver, D. Tuckerman: Demonstration of high performance silicon microchannel heat exchangers for laser diode array cooling. Appl. Phys. Lett. **53**, 1030 (1988)

2.95 L. J Missaggia, J. N. Walpole, Z. L. Liau, R. J. Phillips: Microchannel heat sinks for two-dimensional high-power-density diode laser arrays. IEEE J. QE-**25**, 1988 (1989)

2.96 R. Beach, D. Mundinger, W. Benett, V. Sperry, B. Comaskey, R. Solarz: High-reliability silicon microchannel submount for high average power laser diode arrays. Appl. Phys. Lett. **56**, 2065 (1990)

2.97 D. Mundinger, R. Beach, W. Benett, R. Solarz, V. Sperry, D. Ciarlo: High average power edge emitting laser diode arrays on silicon microchannel coolers. Appl. Phys. Lett. bf 57, 2172 (1990)

2.98 L. J Missaggia, J. N. Walpole: A microchannel heat sink with alternating directions of water flow in adjacent channels. Proc. SPIE **1582**, p. 106 (1991)

2.99 J. P. Donnelly, W. D. Goodhue, C. A. Wang, R. J. Bailey, G. A. Lincoln, G. D. Johnson, L. J. Missaggia, J. N. Walpole: CW operation of monolithic arrays of surface-emitting folded-cavity InGaAs/AlGaAs diode lasers. IEEE Photon. Technol. Lett. **5**, 747–750 (1993)

2.100 R.Beach, M. A. Emanuel, W. J. Benett, B. L. Freitas, N. W. Carslon, J. Skidmore, R. W. Solarz: Improved performance of high average power semiconductor arrays for applications in diode pumped solid state lasers. Proc. SPIE **2148**, (1994)

Chapter 3

3.1 N. S. Kapany, J. J. Burke: *Optical Waveguides* (Academic, New York 1972)

3.2 D. Marcuse: *Theory of Dielectric Optical Waveguides* (Academic, New York 1974)

3.3 R. G. Hunsperger: *Integrated Optics: Theory and Technology*, Springer Ser. Opt. Sci., Vol.33 (Springer, Berlin, Heidelberg 1984)

3.4 P .Yeh: *Optical Waves in Layered Media*: (Wiley, New York 1988)

3.5 T. Tamir (ed.): *Integrated Optics*, 2nd edn., Topics Appl. Phys., Vol. 7 (Springer, New York 1982)

3.6 T. Tamir (ed.): *Guided-Wave Optoelectronics*, 2 edn., Springer Ser. Electron. Photon., Vol. 26 (Springer, Berlin, Heidelberg 1990)

3.7 L. M. Brekhovskikh, O. A. Godin: *Acoustics of Layered Media I: Plane and Quasi-Plane Waves*, Springer Ser. Wave Phenomena, Vol. 5 (Springer, Berlin, Heidleberg 1990)

3.8 L. M. Brekhovskikh, O. A. Godin: *Acoustics of Layered Media II: Point Sources and Bounded Beams*, Springer Ser. Wave Phenomena, Vol. 10 (Springer, Berlin, Heidelberg 1992)

3.9 W. O. Schlosser: Gain-induced modes in planar structures. Bell Syst. Tech. J. **52**, 887–905 (1973)

3.10 H. K. V. Lotsch: Physical-optics theory of planar dielectric waveguides. Optik **32**, 239–254 (1968)

3.11 H. K. V. Lotsch: Reflection and refraction of a beam of light at a plane interface. J. Opt. Soc. Am. **58**, 551–561 (1968)

3.12 D. B. Hall, C. Yeh: Leaky waves in a heteroepitaxial film. J. Appl. Phys. **44**, 2271–2274 (1973)

3.13 R. W. H. Engelmann, D. Kerps: Leaky modes in active three-layer slab waveguides. IEE Proc. J **127**, 330–336 (1980)

3.14 D. D. Cook, F. R. Nash: Gain-induced guiding and astigmatic output beam of GaAs lasers. J. Appl. Phys. **46**, 1660–1672 (1975)

3.15 T. L. Paoli: Waveguide in a stripe-geometry junction laser. IEEE J. QE-**13**, 662–668 (1977)

3.16 M. Abramowitz, I. A. Stegun: *Handbook of Mathematical Functions* (Dover, New York 1965)

3.17 T. Mamine, T. Oda, O. Yoneyama: New class of gain guiding laser with a tapered-stripe structure. J. Appl. Phys. **54**, 4302–4304 (1983)

3.18 T. Mamine: Astigmatism and spontaneous emission factor of laser diodes with parabolic gain, J. Appl. Phys. **54**, 2103–2105 (1983)

3.19 J. Katz, E. Kapon, C. Lindsey, S. Margalit, A. Yariv: Far-field distributions of semiconductor phase-locked arrays with multiple contacts. Electron. Lett. **19**, 660–662 (1983)

3.20 J. Katz, E. Kapon, S. Margalit, A. Yariv: Rate equation analysis of phase-locked semiconductor laser arrays under steady state conditions. IEEE J. QE-**20**, 875–879 (1984)

3.21 J. Katz, E. Kapon, S. Margalit, A. Yariv: Recent developments in monolithic phase-locked semiconductor laser arrays. Proc. SPIE **465**, 175–180 (1984)

3.22 E. Kapon, C. Lindsey, J. Katz, S. Margalit, A. Yariv: Chirped arrays of diode lasers for supermode control. Appl. Phys. Lett. **45**, 200–202 (1984)

3.23 W. Streifer, A. Hardy, R. D. Burnham, R. L. Thornton, D. R. Scifres: Criteria for design of single-lobe phased-array diode lasers. Electron. Lett. **21**, 505–506 (1985)

3.24 P. M. Morse, R. H. Bolt: Sound waves in rooms. Rev. Mod. Phys. **16**, 69–150 (1944)

3.25 L. W. Casperson: Threshold characteristics of multimode laser oscillators. J. Appl. Phys. **46**, 779 (1975)

3.26 W. Streifer, D. R. Scifres, R. D. Burnham: Longitudinal mode spectra of diode lasers. Appl. Phys. Lett. **40**, 305–307 (1982)

3.27 Y. Suematsu, K. Furuya: Theoretical spontaneous emission factor of injection lasers. Trans. IECE Japan E **60**, 467–472 (1977)

3.28 K. Petermann: Calculated spontaneous emission factor for double-heterostructure injection lasers with gain-induced waveguiding. IEEE J. QE-**15**, 566–570 (1979)

3.29 J. Newstein: The spontaneous emission factor for lasers with gain induced waveguiding. IEEE J. QE-**20**, 1270–1276 (1985)

3.30 H. A. Haus, S. Kawakami: On the "excess spontaneous emission factor" in gain-guided laser amplifiers. IEEE J. QE-**21**, 63–69 (1985)

3.31 J. Arnaud, J. Fesquet, F. Coste, P. Sansonetti: Spontaneous emission in semiconductor laser amplifiers. IEEE J. QE-**21**, 603–608 (1985)

3.32 A. Yariv, S. Margalit: On spontaneous emission into guided modes with curved wavefronts. IEEE J. QE-**18**, 1831–1832 (1982)

3.33 J. Arnaud: Theory of spontaneous emission factor in gain-guided laser amplifiers. Electron. Lett. **19**, 798–800 (1983)

3.34 I. H. Deutsch, J. C. Garrison, E. M. Wright: Excess noise in gain-guided amplifiers. J. Opt. Soc. Am. B **8**, 1244–12519 (1991)

3.35 H. G. Danielmeyer: Effects of drift and diffusion of excited states on spatial hole burning and laser operation. J. Appl. Phys. **42**, 3125–3132 (1971)

3.36 W. Streifer, R. D. Burnham, D. R. Scifres: Dependence of longitudinal mode structure on injected carrier diffusion in diode lasers. IEEE J. QE-**13**, 403–404 (1977)

3.37 R. F. Kazarinov, C. H. Henry, R. A. Logan: Longitudinal mode self-stabilization in semiconductor laser. J. Appl. Phys. **53**, 4631–4644 (1982)

3.38 G. P. Agrawal: IEEE J. QE-**23**, 860 (1987)

3.39 F. H. Peters, D. T. Cassidy: Effect of scattering on the longitudinal mode spectrum of $1.3\,\mu m$ InGaAsP semiconductor diode lasers. Appl. Phys. Lett. **57** , 330–332 (1990)

3.40 L. F. DeChiaro: Damage-induced spectral pertubations in multilongitudinal-mode semiconductor lasers. IEEE J. LT-**8**, 1659–1669 (1990)

3.41 J. F. Lucas, L. F.DeChiaro, C. Salla, C. Y. Boisbert: Low coherence reflectometry and spectral analysis for detection of gain anomalies in semiconductor lasers. Electron. Lett. **28**, (1992)

3.42 A. L. Schawlow, C. H. Townes: Infrared and optical masers. Phys. Rev. **112**, 1940–1949 (1958)

3.43 M. Lax: Quantum noise V: Phase noise in a homogeneously broadened maser, in *Physics and Quantum Electronics*, ed. by P. L. Kelly, B. Lax, P. E. Tannenwald (McGraw-Hill, New York 1966)

3.44 R. D. Hempstead, M. Lax: Classical noise VI: Noise in self-sustained oscillators near threshold. Phys. Rev. **161**, 350–366 (1967)

3.45 C. H. Henry: Theory of the linewidth of semiconductor lasers. IEEE J. QE-**18**, 259–264 (1982)

3.46 C. H. Henry: Theory of spontaneous emission noise in open resonators and its application to lasers and optical amplifiers. IEEE J. LT-**4**, 288–297 (1986)

3.47 C. H. Henry: Phase noise in semiconductor lasers. IEEE J. LT-**4**, 298–311 (1986)

3.48 C. H. Henry: Line broadening of semiconductor lasers, in *Coherence, Amplification, and Quantum Effects in Semiconductor Lasers*, ed. by Y. Yamamoto (Wiley, New York 1991)

3.49 J. Arnaud: Natural linewidth of anisotropic lasers. Opt. Quantum Electron. **18**, 335–343 (1986)

3.50 J. Dong, S. Arai, K. Kudo, M. Hotta: Narrow linewidth characteristics of a $1.5\,\mu m$ wavelength single-mode phase-locked laser array. IEEE Photon. Technol. Lett. **5**, 622–624 (1993)

3.51 M. Sargent, M. O. Scully, W. E. Lamb *Laser Physics*: (Addison-Wesley, Reading 1974)

3.52 C. L. Tang, H. Statz: Phase-locking of laser oscillators by injected signal. J. Appl. Phys. **38**, 323–324 (1967)

3.53 M. B. Spencer, W. E. Lamb: Theory of two coupled lasers. Phys. Rev. A **5**, 893–898 (1972)

3.54 R. Lang, K. Kobayashi: External optical feedback effects on semiconductor injection laser properties. IEEE J. QE-**16**, 347–355 (1980)

3.55 S. Kobayashi, T. Kimura: Injection locking in AlGaAs semiconductor laser. IEEE J. QE-**17**, 681–689 (1981)

3.56 L. Goldberg, H. F. Taylor, J. F. Weller: Intermodal injection locking of semiconductor lasers. Electron. Lett. **20**, 809–811 (1984)

3.57 P. Gallion, G. Debarge: Influence of amplitude-phase coupling on the injection locking bandwidth of a semiconductor laser. Electron. Lett. **21**, 264–266 (1985)

3.58 F. Mogensen, H. Olesen, G. Jacobsen: Locking conditions and stability properties for a semiconductor laser with external light injection. IEEE J. QE-**21**, 784 (1985)

3.59 G. R. Hadley, J. P. Hohimer, A. Owyoung: Modeling of injection-locking phenomena in diode-laser arrays. Optics Lett. **11**, 144–146 (1986)

3.60 R. J. Lang, A. Yariv: Local-field equations for coupled optical resonators. Phys. Rev. A **34**, 2038–2043 (1986)

3.61 G. N. Brown: A study of the static locking properties of injection locked laser amplifiers. British Telecom. Technol. J. **4**, 71–80 (1986)

3.62 I. Petitbon, P. Gallion, G. Debarge, C. Chabran: Locking bandwidth and relaxation oscillations of an injection-locked semiconductor laser. IEEE J. QE-**24**, 148 (1988)

3.63 G. L. Abbas, S. Yang, V. W. S. Chan, J. G. Fujimoto: Injection behavior and modeling of 100 mW broad area diode lasers. IEEE J. QE-**24**, 609–617 (1988)

3.64 C. E. Moeller, P. S. Durkin, G. C. Dente: Mapping the injection-lock band of semiconductor lasers. IEEE J. QE-**25**, 1603–1608 (1989)

3.65 Z. Jiang, M. McCall: Phase-locking phenomena in coupled waveguide semiconductor lasers. IEE Proc. J **139**, 88–92 (1992)

3.66 S. A. Shakir, W. W. Chow: Semiclassical theory of coupled lasers. Phys. Rev. A **32**, 983–991 (1985)

3.67 W. W. Chow: Frequency locking in an index-guided semiconductor laser array. J. Opt. Soc. Am. B **3**, 833–836 (1986)

3.68 H. G. Winful, S. S. Wang: Stability of phase locking in semiconductor laser array. Appl. Phys. Lett. **53**, 1894–1896 (1988)

3.69 A. A. Golubentsev, V. V. Likhanskii, A. P. Napartovich: Phase-locking of a two-dimensional laser array with random detuning of eigenfrequencies., Proc. SPIE **1219**, 220–226 (1990)

3.70 A. A. Golubentsev, V. V. Likhanskii: Characteristics of phase-locking of an array of optically coupled lasers with a random scatter of eigenfrequencies. Sov. J. Quantum Electron. **20**, 522–523 (1990)

3.71 G. C. Dente, C. E. Moeller, P. S. Durkin: Coupled oscillators at a distance: applications to coupled semiconductor lasers. IEEE J. QE-**26**, 1014–1022 (1990)

3.72 S. Kobayashi, T. Kimura: Coherence of injection phase-locked AlGaAs semiconductor laser. Electron. Lett. **16**, 668–670 (1980)

3.73 G. C. Dente, D. Bossert, C. E. Moeller: Quantum noise limits for coupled oscillators. unpublished April 1991.

3.74 D. Bossert: *Quantum noise limits for coupled oscillators*. PhD thesis, Oregon Graduate Institute (April 1992)

3.75 J. Katz: Phase-locking of semiconductor injection lasers. TDA progress report 42-66, Jet Propulsion Laboratory (Sept.-Oct. 1981)

3.76 K. Nishi, R. Lang: Lateral mode localization in multistripe laser diode. Jpn. J. Appl. Phys. **24**, L349–L351 (1985)

3.77 E. M. Garmire: Tolerances for phase locking of semiconductor laser arrays. Proc. SPIE **893**, 91–99 (1988)

3.78 C. A. Wang, H. K. Choi, M. K. Connors: Large-area uniform OMVPE growth for GaAs/AlGaAs quantum-well diode lasers with controlled emission wavelength. J. Electronic Mater. **18**, 695–701 (1989)

3.79 R. A. Elliott, R. K. DeFreez, T. L. Paoli, R. D. Burnham, W. Streifer: Dynamic characteristics of phase-locked multiple quantum well injection lasers. IEEE J. QE-**21**, 598–602 (1985)

3.80 R. K. DeFreez, D. J. Bossert, N. Yu, K. Hartnett, R. A. Elliott, H. G. Winful: Spectral and picosecond temporal properties of flared guide Y-coupled phase-locked laser arrays. Appl. Phys. Lett. **53**, 2380–2382 (1988)

3.81 N. Yu, R. K. DeFreez, D. J. Bossert, R. A. Elliott, H. G. Winful, D. F. Welch: Observation of sustained self-pulsation in CW operated flared Y-coupled laser arrays. Electron. Lett. **24**, 1203–1204 (1988)

3.82 N. Yu, R. K. DeFreez, D. J. Bossert, G. A. Wilson, R. A. Elliott, S. S. Wang, H. G. Winful: Spatiospectral and picosecond spatiotemporal properties of a broad area operating channeled-substrate-planar laser array. Appl. Opt. **30**, 2503–2513 (1991)

3.83 K. Otsuka: Self-induced phase turbulence and chaotic itinerancy in coupled laser systems. Phys. Rev. Lett. **65**, 329–332 (1990)

3.84 L. Goldberg, H. F. Taylor, J. F. Weller, D. R. Scifres: Injection locking of coupled-striped diode laser arrays. Appl. Phys. Lett. **46**, 236–238 (1985)

3.85 L. Goldberg, J. F. Weller: Injection-locked operation of a 20-element coupled-stripe laser array. Electron. Lett. **22**, 858–859 (1986)

3.86 G. R. Hadley, J. P. Hohimer, A. Owyoung: High-order ($\nu > 10$) eigenmodes in ten-stripe gain-guided diode laser arrays. Appl. Phys. Lett. **49**, 684–686 (1986)

3.87 L. Goldberg, J. F. Weller: Injection locking and single-mode fiber coupling of a 40-element laser diode array. Appl. Phys. Lett. **50**, 1713–1715 (1987)

3.88 M. K. Chun, L. Goldberg, J. F. Weller: Injection-beam parameter optimization of an injection-locked diode-laser array. Optics Lett. **14**, 272–274 (1989)

Chapter 4

4.1 R. B. Smith, G. L. Mitchell: Calculation of Complex Propagation Modes in Arbitrary, Plane-Layered, Complex Dielectric Strcutures. I. Analytic Formulation II. Fortran Program MODEIG. PhD thesis, University of Washington (December 1977)

4.2 P. Yeh, A. Yariv, C. S. Hong: Electromagnetic propagation in periodic stratified media. I. general theory. J. Opt. Soc. Am. **67**, 423–438 (1977)

4.3 H. Fujii, I. Suemune, M. Yamanishi: Analysis of transverse modes of phase-locked multi-stripe lasers. Electron. Lett. **21**, 713–714 (1985)

4.4 W. K. Marshall, J. Katz: Direct analysis of gain-guided phase-locked semiconductor laser arrays. IEEE J. QE-**22**, 827–832 (1986)

4.5 K. Hayata, M. Koshiba: Direct supermode analysis of phase-locked diode laser arrays. Electron. Lett. **23**, 935–936 (1987)

4.6 P. A. Kirkby, G. H. B. Thompson: Channeled substrate buried heterostructure GaAs – (GaAl) As injection laser. J. Appl. Phys. **47**, 4578–4590 (1976)

4.7 J. K. Butler, J. B. Delaney: A rigorous boundary value solution for the lateral modes of stripe geometry injection lasers. IEEE J. QE-**14**, 507–513 (1978)

4.8 P. M. Asbeck, D. A. Cammack, J. J. Daniele, V. Kelbanoff: Lateral mode behavior in narrow stripe lasers. IEEE J. QE-**15**, 727–733 (1979)

4.9 W. Streifer, E. Kapon: Application of the equivalent-index method to DH diode lasers. Appl. Opt. **18**, 3724–3725 (1979)

4.10 W. Streifer, R. D. Burnham, D. R. Scifres: Analysis of diode lasers with lateral spatial variations in thickness. Appl. Phys. Lett. **37**, 121–123 (1980)

4.11 J. Buus: The effective index method and its applications to semiconductor lasers. IEEE J. QE-**18**, 1083–1089 (1982)

4.12 S. R. Chinn, R. J. Spiers: Calculation of separate multiclad-layer stripe geometry laser modes. IEEE J. QE-**18**, 984–991 (1982)

4.13 M. K. Amann: Rigorous waveguide analysis of the separated multiclad-layer stripe-geometry laser. IEEE J. QE-**22**, 1992–1998 (1986)

4.14 M. Lax, J. H. Batteh, G. P. Agrawal: Channeling of intense electromagnetic beams. J. Appl. Phys. **52**, 109–125 (1981)

4.15 G. P. Agrawal: Fast-Fourier-transform based beam-propagation model for stripe-geometry semiconductor lasers: Inclusion of axial effects. IEEE J. LT-**2**, 537–543 (1984)

4.16 W. W. Simmons, J. T. Hunt, W. E. Warren: Light propagation through large laser systems. IEEE J. QE-**17**, 1727–1743 (1981)

4.17 D. Yevick, B. Hermansson: New fast Fourier transform and finite-element approaches to the calculation of multiple-stripe-geometry laser modes. J. Appl. Phys. **59**, 1769–1771 (1986)

4.18 G. R. Hadley, J. P. Hohimer, A. Owyoung: Comprehensive modeling of diode arrays and broad-area devices with applications to lateral index tailoring. IEEE J. QE-**24**, 2138–2152 (1988)

4.19 Y. Chung, N. Dagli: An assessment of finite difference beam propagation method. IEEE J. QE-**26**, 1335–1339 (1990)

4.20 C. P. Lindsey, D. Mehuys, A. Yariv: Linear tailored gain broad area semiconductor lasers. IEEE J. Quantum Electron. **QE-23**, 775–787 (1987)

4.21 J. K. Butler, D. E. Ackley, D. Botez: Coupled-mode analysis of phase-locked injection laser arrays. Appl. Phys. Lett. **44**, 293–295 (1984)

4.22 E. Kapon, J. Katz, A. Yariv: Supermode analysis of phase-locked arrays of semiconductor lasers. Opt. Lett. **10**, 125–127 (1984)

4.23 J. K. Butler, D. E. Ackley, M. Ettenberg: Coupled-mode analysis of gain and wavelength oscillation characteristics of diode laser phased arrays. IEEE J. QE-**21**, 458–464 (1985)

4.24 H. F. Taylor, A. Yariv: Coupled mode theory for guided wave optics. IEEE J. QE-**QE-9** , 919–933 (1973)

4.25 H. F. Taylor, A. Yariv: Guided wave optics. Proc. IEEE **62**, 1044–1060 (1974)

4.26 H. A. Haus, L. Molter-Orr: Coupled multiple waveguide systems. IEEE J. QE-**19** , 840–844 (1983)

4.27 A. Hardy, W. Streifer: Coupled mode theory of parallel waveguides. IEEE J. LT-**3**, 1135–1146 (1985)

4.28 A. Hardy, W. Streifer: Coupled modes of multiwaveguide systems and phase arrays. IEEE J. LT-**4**, 90–99 (1986)

4.29 W. J. Fader, G. E. Palma: Normal modes of n coupled lasers. Opt. Lett. **10**, 381–383 (1985)

4.30 W. Streifer, A. Hardy, R. D. Burnham, D. R. Scifres: Single-lobe phase-array diode lasers. Electron. Lett. **21**, 118–120 (1985)

4.31 J. Buus: 'Excess' modes in gain-guided laser arrays. Electron. Lett. **22**, 1296–1297 (1986)

4.32 J. E. Epler, N. Holonyak, R. D. Burnham, T. L. Paoli, R. L. Thornton, M. M. Blouke: Transverse modes of gain-guided coupled-stripe lasers: External cavity control of the emitter spacing. Appl. Phys. Lett. **47**, 7–9 (1985)

4.33 J. R. Andrews, T. L. Paoli, W. Streifer, R. D. Burnham: Individual spatial modes of a phase-locked injection laser array observed through spectral selection and selected with an external mirror. J. Appl. Phys. **58**, 2777–2779 (1985)

4.34 D. Mehuys, A. Yariv: Coupled-wave theory of multiple-stripe semiconductor injection lasers. Opt. Lett. **13**, 571–573 (1988)

4.35 G. A. Wilson, R. K. DeFreez, H. G. Winful: Modulation of phased-array semiconductor lasers at K-band frequencies. IEEE J. QE-**27**, 1696–1704 (1991)

4.36 J. M. Verdiell, H. Rajbenbach, J. P. Huignard: Array modes of multiple-stripe diode lasers: A broad-area mode coupling approach. J. Appl. Phys. **66**, 1466–1468 (1989)

4.37 J. M. Verdiell, R. Frey: A broad-area mode coupling model for multiple-stripe semiconductor lasers. IEEE J. QE-**26**, 270–279 (1990)

4.38 C. Zmudzinski, D. Botez, L. J. Mawst: Simple description of laterally resonant, distributed-feedback-like modes of arrays of antiguides. Appl. Phys. Lett. **60**, 1049–1051 (1992)

4.39 C. P. Cherng, M. Osinski: Coupled broad-area mode theory of gain-guided laser arrays. J. Appl. Phys. **70**, 4617–4619 (1991)

4.40 S. R. Chinn, R. J. Spiers: Modal gain in coupled-stripe lasers. IEEE J. QE-**20**, 358–363 (1984)

4.41 P. G. Eliseev, R. F. Nabiev, Yu. M. Popov: Analysis of laser-structure anisotropic semiconductors by the Bloch method. J. Sov. Las. Res. **10**, 449–458 (1989)

4.42 R. F. Nabiev, A. I. Onishchenko: Laterally coupled periodic semiconductor laser structures; Bloch function analysis. IEEE J. QE-**28**, 2024–2032 (1992)

4.43 A. P. Napartovich, D. Botez: Phase-locked arrays of antiguides: Analytical theory. IEEE J. QE-**30** (1994)

4.44 R. F. Nabiev, P. Yeh, D. Botez: Self-stabilization of fundamental in-phase mode in resonant antiguided laser arrays. Appl. Phys. Lett. **62**, 916–918 (1993)

4.45 D. Botez, T. Holcomb: Bloch-function analysis of resonant arrays of antiguided diode lasers. Appl. Phys. Lett. **60**, 539–541 (1992)

4.46 M. W. Austin: Theoretical and experimental investigation of GaAs/GaAlAs and n/n^+GaAs rib waveguides. IEEE J. LT-**2**, 688–694 (1984)

4.47 G. R. Hadley: Two-dimensional waveguide modleing of leaky-mode arrays. Optics Lett. **14**, 859–861 (1989)

4.48 G. R. Hadley: Mode of a two-dimensional phase-locked array of vertical-cavity surface-emitting lasers. Optics Lett. **15**, 1215–1217 (1990)

4.49 R. Amantea, N. W. Carlson, S. L. Palfrey, G. A. Evans, J. M. Hammer, M. Lurie: Network analysis of the modes of two-dimensional grating-surface-emitting diode laser arrays. IEEE J. QE-**26**, 1023–1038 (1990)

4.50 R. Amantea, N. W. Carlson: Network analysis of two-dimensional laser arrays, in *Surface Emitting Semiconductor Laser and Arrays*, ed. by G. A. Evans, J. M. Hammer (Academic, San Diego 1993)

4.51 P. Yeh, C. Gu, D. Botez: Optical properties of dual-state Fabry-Perot étalons. Optics Lett. **17**, 1818–1820 (1992)

4.52 J. Buus: Principles of semiconductor laser modelling. IEE Proc. J **132**, 42–51 (1985)

4.53 G. P. Agrawal: Lateral analysis of quasi-index-guided injection lasers: transition from gain to index guiding. IEEE J. LT-**2**, 537–543 (1984)

4.54 T. Kumar, R. F. Ormondroyd, T. E. Rozzi: A self-consistent model of the lateral behavior of a twin-stripe injection laser. IEEE J. QE-**22**, 1975–1985 (1986)

4.55 G. P. Agrawal: Lateral-mode analysis of gain-guided and index-guided semiconductor laser arrays. J. Appl. Phys. **58**, 2922–2931 (1985)

4.56 J. P. Van De Capaelle, R. Baets, P. E. Lagasse: Self-consistent analysis of waveguiding in phase-locked array lasers. In Integrated Optics, ed. by H. P. Nolting, R. Ulrich, Springer Ser. Opt. Sci., Vol. 48 (Springer, Berlin, Heidelberg 1985)

4.57 Y. Twu, K. L. Chen, S. Wang, J. R. Whinnery, A. Dienes: Eigenmode analysis of phase-locked semiconductor laser arrays. Appl. Phys. Lett. **48**, 16–18 (1986)

4.58 T. Kumar: Steady-state self-consistent analysis of diode-laser arrays. Appl. Phys. Lett. **50**, 877–879 (1987)

4.59 G. R. Hadley, J. P. Hohimer, A. Owyoung: Influence of thermal effects on the eigenmodes of gain-guided diode laser arrays. J. Appl. Phys. **61**, 1697–1700 (1987)

4.60 G. R. Hadley, J. P. Hohimer, A. Owyoung: Free-running modes for gain-guided diode laser arrays. IEEE J. QE-**23**, 765–774 (1987)

4.61 Y. Twu, S. Wang, J. R. Whinnery, A. Dienes: Mode characteristics of phase-locked semiconductor laser arrays at and above threshold. IEEE J. QE-**23**, 788–795, (1987)

4.62 K. L. Chen, S. Wang: Spatial hole burning in evanescently coupled semiconductor laser arrays. Appl. Phys. Lett. **47**, 555–557 (1985)

4.63 M. K. Amann, F. Kappeler: Analytical solution for the lateral current distribution in multiple stripe laser diodes. Appl. Phys. Lett. **48**, 1710–1712 (1986)

4.64 H. S. Carslaw, J. C. Jaeger: Conduction of Heat in Solids, 2nd edn.: (Oxford Univ. Press, Oxford 1959)

4.65 G. R. Hadley: Modeling of diode laser arrays. In Diode-Laser Arrays, ed. by D. Botez, D. Scifres, (Cambridge Univ. Press, Cambridge 1994)

4.66 A. H. Paxton, G. C. Dente: Filament formation in semiconductor lasers. J. Appl. Phys. **70**, 2921–2925 (1991)

4.67 V. I. Beapalov, V. I. Talanov: Filamentary structure of light beams in nonlinear liquids. Sov. JETP Lett. **3**, 307–310 (1966)

4.68 A. J. Campillo, S. L. Shapiro, B. R. Suydam: Periodic breakup of optical beams due to self-focusing. Appl. Phys. Lett. **23**, 628–630 (1973)

4.69 A. J. Campillo, S. L. Shapiro, B. R. Suydam: Relationship of self-focusing to spatial instability modes. Appl. Phys. Lett. **24**, 178–179 (1974)

4.70 W. J. Firth C. Parè: Transverse modulation instabilities for counterpropagating beams in Kerr media. Optics Lett. **70**, 1096–1098 (1988)

4.71 M. Lurie: Coherence and its effect on laser arrays. In Surface Emitting Semiconductor Laser and Arrays, ed. by G. A. Evans, J. M. Hammer (Academic, San Diego 1993)

4.72 B. Saleh: Photoelectron Statistics, Springer Ser. Opt. Sci., Vol. 6, (Springer, Berlin, Heidelberg 1978)

4.73 B. A. Saleh, J. M. Minkowski: On the spatial coherence of laser beams. J. Phys. A **8**, 120–125 (1975)

4.74 P. Spano: Connection between spatial coherence and modal structure in optical fibers and semiconductor lasers. Opt. Commun. **33**, 265–270 (1980)

4.75 G. C. Dente, K. A. Wilson, T. C. Salvi, D. Depatie: Phase and spatial coherence measurements on diode arrays: Comparison to supermode theory. Appl. Phys. Lett. **51**, 9–11 (1987)

4.76 G. A. Evans, J. M. Hammer, R. Amantea, N. W. Carlson, C. G. Dupuy, C. Y. Lai, M. J. Lurie: Design, fabrication, and testing of diode array modules (PILOT4): Final Report Contract No. F29601-88-C-0038, David Sarnoff Research Center, (June 1991)

4.77 V. J. Masin, N. W. Carlson, M. Lurie, J. M. Finlan, S. L. Reinhold, W. T. Walker: Fresnel and Fraunhofer fields of phase locked diode arrays. Conf. on Laser and Electro-Optics 1987 Technical Digest **14**, pp. 54–55 (1987)

4.78 G. A. Hockham: Radiation from a solid-state laser. Electron. Lett. **9**, 389–391 (1973)

4.79 L. Lewin: Obliquity-factor correction to solid-state radiation patterns. J. Appl. Phys. **46**, 2323–2324 (1975)

4.80 Z. Dacic, E. Wolf: Changes in the spectrum of a partially coherent light beam propagating in free space. J. Opt. Soc. Am. A **5**, 1118–1126 (1988)

4.81 D. Faklis, G. M. Morris: Spectral shifts produced by source correlations. Optics Lett. **13**, 4–6, (1988)

4.82 S. K. Liew, N. W. Carlson, R. Amantea: Narrow spectral linewidth in the far-field zone of active-grating surface-emitting semiconductor laser-amplifier. IEEE Photon. Technol. Lett. **5**, 209–211 (1993)

4.83 D.R. Scifres, W. Streifer, R.D. Burnham, T. L. Paoli, C. Lindstrom: Near-field and far-field patterns of phase-locked semiconductor laser arrays. Appl. Phys. Lett. **42**, 495–497 (1983)

4.84 J. E. Epler, N. Holonyak, R. D. Burnham, T. L. Paoli, W. Streifer: Far-field supermode patterns of a multiple-stripe quantum well heterostructure laser operated $\left(\approx 7330\,\text{Å}, 300\,\text{K}\right)$ in an external grating cavity. Appl. Phys. Lett. **45**, 406–408 (1984)

4.85 L. Mandel: Concept of cross-spectral purity in coherence theory. J. Opt. Soc. Am. **51**, 1342–1350 (1961)

4.86 L. Mandel, E. Wolf: Spectral coherence and the concept of cross-spectral purity. J. Opt. Soc. Am. **66**, 529–535 (1976)

Chapter 5

5.1 G. A. Evans, V. J. Masin: Private communication: (February 1985)

5.2 E. Kapon, C. Lindsey, J. Katz, S. Margalit, A. Yariv: Coupling mechanism of gain-guided integrated semiconductor laser arrays. Appl. Phys. Lett. **44**, 389–391 (1984)

5.3 J. E. Epler, N. Holonyak, J. M. Brown, R. D. Burnham, T. L. Paoli, W. Streifer: High-energy $\left(\approx 7330\,\text{Å}, 300\,\text{K}\right)$ operation of single- and multiple-stripe quantum well heterostructure laser diodes in an external grating cavity. J. Appl. Phys. **56**, 670–675 (1984)

5.4 J. P. Hohimer, A. Owyoung, G. R. Hadley: Single-channel injection locking of a diode-laser array with a cw dye laser. Appl. Phys. Lett. **47**, 1244–1246 (1985)

5.5 J. P. Hohimer, G. R. Hadley, A. Owyoung: Interelement coupling in gain-guided laser arrays. Appl. Phys. Lett. **48**, 1504–1506 (1986)

5.6 C. P. Lindsey, E. Kapon, J. Katz, S. Margalit, A. Yariv: Single contact tailored gain phased array of semiconductor lasers. Appl. Phys. Lett. **45**, 722–724 (1984)

5.7 C. P. Lindsey, P. Derry, A. Yariv: Tailored gain broad area lasers with single lobed far-field patterns. Electron. Lett. **21**, 671–673 (1985)

5.8 C. P. Lindsey, P. Derry, A. Yariv: Fundamental lateral mode oscillation via gain tailoring gain in a broad area semiconductor laser. Appl. Phys. Lett. **47**, 560–562 (1985)

5.9 D. F. Welch, D. Scifres, P. Cross, H. Kung, W. Streifer, R. D. Burnham, J. Yaeli, T. L. Paoli: High-power cw operation of phased array diode lasers with diffraction limited output beam. Appl. Phys. Lett. **47**, 1134–1137 (1985)

5.10 A. E. Siegman: Unstable optical resonators for laser applications. Proc. IEEE **53**, 277–287 (1965)

5.11 R. K. DeFreez, Z. Bao, P. D. Carleson, M. K. Felisky: High-brightness unstable resonator semiconductor lasers. Proc. SPIE **1850**, 75–83 (1993)

5.12 J. E. Ungar: High power laser diodes for laser communications: Final technical report RADC-TC-90-247, Rome Air Development Center, October 1990

5.13 G. C. Dente: Unstable resonators for high-brightness semiconductor lasers: January 1993

5.14 G. C. Dente: Private communication (July 1993)

5.15 G. Yao, Y. C. Chen, C. M. Harding, S. M. Sherrick, R. J. Dalby, R. G. Waters, C. Largent: Excess spontaneous-emission factor in unstable-resonator lasers. Optics Lett. **17**, 1207–1209 (1992)

5.16 S. R. Chinn: Unpublished (1993)

5.17 R. Amantea: Private communication (1992)

5.18 R. J. Lang, D. M. Mehuys, A. Hardy, K. M. Dzurko, D. F. Welch: Spatial evolution of filaments in broad area diode laser amplifiers. Appl. Phys. Lett. **62**, 1209–1211 (1993)

Chapter 6

6.1 D. Botez, W. T. Tsang, S. Wang: Growth characteristics of GaAs-Ga$_{1-x}$Al$_x$As structures fabricated by liquid-phase epitaxy over preferentially etched channels. Appl. Phys. Lett. **28**, 234–237 (1976)

6.2 S. Yamamoto, H. Hayashi, S. Yano, T. Sakurai, T. Hijikata: Visible GaAlAs V-channeled substrate inner stripe laser with stabilized mode using p-GaAs substrate. Appl. Phys. Lett. **40**, 372–374 (1982)

6.3 G. A. Evans, J. K. Butler, V. J. Masin: Lateral optical confinement of channeled-substrate-planar lasers with GaAs/AlGaAs substrates. IEEE J. QE-**24**, 737–749 (1988)

6.4 N. W. Carlson, J. C. Connolly: Coherent high-power phased array laser, semiconductor (CHIPPALS): Final Report WRDC-TR-89-5040, Wright Research and Development Center Electronic Technology Laboratory (September 1989)

6.5 W. W. Chow: Effects of spatial variation in and index-guided semiconductor laser array. J. Opt. Soc. Am. B **4**, 324–328 (1987)

6.6 W. Streifer, M. Osinski, D. R. Scifres, D. F. Welch, P. S. Cross: Phased-array lasers with a uniform, stable supermode. Appl. Phys. Lett. **49**, 1496–1498 (1986)

6.7 A. Hardy, W. Streifer, M. Osinski: Chirping effects in phase-coupled laser arrays. IEE Proc. J **135**, 443–450 (1988)

6.8 R. P. Bryan, L. M. Miller, T. M. Cockerill, S. M. Langsjoen, J. J. Coleman: High-power pulsed operation of an optimized nonplanar corrugated substrate periodic laser diode array., IEEE J. QE-**26**, 222–224 (1990)

6.9 S. M. Lee, S. L. Chuang, R. P. Bryan, C. A. Zmudzinski, J. J. Coleman: A self-consistent model of a nonplanar quantum-well periodic laser array, IEEE J. QE-**27**, 1886–1899 (1991)

6.10 V. I. Malakhova, Yu. A. Tambiev, S. D. Yakubovich: Regular integrated array of stripe injection lasers. Sov. J. Quantum Electron. **11**, 1351–1352 (1981)

6.11 I. Suemune, T. Terashige, M. Yamanishi: Phase-locked, index-guided multiple-stripe lasers with large refractive index differences. Appl. Phys. Lett. **45**, 1011–1013 (1984)

6.12 E. Kapon, L. T. Yu, Z. Rav-Noy, S. Margalit, A. Yariv: Phased arrays of buried-ridge InP/InGaAsP diode lasers. Appl. Phys. Lett. **46**, 136–138 (1985)

6.13 N. Kaneno, T. Kadowaki, J. Ohsawa, T. Aoyagi, S. Hinata, K. Ikeda, W. Sasaki: Analysis of the stability of the highest-order supermode in semiconductor laser arrays. Electron. Lett. **21**, 780–781 (1985)

6.14 J. S. Tsang, D. C. Liou, K. L. Tsai, H. R. Chen, C. M. Tsai, C. P. Lee, F. Y. Juang: Fundamental mode operation of high-power InGaAs/GaAs/AlGaAs laser arrays. J. Appl. Phys. **73**, 4706–4708 (1993)

6.15 J. E. A. Whiteaway, G. H. B. Thompson, A. R. Goodwin: Mode stability in real index-guided semiconductor laser arrays. Electron. Lett. **21**, 1194–1195 (1985)

6.16 G. H. B. Thompson, J. E. A. Whiteaway: Analysis of the stability of the highest-order supermode in semiconductor laser arrays. Electron. Lett. **23**, 444–446 (1987)

6.17 D. E. Ackley, R. W. H. Engelmann: Twin-stripe injection laser with leaky-mode coupling. Appl. Phys. Lett. **37**, 866–868 (1980)

6.18 L. J. Mawst, D. Botez, C. A. Zmudzinski, M. Jansen, C. Tu, T. J. Roth, J. Yun: Resonant self-aligned-stripe antiguided diode laser array. Appl. Phys. Lett. **60**, 668 (1992)

6.19 D. Mehuys, J. S. Major, D. F. Welch: High-power, high-efficiency antiguide laser arrays. Proc. SPIE **1850**, 2–12 (1993)

6.20 D. Botez, L. J. Mawst, G. L. Peterson, T. J. Roth: Phase-locked arrays of antiguides: modal content and discrimination. IEEE J. QE-**26**, 482–495 (1990)

6.21 J. S. Major, D. Mehuys, D. F. Welch: 11.5 W pulsed operation of antiguided laser diode array. Electron. Lett. **28**, 1101–1102 (1991)

6.22 R. F. Nabiev, X. Yi, P. Yeh, D. Botez: Self-stabilization of fundamental in-phase mode in resonant antiguided laser arrays. Proc. SPIE **1850**, 23–36 (1993)

6.23 P. D. Van Eijk, M. Reglat, G. Vassilieff, G. J. M. Krijnen, A. Driessen, A. J. Mouthaan: Analysis of the modal behavior of an antiguide diode laser array with Talbot filter. IEEE J. LT-**9**, 629–634 (1991)

6.24 S. Ramanujan, H. G. Winful: Dynamics of resonant optical waveguide semiconductor laser arrays. Appl. Phys. Lett. **62**, 3226–3228 (1993)

6.25 N. W. Carlson, S. K. Liew, R. Menna, P. Gardner, J. Andrews, J. Kirk, J. K. Butler, A. Triano, W. Reichert: Coherent, high-power operation of InGaAsP/InGaAs multiple-quantum well active grating-surface-emitting amplified lasers. Proc. SPIE **1850**, 60–68 (1993)

6.26 N. W. Carlson, S. K. Liew, G. A. Evans J. K. Butler: Observation of self-sustained pulsations from a highly coherent channeled substrate planar diode laser array. CLEO 1989 Technical Digest **11**, pp. 302–303, April 1989

6.27 S. Ramanujan, H. G. Winful, M. Felisky, R. K. DeFreez, D. Botez, M. Jansen, P. Wisseman: The temporal behavior of resonant-optical-waveguide phase-locked diode laser arrays. Appl. Phys. Lett. **63**, April 1994

6.28 L. J. Mawst, D. Botez, M. Jansen, C. Zmudzinski, S. S. Ou, M. Sergant, C. A. Tu, T. J. Roth, G. Peterson, M. Valley, J. J. Yang: IEEE J. QE-**29**, 1906–1917 (1993)

Chapter 7

7.1 S. Wang: Principles of distributed feedback and distributed Bragg-reflector lasers., IEEE J. QE-**10**, 413–427 (1974)

7.2 P. Zory: Corrugated grating coupled devices and coupling coefficients. Integrated Optics Conference, Salt Lake City, Utah (January 1976)

7.3 R. Parke, R. G. Waarts, D. F. Welch, A. F, W. S. Streifer: High-efficiency, high uniformity, grating coupled surface emitting lasers. Electron. Lett. **26**, 125–127 (1990)

7.4 D. F. Welch, R. Parke, R. G. Waarts, W. Streifer, D. Scifres: High power, 16 W, grating surface emitting laser with a superlattice substrate reflector. Electron. Lett. **26**, 757–758 (1990)

7.5 G. Hadjicostas, J.K. Butler, G. A. Evans, N. W. Carlson, R. Amantea: A numerical investigation of wave interactions in dielectric waveguides with periodic surface corrugations. IEEE J. QE-**26**, 893–901 (1990)

7.6 A. Hardy, K. M. Dzurko, D. F. Welch, D. R. Scifres, R. J. Lang, R. Waarts: Design considerations of large aperture perpendicular gratings semiconductor ring lasers. Appl. Phys. Lett. **62**, 931–933 (1993)

7.7 K. M. Dzurko, D. F. Welch, D. R. Scifres, A. Hardy: 1-W single-mode edge-emitting DBR ring oscillators. IEEE Photon. Technol. Lett. **4**, 369–371 (1993)

7.8 P. Zory: Laser oscillation in leaky corrugated optical waveguides. Appl. Phys. Lett. **22**, 125–128 (1973)

7.9 R. D. Burnham, D. R. Scifres, W. Streifer: Single heterostructure distributed-feedback GaAs diode lasers. IEEE J. QE-**11**, 441–449 (1975)

7.10 I. Suemune, Y. Kan, M. Yamanishi: Semiconductor light sources with capabilities of electronic beam-scanning. Electron. Lett. **19**, 1002–1003 (1983)

7.11 G. A. Evans, J. M. Hammer, N. W. Carlson, F. R. Elia, E. A. James, J. B. Kirk: Surface-emitting second order distributed Bragg reflector laser with dynamic wavelength stabilization and far-field angle of 0.25°. Appl. Phys. Lett. **49**, 314–315 (1986)

7.12 N. W. Carlson, G. A. Evans, J. M. Hammer, C. C. Neil: Measurement of effective index and dispersion in an index-guided surface-emitting distributed Bragg reflector laser. Electron. Lett. **23**, 355–357 (1987)

7.13 Y. Kan, Y. Honda, I. Suemune, M. Yamanishi: Electronic beam deflection in a semiconductor laser diode using grating output coupler. Electron. Lett. **22**, 1310–1311 (1986)

7.14 A. A. Zlenko, V. A. Kiselev, A. M. Prokhorov, A. A. Spikhal'skii, V. A. Sychugov: Grating-coupled double-heterostructure AlGaAs diode lasers. Sov. J. Quant. Electron. **4**, 839–842 (1975)

7.15 D. Marcuse: Bell Syst. Tech. J. **52**, 63 (1973)

7.16 D. G. Dalgoutte, C. D. W. Wilkinson: Appl. Opt. **14**, 2983 (1975)

7.17 W. Streifer, R. D. Burnham, D. R. Scifres: Coupling coefficients for distributed feedback single and double-heterostructure diode lasers. IEEE J. QE-**11**, 867–873 (1975)

7.18 R. I. MacDonald, K. O. Hill, V. Makios: Hybrid modes of the passive-core corrugated-waveguide laser. Can. J. Phys. **54**, 849–860 (1976)

7.19 W. Streifer, D. R. Scifres, R. D. Burnham, R. I. MacDonald: On grating-coupled radiation from waveguides. IEEE J. QE-**13**, 67–68 (1977)

7.20 A. A. Hardy, D. F. Welch, W. Streifer: Analysis of second order gratings. IEEE J. QE-**25**, 2096–2105 (1989)

7.21 W. Streifer, D. R. Scifres, R. D. Burnham: Analysis of grating-coupling radiation in GaAs : GaAlAs lasers and waveguides. IEEE J. QE-**12**, 422–428 (1976)

7.22 W. Streifer, D. R. Scifres, R. D. Burnham: Analysis of grating-coupling radiation in GaAs : GaAlAs lasers and waveguides-II: blazing effects. IEEE J. QE-**12**, 494–499 (1976)

7.23 R. I. MacDonald, K. O. Hill: Loss coefficients for blazed waveguide corrugations. Optical and Quantum Electron. **9**, 249–258 (1977)

7.24 M. Matsumoto: Analysis of the blazing effect in second order gratings. IEEE J. QE-**28**, 2016–2023 (1992)

7.25 S. T. Peng, T. Tamir: Directional blazing of waves guided by asymmetrical dielectric gratings. Opt. Commun. **11**, 405–409 (1974)

7.26 L. F. Johnson, K. A. Ingersoll, G. W. Kammlott: An oblique shadow deposition technique for altering the profile of grating relief patterns on surfaces. Appl. Phys. Lett. **34**, 578–580 (1979)

7.27 D. Heflinger, J. Kirk, R. Cordero, G. Evans: Submicron grating fabrication on GaAs by holographic exposure. Opt. Eng. **21**, 537–541 (1982)

7.28 J. B. Kirk, P. K. York, G. A. Evans: Submicron gratings for optoelectronic devices. Optcon '92

7.29 H. Kogelnik, C. V. Shank: Stimulated emission in a periodic structure. Appl. Phys. Lett. **18**, 152–154 (1971)

7.30 D. R. Scifres, R. D. Burnham, W. Streifer: Distributed-feedback single heterojunction GaAs diode laser. Appl. Phys. Lett. **25**, 203–206 (1974)

7.31 H. M. Stoll, D. H. Seib: Distributed feedback GaAs homojunction injection laser. Appl. Opt. **13**, 1981–1982 (1974)

7.32 Y. Uematsu, M. Yamamoto, Y. Unno: Characteristics of grating-coupled GaAlAs-GaAlAs lasers. IEEE J. QE-**13**, 646–651 (1977)

7.33 J. M. Hammer, N. W. Carlson, G. A. Evans, M. Lurie, S. L. Palfrey, C. J. Kaiser, M. G. Harvey, E. A. James, J. B. Kirk, F. R. Elia: Phase-locked operation of coupled pairs of grating-surface-emitting diode lasers. Appl. Phys. Lett. **50**, 659–661 (1987)

7.34 J. H. Weaver: Optical properties of metals, in *Handbook of Chemistry and Physics*: 64th edn., (CRC, Boca Raton 1983-1984)

7.35 R. J. Noll, S. H. Macomber: Analysis of grating surface emitting lasers. IEEE J. QE-**26**, 456–466 (1990)

7.36 S. H. Macomber: Nonlinear analysis of surface-emitting distributed feedback lasers. IEEE J. QE-**26**, 2065–2074 (1990)

7.37 K. Kojima, S. Noda, K. Mitsunaga, K. Kyuma, K. Hamanaka, T. Nakayama: Edge- and suface-emitting distributed Bragg reflector laser with multiquantum well active/passive waveguides. Appl. Phys. Lett. **50**, 227–229 (1987)

7.38 G. A. Evans, N. W. Carlson, J. M. Hammer, M. Lurie, J. K. Butler, S. L. Palfrey, L. A. Carr, F. Z. Hawrylo, E. A. James, C. J. Kaiser, J. B. Kirk, W. F. Reichert: Efficient 30 mW grating surface-emitting lasers. Appl. Phys. Lett. **51**, 1478–1480 (1987)

7.39 G. A. Evans, N. W. Carlson, J. M. Hammer, M. Lurie, J. K. Butler, L. A. Carr, F. Z. Hawrylo, E. A. James, C. J. Kaiser, J. B. Kirk, W. F. Reichert: Efficient high-power ($>$ 150 mW) grating surface emitting lasers. Appl. Phys. Lett. **52**, 1037–1039 (1988)

7.40 N. W. Carlson, G. A. Evans, D. P. Bour, S. K. Liew: Demonstration of a grating-surface-emitting diode laser with low-threshold current density. Appl. Phys. Lett. **56**, 16–18 (1990)

7.41 N. W. Carlson, G. A. Evans, R. Amantea, J. M. Hammer, J. M. Hammer, M. Lurie, L. A. Carr, F. Z. Hawrylo, E. A. James, C. J. Kaiser, J. B. Kirk, W. F. Reichert: Electronic beamm steering in monolithic grating-surface-emitting diode laser arrays. Appl. Phys. Lett. **53**, 2275–2277 (1988)

7.42 J. M. Hammer, G. A. Evans, N. W. Carlson, D. P. Bour, M. Lurie, S. L. Palfrey, R. Amantea, S. K. Liew, L. A. Carr, E. A. James, J. B. Kirk, W. F. Reichert: Lateral beam steering in mutual injection coupled Y-branch grating-surface-emitting diode laser arrays. Appl. Phys. Lett. **56**, 224–256 (1990)

7.43 S. K. Liew, N. W. Carlson, G. A. Evans, R. Amantea, D. P. Bour, J. M. Hammer, J. B. Kirk, W. F. Reichert, R. Stolzenberger: Coherent continuous wave operation of $(10 \times 10 \times 2)$ grating-surface-emitting diode laser array in a ring configuration. J. Appl. Phys. **70**, 7645–7647 (1991)

7.44 D. P. Bour, N. W. Carlson, G. A. Evans, S. K. Liew: Low-threshold current density grating-surface-emitting lasers by OMVPE. In Int'l Electron Devices Meeting 1989. Technical Digest, pp. 857–860

7.45 D. F. Welch, R. Parke, A. Hardy, W. Streifer, D. Scifres: Low-threshold grating-coupled surface-emitting lasers. Appl. Phys. Lett. **55**, 813 (1989)

7.46 N. W. Carlson, G. A. Evans, S. K. Liew, C. J. Kaiser: High-speed switching of monolithic arrays of grating-surface-emitting diode lasers. IEEE J. LT-**7**, 1520 (1989)

7.47 N. W. Carlson, G. A. Evans, S. K. Liew, C. J. Kaiser: Sub-nanosecond electro-optic switching of coherent grating surface emitting diode laser arrays. In 7th Int'l Conf. on Integrated Optics and Optical Fiber Communication, Technical Dig., p. 27 (July 1989)

7.48 S. Noda, K. Kojima, K. Kyuma, K. Hamanaka, T. Nakayama: Reduction of spectral linewidth in AlGaAs/GaAs distributed feedback lasers by a multiple quantum well structure. Appl. Phys. Lett. **50**, 863–865 (1987)

7.49 S. K. Liew, N. W. Carlson, D. P. Bour, G. A. Evans, E. Vangieson: Demonstration of InGaAs/AlGaAs strained-layer distributed feedback grating-surface-emitting lasers with a buried second order grating structure. Appl. Phys. Lett. **58**, 228–230 (1991)

7.50 N. W. Carlson, S. K. Liew, R. Amantea, D. P. Bour, G. A. Evans, E. A. Vangieson: Mode discrimination in distributed feedback grating surface emitting lasers containing a buried second order grating. IEEE J. QE-**27**, 1746–1752 (1991)

7.51 S. K. Liew, N. W. Carlson, R. Amantea, G. A. Evans, J. K. Butler, J. Andrews, J. H. Abeles, N. A. Hughes, P. K. York, J. C. Connolly: Very low threshold InGaAs/GaAs grating-coupled surface-emitting distributed feedback laser. CLEO 1993 Technical Digest **11**, pp. 320–321

7.52 J. K. Butler, W. R. Ferguson, G. A. Evans, P. Stabile, A. Rosen: A boundary element technique applied to the analysis of waveguides with periodic surface corrugations. IEEE J.QE-**28**, 1701–1709 (1992)

7.53 H. Kogelnik, C. V. Shank: Coupled-wave theory of distributed feedback lasers. J. Appl. Phys. **43**, 2327–2335 (1972)

7.54 R. F. Kazarinov, R. A. Suris: Injection heterojunction laser with a diffraction grating on its contact surface. Sov. Phys. Semicond. **6**, 1184–1189 (1973)

7.55 R. F. Kazarinov, Z. N. Sakolova, R. A. Suris: Planar distributed-feedback optical resonators. Sov. Phys.–Tech. Phys. **21**, 130–136 (1976)

7.56 W. Streifer, D. R. Scifres, R. D. Burnham: Coupled wave analysis of DFB and DBR lasers. IEEE J. QE-**13**, 134–141 (1977)

7.57 R. F. Kazarinov, C. H. Henry: Second-order distributed feedback lasers with mode selection provided by first-order radiation losses. IEEE J. QE-**21**, 144–150 (1985)

7.58 S. A. Shakir, T. C. Salvi, G. C. Dente: Analysis of grating-coupled surface-emitting lasers. Optics Lett. **32**, 937–939 (1989)

7.59 C. H. Henry, R. F. Kazarinov, R. A. Logan, R. Yen: Observation of destructive interference in the radiation loss of second-order distributed feedback lasers. IEEE J. QE-**21**, 151–153 (1985)

7.60 W. Streifer, R. D. Burnham, D. R. Scifres: Effect of external reflectors on longitudinal modes of distributed feedback lasers. IEEE J. QE-**11**, 154–161 (1975)

7.61 S. R. Chinn: Effects of mirror reflectivity in a distributed feedback laser. IEEE J. QE-**9**, 574–580 (1973)

7.62 K. Iga: On the use of effective refractive index in DFB laser mode separation. Jpn. J. Appl. Phys. **22**, 1630 (1983)

7.63 S. K. Liew, N. W. Carlson, R. Amantea: Unpublished (1991)

7.64 R. Amantea: Network analysis of uniform grating-surface-emitting ring diode laser arrays. Unpublished (1991)

7.65 H. J. Yoo, A. Scherer, J. P. Harbison, L. T. Florez, E. G. Paek, B. P. Van der Gagg, J. R. Hayes, A. Von Lehman, E. Kapon, Y. S. Kwon: Fabrication of two-dimensional phased array of vertical cavity surface emitting lasers. Appl. Phys. Lett. **56**, 1198–1200 (1990)

7.66 D. G. Deppe, J. P. Van der Ziel, N. Chand, G. J. Zydzik, S. N. G. Chu: Phase-coupled two-dimensional $Al_xGa_{1-x}As$-GaAs vertical-cavity surface-emitting laser array. Appl. Phys. Lett. **56**, 2089–2091 (1990)

7.67 M. Orenstein, E. Kapon, N. G. Stoffel, J. P. Harbison, L. T. Florez, J. Wullert: Two-dimensional phase-locked arrays of vertical-cavity semiconductor lasers by mirror reflectivity modulation. Appl. Phys. Lett. **58**, 804–806 (1991)

7.68 M. Orenstein, E. Kapon, J. P. Harbison, L. T. Florez, N. G. Stoffel: Large two-dimensional arrays of phase-locked vertical cavity surface emitting lasers. Appl. Phys. Lett. **60**, 1535–1537 (1992)

7.69 R. A. Morgan, K. Kojima: Optical characteristics of coherently coupled vertical-cavity surface-emitting laser arrays. Opt. Lett. **18**, 352–354 (1993)

7.70 R. A. Morgan, K. Kojima, M. T. Asom, G. D. Guth, M. W Focht: 1 W (pulsed) vertical-cavity surface-emitting laser. Electron. Lett. **29**, 206–207 (1993)

7.71 R. A. Morgan, K. Kojima, L. E. Rogers, G. D. Guth, R. E. Leibenguth, M. W Focht, M. T. Asom, T. Mullally, W. A. Gault: Progress and properties of high-power coherent vertical cavity surface emitting laser arrays. Proc. SPIE **1850**, 100–108 (1993)

7.72 C. Chang-Hasnain: Two-dimensional vertical-cavity surface-emitting laser arrays, in *Diode-Laser Arrays*, ed. by D. Botez, D. Scifres, (Cambridge Univ. Press, Cambridge 1994)

7.73 D. Vakhshoori, J. D. Wynn, R. E. Leibenguth: Zone lasers, self-focusing high power vertical-cavity surface-emitting lasers with single wavelength operation. CLEO '93 Technical **11**, pp. 140–141

7.74 M. Osinski, W. Nakwaski: Thermal properties of etched-well surface-emitting diode lasers and two-dimensional arrays. Proc. SPIE **1634**, 61–83 (1992)

7.75 C. J. Chang-Hasnain, Y. A. Wu, G. S. Li, G. Hasnain, K. D. Choquette, C. Caneau, L. T. Florez: Appl. Phys. Lett. **63**, 1307 (1993)

7.76 Y. A. Wu, C. J. Chang-Hasnain, R. F. Nabiev: Single-mode emission from passive antiguide region vertical cavity surface emitting laser. Electron. Lett. **29**, 1861–1862 (1993)

Chapter 8

8.1 S. M. Curry, R. Cubeddu, T. W. Hänsch: Intensity stabilization of dye laser radiation by saturated amplification. Appl. Phys. **1**, 153–159 (1973)

8.2 J. L. Emmett, W. F. Krupke, J. I. Davis: Laser R&D at the Lawrence Livermore National Laboratory for fusion and isotope separation applications. IEEE J. QE-**20**, 591–602 (1984)

8.3 D. T. Kyrazis, D. R. Speck, C. Bibeau, R. B. Ehrlich, G. L. Hermes, J. R. Smith, T. L. Weiland, P. J. Wegner: Performance and operation of the upgraded NOVA system. Proc. SPIE **1040**, 169–176 (1989)

8.4 T. Mukai, Y. Yamamoto, T. Kimura: Optical amplification by semiconductor lasers, in *Semiconductors and Semimetals* **22**, ed. by R. K. Williamson, A. C. Beer (Academic, New York 1985)

8.5 Y. Yamamoto, T. Mukai: Fundamentals of optical amplifiers. Opt. QE-**21**, S1–S14 (1989)

8.6 M. J. O'Mahony: Semiconductor laser optical amplifiers for use in future fiber systems. IEEE J. LT-**6**, 531–544 (1988)

8.7 N. A. Olsson: Lightwave systems with optical amplifiers. IEEE J. LT-**7** , 1071–1082 (1989)

8.8 M. Coupland, K. G. Hambleton, C. Hilsum: Measurement of amplification in a GaAs injection laser. Phys. Lett. **7**, 231–232 (1963)

8.9 J. W. Crowe, R. Craig: Small-signal amplification in GaAs lasers. Appl. Phys. Lett. **4**, 57–58 (1964)

8.10 W. F. Kosonocky, R. H. Cornely, F. J. Marlowe: GaAs laser inverter. Int'l Solid-State Circuits Conf., Digest p. 48, (February 1965)

8.11 J. W. Crowe, W. E. Ahearn: Semiconductor laser amplifier. IEEE J. QE-**2**, 283–289 (1966)

8.12 W. F. Kosonocky, R. H. Cornely: GaAs laser amplifiers. IEEE J. QE-**4**, 125–131 (1968)

8.13 D. Schicketanz, G. Zeidler: GaAs-double-heterostructure lasers as optical amplifiers. IEEE J. QE-**11**, 65–69 (1975)

8.14 K. Kishino, Y. Suematsu, K. Utaka, H. Kawanishi: Monolithic integration of laser and amplifier/detector by twin-guide structure. Jpn. J. Appl. Phys. **17**, 589–590 (1978)

8.15 M. B. Chang, E. Garmire: Amplification in cleaved-substrate lasers. IEEE J. QE-**16**, 997–1001 (1980)

8.16 V. N. Luk'yanov, A. T. Semenov, S. D. Yakubovich: Steady-state characteristics of a GaAs injection quantum amplifier receiving a narrow-band input signal. Sov. J. Quant. Electron. **10** , 1432–1435 (1980)

8.17 K. Kono, K. Sakuda: Characteristics of light amplifier of AlGaAs semiconductor diode laser. J. Appl. Phys. **58**, 88–96 (1985)

8.18 G. Eisenstein, N. Tessler, U. Koren, J. M. Wiesenfeld, G. Raybon, C. A. Burrus: Length dependence of the saturation characteristics in $1.5\,\mu$m multiple quantum well optical amplifiers. IEEE Photon. Technol. Lett. **2**, 790–791 (1990)

8.19 G. Sherlock, C. P. Seltzer, D. J. Elton, S. D. Perrin, M. J. Robertson, D. M. Cooper: $1.3\,\mu$m MQW semiconductor optical amplifiers with high gain and output power. Electron. Lett. **27**, 165–166 (1991)

8.20 U. Koren, R. M. Jopson, B. I. Miller, M. Chein, M. G. Young, C. A. Burrus, C. R. Giles, H. M. Presby, G. Raybon, J. D. Evankow, B. Tell, K. Brown-Goebler: High-power laser-amplifier integrated circuit for $1.48\,\mu$m wavelength operation. Appl. Phys. Lett. **59**, 2351–2353 (1991)

8.21 J. M. Verdiell, R. Frey, J. P. Huignard: Analysis of injection-locked gain-guided diode laser arrays. IEEE J. QE-**27**, 396–401 (1991)

8.22 J. M. Verdiell, H. Rajbenbach, J. P. Huignard: Injection-locking of gain guided diode laser arrays: influence of the master beam shape. Appl. Opt. **31**, 1992–1997 (1992)

8.23 J. R. Andrews: Traveling-wave amplifier made from a laser diode array. Appl. Phys. Lett. **48**, 1331–1333 (1986)

8.24 J. R. Andrews, T. L. Paoli, R. D. Burnham: Diffraction effects in a diode array traveling-wave amplifier. J. Appl. Phys. **58**, 2777–2779 (1985)

8.25 J. R. Andrews: Variable focusing due to refractive-index gradients in a diode-array traveling-wave amplifier. J. Appl. Phys. **64**, 2134–2137 (1988)

8.26 S. H. Macomber, P. Akkapeddi: Semiconductor laser power amplifier. Proc. SPIE **723**, 36–39 (1986)

8.27 C. Nabors, R. Aggrawal, H. Choi, C. Wang, A. Mooradian: High power semiconductor optical amplifiers. LEOS Topical Meeting on New Semiconductor Devices and Applications, paper SCF2 (1990)

8.28 L. Y. Pang, E. S. Kintzer, J. G. Fujimoto: Two-stage injection locking of high-power semiconductor arrays. Optics Lett. **15**, 728–730 (1990)

8.29 J. H. Kwon, J. H. Lee, M. D. Williams: Far-field pattern of a coherently combined beam from large-scale laser diode arrays. J. Appl. Phys. **69**, 1177–1182 (1991)

8.30 D. Mehuys, L. Goldberg, D. Hall, D. F. Welch: High-power, diffraction-limited, cw optical amplifiers. Proc. SPIE **1850**, 69–74 (1993)

8.31 L. Goldberg, J. F. Weller, D. Mehuys, D. F. Welch, D. R. Scifres: 12 W broad area semiconductor amplifier with diffraction limited optical output. Electron. Lett. **27**, 927–929 (1991)

8.32 D. Mehuys, D. F. Welch, L. Goldberg, J. Weller: 11.6 W peak power, diffraction-limited diode-to-diode optical amplifier. Appl. Phys. Lett. **62**, 544–546 (1993)

8.33 D. Mehuys, L. Goldberg: Operating characteristics of broad area traveling wave semiconductor amplifiers. Proc. SPIE **1634**, 31–38 (1992)

8.34 G. C. Dente, M. L. Tilton: Modelling broad-area semiconductor optical amplifiers. IEEE J. QE-**29**, 76–88 (1993)

8.35 M. Tamburrini, L. Goldberg, D. Mehuys: Periodic filaments in reflective broad area semiconductor amplifiers. Appl. Phys. Lett. **60**, 1292–1294 (1991)

8.36 M. Jansen, D. Botez, L. J. Mawst, T. J. Roth, J. J. Yang, P. Hayashida, L. A. Dozal, J. Rozenbergs: Appl. Phys. Lett. **60**, 26 (1992)

8.37 M. Jansen, D. Botez, L. J. Mawst, T. J. Roth, J. J. Yang, S. S.Ou, P. Hayashida, L. A. Dozal: Injection locking of leaky-wave coupled resonant optical wavguide arrays. Appl. Phys. Lett. **62**, 547–549 (1993)

8.38 S. O'Brien, D. M. Mehuys, D. F. Welch, R. Parke, D. Scifres: High power monolithically integrated diode laser, preamplifier, and coherent beam expander. Appl. Phys. Lett. **61**, 2638–2640 (1992)

8.39 S. O'Brien, D. M. Mehuys, D. F. Welch, R. Parke, R. J. Lang, D. Scifres: High-power diffraction-limited monolithic broad area master oscillator power amplifier. IEEE Photon. Technol. Lett. **5**, 526–528 (1993)

8.40 S. O'Brien, R. Parke, D. F. Welch, D. M. Mehuys, D. Scifres: High power single mode GaInAs lasers with distributed Bragg reflectors. Electron. Lett. **28**, 1272–1273 (1992)

8.41 J. H. Jacob, M. Rokni, R. E. Klinkowstein, S. Singer: Expanding beam concept for building very large excimer laser amplifiers. Appl. Phys. Lett. **48**, 318–320 (1986)

8.42 R. R. Craig, R. R. Stephans: High power semiconductor laser amplifiers. Proc. SPIE **893**, 25–29 (1988)

8.43 G. Bendelli, K. Komori, S. Arai, Y. Suematsu: A new structure for high-power TW-SLA. IEEE Photon. Technol. Lett. **3**, 42–44 (1991)

8.44 P. A. Yazaki, K. Komori, G. Bendelli, S. Arai, Y. Suematsu: A GaInAsP/InP tapered-waveguide semiconductor laser amplifier integrated with a 1.5 μm distributed feedback laser. IEEE Photon. Technol. Lett. **3**, 1060–1063 (1991)

8.45 W. T. Silfvast, O. R. Wood III: Simple efficient traveling-wave excitation of short-wavelength lasers using a conical pumping geometry. Optics Lett. **14**, 18–20 (1989)

8.46 E. S. Kintzer, J. N. Walpole, S. R. Chinn, C. A. Wang, L. J. Missaggia: High power, strained-layer tapered traveling wave amplifier. Optical Fiber Conf., Digest pp. 44–45 (February 1992)

8.47 D. Mehuys, D. F. Welch, L. Goldberg: 2.0 W CW diffraction-limited tapered amplifier with diode injection. Electron. Lett. **28**, 1944–1946 (1992)

8.48 J. H. Abeles, R. Amantea, et al.: Monolithic high power fanned-out amplifier lasers. Proc. SPIE **1850**, 337–348 (1993)

8.49 R. J. Lang, A. Hardy, R. Parke, D. M. Mehuys, S. O'Brien, J. Major, D. F. Welch: Numerical analysis of flared semiconductor laser amplifiers. IEEE J. QE-**29**, 2044–2051 (1993)

8.50 N. W. Carlson, J. Andrews, S. K. Liew, R. Amantea: Unpublished, (1992)

8.51 N. W. Carlson, G. A. Evans, J. M. Hammer, M. Ettenberg: A design for a monolithic grating-surface-emitting optical amplifier: David Sarnoff Research Center Invention Disclosure 88-267, (October 1988)

8.52 N. W. Carlson, G. A. Evans, J. M. Hammer, M. Ettenberg: A design for a monolithic grating-surface-emitting optical amplifier: United States Patent No. 5,019,787 (May 28, 1991)

8.53 N. W. Carlson: Monolithic semiconductor light emitter and amplifier: United States Patent No. 5,131,001 (July 14, 1992)

8.54 N. W. Carlson: Correction to Design considerations and operating characteristics of high-power active-grating-surface-emitting amplifiers. IEEE J. QE-**29**, 814 (1993)

8.55 D. Mehuys, D. F. Welch, R. G. Waarts, R. Parke, A. Hardy, W. Streifer: Analysis of monolithic integrated master oscillator power amplifiers. IEEE J. QE-**27**, 1900–1909 (1991)

8.56 N. W. Carlson, R. Menna, P. Gardner, S. K. Liew, J. Andrews, A. Triano, J. Kirk, W. Reichert: High-power, single-frequency operation of an InGaAsP/InGaAs active-grating surface emitting amplifier at $\lambda = 1.7\,\mu$m. Appl. Phys. Lett. **62**, 2006–2008 (1993)

8.57 J. H. Abeles, P. K. York, N. W. Carlson, J. T. Andrews, W. F. Reichert, J. B. Kirk, N. A. Dinkel, C. G. Dupuy, J. T. McGinn, J. H. Thomas, T. J. Zamerowski, S. K. Liew, J. C. Connolly, G. A. Evans, J. K. Butler: High-power index-guided distributed out-coupled grating surface emitting laser-amplifiers with narrow spectra and high-quality beams. Appl. Phys. Lett. **62**, 955–957 (1993)

8.58 N. W. Carlson, R. Amantea: Scaling properties of active-grating amplified surface emitting lasers IEEE J. QE-**29**, 2780–2783 (1993)

8.59 N. W. Carlson, G. A. Evans, J. M. Hammer, C. C. Neil: Measurement of the effective index and dispersion in an index-guided surface-emitting distributed Bragg reflecting laser. Electron. Lett. **23**, 355 (1987)

8.60 U. Koren, B. I. Miller, Y. K. Su, T. L Koch, J. E. Bowers: Low internal loss separate confinement heterostructure InGaAs/InGaAsP quantum well lasers. Appl. Phys. Lett. **51**, 1744 (1990)

8.61 K. Magari, S. Kondo, H. Yasaka, Y. Noguchi, T. Kataoka, O. Mikani: A high gain GRIN-SCH MQW optical semiconductor laser amplifier. IEEE Photon. Technol. Lett. **2**, 792–793 (1990)

8.62 Y. Zou, J. S. Osinski, P. Grodzinski, P. D. Dapkus, W. Rideout, W. F. Sharfin, F. D. Crawford: Effects of strain on 1.5 μm quantum well lasers. Proc. SPIE **1850**, 153–164 (1993)

8.63 S. Seki, T. Yamanaka, W. Liu, Y. Yoshikuni, K. Yokoyama: Pure strain effect on differential gain of strained InGaAsP/InP quantum well lasers. IEEE Photon. Technol. Lett. **5**, 500–503 (1993)

8.64 N. W. Carlson: Grating-surface-emitting optical amplifiers. Diode Laser Technology Program, (April 1992)

8.65 N. W. Carlson: High power semiconductor lasers for free-space optical communications. LEOS'92 Annual Meeting Technical Digest (November 1992)

8.66 N. W. Carlson, S. K. Liew, R. Amantea, P. Gardner, R. Menna, E. Vangiseon, J. H. Abeles, G. A. Evans: Spectral linewidth and coherence properties of distributed output coupled master oscillator power amplifier semiconductor laser: In *13th IEEE Semiconductor Laser Conference Proceedings*, (September 1992)

8.67 S. K. Liew, N. W. Carlson, R. Amantea, J. H. Abeles: Coherence and spectral linewidth properties of active grating-coupled, surface-emitting amplified lasers. LEOS '92 Conf., (November 1992)

8.68 S. K. Liew, N. W. Carlson: Method for obtaining a collimated near-unity aspect ratio output beam from a DBR-GSE laser with good beam quality. Appl. Opt. **31**, 2743–2746 (1992)

8.69 R. W. Tkach, A. R. Chraplyvy: Regimes of feedback in 1.5 μm distributed feedback lasers. IEEE J. LT-4, 1655–1661 (1986)

8.70 N. Schunk, K. Peterman: Numerical analysis of the feedback regimes for a single-mode semiconductor laser with external feedback. IEEE J. QE-**24**, 1242–1247 (1988)

8.71 K. Sato, M. Oishi, Y. Itaya, M. Nakao, Y. Imamura: Defect generation due to surface corrugation in InGaAsP/InP DFB laser structures grown by MOVPE on grating-formed InP substrates. J. Crystal Growth **93**, 825–831 (1988)

8.72 D. Botez, M. Jansen, C. Zmudzinski, L. J. Mawst, P. Hayashida, C. Tu, R. F. Nabiev: Flat-phasefront fanout-type power amplifier employing resonant-optical-waveguide structures. Appl. Phys. Lett. **63**, 3113–3115 (1993)

Subject Index

Springer Series in Electrophysics

Editors: D. H. Auston G. Ecker W. Engl L. B. Felsen

Springer-Verlag
and the Environment

We at Springer-Verlag firmly believe that an international science publisher has a special obligation to the environment, and our corporate policies consistently reflect this conviction.

We also expect our business partners – paper mills, printers, packaging manufacturers, etc. – to commit themselves to using environmentally friendly materials and production processes.

The paper in this book is made from low- or no-chlorine pulp and is acid free, in conformance with international standards for paper permanency.